AF810820

Lecture Notes in Intelligent Transportation and Infrastructure

Series Editor

Janusz Kacprzyk, Systems Research Institute, Polish Academy of Sciences, Warsaw, Poland

The series “Lecture Notes in Intelligent Transportation and Infrastructure” (LNITI) publishes new developments and advances in the various areas of intelligent transportation and infrastructure. The intent is to cover the theory, applications, and perspectives on the state-of-the-art and future developments relevant to topics such as intelligent transportation systems, smart mobility, urban logistics, smart grids, critical infrastructure, smart architecture, smart citizens, intelligent governance, smart architecture and construction design, as well as green and sustainable urban structures. The series contains monographs, conference proceedings, edited volumes, lecture notes and textbooks. Of particular value to both the contributors and the readership are the short publication timeframe and the world-wide distribution, which enable wide and rapid dissemination of high-quality research output.

More information about this series at http://www.springer.com/series/15991

Alberto Ochoa-Zezzatti · Diego Oliva ·
Angel Juan Perez
Editors

Technological and Industrial Applications Associated with Intelligent Logistics

Springer

Editors
Alberto Ochoa-Zezzatti
Universidad Autónoma de Ciudad Juárez
Ciudad Juárez, Chihuahua, Mexico

Diego Oliva
University of Guadajalara
Guadajalara, Jalisco, Mexico

Angel Juan Perez
Informática Department
Universtitat Oberta De Catalunya
Barcelona, Spain

ISSN 2523-3440 ISSN 2523-3459 (electronic)
Lecture Notes in Intelligent Transportation and Infrastructure
ISBN 978-3-030-68654-3 ISBN 978-3-030-68655-0 (eBook)
https://doi.org/10.1007/978-3-030-68655-0

© The Editor(s) (if applicable) and The Author(s), under exclusive license to Springer Nature Switzerland AG 2021
This work is subject to copyright. All rights are solely and exclusively licensed by the Publisher, whether the whole or part of the material is concerned, specifically the rights of translation, reprinting, reuse of illustrations, recitation, broadcasting, reproduction on microfilms or in any other physical way, and transmission or information storage and retrieval, electronic adaptation, computer software, or by similar or dissimilar methodology now known or hereafter developed.
The use of general descriptive names, registered names, trademarks, service marks, etc. in this publication does not imply, even in the absence of a specific statement, that such names are exempt from the relevant protective laws and regulations and therefore free for general use.
The publisher, the authors and the editors are safe to assume that the advice and information in this book are believed to be true and accurate at the date of publication. Neither the publisher nor the authors or the editors give a warranty, expressed or implied, with respect to the material contained herein or for any errors or omissions that may have been made. The publisher remains neutral with regard to jurisdictional claims in published maps and institutional affiliations.

This Springer imprint is published by the registered company Springer Nature Switzerland AG
The registered company address is: Gewerbestrasse 11, 6330 Cham, Switzerland

Introduction

This book is a novel, innovative, and adequate source of information that compiles interdisciplinary perspectives about diverse issues related to Intelligent Logistics in Industry 4.0 including Artificial Intelligent applications associated with Logistics 4.0 on different ways about Intelligent Optimization, Industrial Applications on real world, Social applications and Technology applications each one with a different perspective about the correct solution of this kind of methodologies. This book is a collective effort to introduce new ideas and paradigms from a variety of perspectives using innovative techniques related to Bioinspired Algorithms, metaheuristics, and methodologies associated with Logistics 4.0. An innovative and specialized book on optimization considers different aspects to realize this "Intelligent Optimization" and tries to improve with innovative techniques and methodologies different daily aspects of our lives, in each one of them it is possible to understand the necessity to improve scenarios, distances, time, costs, spaces, and a plethora of features associated with the modern life (labor associated with organize, kept, and delivery of goods, materials, people, life issues, or products).

We received manuscripts from renowned researchers from all around the world associated with Theoretical foundations of Intelligent Logistics to understand many paradigms on different Optimization implementation kinds. In addition, we received many manuscripts with expertise on improving optimization related to Logistics 4.0 of deliveries associated with products and services, Optimization of different elements in the time and location, Social Applications to enjoy our life of a better way to improve the life in a Smart City, and finally, Technologies Applications of diverse ways to increase our Life Quality. The book starts with a part entitled Industrial Logistics featuring seven chapters on the theoretical and mathematical ideas related to the correct implementation of a diverse range of Intelligent Logistics optimization algorithms and Logistics 4.0 applications in real world. The first chapter of this section is "Determining and Applying Productive, Environmental and Economical Indicators and Indexes to a Cyber Physical System for Greening Process of Supply Chain" which aboard that a company that transports goods to supply customers usually needs to plan the routes that the fleet must follow, since transportation means a high percentage of the value added to goods. The chapter "Design of Material Delivery Routes with Towing Equipment for the Automotive Sector Using a Von Neumann Topology of PSO Under the Logistics 4.0 Paradigm" suggests that applying Warehousing in profit-oriented environments requires handling complex constraints for planning, organizing, or managing tasks. A large amount of input causes a high degree of complexity, as in the case of the management of customer's time windows to accurately deliver them their products. The chapter "Industrial Accident Induced Coma: A Multi-Disciplinary Perspective Within the Industry 4.0 Paradigm" explains that the warehousing is an essential issue in the logistic supply chain. Among the order picking processes in a warehouse is the picker routing problem which consists in determining the best picking sequence to retrieve the items from locations to satisfy customer demand. The chapter "Coahuila's Future a Perspective Derived from the Growth Per Municipality to Establish a Car Assembler." explains an Order picking model as an activity in a warehouse, carried out by a team, where customer orders are prepared. In chapter, "Layout Problem: Optimization with Material Tour in Open Field. Case of Study", is described also the activity with the highest logistic costs in a company. In this research, a proposal is presented, aimed at improving the order picking process in a warehouse. The chapter "Waste Collection of Touristics Services Sector Residues Vehicle Routing Problem with Time Windows to an Industrial Polygon in a Smart City" detailed that today the problem of product transportation has taken great importance due to the demand of products in the market that has to be delivered, solving the transportation problem is a complex task because it involves the solution of three NP-hard problems. Finally, the last chapter of this part is "Distribution of Merchandise Through Clarke and Wright Heuristic and Mathematical Model: Case Study" which determines that in the purchase of products, a warehouse is an integral part between producers and customers, where the Logistics 4.0 is one of several operations involved. The next part is named Transport and Movements of Vehicles in a Smart City, featuring six chapters related to different comparatives of Metaheuristics used in Intelligent Logistics in the search to improve resources in diverse aspects of our transportations or improve process to this. The chapter "Rethinking the Effects of Fatal Falls on an Italian Scooter in a Smart City: An Approach from Ergonomics and an Ideal and Optimal Helmet: Conditioned Helmet for Improved Security on the Road in a Smart City" explains the use of a metaheuristic algorithm as an optimization method based on swarm intelligence which has demonstrated to be capable of obtaining satisfactory results on a diversity of optimization problems. In the proposed chapter "Design of an Urban Transport Network for the Optimal Location of Bus Stops in a Smart City Based on a Big Data Model and Spider Monkey Optimization Algorithm", is analyzed a novel metaheuristics; however, the implementation of this optimization method has not been much explored on Intelligent Logistics problems, even though Industry 4.0 represents up to 55% of the total operational cost of a typical transportation model. The chapter "Public Urban Transportation in the Smart City: An Exploratory Study in the Northern México" explains the optimization of bus stops in a Smart City, coupled with the improved use of mobile ateliers at some of these bus stops, also called pop-up stores, which sell their products outside their store using a model called pret-a-porter. Therefore, it causes them to go back to it once a product in their mobile store runs out since the customer is waiting for them at the store. The chapter "The Difficulties and Complications of Children When Going to a Zoo and Should Interact with the Colors of the Information in It: An Approach Based on the Use of a Humanoid NAO Robot in an Application for "Smart Cities"", determines as in nowadays that the health industry utilizes critical machines to treat and diagnose illness in patients. In a large majority of hospitals that treat patients with color blindness, especially children, mobile applications are beginning to be used to improve the quality of life of these patients in environments that require information associated with safety colors or the level of extinction of a species, such as in zoos; then, they choose to have many components of their critical machines stored inside a warehouse in the hospital. The chapter "Optimization of Route Planning for the Package Delivery Problem Using Fuzzy Clustering" is explains that nowadays, warehouse operations, specifically order picking process, are receiving close attention of researches due to the need of companies in minimizing operational costs. Finally, the last chapter in this part is "State of the Art for the Creation of a Methodology for the Proper Location of Urban Truck Stops on Route 2A" which determines that in these times, the human factor is key to improve order preparation processes—which is why it is necessary to determine a correct bus stop-. For example, in a Smart City associated with a line of replacement processes for different pieces of equipment or devices—which must be purchased in a specialized store-, it is necessary to find the best bus route in each case to minimize the time consumed, and that this does not affect the continuity of the assembly in small workshops specialized in automotive parts.

The third part is named Humanitarian Logistics, featuring five chapters related to different comparatives of Humanitarian Logistics and Intelligent Logistics Models in the search to improve resources in diverse aspects of companies and to improve our lives. The first chapter of this section is "Financial Analysis Over the Smartest Companies" and in this research chapter is focused the problem of the distribution of escape routes in a public space. The chapter "Simulating Crowd Movements During Emergency Fire Situations: Mexico City Airport Simulation Case" details as in several societies of emerging economies, increasingly large community linked to the scalability and adjustment of components to resolve a problem of Humanitarian Logistics. In the research chapter "Modular Framework for Crowd Simulation "Menge" from a Production Warehouse Simulation Perspective", many industrial sectors use 3D components design customization to improve aspects of competitiveness. In the chapter "Mobile Application for the Detection of COVID19 Suspicious Cases in Mexico Using an Intelligent Model of Virtual Patients". In hospitals with an excessive number of patients and the decrease of medical services associated with other diseases, we propose the use of an intelligent model of virtual patients that allows students of health sciences to practice in an adequate and valid way the detection of various symptoms that affect the health of the population in a

Smart City. The next chapter "Humanitarian Logistics for the Optimal and Timely Evacuation in High Buildings Within a Smart City Using an Adaptive Metaheuristic Context" explains about the analysis concerning the accommodation of safe escapes in a building inside a structure for their respective distribution, verifying the option of finding the correct stow model and accommodation of humanitarian logistics.

In fourth part, which is named E-commerce, Marketing and Mobile Application of Logistics Including Human Factor, are described five chapters related to technologies associated with Intelligent Logistics and Logistics 4.0. The first chapter of this part is "Using Machine Learning to Predict Online Buying Behaviour, Wholesale and Fashion Marketing at Zara, an Analysis Including Z Generation" which presents a problem solved by social modeling, associated with the adequate choice of colors to issues and their distribution in a Fashion Market using a range of 64 colors to specify different features related to the principal attributes of an issue adequate to represent the symbolic capital of a modern society. In the chapter "Analysis of Mental Fatigue Under Delivery Pressure and Considering Creativity and Precision to Organize and Distribute a Diorama to Represent Social Issues Based on Cultural Algorithms" details how mental fatigue is a decisive aspect in the creation of collectible issues, especially in a growing collecting community—mainly from emerging economies—which collects and buys "dioramas and click toys among others", that is, scalable and adjustable collectibles composed of many pieces; this type of paper toys and dioramas have been used for educational purposes. In the chapter "Medicine Inventory Control System Through Fuzzy Logic and Genetic Algorithms: Applied to a Biopharmaceutical" is described a novel research focused on the main economic activities in a Smart City; in another chapter named "Technical Analysis of Shipments in an Automotive Company to Forecast Sales Volumes" is proposed the implementation of an order picking algorithm for the optimization of the packing and distribution of car component products. Finally, in this section is presented "Distributed Programming Applied for the Optimization of Hydraulic Networks Through a Web Application", which proposes new ideas related to web applications by means of an Advanced Selection of systems, which is fundamental for the operational improvement and logistics global supply chain.

And finally, in the part entitled Diverse Kind of Logistics in Amalgamed Application Domains grouped nine different chapters related to solutions derived of specific aspects which try to improve daily activities on Optimization with real applications to amalgamed social topics. In the chapter "What is the Best Location of a Smart Airport in Juarez, Mexico?" is described, as the latter includes, customizing the user interface, as well as the way the system retrieves and processes cases afterward to distribute original products in a novel supply chain and its respective Intelligent Logistics. The purpose of the chapter "Colombian Coffee Price Forecast via LSTM Neural Networks" is to propose a conceptual order picking model to increase the commercialization of coffee in Colombia, through the sequential analysis of activities such as distribution in the warehouse, preparation, packing of orders, and the issuance of orders to final customers. The next chapter "Some Pragmatic Prevention's Guidelines Regarding SARS-CoV-2 and COVID-19 in Latin-America Inspired by Mixed Machine Learning Techniques and Artificial Mathematical Intelligence. Case Study: Colombia" details that due to the world-wide strengthening of the health sector, it presents itself as a challenge for the companies that comprise it to immerse themselves in processes of continuous improvement that contribute to increasing the satisfaction of the needs of its customers, as well as achieving a better positioning in the market. In the chapter "A Drone System for Detecting, Classifying and Monitoring Solid Wastes Using Computer Vision Techniques in the Context of a Smart Cities Logistics Systems" is proposed that the freshness, flavor, good presentation, and nutritional value of fruits and vegetables diminish as time passes until the food begins to lose them completely. That is why the correct implementation of supply chains is a subject of great interest for companies dedicated to the rotation of food marketing. The next chapter "Geo-Referenced Correlation for a Fire in a Smart City Urban Forest Using Hybrid Drone Data and Satellite Images" determines that the purpose of this research is to understand a Multivariable optimization associated with the path of a group of vehicles integrated in an Ecological Community and determine the optimal route involve speed, storage, and travel resources including time of charge for determining the cost-benefit linked to safety in case of a disaster as a wildfire at a Natural Park in a Smart City and considering that most of the drivers in such an ecological community own an electric car, which is coupled with a travel plan associated with the electric power charging point in a Smart City. In the chapter "Evaluation of Drones for Inspection and Control in Industry 4.0" is analyzed that Internet sales have increased exponentially in the last decade. Much of the internet sales are of physical products in urban areas that require product delivery transportation with a tight delivery lead time using drones for this purpose. And finally, in the chapter "Uncertain Analysis Based on Milk-Runs Systems Using Bayesian Networks" is a important considering that one of the most common operations in warehouses of package delivery companies (e.g., UPS or FedEx) is to pack the products in trucks in order to locally deliver them to the customers. The products are generally packed in rectangular-shaped boxes of different dimensions; in addition in the chapter "Implementation of an Intelligent Visual Recognition System for the Proper Classification of Solid Waste Using a Mobile Application in a Smart City" explains a model of ecological support to groutier people with shady hair linked to Greenpeace—which allows a better identification of solid waste and its correct and adequate separation, using a smart mobile application for recycling within the Z generation—to improve the lives of citizens in their environment within a smart city. The last chapter of this part is "Logistics on the Designing of an Electronic Colorblindness Application for Early Colorblindness Detection in Children by Using a Modified Ishihara Test", which describes the social inclusion with colorblindness.

It is important to state that the chapters were selected following a rigorous analysis done by the book editors, and each chapter was double or triple-blind peer-reviewed by at least two experts in the area. This would not have been possible without the valuable help of the Editorial Advisory Board. The content of the chapters included in this book is the sole responsibility of the authors. The views, opinions, or positions expressed by the chapter authors are solely those of the authors, and do not necessarily reflect the views, opinions, or positions of the editors. All trademarks, trade names, service marks, and logos referenced in the chapters of this book belong to their respective companies.

Dr. Diego Oliva
Departamento de Ciencias Computacionales
Universidad de Guadalajara, CUCEI
Guadalajara, México
e-mail: diego.oliva(at)cucei.udg.mx; doliva(at)ucm.es

Dr. Angel Juan Perez
Informática Department
Universtitat Oberta De Catalunya
Barcelona, Spain
e-mail: ajuanp(at)uoc.edu

Alberto Ochoa-Zezzatti
Juarez City University
Juárez, México
e-mail: alberto.ochoa@uacj.mx

Contents

Humanitarian Logistics

E-commerce, Marketing and Mobile Application of Logistics Including Human Factor

Industrial Logistics

Determining and Applying Productive, Environmental and Economical Indicators and Indexes to a Cyber Physical System for Greening Process of Supply Chain

Paula Morella, Dra. María Pilar Lambán, Jesús Royo, Juan Carlos Sánchez, and Mario Enrique Hernández Korner

Abstract Sustainability is taking on increasing value in the industrial world. This article aims to develop and implement in a cyber-physical system two new KPIs capable of transferring the current concept of Supply Chain (SC) to a Green Supply Chain (GSC), which no longer addresses only the productive and economic issues of traditional SC, but also addresses issues of sustainability, relying on industry 4.0 to achieve this goal. Each KPI has a different function within this process. The first one allows selecting the most adequate scenario to be implemented in the SC from all those proposed, attending to these three issues at the same time (productive, economic and environmental). The second allows the monitoring, at the same time, of the machine tool in these three dimensions, so that, it is possible to analyze throughout the production period of the machine its productive, economic and environmental evolution.

Keywords Cyber-physical system · Industry 4.0 · Internet of things · Green supply chain · KPI

P. Morella (✉) · Dra. M. P. Lambán · J. Royo · M. E. Hernández Korner
Design and Manufacturing Department, School of Engineering and Architecture, Universidad de Zaragoza, Maria de Luna 3, 50017 Saragossa, Spain
e-mail: 620453@unizar.es

J. C. Sánchez
Smarts Systems, Tecnalia, Pº Mikeletegi 7, E-20009 Donostia-San Sebastian, Spain

M. E. Hernández Korner
Universidad Tecnológica de Panamá Vía Centenario. Ancón, Campus Victor Levi Sasso, Panamá, Panamá

© The Author(s), under exclusive license to Springer Nature Switzerland AG 2021

A. Ochoa-Zezzatti et al. (eds.), *Technological and Industrial Applications Associated with Intelligent Logistics*, Lecture Notes in Intelligent Transportation and Infrastructure,
https://doi.org/10.1007/978-3-030-68655-0_1

1 Introduction

1.1 Industry 4.0

Industry 4.0 well known as the fourth industrial revolution represents an entire transformation of the traditional production model obtained combining information and communication technologies, and the internet [4]. Industry 4.0 is composed of 9 technologies (Fig. 1) that are the key to its development. However, there is one technology that stands out from the others, cyber-physical systems (CPS), as numerous research studies state that their combination with other technologies such as IoT, Cloud Computing or Big Data, is the key to developing a smart industry [15]. CPS integrate cyberspace and physical processes, turning production machines into a network, which is prepared to make decisions with real-time data and is capable of communicating this data to humans. To do this, it is also necessary to have other types of technologies, such as IoT, which allows data transmission between devices [5].

Fig. 1 Industry 4.0 technology

1.2 Circular Economy: The Way to Green Supply Chain

Circular Economy (CE) is based on an economic system capable of replacing the current 'end of life' model with one in which production, distribution and consumption are reused, recycled and reduced [9]. CE main purpose according to Alamerew and Brissaud [2] is the pursuit of economic prosperity at the same time environmental quality and social equity for a sustainable world for future generations. The economic and environmental prosperity idea added to legal, social, and stakeholder pressures are factors that have made CE a competitive advantage for companies. The GSC concept is born from CE, providing companies with resource optimization as a solution to environmental problems and consumption patterns throughout the SC (SC) [8].

It may seem that GSC only adds an environmental dimension to conventional SC, the reality is that GSC is capable of uniting three dimensions: the productive, the economic and the environmental, making its implementation very attractive for companies. Kazancoglu et al. [8] affirm that GSC does not only focus on the environment, but also on operational, logistical, economic and marketing effi-ciency. Micheli et al. [13] identifies the implementation of GSC practices with improvements in the company from economic, productive and environmental point of view. Vanalle et al. [16] support that there are improvements in the op-erational performance by reducing environmental impact in productive operations.

1.3 Key Performance Indicator for Green Supply Chain

Despite the fact that the GSC proposes an advantageous model for companies, the environmental dimension becomes more complex to manage a balanced score-card, since multiple parameters such as environment, economy, and productivity would be attended at a time [7]. Many authors proposed various indicators (KPIs) for this new model of SC. Kazancoglu et al. [8] proposed various KPIs in different dimensions: environmental, economic, logistic, opera-tions, and organizational. Others proposed multi-criteria decisions through various associated KPIs, as it is the case of Hatim et al. [7] who simulated variations in machining cost, energy consumption and manufacturing time. Alamerew and Brissaud, [2] also proposed one alternative among various simulations but with other indicators such as carbon footprint emission, energy consumption, re-newable energy used, reverse logistics cost, assembly cost, manufacturing cost, and the number of employees. The KPIs obtained from a simulation are only useful for the initial decision making, such as taking a picture of the GSC at the beginning, but they do not allow to continue monitoring the GSC development in real-time.

Moreover, an environmentally-focused SC can be obtained adding CE considerations to the SC concept, that has economic and operational benefits. For that reason, obtaining a right KPI is crucial for the correct operation of a GSC and its performance evaluation [13]. However, several authors, like [6], affirm that GSC indicators, drivers, and conse-quences of GSC implementation have not been developed in depth.

1.4 Advantages of Applying Industry 4.0

Nowadays, GSC is becoming more complex and expensive which makes it dif-ficult to continue using conventional mechanisms and strategies within it [19]. As it is affirmed by Sam Palmisano (2003–2012) IBM Corporation's former president, the use of intelligence in decision making, processes, and management systems is a great improvement to today's speed and capacity approach to SC. Industry 4.0 is able to reduce this complexity through technologies such as IoT or the CPS that allow to interact with the machines in real time, acquiring data and being able to use them, for example, in the calculation of KPIs, decision mak-ing, task optimization or sensorization [5].

Even though GSC present competitive advantages nowadays there are some barriers to implementation, motivated by the lack of sustainable effective indicators and the absence of implemented technology for GSC [1]. This article aims to eliminate these barriers by implementing a cyber-physical sys-tem on a machine tool capable of obtaining real-time data for the calculation of productive, economic and environmental indicators that can be combined, thus en-abling decision making in accordance with the 3 dimensions of GSC. From these KPIs, two indexes with different functions within the GSC are obtained. The first one allows the selection of the most appropriate scenario for the GSC implementa-tion among all considering three dimensions at a time (productive, economic, and environmental). The second one allows real-time monitoring during the whole machine tool process in these three dimensions mentioned, so that it is possible to analyze the integrated productive, economic and environmental evolution of the SKU (Single Key Unit) throughout the production period.

2 Materials and Methods

The previous step to the calculation of these KPIs and the correlational indexes involves the transformation of a machine tool into a CPS. A system that will allow necessary data capture for real-time calculus. Afterward, the KPIs and the testing method for data acquisition are defined. Finally, various software packages are developed for the exact KPI calculus using Python language. Python is a pro-gramming language that allows easy and quick software development [17]. Among the Python remarked features is the large number of libraries for numeric computation, scientific, data analysis, automatic learning, and plots visualization that allows prototypes rapid creation [3].

Our main interest in this software is the analysis of the results provided by the algorithms developed to visualize them in real time, support decision making and solve dynamic problems.

2.1 *Cyber-Physical System Implementation in a Machine Tool*

Cyber-physical technology can obtain real-time data from productive systems and SC to allow efficient resources (materials, energy, water) use [5]. The machine tool used in this research is located in the Uni-versidad de Zaragoza process area laboratory, a vertical milling machine with 5 axes (Haas VF-3), which includes X, Y, Z, and the A and C axes were added using a dual-axis Trunnion 160 rotary table. With the important collaboration of the Tecnalia research center, it was possible to install an industrial computer Beckhoff model C6015 to connect the milling machine through an ethernet cable and a net-work analyzer to be capable of measuring energy consumption. A computer app captures every second the data of 32 variables in real time and stores them, in the cloud every 15 min, being able to access and download them at any time (Fig. 2). Every variable captured by the CPS cyber-physical system was ana-lyzed and selected previously. For the development of this research, relevant vari-ables from the 32 were selected, Table 1 mentions each one.

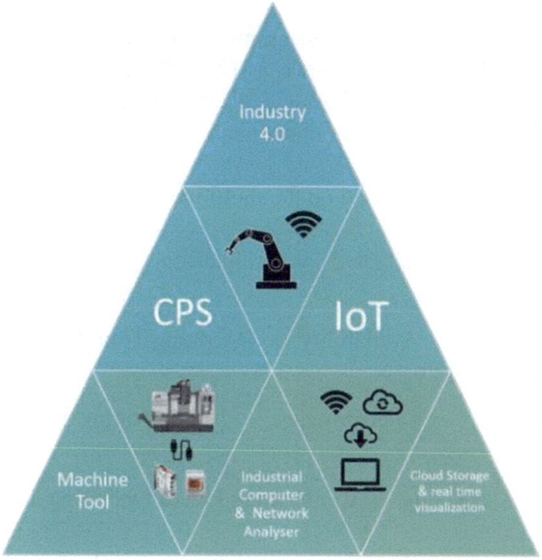

Fig. 2 Cyber-physical system implementation

Table 1 Cyber-physical system parameters

Principal parameters collected from the cyber-physical system
Machine time turned on (s)
Machine time stopped (s)
Machine time milling a piece (s)
Tool usage time (s)
Machining time of each piece (s)
Number of tools used
Number of pieces manufactured
Spindle turn speed (rpm)
Energy (Wh)

2.2 *Key Performance Indicator Definition*

This research seeks, in a single index, to implement and measure productive, economic, and environmental indicators for decision making in a GSC context.

Before moving on to the KPIs definition, a baseline time named operating time has been defined as the manufacturing time without the planned stoppage associ-ated with the productive process to be analyzed.

2.2.1 The Key Performance Indicator for the Productive Dimension

The main idea of measuring productivity, one of the main concepts in SC, is using a quantitative tool, known as Overall Equipment Effectiveness (OEE). The OEE is an indicator that has several years been used for industrial purposes [10]. The OEE indicator will be implemented as a KPI in our CPS to obtain the machine tool productive dimension.

The OEE is based on the multiplication of three parameters (Eqs. 1, 2, 3, and 4):

$$\text{OEE} = \text{Availability} \times \text{Performance} \times \text{Quality} \quad (1)$$

$$\text{Availability} = \text{Operating Time/Planned Operating Time} \quad (2)$$

$$\text{Performance} = \text{Net Operating Time/Operating Time} \quad (3)$$

$$\text{Quality} = \text{Conforming Pieces Operating Time/Net Operating Time} \quad (4)$$

The three parameters calculation is based on the equipment losses classification per operating time [14], described in Fig. 3.

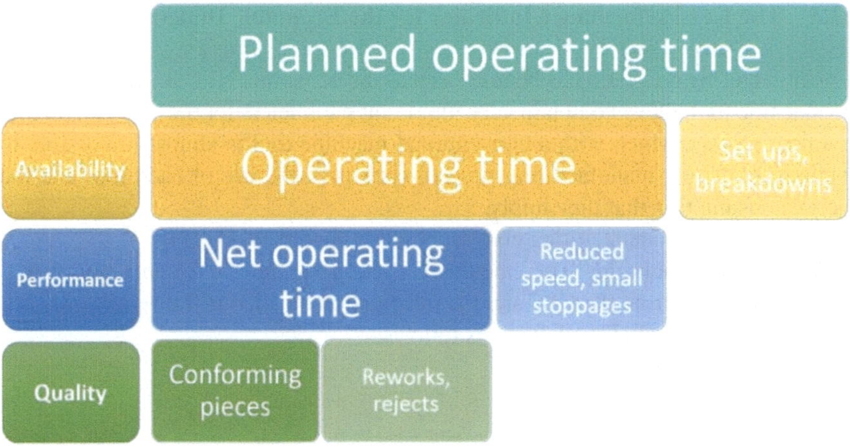

Fig. 3 Overall equipment effectiveness breakdown

CPS implementation allows real-time and accurate OEE calculation. CPS can collect parameter data (Table 1) which is processed by a developed program in Python and processed with other information such as rejected references or maximum time for a minor stoppage. The productive KPI named Productive Index (PI) is based on OEE calculus through Industry 4.0 tools and is calculated to show the machine tool unproductive rate (Eq. 5).

$$PI = 1 - OEE \tag{5}$$

2.2.2 The Key Performance Indicator for the Environmental Dimension

For the environmental dimension, it has been decided to take as an indicator a ratio of the carbon footprint emitted by the machine tool during the production process. Machine tool energy consumption is 75% in the manufacturing industry [18].

The CPS has a network analyzer that can obtain energy consumption during the machine tool operation, which is the operation that consumes the most energy. As in the case of OEE, this consumption parameter is used in a Python program to obtain carbon footprint multiplying the energy consumption by the greenhouse gas emission factor associated. The Environmental Index (EnvI) is calculated as fol-lows:

$$\text{EnvI (Kg CO}_2\text{eq)} = \text{Carbon Footprint/Maximum Carbon Footprint} \tag{6}$$

Considering the Maximum Carbon Footprint (kg CO_2 equivalent) as the maximum carbon footprint emitted continuously among the planned period when operation energy consumption is higher.

2.2.3 The Key Performance Indicator for the Economic Dimension

A manufacturing cost ratio has been chosen as the economic KPI, since moni-toring the machine tool allows to obtain very important parameters for the cost calculation in a precise way, which with a conventional machine tool is almost impossible to obtain, such as the manufacturing time of each part, the use of each tool and the energy consumption that they imply.

We follow the manufacturing cost proposed by Lambán [11] (Fig. 4).

For the manufacturing cost calculus, is required, time and consumption parameters obtained from the CPS and the information detailed in Table 2.

Economic Index (EcI) is defined as follows:

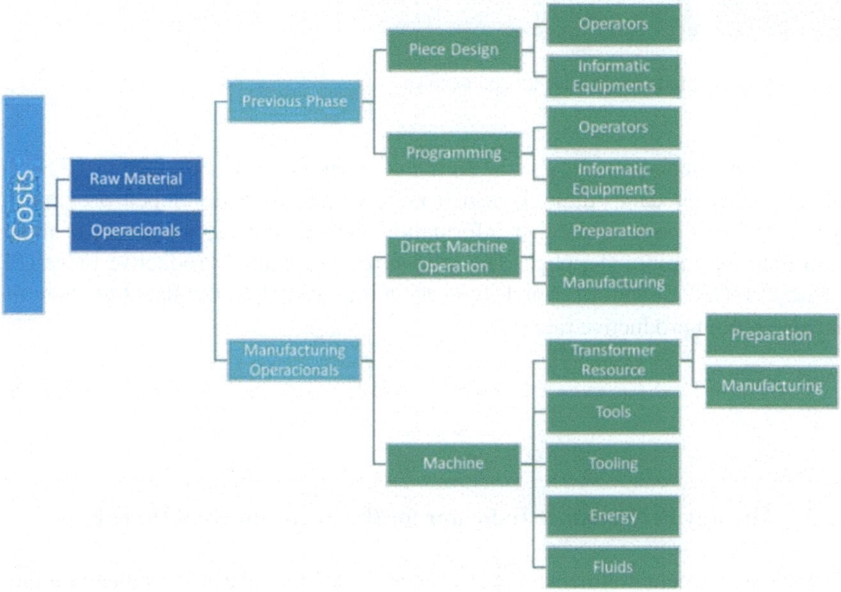

Fig. 4 Cost breakdown structure

Table 2 Parameters to obtain the manufacturing cost

Resource	Required information
Raw material	Cost
Machine operator	Salary
Transformer Resource	Purchase, battery, and mechanical sensors cost
Tools	Cost and lifespan
Tooling	Cost
Fluids	Cost

$$\mathrm{EcI} = \text{Manufacturing Cost}(\in)/\text{Maximum Manufacturing Cost}(\in) \quad (7)$$

Maximum Manufacturing Cost (€) is defined as the maximum manufacturing cost during the worst scenario of the operative costs from the planned operating time explained in Fig. 3.

2.2.4 Definition of the Productive, Economic, and Environmental Index (Green Supply Chain Index) and Decision-Making Index (Green Decision Index)

The objective of this article is to define two indexes that represent the three most important dimensions of the GSC together in a machine tool. To fulfill the goal two different indexes are proposed. Firstly, to collate various scenarios to se-lect one to be implemented in the GSC. Secondly, an index that encompasses all three dimensions and that can be calculated throughout the productive life of the machine tool and compared with itself to see the improvement or possible prob-lems of the GSC.

In the Green Decision Index (GDI) are assumed multiple scenarios (n) (S1, S2, ..., Sn) from which were obtained the productive KPI (**P**roductive **I**ndex **S**cenario **1** (PIS1), **P**roductive **I**ndex **S**cenario **2** (PIS2), ..., **P**roductive **I**ndex **S**cenario **n** (PISn)), environmental KPI (**Env**ironmental **I**ndex **S**cenario **1** (EnvIS1), **Envi**-ronmental **I**ndex **S**cenario **2** (EnvIS2), ..., **Env**ironmental **I**ndex **S**cenario **n** (En-vISn)), and economic KPI (**Ec**onomical **I**ndex **S**cenario **1** (EcIS1), **Ec**onomical **I**ndex **S**cenario **2** (EcIS2), ..., **Ec**onomical **I**ndex **S**cenario **N** (EcISn)).

Once the defined KPIs are calculated for each scenario, the indicators of same dimension and different scenario, such as PI for scenarios 1, 2 and 3, are sorted giving a Value (Vi) from 1 to n, being 1 the best KPI value and n the least relevant from each dimension (Productive Index Value, VPI, Environmental Index Value, VEnvI, and Economical Index Value, VEcI). Every GSC dimension is complemented by a weight (wpr, wenv, wec for productive, environmental and economic dimension, respectively) that should be selected according to the importance given for each dimension and following the strategy of the enterprise, being 1 the sum of the three weights.

Moreover, the GDI is calculated using the weighted sum of values Vi for each scenario associated (Eq. (8)). The scenario with the lower GDI is considered the worst scenario for GSC implementation.

$$\mathrm{GDI}_{\mathrm{Si}} = \mathrm{w}_{\mathrm{pr}} \times (\mathrm{VPI})_{\mathrm{Si}} + \mathrm{w}_{\mathrm{env}} \times (\mathrm{V}_{\mathrm{EnvI}})_{\mathrm{Si}} + \mathrm{w}_{\mathrm{ec}} \times (\mathrm{V}_{\mathrm{EcI}})_{\mathrm{Si}} \quad (8)$$

Calculated scenario Si, (VPI)Si, (VEnv)Si y (VEc)Si values assigned to each KPI scenario i and wpr, wenv and wec to the associated weight respectively for produc-tive, environmental, and economic dimensions in which summing result is 1.

For the productive, economic, and environmental index, GSC Index (GSCI), is calculated as follows, multiplying the indicators of every dimension in the same scenario (Si):

$$GSCI = PI_{Ei} \times EnvI_{Ei} \times EcI_{Ei} \quad (9)$$

2.2.5 Essay Method Definition

Once the KPIs have been theoretically defined, a test for the machine tool transformed into a CPS (Sect. 2.1) has been defined. The test has been divided in 3 cases, during which pieces of the same reference have been manufactured for 1 h. Varying in each case the machine tool speed with respect to the manufac-turer speed recommendation by material to be milled, quality sought, structure, and speed turn of the tool. For case 1 at 100% machine tool speed capacity, for case 2 at 75%, and case 3 at 50%, for each case 1 defective piece is considered. Table 3 resumes case characteristics.

The reference used for the essay is a milled piece from cylindric F-114 steel billet with 50 mm diameter and 75 mm length. For each piece, the following operations were performed: facing, exterior finishing, drilling, and boring (Fig. 5).

3 Results

Below are the results obtained from the tests for the KPIs defined above.

3.1 Productive Dimension—Key Performance Indicator Results

The results obtained by applying the programs developed with Python can be seen in Table 4. The result obtained from this test presents that case 3, with lower speed, has higher OEE because when performing fewer parts, the loss time by mi-nor stoppage and adjustments is lower than in the other two cases. In this case availability and performance are higher, whereas the quality decreases.

Once the OEE value is known the PI can be obtained as shown in Table 5, and the lower PI, the higher productivity.

Table 3 Essay case characteristics summary

Case	Speed (%)	Number of manufactured pieces
Case 1	100	15 pieces
Case 2	75	12 pieces
Case 3	50	8 pieces

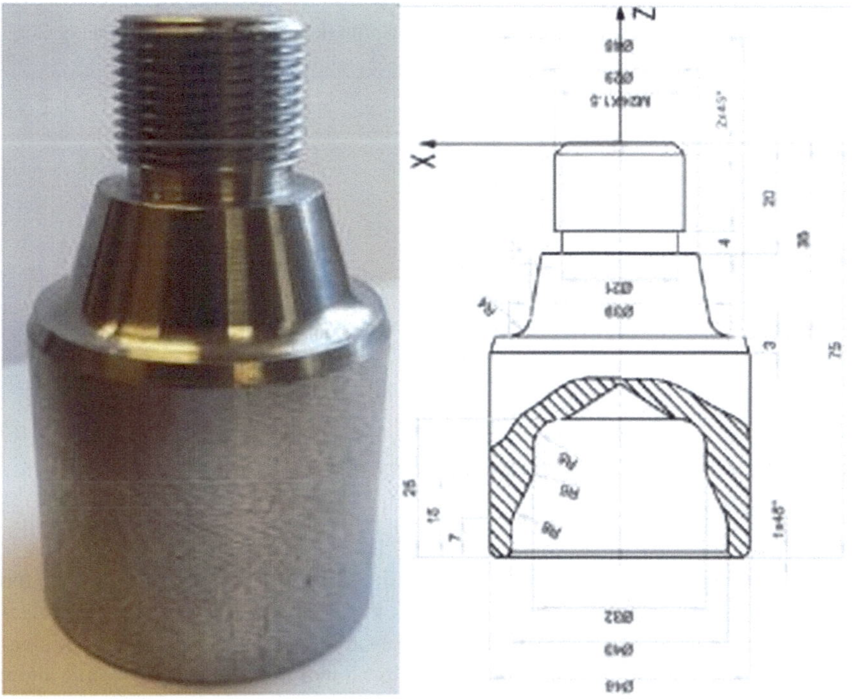

Fig. 5 finished piece (left), piece diagram (right)

Table 4 Overall Equipment Effectiveness results of each case essayed

Case	Availability (%)	Performance (%)	Quality (%)	OEE (%)
Case 1	85.83	72.48	93.33	58.06
Case 2	88.36	72.35	91.67	58.60
Case 3	92.42	74.30	87.50	60.08

Table 5 Productive index results

Case	PI (%)
Case 1	41.94
Case 2	41.40
Case 3	39.92

Table 6 Carbon footprint results for each case

Case	Carbon footprint (kg CO_2 eq)	Carbon footprint (kg CO_2 eq)/piece
Case 1	0.350	0.023
Case 2	0.351	0.029
Case 3	0.292	0.036

Table 7 Calculated EnvI

Case	EnvI (%)
Case 1	39.98
Case 2	39.98
Case 3	32.44

3.2 Environmental Dimension—Key Performance Indicator Results

After obtaining the energy consumption of the CPS in kWh, we multiply by the factor of energy consumption and emissions of Greenhouse Gases from the electricity supply, which AECOC (AECOC, n.d.) establishes at 0.424 kg of CO2 equivalent/kWh to obtain the carbon footprint results shown in Table 6. The results show that the carbon footprint is higher in case 1 than in case 3, however, if the emitted footprint per part is analyzed, case 3 is the one that emits the most carbon footprint per piece.

To calculate the EnvI, it is necessary to know the maximum equivalent CO2 that would be emitted during the entire case, that is, the hour of the operation that consumes the most energy. In the three cases the operation with the highest consumption is machining, which consumes on average 0.25 Wh per second, there-fore, in each case for one hour a maximum of 0.9 kWh would be consumed. With this data the EnvI (Table 7) is calculated according to Eq. (6). The results of the EnvI and the carbon footprint show that case 3 produces lower emission.

3.3 Economic Dimension—Key Performance Indicator Results

The result of the manufacturing cost can be seen in Table 8. As in the environmental case, the total cost is higher in case 1 because more parts are manufactured. The individual cost of each part is less, because the machining time of each part is shorter, causing time-dependent costs to be reduced and fixed costs are divided among a larger number of parts.

The software developed in Python displays various diagrams to analyze manufacturing costs. As an example, the Pareto diagram (Fig. 6, 7, and 8) shows breakdown

Table 8 Economic Key Performance Indicator results

Case	Cost (€/piece)	Total cost	
Case 1	5.96 €	89.40	€ 15 pieces
Case 2	6.89 €	82.68	€ 12 pieces
Case 3	9.22 €	73.76	€ 8 pieces

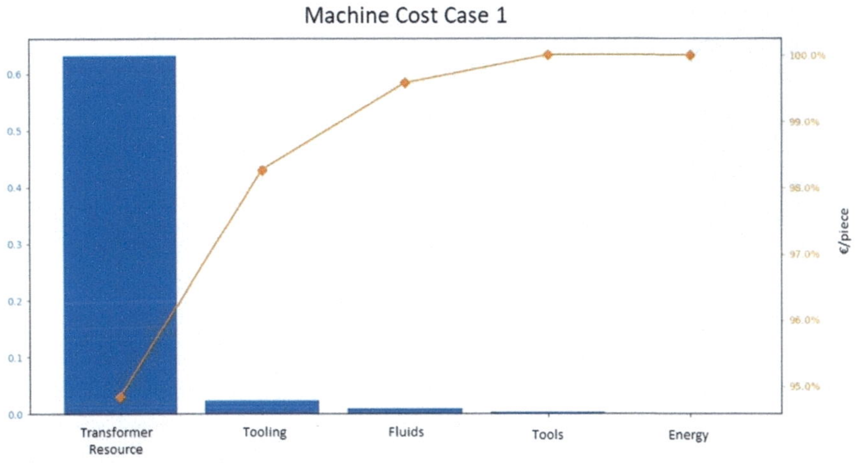

Fig. 6 Machine cost breakdown in case 1

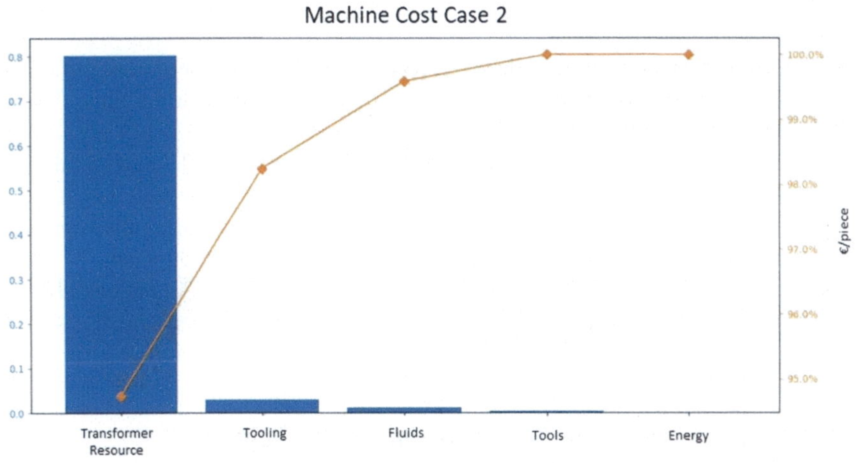

Fig. 7 Machine cost breakdown in case 2

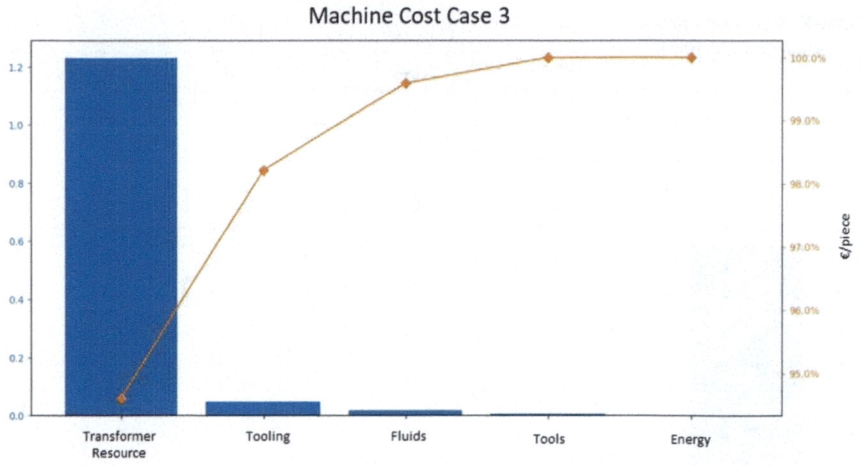

Fig. 8 Machine cost breakdown in case 3

Table 9 Calculated EcI for each case

Case	EcI (%)
Case 1	97.93
Case 2	98.33
Case 3	99.02

of the machine cost and it is possible to observe which element generates higher cost. In this test the transformer resource generates the highest cost in the three cases, followed by the tools cost.

Once the cost of manufacture is determined, the calculation of the maximum cost of manufacture is carried out, in this case it would be the cost of raw material plus the set-up cost plus the maximum cost of manufacturing operations during the hour that lasts each case of the test, being this maximum cost 91.29 €, 84.09 € and 74.49 € respectively for case 1, case 2, and case 3. The result of the EcI is defined in Table 9, the cost of manufacture is close to the maximum cost of manufacture in all cases.

3.4 *Results of the New Key Performance Indicators*

The results of PI, EnvI, and EcI are summarized in Table 10, and they will be used for the calculation of both indexes. The nature of these indexes considers that closer to zero is better.

Table 10 Summary of the calculated indexes for each case

Case	PI (%)	EnvI (%)	EcI (%)
Case 1	41.94	39.98	97.93
Case 2	41.40	39.98	98.33
Case 3	39.92	32.44	99.02

Table 11 Green Decision Index results for each case

Case	VPI	wpr *VPI	EnvI	wenv *VEnvI	EcI	wec *VEcI	GDI
Case 1	3	0.9	2	0.4	1	0.5	1.8
Case 2	2	0.6	2	0.4	2	1	2.0
Case 3	1	0.3	1	0.2	3	1.5	2.0

3.4.1 Green Decision Index Results

The indications given in Sect. 2.2.4. are followed for the calculation of the GDI, so the first step is to select the weights of each index. In this case, the eco-nomic dimension has been considered the most relevant and the environmental the less one. For that reason, the weights are as follows: wpr = 30%, wenv = 20%, and wec = 50%, the total sum of these three weights is 100%. Each indicator is valued between 1 (best scenario) and 3 (most unfavorable scenario), and then multiplied by the weight. For example, focused on the PI results in Table 10, case 3 has the best PI of these three cases, thus, it must be valued as the best scenario (VPI case 3 = 1).

Finally, the factors are sum as seen in Table 11 to obtain de GDI. The case 1 has the lowest GDI, being the most optimal scenario to implement in the GSC.

To validate this KPI, it is necessary the evaluation of results which appear in Table 10. Being one dimension the most important and having the highest ponder-ation, it is expected that the case with the best result of this KPI dimension will obtain the best GDI. To check this hypothesis, different combinations of weights have been proved confirming this hypothesis.

3.4.2 Green Supply Chain Index Results

The GSCI (Eqs. 10, 11 and 12) are obtained following the indications provided in Sect. 2.2.4. Table 10 presents the results of the multiplication of production, environmental and economic indicators.

$$GSCI_{\text{Case 1}} = 0.4194 \times 0.3998 \times 0.9793 = 16.40\,\% \tag{10}$$

$$GSCI_{\text{Case 2}} = 0.414 \times 0.3998 \times 0.9833 = 16.30\,\% \tag{11}$$

$$\text{GSCI}_{\text{Case 3}} = 0.3992 \times 0.3244 \times 0.9902 = 12.82\% \quad (12)$$

It is important to note that the closer gets this index to 0 the better it is. Comparing the GSCI and GDI results of each case were not the same, in case three the GSCI is lower than the other cases. However, the objective of this KPI is not to choose the scenario, but to analyze the situation of the GSC on a continuous basis, for which this GSCI must be calculated continuously and in the same scenario to control the GSC from the three dimensions proposed (productive, environmental, and economic). With the GSCI a problem could be detected if this KPI sudden changes or could check the impact of an improvement implemented in any dimension above them.

This KPI validation has been more complicated, the GSCI has been calculated several times and its results has been compared with the analysis of the conditions of the machine. Checking that the GSCI improve when the conditions of the machine also do it.

4 Conclusions

The industry's new trend towards the GSC has not only productive and economic reasons, but also other reasons such as compliance with environmental regulations, awareness and pressure of stakeholders [8].

Industry 4.0 technologies provided extra value when working with GSC, not only because of the high capacity to monitor and capture machine tool data in real time, but also because it brings speed of response and flexibility to the SC [12]. The strategy of adding "Green SC" and implementation of Industry 4.0 allows the industry to improve its productive, economic and environmental performance [13]. It is interesting to be able to determine a KPI capable of measuring these three perspectives at the same time, obtaining an overview of the GSC. The GDI and GSCI index developed in this article cover two distinct parts of this strategy. On one hand, the GDI would allow the industry to choose the best scenario from the ones proposed at the beginning of the GSC implementation, taking into account the productive, environmental and economic part of the SC. On the other hand, once the GSCI has been implemented, it could be used as an indicator of the evolution of machine tools over time with respect to the three dimensions, allowing to observe whether the improvement decisions in one dimension affect positively or negatively the rest of dimensions.

4.1 Future Research

The pilot scenario of this test has been conducted in a CPS that is used to produce short series. It is considered that the results obtained are linked to a specific study with an established system and under certain conditions. It is proposed as a future line of research to implement these indicators in longer series to analyze the results.

These indicators, which are currently implemented in a CPS, could be integrated into a GSC by particularizing, adapting and implementing them in other machines after study and characterization of the corresponding process and logistics system used.

With a connection to external databases of management software, parameters, such as salaries and other costs that cannot be obtained directly from the CPS, could be automatically fed for the development of the software in Python. Allowing you to work on a system that uses technology 4.0 and software development in Python to obtain values that come from databases, such as the ERP itself used in a certain company.

Acknowledgements The authors would like to acknowledge the participation of TECNALIA RESEARCH & INNOVATIONS by the implementation of the cyber-physical system and for the expertise in electronics and communication. Likewise, to the Precision Mechanic Service of the Universidad de Zaragoza and their contributions during the essays.

References

1. Al Zaabi S, Al Dhaheri N, Diabat A (2013) Analysis of interaction between the barriers for the implementation of sustainable supply chain management. Int J Adv Manuf Technol 68:895–905. https://doi.org/10.1007/s00170-013-4951-8
2. Alamerew YA, Brissaud D (2019) Circular economy assessment tool for end of life product recovery strategies. J Remanufact 9:169–185. https://doi.org/10.1007/s13243-018-0064-8
3. Benítez-Hidalgo A, Nebro AJ, García-Nieto J, Oregi I, Del Ser J (2019) jMetalPy: A Python framework for multi-objective optimization with metaheuristics. Swarm Evol Comput 51:100598. https://doi.org/10.1016/j.swevo.2019.100598
4. Cohen Y, Faccio M, Pilati F, Yao X (2019) Design and management of digital manufacturing and assembly systems in the Industry 4.0 era. Int J Adv Manuf Technol 105:3565–3577. https://doi.org/10.1007/s00170-019-04595-0
5. de Sousa Jabbour ABL, Jabbour CJC, Foropon C, Filho MG (2018) When titans meet—Can industry 4.0 revolutionise the environmentally-sustainable manufacturing wave? The role of critical success factors. Technol Forecast Soc Change 132:18–25. https://doi.org/10.1016/j.techfore.2018.01.017
6. García-Granero EM, Piedra-Muñoz L, Galdeano-Gómez E (2018) Eco-innovation measurement: a review of firm performance indicators. J Clean Prod. https://doi.org/10.1016/j.jclepro.2018.04.215
7. Hatim QY, Saldana C, Shao G, Kim DB, Morris KC, Witherell P, Rachuri S, Kumara S (2020) A decision support methodology for integrated machining process and operation plans for sustainability and productivity assessment. Int J Adv Manuf Technol 107:3207–3230. https://doi.org/10.1007/s00170-019-04268-y
8. Kazancoglu Y, Kazancoglu I, Sagnak M (2018) A new holistic conceptual framework for green supply chain management performance assessment based on circular economy. J Clean Prod 195:1282–1299. https://doi.org/10.1016/j.jclepro.2018.06.015
9. Kirchherr J, Reike D, Hekkert M (2017) Conceptualizing the circular economy: an analysis of 114 definitions. Conserv Recycl Resour. https://doi.org/10.1016/j.resconrec.2017.09.005
10. Kumar J, Soni VK (2015) An Exploratory Study of OEE Implementation in Indian Manufacturing Companies. J Inst Eng Ser C 96:205–214. https://doi.org/10.1007/s40032-014-0153-x

11. Lambán MP (2010) Determinación de costes de procesos de la cadena de suministro, e influencia de factores productivos y logísticos. Universidad de Zaragoza
12. Luthra S, Mangla SK (2018) Evaluating challenges to Industry 4.0 initiatives for supply chain sustainability in emerging economies. Process Saf Environ Prot 117:168–179. https://doi.org/10.1016/j.psep.2018.04.018
13. Micheli GJL, Cagno E, Mustillo G, Trianni A (2020) Green supply chain management drivers, practices and performance: A comprehensive study on the moderators. J Clean Prod 259. https://doi.org/10.1016/j.jclepro.2020.121024
14. Sonmez V, Testik MC, Testik OM (2018) Overall equipment effectiveness when production speeds and stoppage durations are uncertain. Int J Adv Manuf Technol 95:121–130. https://doi.org/10.1007/s00170-017-1170-8
15. Tan Q, Tong Y, Wu S, Li D (2019) Modeling, planning, and scheduling of shop-floor assembly process with dynamic cyber-physical interactions: a case study for CPS-based smart industrial robot production. Int J Adv Manuf Technol 105, 3979–3989. https://doi.org/10.1007/s00170-019-03940-7
16. Vanalle RM, Ganga GMD, Godinho Filho M, Lucato, WC (2017) Green supply chain management: an investigation of pressures, practices, and performance within the Brazilian automotive supply chain. J Clean Prod 151:250–259. https://doi.org/10.1016/j.jclepro.2017.03.066
17. Wagner M, Llort G, Mercadal E, Giménez J, Labarta J (2017) Performance analysis of parallel python applications. Proc Comput Sci Elsevier B.V.,:2171–2179. https://doi.org/10.1016/j.procs.2017.05.203
18. Zeng D, Cao H, Jafar S, Tan Y, Su S (2018) A life cycle ecological sensitivity analysis method for eco-design decision making of machine tool. Proc CIRP Elsevier B.V.: 698–703. https://doi.org/10.1016/j.procir.2017.11.093
19. Zhao J, Ji M, Feng B (2020) Smarter supply chain: a literature review and practices. J Data Inf Manag. https://doi.org/10.1007/s42488-020-00025-z

Design of Material Delivery Routes with Towing Equipment for the Automotive Sector Using a Von Neumann Topology of PSO Under the Logistics 4.0 Paradigm

Alma Luévano, Alberto Ochoa-Zezzatti, Elías Carrum, Darwin Young, Pedro Pérez, and Denise Barzaga

Abstract The supply chain increases its importance in the automotive sector continuously. Therefore, manufacturing systems such as Just in Time, entail a complex implementation, due to the ascending variety of models and components of the vehicles, since currently, the implementations of expert systems within the Logistics 4.0 concept, seek to accelerate the flow synchronized materials in the production lines, using technological tools of the environment Industry 4.0. This project was located in an engine production plant, in the material assortment part from the supermarket (decentralized warehouse of a plant) to the assembly line. A mathematical model of the process was made, and a genetic algorithm programmed in Matlab was used to optimize it, seeking to reduce line stoppages caused by lack of material. The experimentation shown focused on five of the currently implemented assortment routes, which provide 29 different part numbers to 25 workstations. The performance of the current routes and that of the optimization were validated by means of a simulation model.

Keywords PSO algorithm · Material delivery routes · Logistics 4.0

1 Introduction

In recent years, in the operational logistics stage, decision problems have been increasing due to a growing trend of mass customization, which has generated a continuous increase in the variety of vehicles and their components. As a result of this situation, implementing the most used manufacturing methodologies such as "Just in Time" and "Just in Sequence", has become one of the biggest challenges of the current automobile production [1]. One of the steps that have been the most reason

A. Luévano · E. Carrum (✉) · D. Young · P. Pérez · D. Barzaga
COMIMSA, Centro Conacyt, Saltillo, Coahuila, Mexico
e-mail: eliascarrum@comimsa.com

A. Ochoa-Zezzatti
Juarez City University, Doctorado en Tecnología, Ciudad Juárez, Mexico

© The Author(s), under exclusive license to Springer Nature Switzerland AG 2021
A. Ochoa-Zezzatti et al. (eds.), *Technological and Industrial Applications Associated with Intelligent Logistics*, Lecture Notes in Intelligent Transportation and Infrastructure,
https://doi.org/10.1007/978-3-030-68655-0_2

for study, due to the problems that arise in it, is the delivery of productive parts to the assembly line [1]. In addition, due to its complexity and the adoption of production system methodologies such as "World Class Manufacturing WCM" (World Class Manufacturing), used by leading companies in the automotive sector, which does not allow forklifts to enter the stations of work and limits the level of inventory in each station according to the classification that gives the part numbers, that the problems that involve the towing equipment as a means of transport are among the most studied. For all the above, it is precisely the assortment of parts to the assembly line by means of towing equipment, the study area chosen for this project, since to allow a reliable supply of parts with the principle "JIT" (Just in time), the interdependent routing and programming problems need to be solved, being the routing, the assignment of the stations that will be supplied between the towing equipment and the programming, deciding the start times of the trips of each team through its assigned stations. With regard to the resolution of these types of problems, recently in the state of the art it has begun to mention the use of tools of Industry 4.0, mainly the use of simulation. The fourth industrial revolution, better known as Industry 4.0, is currently underway, which was named in 2011 at the Hannover Messe fair in Germany, and is characterized by the integration of digital and physical. The concept of logistics 4.0 starts from these same bases.

2 Formalization of the Problem

Based on the existing literature, we can mention that the most common problems when transporting material with towing equipment are:

- Assign a subset of stations to each towing equipment.
- Determine the number and types of parts to be loaded per trip of the towing equipment.
- Determine a delivery schedule, for each towing equipment and its associated route. [1].

In the present project, the first two problems mentioned above are studied in a real context, in a plant dedicated to engine machining and assembly, located in Saltillo, Coahuila, Mexico, which produces a 3.6–l, 6-cylinder engine, which It has 4 different models.

In the plant there are three types of routes responsible for supplying the productive parts to the assembly line, which are differentiated by the means of transport used and the method of loading and unloading containers with empty material and containers. The first two types use towing equipment as a means of transport, while for the third type of route the main means of transport is the forklift.

There are two subdivisions of the routes used as a means of transport for the towing equipment. The first supplies part numbers with classification of material that allows you to have in stock in process, at each station where they are stocked, enough parts to assemble the engines produced in two hours or more. The part numbers supplied by these routes have among their main characteristics, a high number of pieces per unit container, these are thus sent from the supplier. The cycle time of these routes, that is, the frequency with which they supply material to the line, is two hours. Another important feature is that the loading and unloading of containers with empty material and containers in the warehouse and in the assembly line is done manually by the equipment operator. The project is focused on the following subdivision, which covers the material assortment routes through towing equipment, in which the loading and unloading of material containers and empty containers is done automatically. The process of this type of routes begins in two different material stores, in the first one the material is placed in such a way that the route operators can take the unit containers manually for their subsequent transfer. The second is the area where the complete pallets of material are located, that is, the material loading units (unit containers of material arranged on pallets, commonly protected with rubber and strapping), just as received from the supplier. The material is transported from the warehouses already described to a supermarket area, which is also a warehouse, but decentralized located within the plant and closer to the assembly line, in which the material is placed in such a way that the Person responsible for this area, can take individual containers and place them manually in the supermarket facilities. These facilities are racks specially manufactured for each assortment route, adapted with pneumatic mechanisms so that the material placed in them is automatically transferred to special delivery trucks used for this type of route. These cars are structures made of aluminum profiles and rails, designed and built to load and unload only certain sizes and quantities of material and empty containers. Each structure is fixed to a platform with tires which is anchored to a towing equipment, to be able to transfer it from the supermarket area to the work stations and vice versa. Each towing equipment pulls up to two cars at a time, for which the allocation of material that each one carries was made from the design stage of the routes, currently already implemented in the plant. The transfer of full containers from the supermarket facilities to the cars and the transfer of empty containers from the cars to these facilities, is done at the moment when each car is aligned with its corresponding facility, however, the loading of material to The facilities are done in advance of that time.

Each towing equipment is driven by an operator to the workstations corresponding to its route, where the transfer of material to the station and the transfer of empty containers to the cart are also carried out automatically. In each station where material is supplied, the empty containers accumulated up to that point must be loaded. In both the supermarket area and the assembly line, the loading and unloading of empty material and containers is done for some routes in a single step and for others in two steps for each carriage. It is important to mention that the carriage limit that each towing equipment can pull was given based on its ability to drag and make the necessary turns through the route assigned to each route. Currently, of this type of routes, 5 are implemented and are identified by colors, for reasons of practicality in this work will be identified by numbers. For the design of the routes already implemented, the following variables were mainly considered: production speed of the assembly line (demand for material on each workstation in a given time frame), loading times and unloading of empty material and containers, permitted speed of towing equipment, cargo capacity of delivery carts, number of carts that can be used for each route, whose decision is given mainly based on the possibility of turning in aisles narrow without difficulty.

The production speed considered to perform the allocation and scheduling of the routes already implemented, was taken as a constant data of 105 engines per hour. In the plant you have a production time per shift of 11.5 h, so when multiplied by the two daily shifts you have a total of 23 h of production per day. If the 23 h per day of production is multiplied by the 105 engines per hour, we would have a total production of 2415 engines per day, however, the current daily production target is 1840 engines. Based on this, it would be thought that producing at a rate of 80 engines per hour could reach the daily goal of 1840 engines, so the question arises: What is the reason for assembling the engines at a speed of 105 per hour? On the production line there are several unexpected events, such as quality problems in the production parts, lack of parts due to delays in the delivery time of some supplier, failures in the machines, to name a few, which could cause delays in scheduled production. Because of this, there is a safety margin of 25 engines per hour, to ensure compliance with the daily production target. The problems that affect engine production are so unpredictable and have so many possible causes, that maintaining an invariable production per hour for a day is something that has not been achieved. As examples, it could be that in the early hours of the day engines were produced at a rate of 80 per hour, and that, due to engines rejected in the final quality review, by the middle of the shift the amount of production per hour would increase with the aim of reaching the Goal. On the other hand, it could be the case that, during the first shift due to the absence of problems, they approached the target of 1840 engines per day and therefore, in the last shift the rate of production should be reduced to avoid exceeding the scheduled target, which would also not be suitable for the plant. This point is very important, since one of the main variables to consider for the design of the routes, is precisely the pace of production of the plant. Right after the implementation of the current routes, some issues began to arise. The most serious were line arrests caused by lack of material at workstations. These shutdowns decreased as the learning curve of the towing equipment operators advanced, but unfortunately, they continued to occur. In addition to this, since the implementation to this day, modifications have been made to the design of some tanks of the initial routes, due to the detection of waste in their volumetric capacity. In line with the above, as the learning curve of the operators progressed, downtimes were detected in some of them, contradicting the first recorded observations in which it was described, which due to the cycle time of the assortment process of the routes could not be reached to fill the material in time. In addition, the following observations were made to the assortment process:

- Existence of routes of towing equipment to workstations without unloading material, by decreasing the demand for parts on the production line.
- Problems with automatic transfers, when towing equipment again arrives to supply material to a workstation, before the material that occurred in the previous delivery cycle had been consumed.

Another situation that added difficulty to the design of the routes, is the existence of a narrow corridor (where it is not possible to pass) through which four different material assortment routes have their route, three of them of the type studied in this project. Due to this situation, there have been problems of blockages between material delivery carts, generating the risk of stoppages due to lack of material derived from this condition.

Based on all the features of the problem listed above, it was classified as a routing problem for trained vehicles with time windows. This conclusion is based on the description given in the literature to such problems, which, in order to be classified within this category must meet the following characteristics:

- Have a fleet of vehicles with limited and constant capacity.
- The function of delivering various assigned goods according to customer demand.
- They include the restriction that each customer is only willing to receive orders in the course of a defined time window [2].

In addition to these characteristics, due to the aforementioned WCM philosophy followed by the plant, it is necessary to consider an additional restriction, which states that each workstation has a maximum WPI (Work in process) work inventory allowed, which depends on the classification of the part that is provided in that station, which directly affects the scheduling of the routes.

3 Proposed Methodology

PSO: Particle Swarm Optimization

This method was described around 1995 by James Kennedy and Russell C. Particle Swarm Optimization (PSO) is an optimization/search technique typically used in large-ranging search spaces. This method is inspired by the behavior of insect swarms in the wild. You can think of a swarm of bees because they when looking for pollen look for the region of space where there is more density of flowers, because the probability of pollen is higher. The same idea was moved to the field of computing in the form of an algorithm and is currently used in the optimization of different types of systems. Formally speaking, it is assumed that you have an unknown function f(x,y), which can be evaluated at the points you want but as a black box, so you **cannot** know its expression. The objective is the usual in optimization, to find values of x and y for which the function f(x,y) is maximum or minimum. F(x,y) is often called a "fitness" function, as it will determine how good the current position is for each particle. The idea that follows in PSO begins by placing particles randomly in the search space, but giving them the possibility of moving through it according to rules that take into account the personal knowledge of each particle and the global knowledge of the **swarm.** Providing you with a simple ability to move through the landscape (fitness function) and communication between them, you can discover

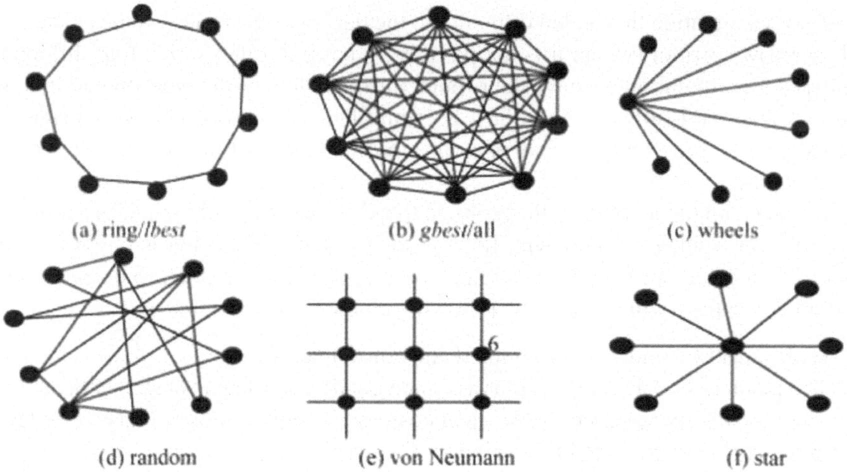

Fig. 1 Topologies used in PSO

particularly high values for f(x,y), spending few computational resources (calculations, memory and time) [3]. A leader can be global from the entire swarm, or local to a swarm neighborhood. Swarm neighborhoods have a structure that defines how information is concentrated and then distributed through their members. The most common organizations and swarms are: Star neighborhood structure, ring neighborhood structure, wheel neighborhood structure, Von Neumann neighborhood structure, and four-cluster neighborhood structure. The organization of the swarm affects search ability and convergence [4] (Fig. 1).

The pseudocode is shown below.

```
N:Number of particles
w:Weight of inertia
c:Cognitive value
pi:Best initial singles
pg:Best Initial Global
x,v ' Assign initial random positions and speeds to the particles (N)
Start Cycle
    For each particle
        Update Speed (w,c,pi,pg)
        x-Update Position (v)
    Calculate new position fitness value (x)
"Upgrade Best Singles"
pg-Update Best Global
The end
```

4 Experimentation of Our Proposed Model

Four years later, in 2011, Nils Boysen and Stefan Bock in the article Programming for the Supply of Fair-to-Time Parts for Mixed Model Assembly Lines [5], raise the problem of finding sequences, according to the philosophy "Just a Time" of boxes of material delivered from a central storage facility. In this case study, the distribution of parts by means of a forklift is bottleneck. Its objective was to improve the efficiency of supply handling for mixed model assembly lines, as well as to minimize the maximum weighted partial inventory that had to be stored near the assembly line of all stations and production cycles. The solution method used was limited dynamic programming and simulated annealed (Figs. 2 and 3).

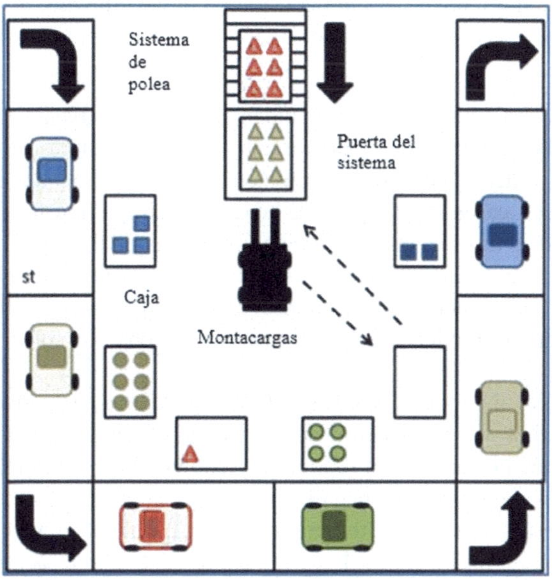

Fig. 2 Problem 2 Scheme. *Source* Boysen and Bock [5]

$$(HSP) \qquad Minimizar\ Z(\pi) = \max_{m \in M;\, t=1,\dots,T} \{w_m \cdot L_{m,t}\}$$

$t^{\pi}_{\pi(s)} = p_{\zeta_{\pi(s)}} + \sum_{s'=1}^{s-1} 2 \cdot p_{\zeta_{\pi(s')}} + s \cdot p' \quad \forall s = 1, \dots, |B|$. Point in time where a box will be available at the workstation.

$\Gamma_{m,t} = \{p \epsilon B_m | t^{\pi}_{p} \le t\} \quad \forall m \in M, t = 1, \dots, T$. Part boxes delivered on time at your station.

$L_{m,t} = A_m + \sum_{p \epsilon \Gamma_{m,t}} a_p - D_{m,t} \quad \forall m \in M, t = 1, \dots, T$. Inventory placed near each station on the assembly line.

Fig. 3 Mathematical Model 2. *Source* Boysen and Bock [5]

The restrictions considered were:

- Variable demand.
- The assortment of a single station per trip.
- Point-to-point deliveries for a single vehicle.
- Each box with a variable number of parts.
- No idle time was allowed between two consecutive deliveries.
- Fixed and determined travel times and forklift manoeuvres.
- Time units normalized to the equidistant length of a production cycle.
- Each box with a predetermined number of single-type units.
- One-part assembly in each station.
- Parts ready to be assembled right after the forklift positioned the box on the workstation.
- The return of empty boxes was not considered in the problem.
- The safety stock was not explicitly considered, but could be included to the demand of the first period.

Nils Boysen and Stefan Bock concluded that limited dynamic programming provided optimal solutions for moderate and medium sized cases (with up to 20 boxes). On the other hand, using simulated annealing, optimal solutions were produced for large cases (with up to 80 boxes). An important feature considered in this investigation was the variable demand of the material.

In the same year, 2011, Nils Boysen but now with Simon Emde in the article Optimal routing and programming of towing equipment for the just-in-time assortment of mixed model assembly lines [6], propose to solve two problems simultaneously.

On the one hand, the routing problem, which is to determine the size of the towing equipment fleet that will be used for the assortment of parts and the distribution of stations among towing equipment (commonly known in the automotive industry as "tuggers"). In addition, the sequence of stops must be decided.

The second problem raised is the programming problem, in which a number of tours have to be determined for each towing equipment, where each tour comprises the loading operation in the supermarket (decentralized storage area within a plant), the cycle through all of its assigned stations, and the arrival time at each stop (Fig. 4).

Continuing with the solution of problems within a line of assembly and/or production, in recent years several authors have continued to work in their study, but now using tools typical of manufacturing 4.0, as are simulation. One of the most relevant advantages of simulation is that it allows for a simpler method of solution when mathematical procedures are complex and difficult. In addition to this, analytical methods are almost always developed relatively straightforward, where a large number of assumptions are made, while in simulation methods it is possible to analyze systems of greater complexity including a better level of Detail. Below are several studies carried out within the area of interest of this project, but with a solution approach more oriented towards the use of simulation.

$\alpha_i = \beta_i - (CP_i - PD_i + dp_i)$ Current inventory

$\delta_i min\left(\frac{K}{D_i},\right)1$ Filling Capacity Rate

$\delta = \prod_{i=1}^{T} \delta_i$ Filling rate for current and previous routes if early loads are used

$K \geq F^{-1}\left(\delta_{EL}, \sum_{i=1}^{N} d_i\right)$ $K \gg D$ Estimate the number of stations supplied by the same towing equipment

$$\min w_i \sum_{i=1}^{TSC-MC} \sum_{j=1+MC}^{TSC} y_{ij}^{+} + w_2 \sum_{i=1}^{TSC-MC} \sum_{j=1+MC}^{TSC} x_{ij} + w_3 \sum_{i=1}^{TSC-MC} \sum_{j=1+MC}^{TSC} x_{ij} \sum_{K=1}^{N} \sum_{i=1}^{j} d_{ku}$$

Minimize the number of towing equipment cycles and the possibility of early deliveries

$$x_{ij} = \begin{cases} 1, \\ 0, \end{cases}$$

1, If there is a towing equipment cycle that covers the cycles of stations from 1 to j

0, If not

$min \sum_{s=1}^{N} \sum_{t=1}^{T} |\varepsilon_{st}|$ Maximize the deviation between line inventory and ideal security stock levels

Fig. 4 Mathematical Model 5. *Source* Alnahhal and Night 2015

Objective function

$$MAX \sum_{t=1}^{N} CWT_t \tag{1}$$

where:

(t) Time measured in the number of engines produced.

N—Number of engines required to be manufactured in a productive day.

- CWT-_t-Inventory in process for all part numbers, at all stations, which will be assorted by towing equipment over time t.

  Calculating the "CWT"_t as follows:

- CWT

$$\begin{aligned} CWT_t &= \sum_{i=1}^{NV} (CWI_i - (s * ups_i)) \\ &+ \sum_{j=1}^{PT} \sum_{i=1}^{NV} \begin{pmatrix} CWp_{ji} - \left(\left(dt_{ji} + rt_{ji} + at_{ji} + tll_{ji}\right) * ups_{ji}\right) \\ + x_{ji} - \left(dif_{ji} * ups_{ji}\right) \end{pmatrix} \\ &+ \sum_{i=1}^{NV} (CWP_i - (MV * ups_i)) \end{aligned}$$

- CWI

$$x_{ji} = \left\{ \begin{pmatrix} Value\ of\ x\_ji,\ IF\ the\ number\ of\ part\ are\ fill\ i\ in\ the\ station\ j. \\ -0\ if\ not\ ,\ j \end{pmatrix} \right.$$

$$0 \leq x_{ji} \leq WSL_i$$

Variable that regulates the material consumption of each part number i.

- Amount of inventory on workstations, part number i, from which the first part number is filled.
- at-ji—Time it takes to align the towing equipment to fill the part number i at station j.
  The time it takes to download part number i on workstation j.
  Time to load empty empty containers of part number i at station j.
  The time it takes for towing equipment to reach the assortment point of part number (i) at station j, considering only distance traveled.

- x_ji—Amount of material of part number i to be supplied by the towing equipment at station j.
- dif_ji—Excess of the quantity of the part number i filled in the workstation j relative to its upper limit of inventory in process allowed.

- MV—Number of engines produced in the time the towing equipment comes out of supplying the last part number on the last workstation, returns to the supermarket and reloads material.
- PT—Total workstations filled by towing equipment.
- NV—Number of part and total numbers, assorted by towing equipment.
- WSL_i—Maximum inventory capacity in process of part number i (Fig. 5).

4.1 Particle Swarm Optimization

The next technique experimented with was particle swarm optimization, better known by its acronym PSO. The code was programmed in MATLAB, using the same evaluation function as in the genetic algorithm, but making some adaptations required by the algorithm itself. The topology used for this algorithm was wheel neighborhood structure, which was mentioned in the theoretical framework. The main code is attached in the annexes section. Tables 1 shows the results of the experimentation performed with PSO varying in the algorithm the parameters of, number of iterations (MaxIt), swarm size (nPop), inertia weight (W), radius of expansion of the weight of inertia (wdamp), these two most recently used to get the feed rate in the algorithm. In addition, variations were made in the individual learning coefficient (c1) and global

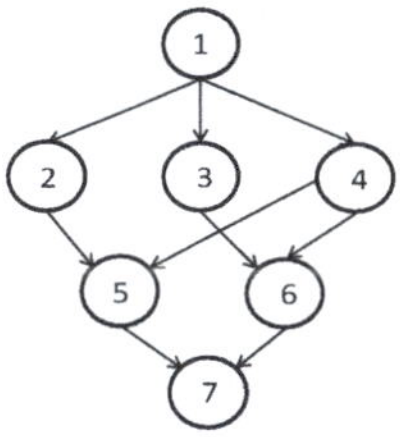

a1	a2	a3	a4
a5	a6	a7	a8
a9	a10	a11	a12

(a) Von Newman PSO Model **(b) Area to explore**

Material Resurshyent Car

MRC	a1	a2	a3	a4	a5	a6	a7	a8	a9	a10	a11	12
1	7	3	9	4	9	4	3	3	1	2	9	9
2	4	7	18	3	2	7	7	2	7	8	7	5
3	8	2	19	4	7	1	2	8	4	4	4	1
4	9	9	17	3	2	7	9	1	2	2	3	5
5	4	8	10	5	9	9	8	8	9	1	4	9
6	5	5	9	8	2	3	5	4	5	8	8	4
7	2	3	11	2	3	3	3	7	3	2	7	3
8	7	5	14	4	7	5	5	5	5	3	3	4
9	3	4	20	3	4	4	4	4	2	4	1	4
10	1	5	2	7	2	5	5	5	10	7	3	2

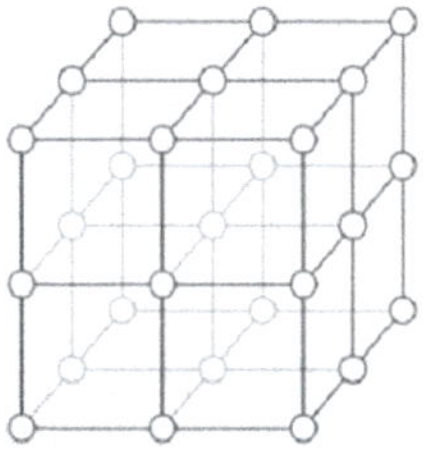

(c) Design of experiments to our proposal Model of Von Newman Topology.

Fig. 5 Process of implementation of our solution

Table 1 Experimentation PSO Algorithm, 105 MH.

PSO Tabla Comparativa 105 MH											
Número de Experimento	Número de iteraciones (MaxIt)	Tamaño del enjambre (nPop)	Peso de Inercia (w)	Radio de expansión del peso de inercia (wdamp)	Coeficiente de aprendizaje individual (c1)	Coeficiente de aprendizaje global (c2)	Valor de función objetivo	Tiempo CPU (seg)	Tiempo Real (seg)	MH	Equipos de remolque
Basal	NA	NA	NA	NA	NA	NA	-1737237.707	NA	NA	105	5
1	500	100	1	0.99	1.5	2	-40999420.47	1.23E+03	1.23E+03	105	3
2	1000	100	1	0.99	1.5	2	-10063176.81	3.12E+03	3.08E+03	105	4
3	2000	100	1	0.99	1.5	2	-261367.8381	7.17E+03	7.14E+03	105	4
4	3000	100	1	0.99	1.5	2	-51059131.31	1.03E+04	1.03E+04	105	5
5	500	200	1	0.99	1.5	2	-52025743.05	3.53E+03	3.53E+03	105	4
6	1000	200	1	0.99	1.5	2	-51025756.56	7.73E+03	7.71E+03	105	4
7	2000	200	1	0.99	1.5	2	-50003199.68	7.88E+03	7.87E+03	105	3
8	3000	200	1	0.99	1.5	2	-12085673.52	2.49E+04	7.07E+04	105	5
9	500	100	1	1	2	3	-42322031.29	1.29E+03	1.29E+03	105	5
10	1000	100	1	1	2	3	-15250043.68	1.66E+03	1.65E+03	105	4
11	2000	100	1	1	2	3	-42398393.92	5.13E+03	5.10E+03	105	5
12	3000	100	1	1	2	3	-10632227.33	4.42E+03	4.38E+03	105	5
13	500	200	1	1	2	3	-9369976.68	2.04E+03	2.04E+03	105	5
14	1000	200	1	1	2	3	-1823862.003	7.83E+03	1.06E+04	105	6
15	2000	200	1	1	2	3	-50015404.96	1.56E+03	1.55E+03	105	3
16	3000	200	1	1	2	3	-18177578.49	1.39E+04	1.38E+04	105	5

Source Own elaboration

learning coefficient (c2) parameters. The experimentation was carried out considering a production speed of 105, 90 and 80 engines per hour. The experiments were conducted based on a custom orthogonal design with 16 runs. The results are shown

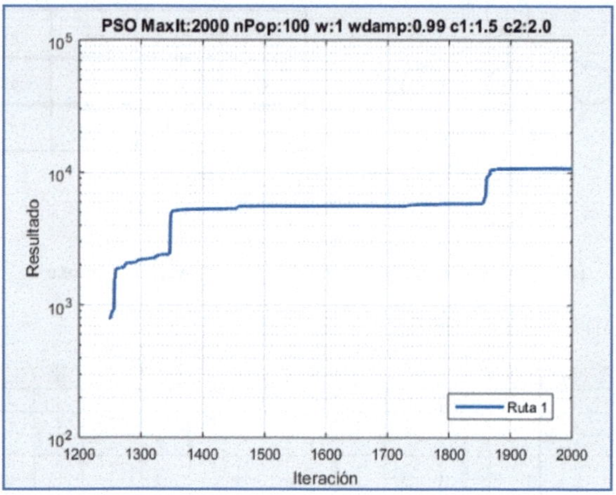

Fig. 6 Convergence Graph, PSO Experimentation, 105 MH, Experiment 3, Route1. *Source* Own elaboration

below and highlighted in green in each table, the best result of each experiment group, corresponding to a different production speed (Fig. 6).

5 Conclusions and Future Research

- The maximum value of the target function, for the material assortment process with a production speed of 105 engines per hour, was found using a Genetic Algorithm with general crossover type, population of 200 individuals, 300 generations. The time it took to run the algorithm was 22.28 min.
- The maximum value of the target function, for the material assortment process with a production speed of 90 engines per hour, was found using a Genetic Algorithm with general crossover type, population of 200 individuals and 100 generations. The time it took to run the algorithm was 5.13 min.
- The maximum value of the target function, for the material assortment process with a production speed of 80 engines per hour, was found using a Genetic Algorithm with general crossover type, population of 200 individuals and 200 generations. The time it took to run the algorithm was 9.49 min.
- It could therefore be concluded that the best optimization technique for the mathematical model developed was Genetic Algorithm.
- Validation by means of simulation, of the results obtained when performing the optimization, considering a production speed of 105 engines per hour, showed a reduction of line stops from 2.42 to 1.30%, using the same 5 towing equipment, but decreasing the total number of kilometers travelled over a 23-hour period, from 448.8 to 410.71 km.
- In the current process the amount of material that is sent to the assembly line remains fixed, even if the production speed of 105 motors per hour decreases, so it was expected that in experiments where production speeds of 90 and 80 mots

are considered you decrease the amount of material that is sent per cycle, the impact of optimization was greater. However, although the optimization results for lower speeds decreased the use of 1 towing equipment, the validation by means of the simulation did not show good statistics, since, although the value of the target function increased with compared to baseline, a higher percentage of line stops than the current process was obtained in the simulation at the same speeds. An important detail observed when simulating, was that the solution generated, gave a part number a low amount of material to be sent per cycle (see Fig. 93 and 98), which affected the total result of the percentage of line stops of all workstations considered in this study. The above gives rise to consider continuing experimentation with other optimization techniques, or even with it, but with a variety of different parameters, in a future work.

- The implementation of the results obtained would require changes in the structure of the facilities of the supermarket area, delivery carts, and special facilities for automatic transfers found at the workstations, for which they are would have to do a cost-benefit analysis that would justify the change considering the cost of the changes.
- Simulation of individual assortment routes showed no greater difficulty. The problem arose when testing all the routes running at the same time, since the interference of each other in corridors where it is not possible to go over, cause the simulation of the 5 routes, in the simulation of the process with waits, to stop 2 h with 52 min. The complexity was found in representing the decisions that towing route operators make so as not to block each other. Another practice they perform to avoid these blockages is to return to the supermarket area with the material that was loaded at the beginning of the cycle without unloading it at the workstation, seeing that the material from the previous cycle has not yet been consumed, which occurs when the Production speed decreases, this causes unnecessary material handling and is a very bad practice carried out in the current process.

5.1 Future Research

- Perform the optimization of material assortment routes using the simulation model.
- Develop in the Plant Simulation software a graphical interface to facilitate the use of the simulation model of the current process and subsequently perform the optimization of that model.
- Improvement of the developed mathematical model, adding considerations such as, the percentage of material damaged by part number present in the workstations, storage capacity of supermarket facilities or loading times and different download per workstation, to name a few.

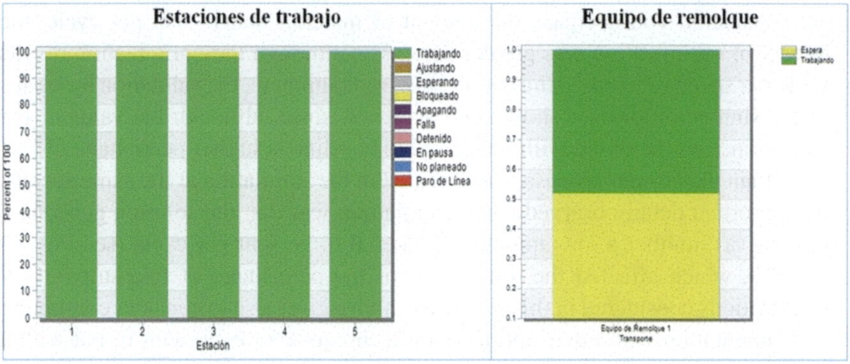

Fig. 7 Simulation software proposed

- Experiment incasing with optimization techniques other than those studied in this research, such as ant colony, simulated annealing or dynamic programming, to mention some of the most outstanding techniques in the study of state of the art.
- Perform in Matlab a graphical interface, in order to facilitate the use of the programmed genetic algorithm, to obtain the routes of assortment of material.

5.2 Simulation

This research still needs to be complemented with simulation and testing prior to implement it in a real environment (Fig. 7). Beside the engineering challenges of a physical Material Resurshyent Cars, the developed algorithm described in Sect. 4 will be simulated for fine tuning and validation. For this, the above is why in the mathematical model wait times are sanctioned, so that the solution that searches for the algorithm avoids them.

References

1. Boysen N, Emde S, Hoeck M, Kauderer M (2015) Part logistics in the automotive industry: decision problems literature review and research agenda. Eur J Oper Res:107–120
2. Sánchez Barrientos M, Carrum Siller E, Cedillo Campos M, Peña Reyna L (2016) Revisión del estado del arte de los problemas de ruteo de vehículos en redes de transporte LTL. Congreso Internacional de Logística y Cadena de Suministro
3. Sancho Caparrini F (2016) PSO: Optimización por enjambres de partículas
4. Ortiz-Zezzatti O, Rivera A, Gilberto G (2019) Handbook of research on metaheuristics for order picking optimization in warehouses to smart Cities, 2019 ómez-Santillán, ClaudiaSánchez–Lara, Benito
5. Boysen N, Bock S (2011) Scheduling just-in-time part supply for mixed-model assembly lines. Eur J Oper Res 211:15–25

6. Emde S, Boysen N (2012) Optimally routing and scheduling tow trains for JIT-supply of mixed-model assembly lines. Eur J Oper Res 217:287–299
7. Boysen et al (2015) Part logistics in the automotive industry: decision problems, literature review and research agenda. Eur J Oper Res:107–120

Industrial Accident Induced Coma: A Multi-Disciplinary Perspective Within the Industry 4.0 Paradigm

Ismael Rodriguez and Alberto Ochoa-Zezzatti

Abstract In this work, we employed TOPSIS (Technique for Order of Preference by Similarity to Ideal Solution) methodology to assess which accidents could be treated through medically induced coma in a society under the paradigm of industry 4.0. We determined 7 variables: Laboral age, previous accidents, incidence by race, general health status, consciousness, collateral effects, and in-hospital complications, besides we took into account 7 different work-related accidents: Electrocution, burn, head injury, body trauma, amputation, chemical inhalation and drinking a dangerous substance. Through the TOPSIS analysis, we determined that drinking a dangerous substance and electrocution are at the top of the list of priorities, while amputation and head injury are at the bottom of the list. Considering the hospital requirements, we also found that drinking a dangerous substance was at the top of the list while burns were at the bottom. Furthermore, when we analyzed the human factor or human error within the TOPSIS method, we found that electrocution is the top priority, while head injury appears at the bottom of the list. This analysis represents one of its kind in determining whether patients suffering from work-related accidents could be treated with a revolutionizing therapy that involves medically induced coma while attending other manufacturing and information requirements in the medical field to improve the quality of life of workers from an industry 4.0.

Keywords TOPSIS · Industry 4.0 · Induced coma

I. Rodriguez
Jagiellonian University, Kraków, Poland

A. Ochoa-Zezzatti (✉)
Universidad Autónoma de Ciudad Juárez, Ciudad Juárez, Mexico
e-mail: alberto.ochoa@uacj.mx

© The Author(s), under exclusive license to Springer Nature Switzerland AG 2021

A. Ochoa-Zezzatti et al. (eds.), *Technological and Industrial Applications Associated with Intelligent Logistics*, Lecture Notes in Intelligent Transportation and Infrastructure,
https://doi.org/10.1007/978-3-030-68655-0_3

1 Introduction

Industrial accidents statistics indicate that there is an increasing trend with 4.49 fatal accidents per 100 000 people employed in Rumania and 1.65 across whole Europe in 2017 and a total of 1558 non-fatal accidents per 100 000 employed people in the same year according to Eurostat 2017 [1]. In Mexico, in 2019, according to data published by the National Secretariat of Labor and Social Welfare, there were a total of 200,763 industrial accidents of which 285 were fatal [2]. The fatalities in the labor field are related to sociodemographic variables such as age, sex, and occupation, the work environment and workplace conditions [3]. The prevention of the fatalities must play a critical role within a society adapted to the industry 4.0 paradigm, where fatalities can be reduced through treatments such as induced coma.

Industrial accidents prevention and reduction are one of the main goals of industry 4.0, this will benefit both companies and employers in multiple aspects such as economic and life quality. It has been estimated by the Health and Safety Executive that in Gran Britain total costs of workplace injuries in 2017/18 was £15.0 billion, being the individuals who bear the majority of the costs [4], and according to the National Health Council of the USA, the total economic burden of industrial accidents was of $170.8 billion and a total of 103,000,000 days lost just in 2018 [5]. In 2017, it was published by the European Agency for Safety and Health at Work that in Europe, work-related accidents, and injuries cost €476 billion a year and that worldwide these account for a loss of 3.9% of GDP [6].

Coma is a state of prolonged unconsciousness and lack of response to noxious stimuli, it can be either induced or due to a pathological condition [7]. It is known that the state is a mechanism of defense of the brain, in which it profoundly shuts down. Drugs that are used to induce coma include Propofol, Barbiturates, Benzodiazepines, and Opioids; following their administration a reduced blood flow and metabolism of the brain are induced achieving large-scale brain inactivation [8] which can be monitored on an electroencephalogram (EEG) by a burst suppression pattern (BSP) or a completely isoelectric EEG [9]; usually, the EEG pattern needs to be continuously monitored by a clinician and up to this day, no guidelines specify the level of BSP required to achieve a medically induced coma state [8], although some efforts have been made to standardize this therapy with highly variable results that may arise from inter- and intra- individual pharmacokinetic, pharmacodynamic and pharmacogenetic differences [10].

A comatose state has been divided in three different types according to the conditions that generated it [11]:

(a) Medical coma: coma generated by an underlying medical condition
(b) Induced coma: pharmacologically induced coma
(c) Multifactorial coma: coma generated by the combination of administered drugs and underlying medical conditions

There are multiple scales to classify coma, some of the most used are: the Glasgow Coma Scale, the Ramsay scale, the Sedation—Agitation Scale (SAS), and the Richmond Agitation Sedation Scale (RASS), some other scales are further described in

Table 1 Scales used to identify the comatose state

Sedation scale	Coma	Deep coma
Ramsay scale	5 or 6	6
Sedation – Agitation scale (SAS)	1 or 2	1
Richmond Agitation Sedation Scale (RASS)	3–5	5
Glasgow coma scale	9–12	Less than 8

Source [12, 13]

Table 1. As presented, each coma scale presents a different system to identify the medical status in which a coma is present in a patient, such different scales can vary according to different medical characteristics and reactions such as motor responsiveness, verbal performance, and eye-opening to appropriate stimuli that are deeply described in their respective guidelines provided by diverse medical institutions such as the CDC (Center for Disease Control), among others. A brief explanation for each scaling system is provided further in the text.

Ramsay Scale

This scaling system is comprised of six different descriptions that account for diverse sedation states that a patient can be presented with, these descriptions can be found in Table 2. As previously described, it is considered that a patient is under a coma state if the result of the medical evaluation indicates a state number 5 or 6, being 6 an indication of a deep coma state.

Sedation Agitation Scale (SAS)

This sedation scale is comprised of seven descriptors for diverse possible responses or activities performed by patients in the Intensive Care Unit (ICU), these descriptions, their corresponding terms, and scores are presented in Table 3.

Richmond Agitation Sedation Scale (RASS)

This scale is used to assess patients' level of sedation and agitation in the ICU, the specific scores are presented in Table 4.

Table 2 Ramsay sedation scale

Score	Description
1	Patient is anxious and agitated or restless, or both
2	Patient is co-operative, oriented, and tranquil
3	Patient responds to commands only
4	Patient exhibits brisk response to light glabellar tap or loud auditory stimulus
5	Patient exhibits a sluggish response to light glabellar tap or loud auditory stimulus
6	Patient exhibits no response

Table 3 Sedation Agitation Scale

Score	Term	Description
7	Dangerous agitation	Pulling at endotracheal tube, trying to remove catheters, climbing over bed rail, striking at staff, thrashing side to side
6	Very agitated	Does not calm, despite frequent verbal reminding of limits; requires physical restraints, biting endotracheal tube
5	Agitated	Anxious or mildly agitated, attempting to sit up, calms down to verbal instructions
4	Calm and cooperative	Calm, awakens easily, follows commands
3	Sedated	Difficult to arouse; awakens to verbal stimuli or gentle shaking, but drifts off again; follows simple commands
2	Very sedated	Arouses to physical stimuli, but does not communicate or follow commands, may move spontaneously
1	Unable to rouse	Minimal or no response to noxious stimuli, does not communicate or follow commands

Table 4 Richmond sedation agitation scale

Score	Term	Description
+4	Combative	Overtly combative, violent, danger to staff
+3	Very agitated	Pulls or removes tubes or catheters; aggressive
+2	Agitated	Frequent nonpurposeful movement, fights ventilator
+1	Restless	Anxious, but movements not aggressive or vigorous
0	Alert and calm	–
−1	Drowsy	Not fully alert, but has sustained awakening (eye opening/eye contact) to voice (>10 s)
−2	Light sedation	Briefly awakens with eye contact to voice (<10 s)
−3	Moderate sedation	Movement or eye opening to voice (but no eye contact)
−4	Deep sedation	No response to voice, but movement or eye opening to physical stimulation
−5	Unable to rouse	No response to voice or physical stimulus

Glasgow Coma State

This scale is based on motor responsiveness, verbal performance, and eye-opening to appropriate stimuli. Three different behaviors are accounted for, and the responses presented by the patients are given a score, those patients with a final score of 8 or less are classified as under a comatose scale. The specific behaviors, responses, and scores are presented in Table 5.

Table 5 Glasgow coma scale

Behavior	Response	Score
Eye opening response	Spontaneous	4
	To verbal stimuli, command, speech	3
	To pain only (not applied to face)	2
	No response	1
Verbal response	Oriented	5
	Confused conversation, but able to answer questions	4
	Inappropriate words	3
	Incomprehensible speech	2
	No response	1
Motor response	Obeys commands for movement	6
	Purposeful movement to painful stimulus	5
	Withdraws in response to pain	4
	Flexion in response to pain	3
	Extension response in response to pain	2
	No response	1
Total score	Best response	15
	Comatose patient	8 or less
	Totally unresponsive	3

Medically Induced Coma as a Treatment

Medically induced coma is effective in the treatment of specific conditions that involve brain injuries, drug overdose, strokes, and status epilepticus, among others [10].

Head Injury, Vascular Neurosurgical Procedures and Strokes

Barbiturates are used to manage intractable intracranial pressure (ICP) elevations after procedures such as Arteriovenous Malformation (AVM) resection, extracranial—intracranial bypass, aneurysmorrhaphy, after significant bleeding due to AVMs or ruptured aneurysms and they are mainly used due to their neuroprotective effect in vascular neurosurgery during episodes of temporal or permanent focal ischemia [14].

Kim et al. analyzed in a 10-year study 92 semicomatose and comatose patients with severe intracranial hypertension who either received barbiturate coma therapy (40 patients) or those who didn't (52 patients), indicating demographic data such as age, sex ratio and the initial Glasgow coma scale (GCS) and variables such as the types of brain insult and the types of conducted operations and demonstrated that barbiturate coma therapy is an effective treatment method for this disease especially in younger patients [15].

To cause EEG burst suppression Wassman et al. administered short-acting barbiturates to sixty patients undergoing carotid endarterectomy during cross-clamping of the internal carotid artery, during the procedure, the arterial and venous blood concentration of lactate were measured, they found that the blood lactate levels rise above normal only in those patients with the shortest burst suppression times, being an indicator of how effective barbiturates are in preventing ischemia during carotid artery occlusion [16].

Cardiac Arrest and Different Cardiac Procedures

Propofol was used by Kortelainen et al. to demonstrate that the induction of slow waves (less than 1 Hz) after the administration of this drug was correlated with good neurologic outcome 48 h after cardiac arrest [17].

Treatment-Resistant Depression And Other Mental Disorders

Elowe et al. provided evidence of a potential anti-maniac effect of a benzodiazepine, after treating a 46-year-old woman admitted to the psychiatric unit under a severe manic episode whose symptoms almost completely disappear after the administration of 0.07 mg/kg/h of intravenous midazolam for 5 days [18].

About one-third of the patients suffering from depression develop treatment-resistant depression, some patients opt for electroconvulsive therapy (ECT), but still many evade this treatment due to the cognitive secondary effects; Mickey et al. proposed that propofol has antidepressant properties and demonstrated that high-dose treatment with this drug is feasible and well tolerated by patients with treatment-resistant depression, similarly to ECT with reduced side effects [19].

Benzodiazepines have been used in combination with antidepressants to treat anxious depression and the initial symptoms of anxiety-related depression [20], it has been found in some studies that the combinatorial therapy of both drugs produces better therapeutic outcomes in the treatment of major depression [21], however, it is recommended to consider different alternative drugs for the treatment of depression due to secondary effects that could be present after a long treatment with benzodiazepines, such as drug dependence.

Status Epilepticus

This condition is defined as more than 30 min of either continuous seizure activity or two or more consecutive seizures without full consciousness recovery between them [22]. There are three recognized stages of this condition in which different drugs are used in order to treat it [23].

(1) Early Stage—Therapy is with benzodiazepines
(2) Established Status Epilepticus—Therapy is with intravenous anti-epileptic drugs such as phenytoin, phenobarbital or valproate
(3) Refractory Status Epilepticus - General anesthesia is usually recommended, at doses resulting in EEG burst suppression, at which level all seizure activity remains controlled.

Despite the great efforts to establish a line of action to treat status epilepticus or refractory status epilepticus, many authors have tried to establish protocols that have been used in adults and children; however, continuous monitoring of the EEG is of critical importance once the clinical treatment has initiated.

Glauser et al. developed an evidence-based treatment algorithm for children and adults with convulsive status epilepticus, they found intramuscular midazolam, intravenous lorazepam, intravenous diazepam and intravenous phenobarbital to be effective as initial therapy in adults, while in children, they reported intravenous lorazepam and diazepam as effective at stopping seizures lasting at least 5 min [22].

Newey et al. identified the complications (peripheral neuropathy, cerebral atrophy, volume overload, renal/metabolic, gastrointestinal, endocrine, cardiac/hemodynamic/vascular, respiratory and infectious) that appeared in patients who were administered pentobarbital and phenobarbital for the treatment of refractory status epilepticus, demonstrating that it is necessary to consider such complications to improve survival and functional outcomes [24].

McGinn et al. reported that the administration of ketamine in conjunction with pentobarbital reduced the frequency of epilepsy and maintained a state of burst suppression in a 56-year-old patient; something similar was observed in a 57-year-old patient, for whom in order to minimize the dose of pentobarbital needed to reach burst suppression, a continuous ketamine infusion was initiated as an adjunct to pentobarbital [25].

Claassen et al. carried out a systematic review to compare the efficacy of midazolam, propofol, and pentobarbital for terminating seizures and improving outcomes in refractory status epilepticus patients; they found that compared with treatment with midazolam or propofol, pentobarbital treatment was associated with a lower frequency of short-term treatment failure, breakthrough seizures and changes to a different infusion antiepileptic drug, furthermore, they reported that compared with seizure suppression, titration of treatment to EEG burst suppression was associated with a lower frequency of breakthrough seizures and a higher frequency of hypotension [26].

Mechanism of Action of the Drugs Used to Clinically Induce Coma

Most of the drugs previously mentioned alter the neurotransmission of GABA (gamma-aminobutyric acid), specifically binding to the GABAA receptor at multiple allosteric sites in the hypothalamus, the frontal lobe, the temporal lobe [27], disrupting the communication between the medial prefrontal cortex and the thalamus in the case of Propofol [28], selectively acting in different brain regions such as the neocortex and the thalamus in the case of barbiturates [29], and benzodiazepines providing their therapeutic effects through receptors highly concentrated in the cortex, thalamus, and cerebellum [30]. A brief description of the mechanism of actions can be seen in Table 6.

Despite the broad knowledge of the acting mechanism of these drugs and their therapeutic usability, little is known about the settings required to generate positive outcomes by protecting the brain in specific industrial accidents that derive in medical complications; industry 4.0 and artificial intelligence will provide the required aid

Table 6 Mechanism of action of the drugs employed to induce coma

Drug	Examples	Mechanism of action
Propofol	–	It is an alkylphenol sedative and hypnotic agent with a rapid onset and offset that was developed in 1975. The mechanism of action of this drug has not yet been clearly elucidated, although is known to have a non-specific on the GABAA receptor complex in a different location than barbiturates and benzodiazepines, increasing the conductance of chloride; it enhance the action of the GABA neurotransmitter, which inhibits the transmission by a "hyperpolarization mechanism" caused by the opening of the chloride channels. It also influences the presynaptic mechanisms of GABAergic transmission such as GABA uptake and GABA release and it has been shown to interact with different neuro transmitter receptors such as glycine receptors, nicotinic acetylcholine receptors and glutamate receptors [31]
Barbiturates	Thiopental Pentobarbital Methohexital Phenobarbital	This class of drugs act by potentiating the action of GABA, specifically binding to an allosteric site of the GABAA receptors, being critical for its activity the binding to the second and third transmembrane domains of the β subunit indirectly opening the chloride channels or at high doses directly activating them. Additionally, they block the AMPA (α-amino-3-hydroxy-5-methyl-4-isoxazolepropionic acid receptor)/kainite receptors and block glutamate release through the P/Q type high voltage activated calcium channels [32]
Benzodiazepines	Midazolam	Although benzodiazepines produce their effect binding to the GABAA receptors similarly to barbiturates and propofol, studies have shown that the molecular interplay underlying their interaction are different, since it occurs at different allosteric sites. Specifically, this class of drugs bind to the pocket created by the α and γ subunits of the GABA receptor and induce a conformation change in the chloride channels that hyperpolarizes the cell provoking their inhibitory effect [30]

(continued)

by offering new algorithms that will eventually consider all the variables and factors that play important roles in determining the faith of patients after they have suffered an accident.

Table 6 (continued)

Drug	Examples	Mechanism of action
Opioids	Morphine Fentanyl	Opioids act on G-protein coupled receptors named DOP, KOP, MOP and NOP which produce analgesia and other effects according to their location whenever they are activated. Binding of an opioid agonist to the G protein, causes the α subunit of the protein to exchange GDP (guanosine diphosphate) for GTP (guanosine triphosphate). This causes the α subunit to dissociate from the βγ complex and both complexes are then able to interact with their target proteins. The binding results in the reduction in intracellular cyclic adenosine monophosphate (cAMP) levels by inhibiting adenylylcyclase; in addition, the interaction with ion channels cause the activation of potassium conductance and an inhibition of calcium conductance causing a cell hyperpolarization and reduced neurotransmitter release [33]

2 Implementation of the Intelligent Application

In this work, we will focus in the development of a predictive algorithm that determines whether a patient can be aided with a medically induced coma following an industrial accident, for that, we will consider the following variables: age, gender, race, general health status, consciousness status according to the Glasgow Coma Scale, in-hospital complications and the type of injury (Head, Heart. Mental Illness or derived conditions). The variables have been previously shown to be directly associated with the comatose state outcome in the earlier described studies. To start analyzing the odds of a patient recovering from an industrial accident, the TOPSIS decision-making model will be employed. A general scheme of the decision-making process for this work is described in Fig. 2. In Fig. 1, a woman in a comatose state can be observed.

Multicriteria decision-making techniques are tools that aid those researchers to select diverse options mainly on discrete problems, this kind of tools are usually aided by the use of computers allowing the velocity for researchers to be able to take well-planned decisions based on different available methods such as MAXMIN, MAXMAX, SMART, TOPSIS, among others.

Some of the presented advantages by this method have been previously described, and are as follows:

- Simple, rational, comprehensible concept,
- Intuitive and clear logic that represent the rational of human choice,
- Ease of computation and good computational efficiency,
- A scalar value that accounts for both the best and the worst alternatives
- Ability to measure the relative performance for each alternative in a simple mathematical form,

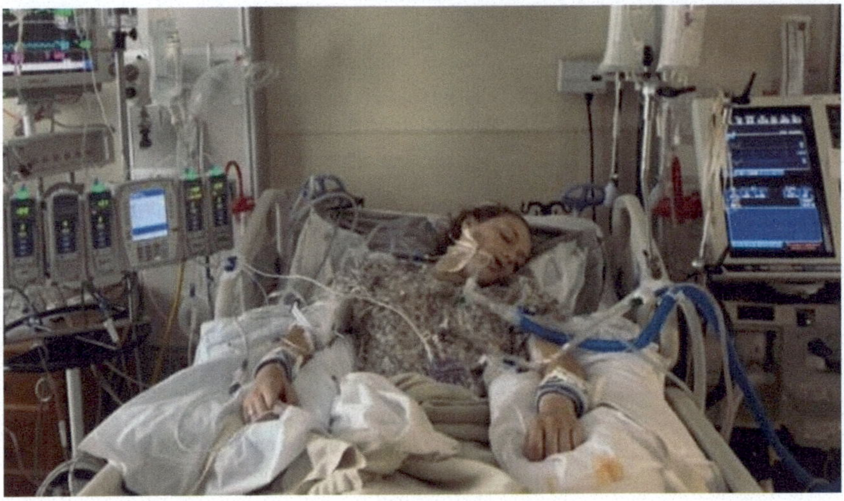

Fig. 1 Medically induced coma at the intensive care unit

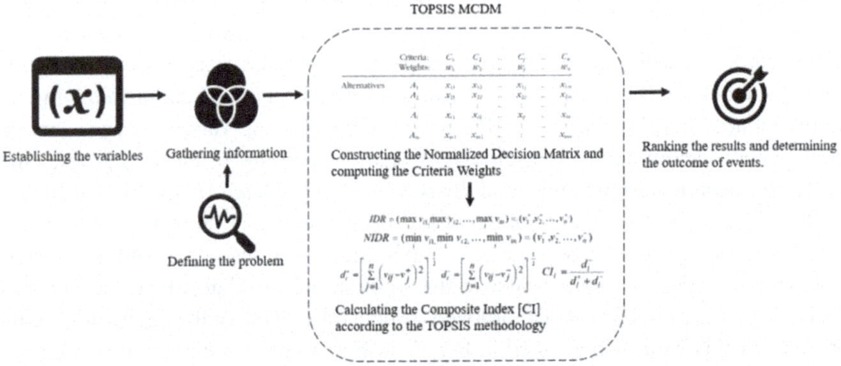

Fig. 2 Topsis methodology applied for this work

– Possibility for visualization

As it has been previously well described by [34], the TOPSIS algorithm starts by forming the decision matrix representing the satisfaction value of each criterion with each alternative. Next, the matrix is normalized using criteria weights, subsequently, positive-ideal and negative-ideal solutions are calculated, and the distance among the alternatives to each of these solutions is finally calculated and ordered, maintaining those closest to the ideal solutions at the top of the ranking.

The first step of the analysis consists of providing to each of the variables a weight, which can be seen in Table 7; the weights presented in this table have been assigned according to the authors expertise.

Table 7 Variables associated with the clinical outcome after suffering and industrial accident and their assigned weight

Variable	Weight
Laboral age	0.12
Previous accidents by gender	0.11
Incidence by race	0.09
General health status	0.18
Consciousness status according to the Glasgow coma scale	0.19
Collateral effects (head, heart, mental illness, or derived conditions)	0.20
In-hospital complications	0.11

We considered seven different accidents that could occur to workers of an industry, these are: electrocution, burn, head injury, body trauma, amputation, chemical inhalation, and drinking a dangerous substance.

3 Multicriteria Analysis

For the multicriteria analysis using TOPSIS, matrices were created to get the optimal solution considering the previous variables. For the selection of the accident with the highest priority to be attended by means of the induced coma, the following table of weights was carried out, the positive ideal row (PIR) and the negative ideal row (NIR) were also determined (Table 8). Three different methods were applied in order to calculate the rankings, the first method is the common TOPSIS using the previously established weights for each alternative (Table 9), secondly the point method was employed using the point's rank of each alternative set by the industry 4.0 bureau according to each criterion based on hospital requirements (Table 10), and finally, the AWM method to consider the human factor (Table 11) [35].

4 Multivariable Analysis of Our Results

According to the multicriteria analysis of the results, we found out that drinking a dangerous substance is the accident that must be given the highest priority in comparison with all the others, as the TOPSIS determined that this accident was ranked as 1, while the accident that must be given the lowest priority according to our results is head injury, as there might be other factors that could determine the outcome of patients with head injury, besides, when we make the analysis and consider the hospital requirements to treat such conditions we found out that once again, drinking a dangerous substance must be given the highest priority, followed

Table 8 TOPSIS methodology

Accident	Laboral age	Previous accidents	Incidence by race	General health status	Conscious ness according to the Glasgow Coma Scale	Collateral effects (head, heart, mental illness, or derived condition)	In hospital complications	Sum of weights
Electrocution	6.03	5.23	6.79	1.05	1.5	3.26	4.06	–
Burn	2.57	5.85	6.78	6.75	5.79	5.3	1.83	–
Head injury	4.2	4.65	5.9	6.63	7	5.24	5.72	–
Body trauma	3.57	4.98	1.63	3.24	5.13	5.09	3.94	–
Amputation	1.81	6.62	1.32	6.26	4.58	5.55	6.18	–
Chemical in halation	3.19	5.09	5.09	5.09	3.81	2.2	2.51	–
Drinking of a dangerous substance	2.91	1.87	1.56	1.95	1.03	4.97	3.12	–
Weights	0.12	0.11	0.09	0.18	0.19	0.2	0.11	1
PIR	1.81	1.87	1.32	1.05	1.03	2.2	1.83	–
NIR	6.03	6.62	6.79	6.75	7	5.55	6.18	–

Table 9 Rank method using only the rank of each alternative under each criterion

Accident	Laboral age	Previous accidents	Incidence by race	General health status	Consciousness according to the Glasgow Coma Scale	Collateral effects (head, heart, mental illness, or derived condition)	In hospital complications	Weighted rank	Conclusion
Electrocution	7	5	7	1	2	2	5	1.51	2
Burn	2	6	6	7	6	6	1	3.71	5
Head injury	6	2	5	6	7	5	6	4.07	7
Body trauma	5	3	3	3	5	4	4	2.73	4
Amputation	1	7	1	5	4	7	7	3.83	6
Chemical in halation	4	4	4	4	3	1	2	1.71	3
Drinking of a dangerous substance	3	1	2	2	1	3	3	1.48	1

Table 10 Point method considering the hospital requirements

Accident	Laboral age	Previous accidents	Incidence by race	General health status	Consciousness according to the Glasgow Coma Scale	Collateral effects (head, heart, mental illness, or derived condition)	In hospital complications	Weight sum of points	Conclusion
Electrocution	5.07	9.73	8.09	1.27	5.54	6.02	2.09	2.715	5
Burn	6.16	10	6.14	2.2	2.18	3.47	1.45	1.664	7
Head injury	7.59	8.73	9.55	2.81	6.64	7.35	1.93	3.450	4
Body trauma	7.44	8.18	9.73	3.17	3.42	1.9	4.23	2.066	6
Amputation	6.71	4.26	1.2	2.58	9.95	7.51	8.14	4.752	3
Chemical in halation	8.36	8.11	5.79	7.3	6.21	8.29	7.97	5.029	2
Drinking of a dangerous substance	5.94	7.16	2.53	3.9	8.37	9.36	8.76	5.128	1

Table 11 AWM method considering the human factor

Accident	Laboral age	Previous accidents	Incidence by race	General health status	Conscious ness according to the Glasgow Coma Scale	Collateral effects (head, heart, mental illness, or derived condition)	In hospital complications	Weighted sum	Conclusion
Electrocution	0.000	0.293	0.000	1.000	0.921	0.684	0.487	0.545	1
Burn	0.820	0.162	0.002	0.000	0.203	0.075	1.000	0.163	5
Head injury	0.434	0.415	0.163	0.021	0.000	0.093	0.106	0.034	7
Body trauma	0.583	0.345	0.943	0.616	0.313	0.137	0.515	0.254	4
Amputation	1.000	0.000	0.086	0.405	0.000	0.000	0.000	0.092	6
Chemical in halation	0.0673	0.322	0.172	0.239	0.534	1.000	0.844	0.437	3
Drinking of a dangerous substance	0.739	1.000	0.956	0.842	1.000	0.173	0.703	0.454	2

by chemical inhalation, amputation, head injury, electrocution, while burns and body trauma are at the bottom of the list.

A dangerous or a toxic substance can be defined as a substance that can cause poisonous or negative health effects, such effects can be diverse such as vomit, dizziness, headache etc. and must be treated immediately after the exposure [36]. There are certain factors that must be considered when talking about toxic substances, such as the toxicity, the route of exposure, the dose, the duration of the exposure, the reaction and interaction with the different organs and systems of the human body and the sensitivity of the body for the substance. Although there are multiple routes of exposure, we consider that the ingestion of toxic substances is one of the most harmful and hard to treat in the intensive care unit since this route of exposure is rarely considered and usually is considered unimportant due to a series of reasons [37].

Regarding the hospital requirements, since we also found that drinking a dangerous or a toxic substance is the most important accident and it must be treated in some specific serious patient conditions, it is necessary that the hospitals and the personnel is well trained to proceed with a treatment such as induced coma to treat these kind of accidents; so far there is no evidence that this work-related incident has been treated with this therapy, however, there is the possibility that the toxic substance might, in some specific cases, be related to drugs that can at some point induce a state of unconsciousness similar to a medically induced coma, so for that it is important that the hospital and the personnel are well prepared for this cases.

Considering the human factor, we found that electrocution is the top priority, followed by drinking a dangerous substance, chemical inhalation, body trauma, burns, amputation and head injury, meaning that there is a high-degree of human error involved in electrocution accidents. These accidents occur due to multiple reasons such as inadequate wiring, overloads, ground faults, among other, commonly resulting in high-grade burns, multiple tissue and organs malfunction, commonly associated with a high morbidity and mortality.

Victims of electrocution or electrical injuries must be approached at the intensive care unit as both victims of trauma and cardiac affections, commonly the treatment involves a constant monitoring of the cardiac function through an EEG until the patient is stabilized and normal cardiac functions are restored. Some cardiac traumas have been involved in a pathological induced coma, as it was previously noted in the introduction section, it is important to consider that the outcomes for a pathological and a medically induced coma are different since patients can recover in almost all the cases immediately after the medication has stopped during a medically induced coma, however, during a pathological induced coma, the recovery time is not well stablished and is typically associated with the degree of the injury, sometimes resulting in a fatality.

We found head injury at the bottom of list considering the human factor, this indicates that there is little human error associated with this kind of injuries. Common head injuries include concussions, skull fracture, and scalp wounds, the treatment of such injuries varies greatly depending on its severity. Generally, those patients that suffer high-degree head injuries such as concussions are left with long-lasting damage to the brain. It is in all the cases important to receive a treatment from a specialized health care professional to evaluate the damage, the initial treatment commonly involves checking if the airway is clear, stabilizing the neck and spine and stopping any severe bleeding followed by an observation period to assess the level of consciousness, the size of the pupils and their reaction to light, the movement of arms and legs and breathing, heart rate, blood pressure among others to finally analyze a CT scan to monitor any kind of swelling or bleeding inside the skull [38].

5 Conclusions and Future Research

Fourth Industrial Revolution or Industry 4.0 has demonstrated its capacity to fulfill the requirements of the medical field in many aspects, it is expected that with this broad technology, work-related accidents will be treated in revolutionizing ways so far not foreseen in the field. There are multiple ways in which accidents can be treated and prevented employing interconnected systems that offer multidisciplinary ways of using information, manufacturing improving by this means quality of life.

In this work we demonstrated that medically induced coma can be an alternative treatment in some specific work-related accidents such as drinking a toxic substance, inhaling and poisonous substance, burns or head trauma, among others. We consider that this kind of therapies in conjunction with better manufacturing systems will improve treatment practices at the intensive care unit in these and other accidents. Prevention is the best way of treatment for any disease, however, in some cases it is not possible to completely stop accidents from happening, thus resulting in increasing numbers of people that every year suffer devastating consequences.

As a future research, we would like to apply this TOPSIS methodology for the selection of other kind of accidents that might happen in a society under the paradigm of industry 4.0, this would allow the hospitals and their personnel to acquire the necessary knowledge, techniques and equipment for improving the standard of care for specific injuries, such as those analyzed in this research work.

References

1. Eurostat (2019) Accidents at work statistics. https://ec.europa.eu/eurostat. Accessed 12 April 2020
2. Secretaria del Trabajo y Prevision Social (2019) Estadisticas Del Sector. https://www.gob.mx/stps/acciones-y-programas/estadisticas-del-sector. Accessed 12 April 2020
3. Gonzalez-Delgado M, Gómez-Dantés H, Fernández-Niño J, Robles E, Borja V, Aguilar M (2015) Factors associated with fatal occupational accidents among Mexican workers: a national analysis. PLoS ONE 10(3):e0121490
4. Health and Safety Executive (2019) Costs to great Britain of workplace injuries and new cases of work-related Ill health—2017/18. https://www.hse.gov.uk/statistics/cost.htm. Accessed 12 April 2020
5. Injury Facts (2018) Work injury costs. https://injuryfacts.nsc.org/. Accessed 13 April 2020

6. van den Heuvel S, van der Zwaan L, van Dam L, Oude-Hengel K, Eekhout I, van Emmerik M (2017) Estimating the costs of work-related accidents and Ill-health: an analysis of European data sources. https://osha.europa.eu/. Accessed 13 April 2020
7. Pavel B (2018) Important issues in coma and neuromonitoring. In: Current topics in intensive care medicine. IntechOpen
8. Ching S, Liberman M, Chemali J, Westover M, Kenny J, Solt K, Purdon P, Brown E (2013) Real-time closed-loop control in a rodent model of medically induced coma using burst suppression. Anesthesiology 119(4):848–860
9. Trapani G, Altomare C, Sanna E, Biggio G, Liso G (2000) Propofol in anesthesia. Mechanism of action, structure-activity relationships, and drug delivery. Curr Med Chem 7(2): 249-271
10. An J, Jonnalagadda D, Moura V, Purdon P, Brown E, Westover M (2018) Variability in pharmacologically-induced coma for treatment of refractory status epilepticus. PLoS ONE 13(10):e0205789
11. Ouimet S, Kavanagh B, Gottfried S, Skrobik Y (2006) Incidence, risk factors and consequences of ICU delirium. Intensive Care Med 33(1):66–73
12. Bordini A, Luiz T, Fernandes M, Arruda W, Teive H (2010) Coma scales: a historical review. Arq Neuropsiquiatr 68(6):930–937
13. Devlin J, Fraser G, Riker R (2012) Drug-induced coma and delirium. In: Papadopoulos J, Cooper B, Kane-Gill S, Corbett S, Barletta J (ed) Drug-induced complications in the critically-Ill patients, 1st ed. Society of Critical Care Medicine, pp 107–116
14. Clark J, Ellens N, Figueroa B (2015) The use of barbiturate-induced coma during cerebrovascular neurosurgery procedures: a review of the literature. Brain Circul 1(2):140
15. Kim Y, Park S, Nam T, Park Y, Min B, Hwang S (2008) The effect of barbiturate coma therapy for the patients with severe intracranial hypertension: a 10-year experience. J Korean Neurosurg Soc 44(3):141
16. Wassman H, Fromm G, Nadstawek J, Bannister C, Hartmann A, Pavlidis C (1989) The influence of barbiturates on cerebral metabolism in patients with borderline cerebrovascular reserve during intraoperative transient carotid occlusion. Br J Neurosurg 3(4):429–434
17. Kortelainen J, Väyrynen E, Huuskonen U, Laurila J, Koskenkari J, Backman J, Alahuhta S, Seppänen T, Ala-Kokko T (2017) Pilot study of propofol-induced slow waves as a pharmacologic test for brain dysfunction after brain injury. Anesthesiology 126(1):94–103
18. Elowe J, Zimmermann M, Hasselmann M, Danion J (2014) Benzodiazepine-induced coma in the treatment of severe acute mania. J Neuropsych Clin Neurosci 26(1):E09–E10
19. Mickey B, White A, Arp A, Leonardi K, Torres M, Larson A, Odell D, Whittingham S, Beck M, Jessop J, Sakata D, Bushnell L, Pierson M, Solzbacher D, Kendrick E, Weeks H, Light A, Light K, Tadler S (2018) Propofol for treatment-resistant depression: a pilot study. Int J Neuropsychopharmacol 21(12):1079–1089
20. Benasi G, Guidi J, Offidani E, Balon R, Rickels K, Fava G (2018) Benzodiazepines as a monotherapy in depressive disorders: a systematic review. Psychother Psychosom 87(2):65–74
21. Ogawa Y, Takeshima N, Hayasaka Y, Tajika A, Watanabe N, Streiner D, Furukawa T (2019) Antidepressants plus benzodiazepines for adults with major depression. Cochrane Database of Systematic Reviews
22. Glauser T, Shinnar S, Gloss D, Alldredge B, Arya R, Bainbridge J, Bare M, Bleck T, Dodson W, Garrity L, Jagoda A, Lowenstein D, Pellock J, Riviello J, Sloan E, Treiman D (2016) Evidence-based guideline: treatment of convulsive status epilepticus in children and adults: report of the guideline committee of the American epilepsy society. Epilepsy Curr 16(1):48–61
23. Shorvon S, Ferlisi M (2011) The treatment of super-refractory status epilepticus: a critical review of available therapies and a clinical treatment protocol. Brain 134(10):2802–2818
24. Newey C, Wisco D, Nattanmai P, Sarwal A (2016) Observed medical and surgical complications of prolonged barbiturate coma for refractory status epilepticus. Therapeutic Adv Drug Saf 7(5):195–203
25. McGinn K, Bishop L, Sarwal A (2016) Use of ketamine in barbiturate coma for status epilepticus. Clin Neuropharmacol 39(1):62–65

26. Claassen J, Hirsch L, Emerson R, Mayer S (2002) Treatment of refractory status epilepticus with pentobarbital, propofol, or midazolam: a systematic review. Epilepsia 43(2):146–153
27. Zhang H, Wang W, Zhao Z, Ge Y, Zhang J, Yu D, Chai W, Wu S, Xu L (2010) The action sites of propofol in the normal human brain revealed by functional magnetic resonance imaging. The Anatomical Record: Advances in Integrative Anatomy and Evolutionary Biology 293(12): spc1-spc1
28. Flores F, Hartnack K, Fath A, Kim S, Wilson M, Brown E, Purdon P (2017) Thalamocortical synchronization during induction and emergence from propofol-induced unconsciousness. Proc Natl Acad Sci 114(32):E6660–E6668
29. Mathers D, Wan X, Puil E (2007) Barbiturate activation and modulation of GABAA receptors in neocortex. Neuropharmacology 52(4):1160–1168
30. Griffin C, Kaye A, Rivera Bueno F, Kaye A (2013) Benzodiazepine pharmacology and central nervous system-mediated effects. Ochsner J 13(2):214–223
31. Brown E, Lydic R, Schiff N (2010) General anesthesia, sleep, and coma. N Engl J Med 363(27):2638–2650
32. Löscher W, Rogawski M (2012) How theories evolved concerning the mechanism of action of barbiturates. Epilepsia 53:12–25
33. Pathan H, Williams J (2012) Basic opioid pharmacology: an update. Br J Pain 6(1):11–16
34. Roszkowska E (2011) Multi-criteria decision-making models by applying the topsis method to crisp and interval data. University of Economics in Katowice, pp 200–230
35. Zulqarnain M (2017) Choose best criteria for decision making via fuzzy topsis method. Math Comput Sci 2(6):113
36. Department of Health (2013) What you know can help you - an introduction to toxic substances. https://www.health.ny.gov/environmental/chemicals/toxic_substances.htm. Accessed 15 April 2020
37. The Annals of Occupational Hygiene (2006) How important is inadvertent ingestion of hazardous substances at work?
38. NHS (2018) Severe head injury. https://www.nhs.uk/conditions/severe-head-injury/treatment/. Accessed 15 April 2020

Coahuila's Future a Perspective Derived from the Growth Per Municipality to Establish a Car Assembler

Cynthia Rodríguez, Moisés Sarmiento, Jair Martinez, Daniel Castro, Francisco Tarango, and M. A. Gerardo Yáñez

Abstract In this document an investigation is developed that used the TOPSIS Methodology to evaluate by means of Socio-Economic Attributes the economic power of the most salient communities of the Coahuila State, based on the Database collected from the INEGI (National Institute of Statistic and Geography) Census of 2010, This being the last census carried out in the country, the next one is in the data collection stage during the year 2020. Elaborating a first review, 5 attributes related to 5 economic aspects were determined, which are: 1. HEALTH 2. EDUCATION 3. HOUSING 4. INTERNET 5. POPULATION. Each of these elements, represented by their attributes, were analyzed against the mass of the total population or economically active population; With this, positioning of the municipalities was obtained, where the 5 best were selected after Saltillo and Torreón, looking in them for a candidate to achieve outstanding economic growth and position themselves in a privileged position within the state and at the national level. These two localities were excluded from the evaluation considering their economic development considerably in advantage over the others. Torreon is part of a multi-developed area that is supported by the Comarca Lagunera de Durango, with the City of Gomez Palacio and Lerdo, strengthening the attributes of social, economic and educational development of its inhabitants; it already has national and international investments. On the other hand, the city of Saltillo as the capital of the state is favorable to investments in the automotive sector in addition to indices of social and economic development. By considering all these characteristics, and avoiding a bias in the information of the crossings; these two locations were withdrawn from the evaluation. The main objective of this work is to know which of the competing cities is the most suitable for the construction and development of a new automobile assembly plant; taking into account the infrastructure in operational, logistical, economic and social demand that this implies; that precisely these characteristics are associated with the evaluated attributes. After performing the evaluations in the software and after the evaluation with the multicriteria method, the 5 municipalities with the best economic indicators

C. Rodríguez · M. Sarmiento · J. Martinez · D. Castro · F. Tarango · M. A. Gerardo Yáñez (✉)
Maestría en Administración y Alta Dirección, Universidad Iberoamericana de Torreón, Torreón, Mexico
e-mail: gerardo.yanez@iberotorreon.edu.mx

© The Author(s), under exclusive license to Springer Nature Switzerland AG 2021
A. Ochoa-Zezzatti et al. (eds.), *Technological and Industrial Applications Associated with Intelligent Logistics*, Lecture Notes in Intelligent Transportation and Infrastructure,
https://doi.org/10.1007/978-3-030-68655-0_4

were determined, these were: **1. Monclova 2. Matamoros 3. Piedras Negras 4. San Pedro 5. Ramos Arizpe 6. Acuña.** Once the ideal municipalities had been identified and whose attributes meet the optimal values to be considered in our analysis, it was taken into consideration that a triggering factor for the development of a municipality is investment in the automotive industry such as an automobile assembly plant, this type of Over the years the industry is subject to new regulations and certifications as it is undergoing a technological revolution where consumer demand means that the product has greater digital requirements, software, innovations in infotainment etc. That is why, in addition to complying with the necessary attributes, the municipality that proves to have outstanding economic and infrastructure potential in this sector is Ramos Arizpe.

Keywords Car assembler · Automotive industry · Car manufacturing · Economic growth · Industrial development · Industry

1 Introduction

The state of Coahuila is the third largest in the country with an area of 151,572 km^2 that corresponds to 7.8% of the country's total. It limits to the north with the state of Texas of the United States of North America having the Rio Bravo as its border; to the south with the state of Zacatecas; to the southwest with Durango; to the east with Nuevo León and to the west with Chihuahua. It has 38 municipalities and over the last 20 years, Coahuila has increased its population, in absolute terms, by just over 776 thousand inhabitants, to register by 2010 a population figure of the order of 2,748,391 inhabitants according to the INEGI 2010 Population and Housing Census. The population weight that the entity has within the national context has remained stable throughout these last two decades, concentrating 2.4% of the national population. It also has a development that places it in privileged places in the socioeconomic structure of the country. 76% of this total population is concentrated in seven municipalities: Saltillo, Torreón, Monclova, Piedras Negras, Acuña, Matamoros and San Pedro; being the municipalities of Saltillo and Torreón those that concentrate almost 50% of the total population. Coahuila has an important population dynamics since only in the last twenty years of the entity's history (between 1990 and 2010), the number of residents living in a municipality with a population greater than 100,000 inhabitants practically doubled. In the state economy, the main sector is the tertiary which contributes 50.5% of GDP (Gross domestic product), closely followed by the secondary sector with 46.7% and finally the primary sector with 2.8% (INEGI). There are important activities such as the production of mineral coal, liquid steel, agricultural and mining activities, milk production and crops such as forage, cotton, walnut, apple tree and vegetables; in addition to the establishment of the maquiladora industry and electricity generating companies, as well as the production of automobiles and trucks for export. The state has an important urbanization process, according to SEDESOL (Secretaria de Desarrollo Social, Ministry of Social development), there are four metropolitan areas: Saltillo-Ramos Arizpe-Arteaga, Torreón-Matamoros, Monclova-Frontera-Castaños and Piedras Negras-Nava. The Government of Coahuila also decreed on June 15, 2011 a fifth Metropolitan area, La Carbonífera, made up of the municipalities of Sabinas-San Juan de Sabinas-Múzquiz [1]. In accordance with the population concentration that Coahuila lives in its metropolitan areas, the economically active population is concentrated significantly in these urban centers, since they concentrate 77% of the state's EAP. The economically active population (EAP) of Coahuila has had a significant increase, of just over 251 thousand people, registering an average annual growth rate of 2.5%, well above the population growth registered in that decade. In terms of the age and sex of the employed population, the data reveals that of the 1,037,924 employed persons that exist in Coahuila for 2010, 65.8% are men, while the remaining 34.2% are women. It is striking how young the employed population is in the state; for both sexes, 30% of those employed are under 30 years old and 65% are between 30 and 64 years old. Only 5% of the employed population is 65 years old and over; 35,700 employed [2]. According to the data from the Population and Housing Census in 2010, Coahuila has a municipal average of access to basic services (water, drainage and electricity) of 91.28% of its homes. In the case of piped water, 96.53% of the homes have piped water from the public network; However, there are some municipalities with serious access problems such as Jiménez, in which only 70% of the homes have access to water; Ocampo and Escobedo, in which around 80% of homes have access. Similarly, 93.58% of the homes have drainage in the state, however, the municipality of Hidalgo only registers 68% of its homes with this service. The electricity service reports less disparity in its availability (97.43% of the homes in the state have electricity), the municipality with the least access to electricity service (Hidalgo) registers 89.34% of its homes with access to the service [3]. The state shows an average of schooling higher than the national average (9.47 and 8.25 years respectively). On average, attendance at an education center by age group in the state follows the national trend: while a high percentage of the population (95%) in the age group of 6–14 years corresponding to the primary level, attends the school; only 5% of the age group 25–29 corresponding to postgraduate studies, attend an education center.

The state of Coahuila has limited access to higher education in most of its municipalities since the concentration of universities and higher education centers in only four municipalities (Saltillo, Torreón, Acuña and Monclova), where 41 of the 60 universities are located of the state to 2010. 76.5% of the population of the state of Coahuila has the right to health care in an institution, whether public or private. It is important to consider that the vast majority of beneficiaries are affiliated with at least one public institution and only a low percentage (less than 5%) resort to private institutions. The percentage of eligibility by sex is slightly higher among women (78%) than in men (75%), this means that of the total of women and of the total of men living in Coahuila, 3 out of 4 are beneficiaries approximately [4]. According to figures provided by the National Institute of Statistics and Geography (INEGI) and the Mexican Association of the Automotive Industry (AMIA), 10 States concentrate 97.5% of the production of cars and trucks in Mexico, among which Coahuila ranks second rung at the national level. In the creation of auto parts, Coahuila leads this area by far, by producing 18.1% of the auto parts manufactured nationwide; surpassing states like Nuevo León, which contributes 11.4%; and Puebla, with 10.4%, being the only entity located in the first places in the manufacture of vehicles and auto parts.

Coahuila is considered one of the entities with the best economic and educational development in Mexico, for this reason it is not difficult to understand that it is home to some of the multinational companies in the automotive sector with the greatest economic spillover; considering its position is that this study is based on evaluation, directing the study of certain population and economic attributes. This seeks to know which would be the best area or locality in which a following automotive company could build an assembly plant.

During this study, evaluations are carried out in the candidate populations through multicriteria analysis, software analysis and field investigations. The software used in this analysis was WEKA wich is intended for automatic learning and data mining based JAVA. WEKA contains a collection of algorithms to perform data analysis and predictive modeling, it also has tools for visualizing this data, it also provides a graphical interface that unifies the tools so that they are better available. The data to be evaluated are taken from a database, the source of the database was from INEGI which contains a considered number of parameters or attributes to be studied.

The attributes allow knowing the situations of the localities in the areas of connectivity, schooling, total population, services, social development and economically active population; All these parameters were considered important for the good reception, construction, development and profitability of a car assembler. For the study and analysis of interest, the research team reviewed literature and research similar to the topic presented to allow building a solid base of the present analysis, allowing added value to the interested parties.

The document found that contributed much to this work was "The new factors of industrial location in Mexico. The experience of automotive export complexes in the north" (Jose Carlos Ramirez).

In that research work seeks to offer evidence on the new location factors in northern Mexico beyond the traditional industrial geolocation factors called Weberian factors. An interesting fact that is mentioned is that an important part of the manufacture of cars of the center of the country moved to the north and although this situation has already been studied by many authors, there is no consensus on the existence of a pattern common behavior.

This author´s analysis is base on making an factorial model and solving the model using the principal components method. The factors that he weighs and classifies were taken according to the considerations that the plants have to settle in Mexico, it was important fo us to know what these factors were and have a reference to relate them to the attributes available in the database in which we work [5].

2 Selected Attributes

1. PEA: People aged 12 and over who worked; they had work, but did not work or; they looked for work in the reference week. This variable allows us to know the potential mass of human resources available to cover the demand for human resources for administrative and operational roles.
   This figure speaks of the inhabitants with the ability to generate economic value through the fulfillment of a work activity that contributes to the value and supply chain of the assembly plant or any other work center.
2. VPH_C_SERV: Inhabited private homes that have electricity, piped water inside or outside the home, but within the land, as well as drainage. It includes private dwellings for which the characteristics of the dwelling were captured, classified as independent house, apartment in a building, dwelling or room in the neighborhood and dwelling or room on the roof, and for which no class of dwelling was specified. This variable was considered as a reflection of the socio-economic level that the potential human resource that can participate within the assembly plant's infrastructure needs to count. It is important to highlight that the manufacturing process in the automotive sector demands specific hard and soft skills; so the human resource must be evaluated in various skills.
3. GRAPROES: Result of dividing the amount of school grades approved by people between 15 and 130 years of age by people of the same age group. Exclude people who did not specify approved grades.
4. TOTAL, POP: Total of people who habitually reside in the country, state, municipality and locality. Includes the estimate of the number of people in private homes without information on occupants. Includes the population that did not specify their age.
   For administrative positions certain skills and knowledge are essential; Therefore, this variable allows us to know the volume of inhabitants with an educational level necessary to cover the demanded places of the assembly plant; therefore, it is important to study this data.
   Variable that shows us the total population of the community, allows us to evaluate it against population with level of education, population with housing and the age range evaluated in the first variable.
   It is a variable for a complementary evaluation for decision making.
5. PDER_SS: Total number of people who have the right to receive medical services in a public or private health institution such as: the Mexican Institute of Social Security (IMSS), the Institute of Social Security and Services for State Workers (ISSSTE and ISSSTE state), Petróleos Mexicanos (PEMEX), the Secretariat of National Defense (SEDENA), the Secretariat of the Armed Navy of Mexico (SEMAR), the Social Protection System in Health or another. The health infrastructure is one of the most prominent variables since an assembler has a minimum staff of 500 people; Taking this figure into account, the hospital infrastructure must correspond to it, so that a viable location to receive a car assembler must have a favorable index in this attribute.

6. VPH_INTER: Private inhabited homes that have internet service. It includes the private dwellings for which the characteristics of the dwelling were captured, classified as independent house, apartment in a building, dwelling or room in the neighborhood and dwelling or room on the roof and for which no class of dwelling was specified [6]. This variable was considered for its interpretation of the socio-economic level of the population; The population with internet access gives meaning to the level of income, communication and accessibility.

2.1 CROSSING I. Impact of the Economically Active Population on Housing with Basic Services (PEA—VPH_C_SERV)

Cross Justification

The concept of the Economically Active Population was introduced by the United Nations (UN) in 1958 in its principles and recommendations regarding the 1960 population censuses as specified in its Latin American Demographic Center (Elizaga and Mellon, 1971), because most of the economic characteristics in a census correspond only to economically active people; For this reason, the elements used to identify the employment situation are key to the entire block of economic elements and are fundamental to formulate almost all economic and social policies.

The hypothesis we propose is that there is a direct relationship between the economically active population and the number of homes that have all the basic services in Coahuila, assuming in this way that the greater the relationship between these two factors, the quality of life in said municipality will be higher.

Results

In a preliminary analysis of the relationship between PEA and VPH_C_SERV, it is observed that Monclova, Piedras Negras and Torreón are above the state average (62%), while Acuña, Matamoros, Saltillo and San Pedro are below it (Fig. 1).

Prior to performing the crossing, a linear regression is performed to determine the correlation of the PEA and VPH_C_SERV attributes, finding it is 99.96%, with an absolute relative error of 3.31% (Table 1).

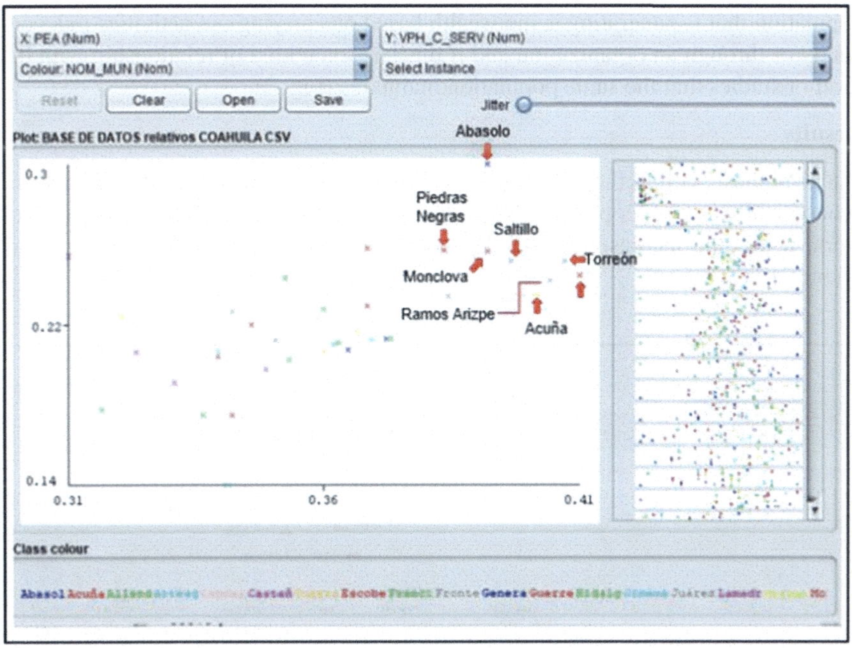

Fig. 1 Graphic results of the crossing PEA—VPH_C_SERV

Table 1 Total, population per municipality with basic services [2]

NOM_MUN	POBTOT	VPH_C_SERV
Saltillo	725123	0.244
Torreón	639629	0.254
Monclova	216206	0.259
Piedras Negras	152806	0.254
Acuña	136755	0.247
Matamoros	107160	0.217
San Pedro	102650	0.203
Ramos Arizpe	75461	0.237

2.2 *CROSSING II. Impact of the Average Level of Schooling with the Economically Active Population (PEA—GRAPROES)*

Cross Justification

The analysis of the Economically Active Population (PEA) and the Average Level of Education (GRAPROES) was carried out, with the intention of identifying that population that is generating some wealth based on carrying out activities independently or through an organization and if this has some kind of relationship with the level of studies that the same population maintains (Fig. 2).

Results

Our hypothesis is that the higher the level of schooling, the greater the number of economically active people and that is that, although studying is not a guarantee of work, if it could increase the chances of getting one or being able to self-employ (Table 2).

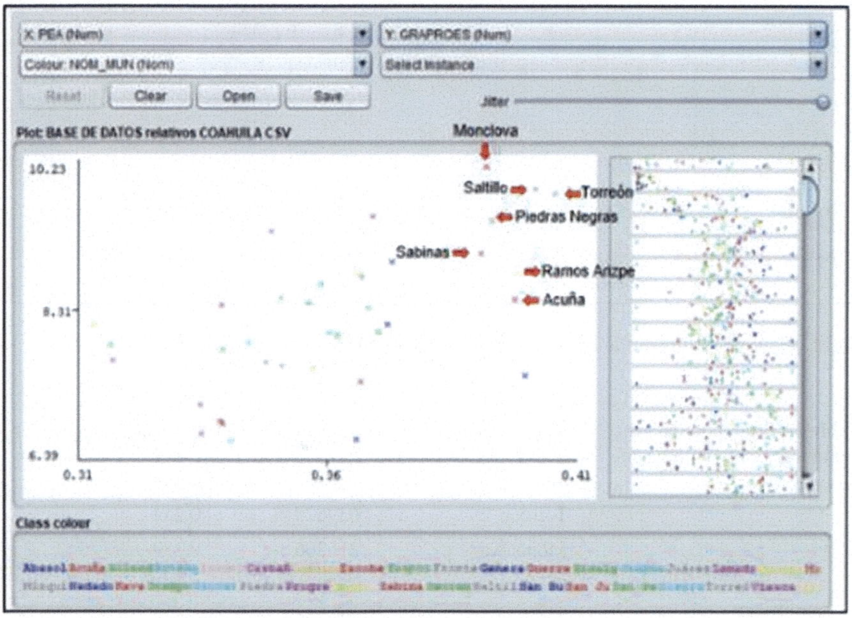

Fig. 2 Graphic results of the crossing PEA—GRAPROES

Table 2 Total population per municipality with average grade of schooling [2]

NOM_MUN	POBTOT	GRAPROES
Saltillo	725123	10.14
Torreón	639629	10.23
Monclova	216206	10.07
Piedras Negras	152806	9.24
Acuña	136755	8.46
Matamoros	107160	8.13
San Pedro	102650	7.9
Ramos Arizpe	75461	9.28

2.3 CROSSING III. Impact of the Total Population on the Economically Active Population (PEA—TOTAL POP)

Cross Justification

These indicators or attributes were selected to analyze the portion of the economically active population on the total population of the municipalities of Coahuila, which the INEGI defines as those 12 years and over who worked; they had work but did not work or; they looked for work in the week of reference to the census. The importance of this crossing is given since it is considered that the prosperity and growth of a Municipality is largely determined by the number of population and the population that has an income (Fig. 3).

Results

The 5 prospective municipalities based on this analysis and without considering the municipalities of Saltillo and Torreón are: Acuña, Ramos Arizpe, Piedras Negras, Monclova and Matamoros below, the results of the analysis are shown and a table showing the total population and the economically active population of each prospective municipality and only as a reference to Saltillo and Torreón.

A significant reference is that these prospective municipalities together with Torreón and Saltillo concentrate almost 76% of the population of Coahuila and also

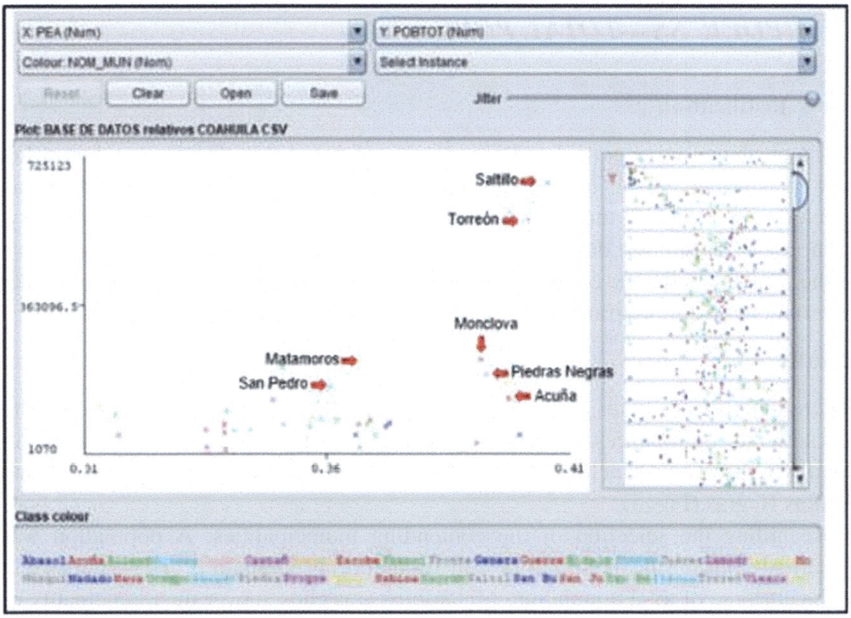

Fig. 3 Graphic results of the crossing PEA—TOTAL POP

Table 3 Total population per municipality and PEA [2]

NOM_MUN	POBTOT	EAP
Acuña	136755	0.412
Torreón	639629	0.409
Saltillo	725123	0.406
Ramos Arizpe	75461	0.403
Piedras Negras	152806	0.398
Monclova	216206	0.393
Abasolo	1070	0.393
Frontera	75215	0.385

belong to 4 of the 5 metropolitan areas of the state with the exception of Acuña, which is a border reference (Table 3).

In the state, the highway axes that interconnect the metropolitan areas with the most important population centers of the country such as Durango-Mazatlán, San Luis Potosí-Querétaro, Mexico City, to the north with Eagle Pass and Del Río Texas, to the north stand out. Northwest with the metropolitan area of Monterrey and the main border cities of Tamaulipas.

2.4 CROSSING IV. Impact of the Total Population with the Population Entitled to Health Services (PDER_SS—TOTAL POP)

Cross Justification

These attributes were considered since it is estimated that there is a relationship between the percentage of the population entitled to health services and human development; which in turn, is reflected in the economic power of the community; furthermore, as support data, the segmentation of the eligible individuals is known in the 3 health schemes of the country which are IMSS, ISSSTE [7] and popular government health programs (DIF and Seguro Popular) [8] (Fig. 4).

Results

By crossing the attributes of Total Population and Population with access to Health services, the communities of Torreón and Saltillo stand out for their total mass; after them we observed Monclova, Matamoros, San Pedro, Acuña, Ramos Arizpe and Piedras Negras (Fig. 5).

Regarding the selection of the contending municipalities; A population was discarded, considering its similarity with a neighboring community and giving weight to the attribute of total population, Matamoros was chosen over the municipality of San Pedro, leaving the latter out of the analysis. On the other hand, Ramos Arizpe

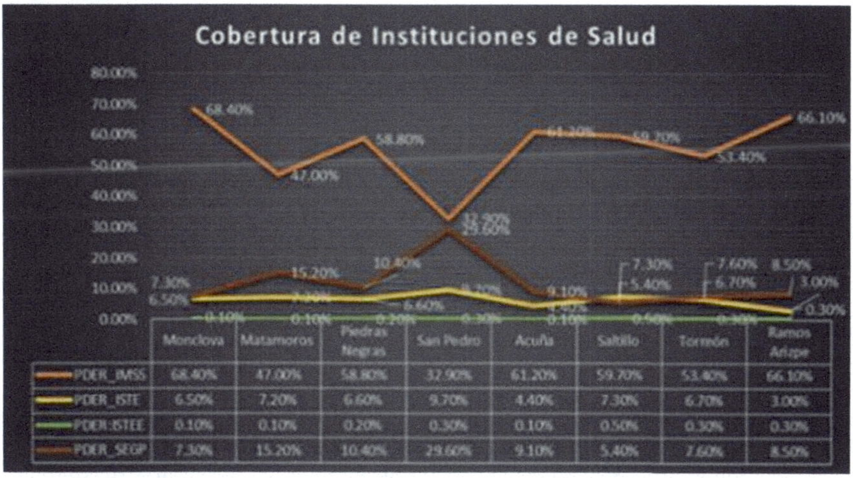

Fig. 4 Coverage of Health Institutions [9]

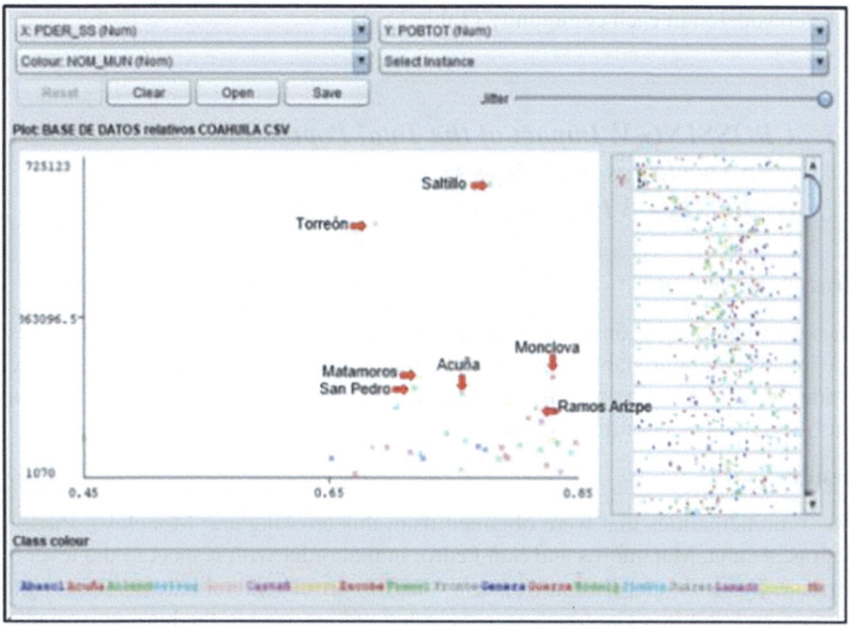

Fig. 5 Graphic results of the crossing PDER_SS—TOTAL POP

Table 4 Total population per municipality and PDER_SS [2]

NOM_MUN	POBTOT	PDER_SS
Saltillo	725123	0.774
Torreón	639629	0.709
Monclova	216206	0.831
Piedras Negras	152806	0.772
Acuña	136755	0.759
Matamoros	107160	0.724
San Pedro	102650	0.728
Ramos Arizpe	75461	0.81

is entered, which although it has a lower total population volume, its economic, strategic, logistical and operational benefits make it a strong contender; in addition to adding a competitive percentage of the eligible population. We observe the segmentation of health services in the following image, in this perspective, the percentage of the population entitled to government-type popular health services reflects a reality to be considered in each community (Table 4).

2.5 CROSSING V. Impact of the Total Population with Dwellings that Have Internet Use (POBTOT—VPH_INTER)

Cross Justification

These attributes were considered to analyze the total population of all municipalities, a degree of economic stability in their inhabitants, since it follows that having internet inside the home, not being a basic need, is considered to be an extra resource and thus represents a better quality of life.

Results

The 5 municipalities that were obtained from this crossing are: Monclova, Piedras Negras, Acuña, Matamoros and San Pedro, in that order with respect to the level of attribute 2, since in population we know that they are considerably less compared to Torreón and Saltillo (The first 5 were selected with respect to the housing variable, since it is in proportion to the number of inhabitants) (Fig. 6).

According to the objective of the research, Piedras Negras and Acuña were selected since they are smaller urban centers compared to Tijuana or Ciudad Juárez, if the perspective of the intensity and continuity of commercial transactions between Mexico and the United States is considered, for which there is An area of opportunity since on the other side of the Coahuila border is located one of the poorest regions of that country. It should be added that the proximity of Nuevo Laredo has a significant

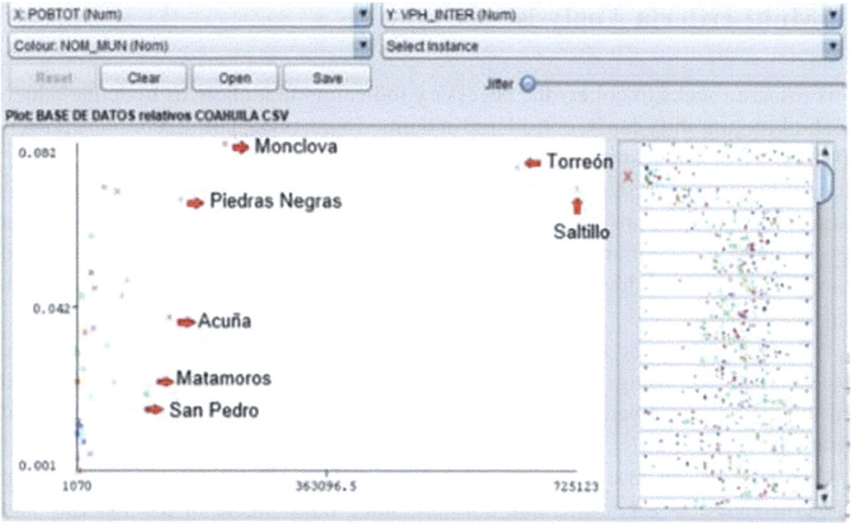

Fig. 6 Graphic results of the crossing POBTOT—VPH_INTER

impact on the formation of those weak ties of commercial transaction generated from the border cities of Coahuila (Table 5).

According to the distribution of the population, most of the houses in Coahuila are in Torreón and Saltillo. Between both localities they accumulate almost 50% of the homes, while Monclova and Piedras Negras follow in importance. Most of the services were also concentrated in the mentioned populations. For Coahuila, 92.9% of every 100 homes have electricity, water and drainage services. Meanwhile, at the state level, 77.8% of the homes have roofs made of resistant materials and the percentage of homes with dirt floors is 1.5. On the other hand, the proportion of households that have access to broadband internet is 34.56% [10].

Table 5 Total, population per municipality and wifi [2]

Municipio	Población total	Viviendascon internet
Saltillo	725123	0.071
Torreón	639629	0.076
Monclova	216206	0.082
Piedras Negras	152806	0.068
Acuña	136755	0.039
Matamoros	107160	0.022
San Pedro	102650	0.02
Ramos Arizpe	75461	0.052

3 Multi-criteria Analysis

This research seeks to collect the necessary indicators that allow us to define which Coahuila community can be considered or defined as an outstanding economic power; For this, we consider it important to obviate that cities such as Saltillo and Torreón already have such recognition, so the research was concentrated on those municipalities that are, perhaps, achieving the same competitive advantages of growth and development. As mentioned previously, the towns of Torreon and Saltillo are omitted from the evaluations due to their characteristics outside the average (Tables 6, 7, 8 and 9).

After generating the multi-criteria evaluation, we obtain the localities with the best scores; According to what was explained above, we removed Torreon and Saltillo from the competition for being out of the average and 5 locations with good parameters stand out.

Although we can observe 3–4 more localities in the evaluation, the total population was considered as a considerable attribute since the demand of the working and economically available population will be a weighting factor for the construction of the automobile assembly plant, from there generated a second filter, so populations with a higher volume of inhabitants by then were taken into account and evaluated with respect to the selected attributes. Thus, the communities of Monclova, Ramos Arizpe, Piedras Negras and Acuña stand out. As a complementary investigation, the geographical location of the town is taken into account, in addition to the viability of routes and logistics points for the best use of supply chains that are of vital importance in the automotive sector, said this, the town of Ramos Arizpe It is the one that is shaping up to be the ideal one to house an automobile assembly company and meet its global demand.

4 Results

After determining the attributes or indicators, we have been able to obtain a select and, almost homogeneous, group of 5 municipalities with outstanding characteristics and that we will concentrate the following conclusions on them. The municipalities are: 1. Acuña 2. Piedras Negras 3. Monclova 4. Ramos Arizpe 5. Matamoros (Fig. 7).

Monclova

The municipality has had one of the highest GDP per capita in the state of Coahuila; It is located in the central part of the state with the 3rd largest population, in the HDI of 2014 it achieved a value of 0.8064 considering a High Human Development Index [11]; It has more than 60% of its population with the right to health services affiliated with the Mexican Institute of Social Security, considering then, that most of the economic strength of the town comes from employers of small, medium and large companies. It is a community that has grown with investment and development.

Table 6 TOPSIS Methodology

NOM_MUN	POBTOT	PEA	GRAPROES	VPH_C_SERV	PDER_SS	VPH_INTER
Town	Total population	Economically active population	Average level of aducation	Inhabited private homes that have electricity, public water and sewer system	Population with access to public health services	Inhabited private homes that have internet
	Population	Economic strength	Education level	Quality of life	Health	Communication
Monclova	216206	0.393	10.07	0.259	0.831	0.082
Saltillo	725123	0.406	10.14	0.244	0.774	0.071
Torreón	639629	0.409	10.23	0.254	0.709	0.076
Ramos Arizpe	75461	0.403	9.28	0.237	0.81	0.052
Piedras Negras	152806	0.398	9.24	0.254	0.772	0.068
San Juan de Sabinas	41649	0.368	9.29	0.261	0.851	0.071
Sabinas	60847	0.384	9.22	0.26	0.796	0.07
Frontera	75215	0.385	9.05	0.236	0.827	0.048
Acuña	136755	0.412	8.46	0.247	0.759	0.039
San Buenaventura	22149	0.376	9.2	0.248	0.782	0.05

Table 7 Rankmethod using only the rank of each alternative under each criterion

NOM_MUN	POBTOT	PEA	GRAPROES	VPH_C_SERV	PDER_SS	VPH_INTER	Weighted range	Conclusion
TOWN	Total population	Economically active population	Average level of aducation	Inhabited private homes that have electricity, public water and sewer	Population with access to public health services	Inhabited private homes that have internet		
	Population	Economic strength	Education level	Quality of life	Health	Communication		
Rank method we use only rank of each alternative under each criterion								
Monclova	36	32	36	35	33	38	34.87	1
Saltillo	38	36	37	28	16	35	32.99	2
Torreón	37	37	38	32	7	37	32.7	3
Ramos Arizpe	31	35	34	27	26	31	31.31	4
Piedras Negras	35	34	33	32	15	33	31.21	5
San Juan de Sabinas	25	24	35	37	38	35	31.02	6
Sabinas	28	30	32	36	22	34	30.06	7
Frontera	30	31	30	26	32	29	29.88	8
Acuña	34	38	27	30	14	25	29.37	9
San Buenaventura	20	29	31	31	19	30	26.38	10

Table 8 AWM method—The normalization supposed linear function of utility

NOM_MUN	POBTOT	PEA	GRAPROES	VPH_C_SERV	PDER_SS	VPH_INTER	Weighted range	Conclusion
TOWN	Total population	Economically active population	Average level of aducation	Inhabited private homes that have electricity, public water and sewer system	Population with access to public health services	Inhabited private homes that have internet		
	Population	Economic strength	Education level	Quality of life	Health	Communicat1on		
AWM method—the normalization supposed linear function of utility								
Saltillo	1.00	0.94	0.98	0.63	0.81	0.86	0.90	1
Torreón	0.88	0.97	1.00	0.69	0.65	0.93	0.88	2
Monclova	0.30	0.82	0.96	0.72	0.95	1.00	0.75	3
Piedras Negras	0.21	0.87	0.74	0.69	0.80	0.83	0.66	4
Ramos Arizpe	0.10	0.92	0.75	0.59	0.90	0.63	0.62	5
Sabinas	0.08	0.74	0.74	0.73	0.86	0.85	0.61	6
San Juan de Sabinas	0.06	0.58	0.76	0.74	1.00	0.86	0.60	7
Acuña	0.19	1.00	0.54	0.65	0.77	0.47	0.59	8
Frontera	0.10	0.75	0.69	0.59	0.94	0.58	0.58	9
San Buenaventura	0.03	0.66	0.73	0.66	0.83	0.60	0.55	10

Table 9 Point method considering the requirements

NOM_MUN	POBTOT	PEA	GRAPROES	VPH_C_SERV	PDER_SS	VPH_INTER	Weighted range	Conclusion
Town	Total population	Economically active population	Average level of education	Inhabited private homes that have electricity, public water and sewer system	Population with access to public health services	Inhabited private homes that have internet		
	Population	Economic strength	Education level	Quality of life	Health	Communication		
Point method								
Saltillo	**10**	**9**	**8**	**8**	**9**	**9**	**8.91**	**1**
Torreón	**9**	**9**	**8**	**9**	**9**	**7**	**8.58**	**2**
Monclova	8	7	9	7	8	8	7.87	3
Matamoros	7	8	6	9	9	9	7.73	4
Piedras Negras	7	8	7	8	8	8	7.57	5
San Pedro	6	7	8	7	9	8	7.34	6
Ramos Arizpe	4	9	7	8	8	9	7.2	7
Acuña	7	8	7	7	9	4	7.14	8
Sabinas	4	7	8	7	9	7	6.77	9
San Buenaventura	5	7	6	8	7	6	6.35	10

Fig. 7 Best municipalities in socioeconomic attributes performaces [2]

According to the latest economic census conducted by INEGI, the total number of economic units in the municipality amounted to 9,037; 7,807 of them were commercial establishments and 1,041 from the industrial sector, which represented 86% and 11.5% of the total, respectively. The economically active population amounted to 84,977 people. There are 8 legally constituted ejidos, with an area of 16,537 ha [12].

The city of Monclova has a 66% relationship between the economically active population and that has housing with the basic services considered; regarding the schooling index we have a good index with 10.07 units, first place without taking into account Torreón and Saltillo. This should not surprise us since we have a community in which 39% contribute economically; however, more than 80% of this population has the right to a health service. Furthermore, with a fairly good index we know that 8% have internet access in their home (Table 10).

Table 10 Popultation per Square Kilometer [2]

Municipality	Population Habitant	Area (km^2)	Population Density Population per Square Kilometer
Torreón	679288	1255.98	540.84
PIEDRAS NEGRAS	163595	474.52	344.76
MONCLOVA	231107	1251.32	184.69
SALTILLO	807537	5652.98	142.85
MATAMOROS	108950	806.17	135.15
RAMOS ARIZPE	92828	6769.52	13.71
Acuña	147809	11467.84	12.89

Taking into account the population density, we observe that the city of Monclova has a higher number than Saltillo; it still has the capacity to grow in population.

Piedras Negras

It belongs to the economic region of northern Coahuila, considering mostly the secondary and services sector. There are eight coal sub-accounts in the Sabinas-Monclova area, one sub-basin in Piedras Negras, and one basin in the Hidalgo-Colombia-San Ignacio area. Currently the city is in a privileged position in terms of economy and standard of living for its inhabitants; Due to its border position, it has become a very clear target for foreign investment, mostly American.

The economically active population is fully employed, with the secondary and service sectors standing out in sectors such as metal, textile, industrial and automotive. Within the Human Development Index [11] it undoubtedly has a very high category, understanding that it offers its population a range of development possibilities.

It has in its workforce a mining company such as Central Termoeléctricas Carbón 1 and Carbón 2; Coca cola plant; Grupo Modelo and countless transformation companies; This range of job options is related to the percentage of 77% of the population with access to health services, derived from job options; of this total, 58% are affiliated with the IMSS and only 10% with Seguro Popular [13, 14], so the formal offer of economic activity is a positive and too strong indicator for this municipality of Coahuila [15, 16].

According to INEGI, the total number of economic units in the municipality amounted to 5,889, 5,288 of them were commercial establishments, 429 from the industrial sector, which represented 90% and 10% of the total, respectively. The economically active population amounted to 60,875 people. There are 5 legally constituted ejidos, with an area of 17,665 ha [2].

Piedras Negras stands out as an important economic source; It has a ratio of 64% of the economically active population against the total; we can conclude that it is considerably attractive. Definitely, this gives its population a future of certainty with development strategies, and in education, it is not so bad, with an index of 9.24, it stands out in the education options for its individuals. With this, it builds a 39% of the population that contributes economically, of which more than three quarters are in the program of the Mexican Institute of Social Security, here we speak of formal labor supply [17]. Within the communication index, being a border area, definitely the communication infrastructure is so outstanding, if we talk about internet access, the 7% index is its indicator; low in consideration of other nations; but profitable if we do the state comparison.

Although the state's border line extends for more than 650 km, trade with the United States through the checkpoints of the Coahuila cities on the south bank of the Bravo does not stand out in the context of commercial transactions of the entire northern border. Piedras Negras and Acuña are smaller urban centers compared to Tijuana or Ciudad Juárez, if the perspective of the intensity and continuity of commercial transactions between Mexico and the United States is considered. It should be added that the proximity of Nuevo Laredo has a significant impact on the formation of those weak ties of commercial transaction generated from the border cities of Coahuila [18].

Ramos Arizpe

This municipality located in the southeast of the state contains one of the most productive areas of the state and the country; about 1000 companies; According to the IMSS registry, between micro, medium and large companies are operating in this municipality, this fact has caused Ramos Arizpe to be considered as one of the communities that contribute the most GDP at the state and federal level.

With a population of more than 90,000 inhabitants, the municipality has been awarded a high level of social development; This translates into opportunities for individuals, such as job options, educational development, housing, and service infrastructure.

It is a main source of labor supply in the northern part of the country, due to its geographical location, it offers investors benefits in transportation costs, logistics, imports, exports and an immense variety of access to supply chains in the area.

Its population with entitlement is 81%, which belongs to 40% of the currently economically active population; quite competitive indicators; Of course, this is related to the job offer, which are transnational or national companies with a broad social security scheme that provide their employees with an accurate social development panorama; perhaps the only disadvantage with Torreón and Saltillo is their total population mass; although in recent years it has maintained a growth rate of 152%; but definitely Ramos Arizpe is a strong contender for a favorable economic position for much longer. The labor supply has an effect on the population's degree of study index, the 9.28 obtained in the last 2010 census generates certainty in the competences of the labor force generated in the locality; access to networks and communication diagrams through digital means provides an important factor for investments.

Acuña

Municipality located in the north of the country, it has a considerable extension of border territory with the United States of America; Its population exceeds 145,000 inhabitants and with a balanced economic turn with a primary sector of 40%; secondary sector 40% and tertiary sector 20%. Acuña managed to qualify with a high index in Social Development according to the latest study carried out. The economic activity is divided into agriculture, livestock, industry, forestry, fish farming, tourism, trade and services; with a considerable participation; in contrast, although the economic and border movement is moderately considerable, the percentage of the population with internet access is low, with only 4%.

It has 75% of the population with access to health services, of which 61% is affiliated with the Mexican Institute of Social Security and only 9% with popular security, health services derived from formal activities are predominant [15, 16]. In recent years, the transformation, livestock and agriculture services industry has increased; in addition to a boom in the tourism sector; that generates a virtuous circle with the job offer. While there are opportunities for work and development, crime decreases, therefore, there is peace and cordiality in the population and that brings opportunities in sectors such as tourism. Acuña is emerging as a town with economic potential with an economically active population percentage of 41%. However, its degree of studies index is below municipalities such as Monclova, Piedras Negras and Ramos Arizpe with an 8.46.

According to the latest economic census, carried out by INEGI, the total number of economic units in the municipality amounted to 4,387, 3,931 of them were commercial establishments, 316 from the industrial sector, which represented 89.60% and 7.0% of the total, respectively. The economically active population amounted to 56,311 people. There are 13 legally constituted ejidos, with a surface area of 331,380 ha. Matamoros

It is located in the northwest of the state; being 248 km from the current capital, the current location gives it a strategic advantage to detonate the development in the Coahuila lagoon as if to the state in its entirety; Taking into account the characteristics of its soil, it is benevolent with livestock and agricultural activities; We observe mineral activity for extraction of construction component.

The population issue may be its strong point, in addition to its geographical advantage; In this community, 72% of the economically active population has medical services; less than half are with IMSS services and 15% with popular insurance; there is still an informal job offer that does not provide stability in health services; Along with all this, the community is considered to rank high on the Human Development Index; although the activities are of the first economic sector [15, 16].

According to the latest economic census, carried out by INEGI, the total number of economic units in the municipality amounted to 2,681, 2,365 of them were commercial establishments, 259 from the industrial sector, representing 88% and 10% of the total, respectively. The economically active population amounted to 39,240 people. There are 64 legally constituted ejidos, with a surface area of 53,172 ha. The municipality presents indicators that are not as favorable as the populations we have already analyzed; its percentage of index of Housing with services against the economically active population is 59%, a number with an opportunity area in quality of life. Only 36% have an economically active capacity and only 2% of its population has a digital communication system with internet at home.

5 Growth Strategies

Monclova

The conditions of the City of Monclova allow us to suggest that you could develop an exponential growth; Although it gives us very favorable indicators, we suggest conditions of "Private Foreign Investment" as a viable strategy; The arrival of some transnational companies could allow the creation of a new industrial nest in the state, with this, it would cause population movement to the community that would consequently increase the total population, a population with an economic contribution and therefore an optimal human development [19].

On the other hand, "the development of competitive jobs, as part of a Human Capital Development strategy" could create a development of human capital that places Monclova as a job destination for new generations.

Without hesitation, the indicators of competitive salaries, employment conditions and the formal job offer associated with private or government employers could be developed as new studies to monitor the growth of Monclova compared to other Coahuila communities.

Piedras Negras

The city of Piedras Negras is the most important border area in Coahuila; however, it is far below in tariff demand compared to Tijuana and Ciudad Juárez; even with its closest cousin city, Laredo, it is at a disadvantage [17].

We observe that we have good indicators in social development, labor supply and competitive human resources; but definitely an important strategy can be the "Investment in road infrastructure", where we can place Black stones as an important connection matrix; an important logistics center for the cities of the American Union; but with cost savings in transportation and packaging.

Studies such as the Tariff Demand Index; Export and Import Index; In addition to the Labor Supply Index in Mexico and the US, we could open the gap for new socio-economic and political strategies where the appropriate factors are worked on and the City of Piedras Negras can be positioned as a faithful competitor of the Large Cities-Crossings that our country has.

Ramos Arizpe

With an exorbitant population growth, Ramos Arizpe continues to give more reasons to occupy a preferential position in a select group of municipal powers in the state of Coahuila. It has a job offer, a social development index, a competitive degree of study in its population and an excellent health infrastructure with 81% of its affiliated EAP; It is only a matter of Ramos Arizpe betting on "Social Development Level Strategies" for its population; With this, it will be able to offer the country's workforce opportunities for professional growth accompanied by a community with added value and an excellent quality of life.

Acuña

With a positioning behind Ramos Arizpe; Acuña is betting on "Security and investment strategies"; relying on transnational companies for services and products; and with an important boom in the maquilero sector for more job offers as they have been in recent years.

National and international chain companies are considering their expansion in the northern community of Coahuila due to its border position and interconnection for our northern country. This gives the population a better opportunity for resources for development. roll and quality of life. Evaluation in indicators such as foreign and national investment and educational infrastructure could generate rational strategies to boost the border city.

Matamoros

Matamoros is a population with economic strength; however, we see performance in the first sectors or primary activities. "Social development programs with their impact evaluations on the population could trigger progress in the community; the recent openings of franchise companies", accommodation options and job offers have allowed Matamoros to be seen as a new accommodation option for the population of the municipalities of La Laguna de Durango and Coahuila; this helps in total population and new development zest.

The economy of Coahuila is strong, and proof of this is that the entity maintains the national leadership in Foreign Investment Per Capita; the first in labor formality and the sixth in Competitiveness. It has excellent levels of competitiveness that are strategic factors for potential investors, such as a qualified workforce, educational level, skill development or skills of its workers. Of the 80 investment projects confirmed for 2020 in the entity, estimated at 3,371.76 million dollars, 35 correspond to the Southeast Region for an amount of 1,573.27 million dollars, which will be the source of employment for at least 13 thousand 57 people. According to information from the Ministry of Economy in Coahuila, 59 of the projects that confirm their installation in Coahuila territory, is their first step through the entity, while in 21 of the cases they are investments for the expansion of activities.

The Southeast Region, with 26% of the investments made in 2019, was followed by La Laguna, with 21 confirmed projects and a spill of 886.98 million dollars. This may directly affect the growth of the northern zone, which in previous years has presented an important development due to the strong investments represented by the acquisition of the Grupo Modelo brewery in Piedras Negras by Constellation Brands, which in turn implied the installation of a glass plant and a cardboard plant, which represented an investment of 4.738 million dollars for the year 2015.

Throughout the analysis, we found various patterns, where some municipalities are distinguished as those with greater opportunities for growth and that undoubtedly with a good strategy could rapidly increase the values of each attribute and thus continue generating towards the state and towards the country like Torreón and Saltillo do.

Considering the analysis, the strategies and the results that were generated in the research, it is possible to contemplate different approaches in certain municipalities to stimulate the economy and growth.

6 Conclusions

Investment In The Automotive Industry

According to the results, in this approach Ramos Arizpe is the ideal candidate for exponential growth as a municipality in addition to already having a trajectory within this sector since it belongs to one of the most important metropolitan areas of Coahuila (Saltillo-Ramos Arizpe- Arteaga) and is part of the Saltillo-Monterrey automotive industrial corridor where more than 50% of the companies in this sector are located in Mexico, it has become one of the points with the most offer of well-paid jobs in the country. To determine that Ramos Arizpe is the ideal place for an assembly plant, in addition to the attributes included in our analysis, other factors that various authors consider of vital importance were taken into consideration as Jose Carlos Ramirez in "The new factors of industrial location in Mexico", such as the application of the just-in-time (JAT) system which has been determined the new patterns of localization of the country's industry at least in northern Mexico.

The companies with the JAT system cannot be installed or function independently from the rest of the component or material supplier companies since in this model the suppliers are within a delimited geographic space and work as if they were modules from the same factory, in the area metropolitan area of Saltillo-Ramos Arizpe there are 10 of the most important industrial parks in the state with more than 200 companies installed.

Another important factor is the strategic location with the United States, so in general it is a benefit since Coahuila is located in the northern part of the country, although Ramos Arizpe is not a border, it has easy access to road and rail infrastructure in addition to 1 airport international.

The qualified and specialized workforce is another important factor to consider in this approach, Coahuila is the first national in manufacturing productivity in addition to having more than 88 universities, 6 specialized research centers and 2 technological innovation centers of the of which a significant number are found in the southeast area of the state. In addition, something that is very attractive and that has previously helped to consolidate the automotive sector in Ramos Arizpe has been the subsidies granted by the government, infrastructure qualification, credits, exemptions from the payment of taxes and telecommunications services and infrastructure.

References

1. Romo, L. (2012). Programa de Desarrollo Urbano y Ordenamiento Territorial del Estado de Coahuila Síntesis Ejecutiva. 1–16. http://www.coahuilatransparente.gob.mx/otrainfonew/documentos_otrainfo/SSAProgramaSectorial v.completa17.pdf
2. Censo de Población y Vivienda 2010. (n.d.). Retrieved February 19, 2021, from https://www.inegi.org.mx/programas/ccpv/2010/default.html#Datos_abiertos

3. Secretaría de Gestión Urbana, Agua y Ordenamiento Territorial, Programa Es-tatal de Desarrollo Urbano y Ordenamiento Territorial de Coahuila de Zaragoza
4. Fajardo-Dolci G, Gutiérrez JP, García-Saisó S (2015) Acceso efectivo a los servicios de salud: operacionalizando la cobertura universal en salud. Salud Pública de México 57(2):180–186. Recuperado el 24 de junio de 2020. http://www.scielo.org.mx/scielo.php?script=sci_arttext&pid=S0036-36342015000200014&lng=es&tlng=es
5. Ramirez JC (1999). https://www.researchgate.net/publica-tion/227428784_Los_nuevos_factores_de_localizacion_industrial_en_Me-xico_La_experiencia_de_los_complejos_automotrices_de_exporta-cion_en_el_norte
6. Destinos México Arduinna SA (2018). https://www.mexicoindus-try.com/es/news/coahuila/industria-automotriz-coahuilense-se-mantiene-entre-las-ms-productivas
7. Centro de Estudios Sociales y de Opinión Pública, "Políticas públicas y gestión gubernamental de la administración vigente", en Seguridad Social [Actualización: 17 de febrero de 2006], en https://www.diputados.gob.mx/cesop/
8. Monclova (2020) In Wikipedia, 19 Sept 2020. https://es.wikipedia.org/wiki/Monclova
9. Montiel L (2004) Derecho a la salud en México. Un análisis desde el debate teórico contemporáneo de la justicia sanitaria. Revista IIDH 40:291–313. https://dialnet.unirioja.es/ejemplar/530338
10. Vazquez G (2020) Ramos Arizpe es atractivo para la llegada de nuevas inversiones. Mexico Industry. Recuperado el 28 de junio de 2020. https://mexicoindustry.com/noticia/ramos-arizpe-es-atractivo-para-la-llegada-de-nuevas-inversiones
11. Subsecretaría de Prospectiva, Planeación y Evaluación (2010) Informe anual sobre la situación de pobreza y rezago social. Recuperado el 29 de junio de 2020. https://www.gob.mx/cms/uploads/attachment/file/32675/Coahuila_027.pdf
12. Instituto Mexicano del Seguro Social (2010) Asegurados. Recuperado el 24 de junio de 2020. http://datos.imss.gob.mx/dataset/asegurados-2010/resource/d74712d8-cb08-4ed4-b37a-52ad399a8972
13. Secretaría del medio Ambiente (s.f). Sistema Integral de Información Am-biental del Estado de Coahuila. Disponible en. https://www.sema.gob.mx/SRN-SIIAECC-DG-MSE-DH.php
14. Índice de Desarrollo Humano Municipal en el estadode Coahuila. https://www.sema.gob.mx/
15. ISSSTE gasta casi cinco veces más de lo que ingresa Cerró con superávit, favorecido por transferencias y subsidios del gobierno. https://www.elecono-mista.com.mx/sectorfinanciero/ISSSTE-gasta-casi-cinco-veces-mas-de-lo-que-ingresa-20180612-0131.html
16. Plan municipal de Desarrollo de Matamoros Coahuila. http://www.matamo-roscoahuila.gob.mx/Go-bierno/PDI/PMD/PMD%20Matamoros%20Coahuila%202019-2021.pdf
17. Tiene Acuña potencial económico. http://www.zocalo.com.mx/seccion/ar-ticulo/tiene-acuna-potencial-economico
18. Piedras Negras | AAAPN. (n.d.). Retrieved February 19, 2021, from http://www.aaapn.mx/aniv50/piedras/ciudad.php
19. Conoce Monclova – Presidencia Municipal de Monclova Coahuila México. (n.d.). Retrieved February 19, 2021, from https://monclova.gob.mx/sitio/conoce-monclova/
20. Sistema Integral de Información Ambiental (s.f.). https://www.sema.gob.mx/SRN-SIIAECC-DG-MSE-VIV.php

Layout Problem: Optimization with Material Tour in Open Field. Case of Study

Victor Manuel Valdes Flores, Pedro Henoc Ireta Sánchez, Ricardo Martínez López, Marco Aurelio Jiménez Gómez, Adriana Mexicano Santoyo, and Alan Tijerina de la Rosa

Abstract World-class companies are always looking to minimize their production costs. One of these costs is the distribution of materials in the production lines. This article addresses a real problem of material handling in a chemical company through 14 reactors where each reactor has a specific demand. The case study uses a hydraulic skid and a forklift to move material through the plant without a specific route, resulting in a high cost for material handling because when the operator cannot meet the demand at one of the reactors, he has to return to the warehouse to pick up material and continue with delivery. To solve the problem, we use the Traveling Agent Problem (TSP) as a mathematical model to create an a priori route that helps make sub routes using two solution methods: Branch and Bound and Simulated Annealing. The results show that Branch and Bound works better at finding a solution faster than the Simulated Annealing model.

Keywords Layout · Optimization · Branch and bound · TSP and simulated annealing

V. M. Valdes Flores (✉) · P. H. Ireta Sánchez · R. Martínez López · M. A. Jiménez Gómez · A. Tijerina de la Rosa
Tecnólogico Nacional de México/I.T. de Saltillo, Saltillo, Mexico
e-mail: vvaldes@itsaltillo.edu.mx

P. H. Ireta Sánchez
e-mail: pireta@itsaltillo.edu.mx

R. Martínez López
e-mail: ricmartine@itsaltillo.edu.mx

M. A. Jiménez Gómez
e-mail: mjimenez@itsaltillo.edu.mx

A. Tijerina de la Rosa
e-mail: altijerina@itsaltillo.edu.mx

A. Mexicano Santoyo
Tecnológico Nacional de México/I.T. de Ciudad Victoria, Ciudad Victoria, Mexico
e-mail: mexicanoa@gmail.com

© The Author(s), under exclusive license to Springer Nature Switzerland AG 2021

A. Ochoa-Zezzatti et al. (eds.), *Technological and Industrial Applications Associated with Intelligent Logistics*, Lecture Notes in Intelligent Transportation and Infrastructure,
https://doi.org/10.1007/978-3-030-68655-0_5

1 Introduction

A material path layout problem seeks to ensure the entry of material in the appropriate locations by handling the materials and can be transported by belt, roller wheel, automated vehicle (robot), forklifts and skates hydraulic [1]. Two dependent design issues are considered: facility design and equipment selection. Many types of layouts exist for example the layout of a row, multiple rows, double row, parallel row order, loop open field, and multi-floor [2].One of the objectives of the layout problem is the minimization of routes, the cost of handling materials and the distribution of fixed-dimensional installations and unequal areas; therefore there is a need for the construction of mathematical models that represent and serve to solve exactly the problem of optimizing the material layout of open-field fixed areas that aims both to optimize the cost of handling materials between stations based on criteria independent of flows and costs between stations [3]. Some authors focus on the journey of materials to find the minimum distance between the routes that the forklifts have to make when collecting the goods from the areas and taking them to the dispatch meeting labor journeys and satisfying customer requests with the use of exact methods such as the Branch and Bound [4]. Metaheuristics are generally applied to problems that do not have an algorithm or heuristics specifies that provide a satisfactory solution; or when it is not possible to implement that optimal method. Most solution methods target optimization problems and apply to any problems that may be reformulated in heuristic terms [5]. The next approach is the use of Solution Methods such as Simulated Annealing where this process starts with a random estimate of the values of the variables of the cost function and allows to randomly modifying the values of the variables and involves best random variations [6]. To solve the material path layout problem, TSP is considered as set graphs whose edges are the possible paths that the entity can follow to visit all nodes, where the routes are searched and shorter sub routes for the best routes [7]. Also these methods are used for troubleshooting optimization problems that need to substantially improve the compute times for the solution of a specific application as well as the creation of the distance matrix to determine the costs of the materials and different types of vehicles to be used for fixed operations [8].

The objective of optimization is to propose and develop sets of sub-routes that allow the transfer capacity of materials and vehicles that optimize the best costs, taking into account distances in operations and vehicles to be used [9]. It is the use of layout design by classification to solve the optimization problem by looking for all the types of layout to identify the type of design that helps solve the case study with solution methods [10]. Redesigning reactor material delivery routes into an open field layout problem will reduce material routing costs by improving the work process. The key is to develop an optimization model in an open field layout design with material routing that reduces the cost of operations paths using solution methods and seeks to define the fastest solution method to solve the problem.

Below this document is organized as follows: Sect. 1 is motioned in the introduction, Sect. 2 describes Workshop characteristics, Sect. 3 describes Mathematical model, Sect. 4 describes Solution methods, Sect. 5 Branch and Bound, Sect. 6 Simulated annealing, Sect. 7 Application, Sect. 8 Methodology, Sect. 9 State of the art, Sect. 10 Data collection, Sect. 11 Results, Sect. 12 Conclusion, Sect. 13 Future work, Sect. 14 acknowledge and in Sect. 15 bibliographic references.

2 Workshop Characteristics

The theory for solving the case study and the tools to track layout optimization research is presented. A layout commonly shows the spaces and working distances that a certain organization has, variables and areas of opportunity are detected with the naked eye from the company location, difficulties in working, problems of the material path and even the type of office design. Layout designs are found in various types of manufacturing systems, usually related to the location of a plant's facilities such as the location of operations in an area of the plant, determined that the most efficient accommodating according to objectives or criteria under some restrictions are in shape, size and orientation of materials between work facilities with the use of mathematical models such as TSP among others [10]. The production of materials or service delivery and a layout design states that the performance of any type of work, machines, and spaces of any system can improve significantly if costs are attacked in terms of distances [11].

Layouts are directly influenced by the specifications of production systems, variety of products, production volumes. They can be divided into four categories in form and dimensions, manufacturing system, material handling and flow handling, at this point the line to be investigated will be "Material Handling" and shown in Fig. 1[10].

(a) *Material Handing*
It is defined as the selection and arrangement of handling and handling materials in order to reduce time, costs and efforts. It can include conveyors, cranes, elevators, auxiliary equipment, industrial trucks and automated vehicles as they play a very important role in the manufacturing system. [10].

(b) *Layout Configuration*
The classifications of the seven layout types are: Open-field Layout, Loop Layout, Multi-rows Layout, Single Row, Double Row, Parallel Row Ordering Problem and Multi-floor Layout [10].

(c) *Multi-floor Layout*
In cities the spaces are insufficient and the exorbitant costs of providing living space make designers and engineers consider multi-floor designs instead of single-floor designs [12]. Multi-floor factories are preferable to save spaces for future extensions, Fig. 2 [13].

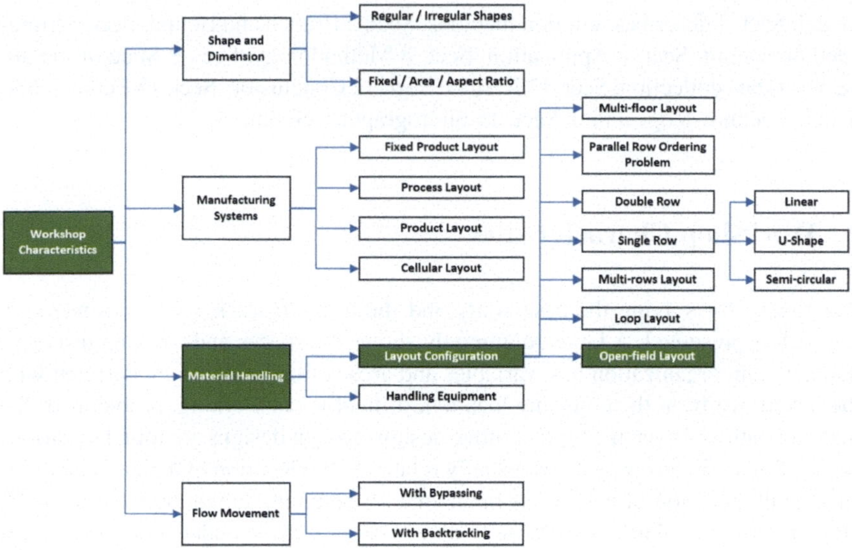

Fig. 1 Workshop characteristics

Fig. 2 Multi-floor layout

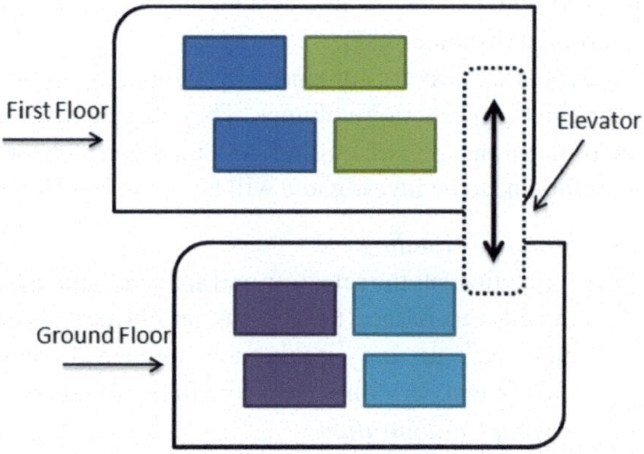

(d) *Parallel Row Ordering Problem*
Installations with some common features are arranged along a row, with the remaining installations left to be fixed a parallel row. Arrangements in both rows start from a common point that does not allow space between two adjacent installations, Fig. 3 [14].

(e) *Double Row*
It involves a series of rectangular installations with variables on both sides of a straight line to minimize the total cost of handling materials between installations and operate along a corridor to move materials from one facility to another, Fig. 4 [15].

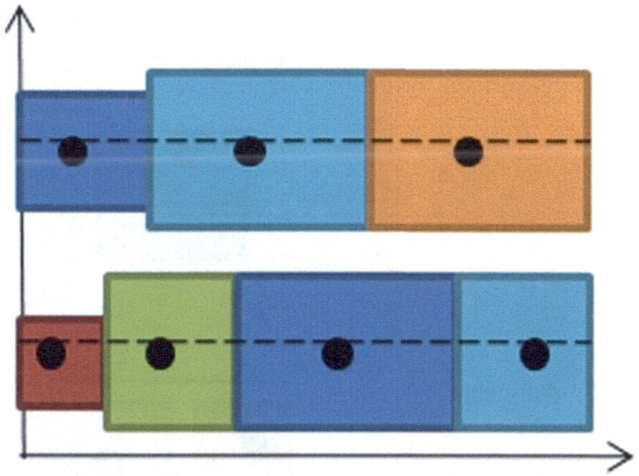

Fig. 3 Parallel row ordering problem

Fig. 4 Double row

(f) *Single Row*
Refers to a certain number of rectangular instances along a line to minimize the total cost of the layout and the sum of the product and the distances between all instances [16]. The shape of this layout can be straight line, semicircular, or U-shaped, Fig. 5 [17].

(g) *Multi-rows Layout*
Install a system of rectangular installations in each number in rows in a two-dimensional space between all pairs of installations assigned to any of the given rows to reduce the minimum cost, Fig. 6 [18].

(h) *Loop Layout*
This type of design aims to find an assignment of facilities to certain selected locations in a closed ring network, so that the total cost of handling is minimized. [18]. A Loop Design is incorporated by a load and unload i.e. a part location enters and leaves the loop, Fig. 7 [17].

Fig. 5 Single row

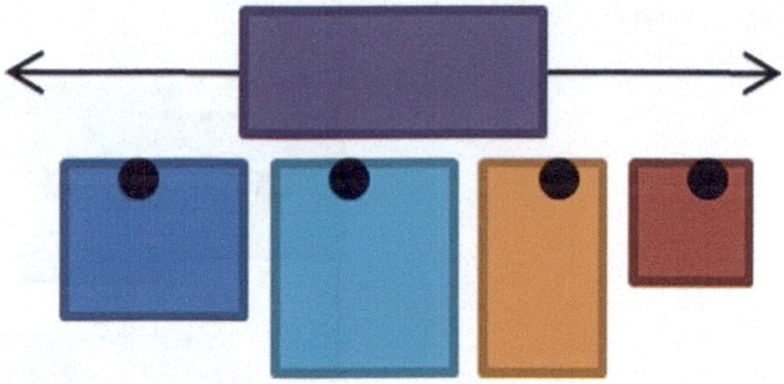

Fig. 6 Multi-rows layout

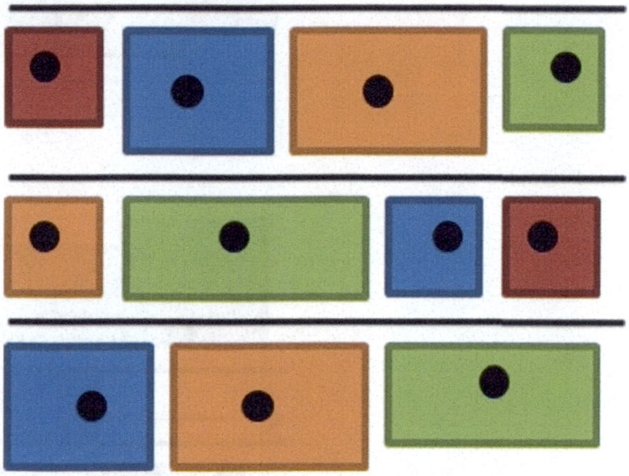

Fig. 7 Loop layout

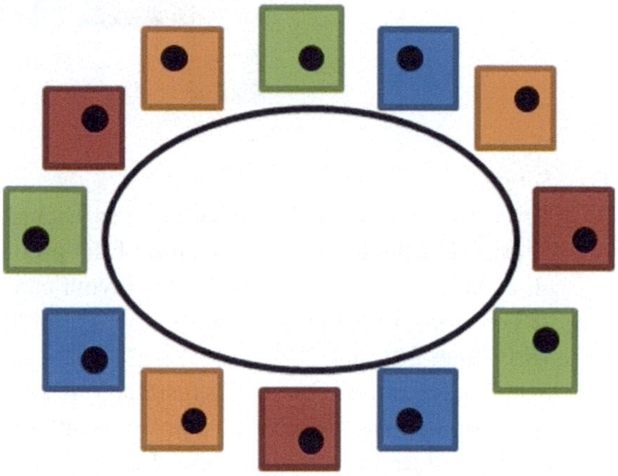

(i) *Open-field Layout*
The open field layout corresponds to situations in which the installations can be placed are the constraints induced by such arrangements as the layout of a row, double row, parallel row, multi-row or loop and consists of multiple

Fig. 8 Open-field layout

loops and can include side paths, this type of layout is normally appropriate for processing large numbers of part families, Fig. 8 [17].

A new circular open field layout composed of a vehicle-based material handling system and multiple circular loops is proposed. The suggested layout offers more flexibility for high-mix low- volume production systems compared to other layouts as it maximizes the number of equidistant machines [19].

For manufacturers to stay competitive in today's market, they must reduce, if not eliminate, the long setup time between jobs, increase productivity and enhance the flexibility to produce a wider variety of products. A Flexible Manufacturing System can produce a variety of products for a limited range of part families without the need to produce in batches or any time lost in reprogramming. Enhancing the layout of the system can significantly affect the cost, throughput, and productivity of the facility. The effect of workshop characteristics specifically layout configuration was investigated by researchers but a major drawback of most of the proposed layouts is that they are biased towards a certain processing order [20].

3 Mathematical Model Applied for the Case Study

The (TSP) is defined, given an integer $n > 0$ and the distances are given by the matrix (dij) of dimension $n \times n$ where dij an integer is greater than or equal to zero. A tour is a journey that visits all cities exactly once; the problem is to find a route with minimum total length [10].

TSP consists of finding the shortest possible path among a variable number of villas or sites to visit and for which it should only be passed once, with the condition that at the end of the tour you must return to the point of origin. It has been shown that the complexity of the problem is n![21].

The TSP has a single seller that customers must visit on a single route at minimum cost. There is usually no deposit (and if there is, it would not be distinguished from customers), there is no demand associated with customers and no temporary restrictions [11].

- For all $X_{ij} = 0ó1$, *for all* $U_i \geq 0$ and is set of clients.
- Xij : Equals 1 if all clients are visited.
- Cij: Is the cost to visitclient i to client j.

The mathematical model is described below:

Subject:

$$\min \sum\nolimits_{(i,j) \in E} cijxij \tag{1}$$

$$\sum\nolimits_{j \in \Delta+(i)} xij = 1 \quad \forall i \in V \tag{2}$$

$$\sum\nolimits_{i \in \Delta-(j)} xij = 1 \quad \forall j \in V \tag{3}$$

$$\sum\nolimits_{i \in S, j \in \Delta+(i) \setminus S} xij \geq 1 \quad \forall S \subset V \tag{4}$$

(1) Is the objective function that is to minimize the travel distance from customer i to customer j, (2) and (3) indicate that all customers should be visited and must be within a route and (4) indicates that no sub routes are allowed and the problem can formulate mathematically by full linear programming as mentioned [16].

This problem is one of the most studied in the combination of optimization, in general this problem can be defined how the search to minimize the distance from the cities to visit, started and ending in the same city, visiting only once the city [22]. The TSP is one of the most famous and complex problems of computer science have been addressed by several branches of engineering for different reasons, its main application is to route from different perspectives either a process that carries a sequence específica or a distribution of logistics involving elements of transport seeking the best possible route with criteria of economy in distance or cost as well as providing solutions that contribute to improving tasks and processes in different areas [22]. Given a number n of cities, which have cost or distance characteristics, the seller starting from one city must go through all the others exactly once until returning to his starting point. This type of travel under graph theory is known as a Hamiltonian travel [23].

When adding the vehicle capacity restriction, the case study can be seen as a VRP (Vehicle Routing Problem). To solve this problem, it is necessary to take into account three elements: location of the warehouse to collect sacks and the load capacity of the vehicles. The theory of the mathematical model is presented to solve the case study as is the use of: Distance between two points, a priori route and Sub-routes and Distance.

i *Distance between two points*
From the location of the coordinates of the distances of two points [16] it is possible to calculate the distance between them, when the points are located on the x-axis (of the abscissa) or in a line parallel to this axis, the distance between the points corresponding to the absolute value of the difference of their abscissa $(x2 - x1)$ [24].When the points are located on the y-axis (of the ordered) or in a line parallel to this axis, the distance between the points corresponding to the absolute value of the difference of their sorts where the points are anywhere in the coordinate system [24].

ii *Route* a priori
Under the a priori route approach, routes are established before the customer demands are known, so that the vehicle distributes the goods following the customer route established before the start of the journey. Given that customer demand is only known at the time of service, a failure in the route may occur, which means that the vehicle does not have the necessary merchandise to satisfy the demand of the current node; for this reason, different strategies are presented that avoid modifying the sequence proposed a priori [25].

iii *Sub-routes and Distances*
The sub-routes considered have a common city called base. The procedure for joining different sub-routes is based on eliminating the edges that connect two cities of different sub-routes with the base city, later joining the cities with each other. The difference in cost between the eliminated edges and the added one is known as a saving. The aim is to minimize the sum of the distances of the sections of the complete route that allow all the cities to be covered starting from a source city [26].
Each reactor has its own demand. If the sum of the demand of the reactors visited on the route a priori is greater than the loading capacity of the vehicle, it will return to the warehouse to reload and continue with the route where it was left with the unattended reactor, where n is the number of customers in the subroutines that must start and end at the warehouse, V is the distance value between the reactor and the reactor including the distance matrix warehouse [26].

iv *Mathematical model to obtain cost factor*
To obtain the $Tcost$, it is necessary to have the cost per meter of the use of the forklift and the hydraulic skid in order to trace how much it costs to carry sacks to each of the reactors and remembering that the return to the warehouse does not generate any cost, below is Eq. (5) called the mathematical cost model [27].

$$Tcost = \left[\frac{\sum_{i=1}^{n}\sum_{j=1}^{n} f_{ijd_{ij}}}{V_g}\right] \tag{5}$$

v *Mathematical model to divide routes and distances*
Equation (6) is known as an objective function since it is the sum of each of the distances of the sub-routes created, Eq. (7) if the sum of the capacity of the reactors is greater than the load capacity of the hydraulic skid and forklift at that time the skid or forklift will return reactor to the warehouse to reload material and continue with the next reactor, Eq. (8) indicates that each sub-route must begin and end at the material storage and Eq. (9) mentions that the reactors must be incorporated into the sub-routes [28].

$$\min V_g \sum_{i}^{m} X_{(y_i,(y_i+1))} * Tcost \tag{6}$$

Sujeto a:

$$\sum_{i=1}^{n} V_g \geq Q.:\left[y_i, \ldots, 0\right]\forall\left[y_0, \ldots, y_{i+1}\right] \tag{7}$$

$$V_g = [0, \ldots..0] \tag{8}$$

$$i \in V_g \tag{9}$$

4 Solution Methods

Various solution methods and procedures have been developed to solve a problem in layout design and can be divided into exact, heuristic and metaheuristic, mentioning various methods and procedures to solve an open field layout design problem [10]. Regarding solution methods, it was concluded that three major categories have been addressed, which can be grouped as follows: Exact, Heuristics, Metaheuristics in Fig. 9 [10].

1. *Exact*
   Accurate methods are efficient in problems with few constraints and are procedures that provide quality acceptable solutions through limited search scanning. A considerable amount of research was done to develop the optimal

Fig. 9 Solution Methods

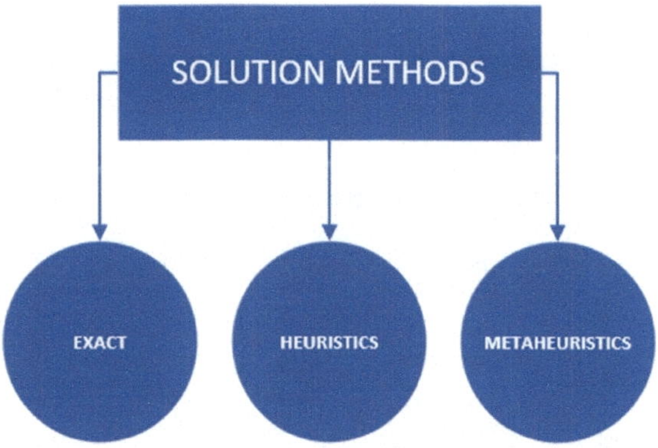

algorithms; the exact methods are useful approaches to find optimal solutions for small layout designs and can be classified into three groups: direct tree search, dynamic programming, linear programming and integer [29].Dynamic programming, the Branch and Bound method, the flat cutting algorithm, and Simplex Method are examples of exact approaches [10]. Local search procedures are applied to improve an already obtained solution. These procedures define a set of neighboring solutions and part of a primary solution and then replace it with a neighboring solution with lower cost [29].

2. *Heuristics*
   Heuristics are simple procedures that make a limited explosion of the search space and give solutions of acceptable quality in generally moderate calculation times [23]. Solutions obtained with this kind of procedure can, in general, be improved by using more sophisticated search methods, but incurring high execution times. Many of these heuristics can be extended to handle additional constraints to the VRP Vehicle Routing Problem [30].
3. *Metaheuristics*
   Metaheuristics are generally applied to problems that do not have a specific algorithm or heuristic that provides a satisfactory solution; or when it is not possible to implement that optimal method [7]. Most metaheuristics target combinatorial optimization problems and can be applied to any problem that can be reformulated [5]. The metaheuristics for solving layout design cases and the most well-known techniques are the Genetic Algorithm, Taboo Search, Simulated Annealing and the optimization of Ant Colonies among others [10] is seen in Fig. 10.

5 Branch and Bound

Branch and Bound's philosophy is to solve successive solutions to linear programming problems, initially solving the original problem by relaxing the integrity of their entire variables, that is, allowing integer variables to have real values [17]. If

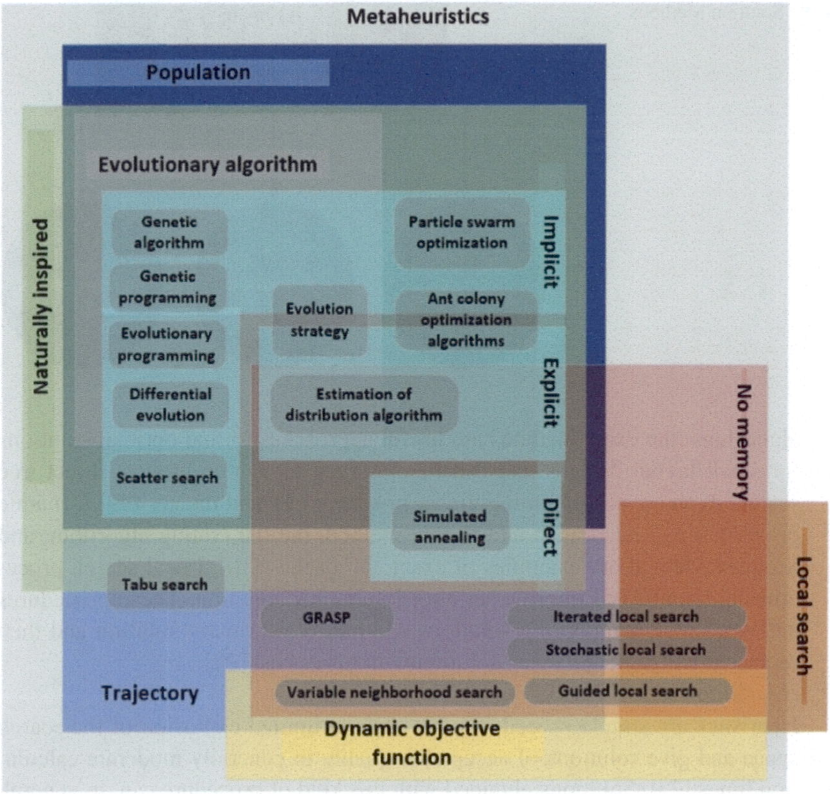

Fig. 10 Most popular metaheuristics

the problem has a complete solution across all variables, it means that the optimal global solution has been found. Otherwise, one of the non-integer solution variables is randomly chosen and an iterative branching process is applied to find an integer value. The Branch and Bound algorithm is an accurate method for finding the solution to linear problems with whole and mixed solution [17].Combinatorial optimization issues are about finding an optimal solution within a finite solution space. If the solution space is small, you can list all possible solutions, however, if the solution space is too long, the solution enumeration becomes more complicated [18].

The Branch and Bound algorithm helps to find the optimal solution within the solution space. It can be represented as a tree whose root is the original problem, and each branch represents a division of solution space [18].Each node in the tree is responsible for finding the optimal solution in the solution space. The nodes of the tree are pruned to find the solution, if the dimension is no better than the previously obtained solution, it is not necessary to explore that zone otherwise it must be done [18]. The main feature is that this technique is responsible for detecting which branching the optimal solution is in and where the solutions found are moving away from the optimal solution [18].There are numerous Branch and Bound strategies that solve the TSP problem. The cost of nodes is the distance of the path of a complete graph. The adjacency matrix and its given elements between two vertices is not necessarily symmetrical although it is symmetrical, although it is symmetrical with respect to its non-negative elements [15].

6 Simulated Annealing

Simulated Annealing is a bioinspired metaheuristic in metal smelting and was proposed by researchers Kirkpatrick in 1983 and V. Cerny in 1985 [14]. Most metaheuristics target combinatorial optimization issues and can be applied to any reformulated problem [7]. Simulated Annealing is an optimization method inspired by the metal tempering process. The metal tempering process consists of three phases: a heating phase at a given temperature; in the second phase the high temperature is held which allows the molecules to settle in states of minimum energy and follows a controlled confrontation phase to increase the size of their crystals and reduce their defects applies to optimization problems (TSP) with a relatively large number of cities [12].

The algorithm works by mimicking the physical process by which a solid is slowly cooled, until eventually its molecular structure is frozen, and it passes to a minimum energy configuration. The algorithmic version of this initial process with a random estimate of the values of the variables of the cost function [6].

Simulated Annealing (SA) is an iterative method that starts with a certain s state. Using a particular process generates a neighbor state s' to the current state. If the power or evaluation of s' is less than that of the s state, the s state is changed to $s^{'}$. if the evaluation of $s^{'}$ is greater than that of s then it can be worsened by choosing s' instead of s with a certain probability that depends on the differences of the evaluations $\Delta f = f(s) - f(s')$ and the current temperature of the T system mentioned in Eq. (10) [12].

$$P(\Delta f, T) = e\Delta f/T \tag{10}$$

If the energy difference, $E_u = E_j - E_i$, is less than zero $E_u \leq 0$), the solution S_j is accepted as the best one, that is, it substitutes S_i, otherwise, if $E_u > 0$, the solution is not rejected and an acceptance criterion known as Boltzmann probability is applied. Where Eu is the energy difference, K_B is the Boltzmann constant whose value is $1.38x10^{-16}ergs/K\ T$ is the temperature of the exchanged value. When $P(E_u)$ greater than a random number between *0* and *1* is, S_j takes the place of S_i, otherwise S_i remains the best solution. The initial temperature is initialized to a value, k represents the number of iterations. The outdoor cycle is responsible for reducing the temperature with the temperature reduction parameter α. In thermal equilibrium, that is, after the material has been at a constant temperature for a sufficiently long time, the probability that the material is in the state u with energy E_u at temperature T

is where the sum extends over all possible states, the Eq. (11) Boltzmann Probability [12] is presented.

$$PT\{X = u\} = \frac{exp\left(-\frac{E_u}{k_{BT}}\right)}{\sum_v exp\left(-\frac{E_v}{k_{BT}}\right)} \tag{11}$$

7 Application

This section shows the case study location and the reason for optimizing layout paths. The case study presents a design of layout in a chemical company that is located in the city of Ramos Arizpe, Coahuila de Zaragoza and where a continuous operation is carried out within the same process which the material is moving in different reactors chemicals and material shipping routes are not properly established by making this a journey in question of customer requirements. It is important to highlight the reason for working in this case study is to generate sub routes through an a priori route to calculate the cost of transporting materials that there are no established registrars and waste of time, improper use of materials.

A way of transporting sacks of material to each reactor with the use of a forklift and a hydraulic skid is shown below in Fig. 11.

8 Methodology

This selection explains the information needed to the case study and the ways of working for the research are: Collection, Data Analysis, Creation, Validation, Selection and Implementation. Because it is an essential part of solving the problem raised each of them is explained below and a diagram of the methodology is shown in Fig. 12.

(1) *Collection*

- Location of chemical reactors: To obtain the distance matrix between each reactor.
- Demand for reactors for material routes: Each of the reactors has different chemical sacks loading needs because of the size of each of them since some reactors are larger than others and require more sacks to produce.
- Vehicle load capacity when transporting materials: As a restriction to perform the solution algorithm since the sum of the demand for the material routes for each reactor may not be greater than load capacity.
- Vehicles for material routes: They are represented as another restriction of the Solution algorithm. Identify the difference in vehicle efficiency by using

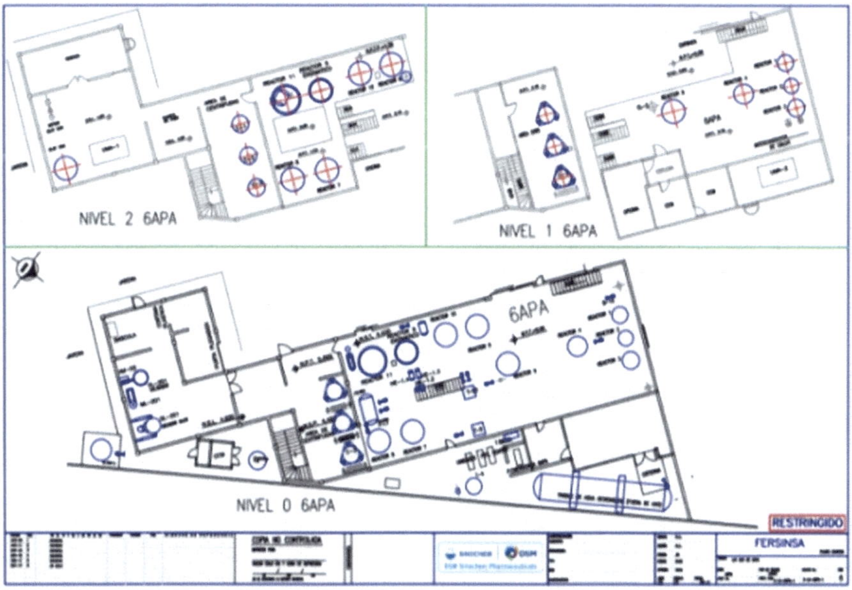

Fig. 11 Layout of a chemical company

a forklift to traverse the layout and a hydraulic skid to navigate the same layout of the chemical company.

(2) *Analysis*

- Data analysis: Distances and capacity of the 14 reactors and the warehouse to create the symmetrical distance matrix of the design of a chemical company's layout in order to generate an a priori route with the help of solution methods.

(3) *Creation*

- Creation of the a priori route: As the basis of the mathematical model of Eq. (4) the route will be obtained by minimizing the total distances of the routes. The x value will be the distance between the reactors of the a priori route of the distance matrix in Eq. (12) an in Eq. (13) distance between reactors.

$$y_0 = \left[0, y_{1,} y_{2,} y_3, \ldots, y_{14}, 0\right] \tag{12}$$

$$\sum_{i=1}^{14} X_{i,(i+1)} \tag{13}$$

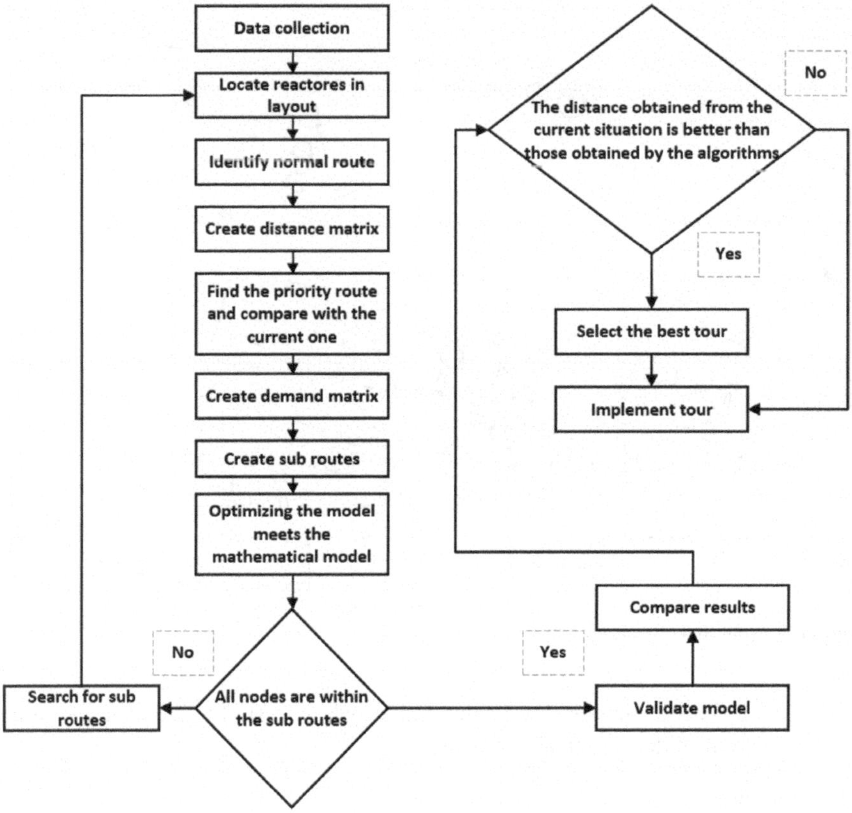

Fig. 12 Methodology diagram

- Create sub route and distance: Each reactor has its own demand for materials; the sum of the demands of the chemical reactors visited on the route is a priori greater than the load capacity of the forklift and the hydraulic skid, which will return to the warehouse to reload chemical sacks and follow the route where the reactor was left unattended. The re-reload generates a sub routes called X_g where $g = \{1, 2, .., n\}$ and will be named as the sub routes number $X = 1$.
  - n = number of reactors in the sub routes.
  - Every sub routes must start in the warehouse.
  - X = the distance between the reactor and reactor of the distance matrix.

(4) *Validation*

- Validate sub route: Accept sub routes that meet the results of optimization models. When they are valid, the other paths of the results obtained from the proposed algorithms are checked.

(5) *Selection*

- Select the best route: Choose the best algorithm that minimizes the total path of sub routes obtained from the layout.

(6) *Implementation*

- Implement tour: It shows how to perform the best journey of Branch and Bound results as well as Simulated Annealing.

9 State of the Art

This chapter will present important contributions in layout design from certain authors where they mention the diversity and importance of material path and open field layout designs as well as the different ways to solve the case study being presented in this research.

"Classification of facility layout problems: a review study" The authors considered in their research that layout design in general is a classic and most important problem of production management and industrial engineering that has attracted the attention of many researchers in the literature during the last decades. Despite the research on many aspects of the problem is still in its initial stage; therefore, it seems interesting and promising to work on these studies, a comprehensive review of several articles on layout design was presented, an attempt was made to present an extensive and comprehensive review of the types of layout, classifications and forms were considered many references that considered layout evolution, layout, problem formulation and resolution approaches. Based on this analysis, layout design and its classification continue to be published in the main academic journals, being a subject of open research. As results and comparisons, the types of layout and their derivations are broken down, as a future work we wish to investigate more deeply with each of the types of layout, among the many references listed at the end of this article we find more than 180 publications since 1987 that are considered in different aspects to be questioned and comparing different authors [10].

"Investigation of the Layout Problem of Different Types of Floors" The authors presented a comprehensive survey related to layout design problems of multi-rows facilities along with modeling and resolution approaches. For this study, all related and compared documents in the literature are collected and analyzed from different aspects, due to the increasing importance of layout over time and also the lack of attention. Firstly, allocating efforts to benefit other areas such as markets, ships, hospitals, administrative organizations and schools would be of valuable importance. As a final point, a more in-depth direction for future research can be identified by using a tree representation of the layout of different floors for each branch of the tree. In addition, it is noted that some topics have not been researched in the literature and have received less attention than the 10% research on the topic. Therefore, research is needed on high quality solutions to the problem of layout design using comparisons

of ant colonies and particle swarms that demonstrate their robust performance with the use of these metaheuristics, finding results and comparing fences to the local optimum at all times [31].

"Research on the design of the facilities" The information studied and compared by the authors in this article focuses on manufacturing system applications. However, layout design problems may also apply, ports, supermarkets, airports, etc. These are interesting areas that could benefit from advances made in the specific area of manufacturing especially material path designs the key is to find a suitable metaheuristic that fits. As future work the authors will consider that the commercial software tools available on the market to assist in the overall design of manufacturing are currently limited, it is likely that it will be necessary to try procedures that supersede such tools. The results and methods used to solve the problem of layout design can obviously be seen that the use of metaheuristics, evolutionary algorithms seem to be among the most popular approaches since they serve to solve complex problems such as Simulated Annealing and provide more realistic solutions for example connecting evolutionary principles with simulation and artificial intelligence based approaches are now rarely published since they are difficult to solve with the use of hybrid methods capable of optimization [17].

"Model of optimization of the plant distribution" When implementing any of the different distribution models, the authors should propose different alternatives on which an optimization model should be applied to determine the best option. Based on qualitative criteria, it provides good criteria to be taken into account in the plant distribution regardless of the nature of the plant. Additionally, through an adequate methodology, this process evidences improvement in the quantitative variables that are taken into account and complemented with the application of an adequate software evidences an important optimization tool on which the material path can be reliably based. As a future proposal it is considered that the layout will have further research and new methods proposed to future researchers who start this path with too much area of opportunity to experiment, evaluate and compare different decision alternatives and identify system problems generating concepts with the use of simulation software [9].

"A non-capacity layout problem with stochastic demands" In this research the authors have formulated models to maximize and minimize, they have studied during half a century only in specific environments this type of work without investigating problems without capacity with stochastic demands, to solve problems of programming and simulation models it is necessary to use numerical examples and compare to show efficiency in all the proposed algorithms, as future proposal it is intended to use hybrid intelligence algorithms to maximize and minimize layout problems with different mathematical models and solution methods as Simulated Annealing and Ant Colony among others, The steps taken were the generation of input and output data for stochastic simulations, neural networks to approximate uncertain functions according to the data, size parameters and as a result the calculation of the target values of the routes through the neural networks to verify the variability and report the best route as the optimal solution and as future work to solve 80% of the cases proposed by the authors [31].

"A dynamic, multi-objective approach to the layout design problem with multiple reconfigurable installations and paths" The authors studied the problems of layout design with two different situations: tours of the departments within a facility and the location of the departments within each facility to each other. The objectives and results of the model are to minimize the costs of materials handling within facilities and between facilities by comparing solution models such as Simulated Annealing and Branch and Bound. As future work, add travel time between facilities taking into account transportation costs and use simulation software to recreate the work more dynamically as well as use some metaheuristics known as Simulated Annealing and Ant Colony to compare results finding the optimum location with 100% of the solutions in their proposals [32].

"Diseño robusto de las instalaciones bajo demandas de productos inciertos" el diseño de layout tiene un impacto significativo por parte de los autores en los costos de fabricación y transporte de materiales, trabajo en proceso, tiempo de ejecución y la productividad. Implica la determinación de la colocación óptima de materiales, de acuerdo con la declaración del problema de un modelo matemático del layout dinámico y encontrar un diseño robusto definiendo varias variables a trabajar. En este trabajo se propone un nuevo modelo robusto y novedoso para resolver problemas de diseño de layout dinámico se analiza los principales factores que causan las demandas sobre una teoría difusa, como posibles trabajos futuros es presentar el modelo en un caso de estudio industrial realista y enfocarse en múltiples objetivos como encontrar la mínima ruta en recorridos de materiales y tiempos. Como resultado un modelo dinámico de diseño de instalaciones con área desigual en un entorno aleatorio difuso establecido en este documento busca encontrar un diseño sólido y resolver los casos propuestos y futuros de los investigadores [33].

"A New Algorithm for Structure Optimization: Simulated Annealing" The authors investigated that in their research it is possible to exploit the physical knowledge of such processes with a view to searching for the absolute minimum of any mathematical function. Simulated annealing can only be considered new as far as the search for optimal values of mechanical structures is concerned, although as a general-purpose algorithm it is already a decade old. In order to highlight some salient features of the general problem of optimizing a function, a paradigmatic problem known as the traveling agent (TSP) can serve as an illustration, since it can be posed very briefly. As a result, the problem here is not so much to optimize a function, but rather to find the mathematical function that faithfully represents the objectives of the design, which in activities such as engineering and architecture cannot always be made explicit, much less given an algebraic form. For new or unexplored problems and future work you may be able to suggest new combinations of mathematical models as well as design an experiment that can use these properties to obtain excellent approximations or when other objectives demand it and predict how far your design will deviate from the theoretical optimum. And, of course, without needing powerful machines to do so. However, for new or unexplored problems they may be able to suggest designing theory which synthetic properties should be investigated and their relative importance in solving difficult problems by purchasing genetic algorithms [21].

"A vehicle routing problem with stochastic demand sub-routes a priori*"* The author considers that the probabilistic description of the demand is necessary to design routes for the applications in the strategic planning of the distribution systems and instead of solving the problem he proposes to build an a priori sequence among all the customers of the minimum expected total. We found algorithms to compute the expected length of an a priori sequence in which the local optimum is held, analytical evidence is provided and it is based on finding an a priori sequence among all the customers set, it works quite well from the worst case perspective, especially if the distribution of customer demand is the same, although the limits are proposed in the worst case. In conclusion, they provide a sound alternative to the re-optimization strategy and can therefore be useful in practice, especially in the absence of intense computing power. In relation to this point, an interesting question that deserves further research is to compare computationally the a priori and re-optimization strategies [12].

"Simulated Annealing Cooling Models" The authors make a compilation of the most used cooling models in The Simulated Annealing Algorithm. A comparison of model performance in the context of the combinatorial problem of quantitative data distribution is shown. In addition, an alternative empirical model is proposed to accelerate the geometric model, which is the most used in practice. The algorithm of Simulated Annealing and performing the cooling geometrically, with the results exposed this experiment the simplicity of the geometric model and its high performance. However, the existence of other models that also presented a performance like the geometric one was noted. In particular, the Lundy & Mees model was refined as a very suitable alternative to perform the cooling and comparison. In the experiment it was found that the logarithmic model has little practical importance, since it performs the cooling in a very significant number of iterations and using long execution times [34].

10 Data Collected

For each reactor you have different capacities as sizes range from large reactors to small reactors whose quantity is essentially defined by the size of the reactor.

The material travel of 14 reactors and a material warehouse with capacity of the forklift of 120 sacks and the hydraulic skate 70 sacks, data to be worked in Table 1.

Table 1 Demands and reactor capabilities

Chemical reactor	1	2	3	4	5	6	7	8	9	10	11	12	13	14
Demand for sacks for each reactor	30	40	28	31	32	15	28	13	29	40	12	18	22	17
Forklift capacity	120													
Hydraulic skate capacity	70													

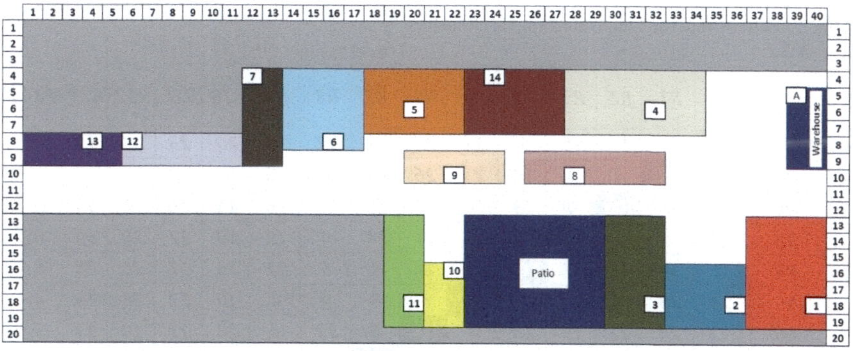

Fig. 13 Reactor layout of a chemical company

Figure 13 shows 14 reactors and warehouses identified by different colors to distribute sacks with a certain chemical and are found by coordinates in *x* and *y* the closest to the actual distances but which do not affect the result obtained. To improve the only way is to control the chemical sack materials for each of the reactors by creating efficient sub-routes to supply each of the reactors without generating so many trips for the forklift and the hydraulic skid and to improve the distance times and the ease of loading the materials.

In Table 2 table *x* and *y* of each reactor including the warehouse using Fig. 3 to identify the distances of each reactor for forklift and hydraulic skid paths.

Table 2 Reactor and warehouse coordinates

Coordinates	X	Y
Warehouse	39	5
Reactor 1	40	18
Reactor 2	36	18
Reactor 3	32	18
Reactor 4	32	6
Reactor 5	20	6
Reactor 6	16	8
Reactor 7	12	4
Reactor 8	28	10
Reactor 9	22	10
Reactor 10	22	16
Reactor 11	20	18
Reactor 12	6	8
Reactor 13	4	8
Reactor 14	24	4

Table 3 Path of reactor materials and warehouse

Travel table

	dij	A	R1	R2	R3	R4	R5	R6	R7	R8	R9	R10	R11	R12	R13	R14
0	A	100	13	13	15	7	19	23	27	12	18	20	23	33	35	15
1	R1	13	100	4	8	14	23	26	31	14	20	18	20	35	37	21
2	R2	13	4	100	4	13	20	22	28	11	16	14	16	32	34	18
3	R3	15	8	4	100	12	17	19	24	9	13	10	12	28	30	16
4	R4	7	14	13	12	100	12	16	20	6	11	14	17	26	28	8
5	R5	19	23	20	17	12	100	4	8	9	4	10	12	14	16	4
6	R6	23	26	22	19	16	4	100	6	12	6	10	11	10	12	9
7	R7	27	31	28	24	20	8	6	100	17	12	16	16	7	9	12
8	R8	12	14	11	9	6	9	12	17	100	6	8	11	22	24	7
9	R9	18	20	16	13	11	4	6	12	6	100	6	8	16	18	6
10	R10	20	18	14	10	14	10	10	16	8	6	100	3	18	20	12
11	R11	23	20	16	12	17	12	11	16	11	8	3	100	17	19	15
12	R12	33	35	32	28	26	14	10	7	22	16	18	17	100	2	18
13	R13	35	37	34	30	28	16	12	9	24	18	20	19	2	100	20
14	R14	15	21	18	16	8	4	9	12	7	6	12	15	18	20	100

The distances of two points in Eq. (14) are observed and the distance matrix is obtained by obtaining the distances of all the routes in Table 3.

$$d = \sqrt{(x_2 - x_1)^2 + (y_2 - y_1)^2} \tag{14}$$

To create the symmetrical distance matrix of the case study, the quantities of chemical sacks of the chemical reactors were considered the distances from one reactor to another considering the 14 reactors and the warehouse in Table 4.

11 Results

The route of the cargo vehicles was carried out with the Branch and Bound to obtain the route a priori minimizing the distances and capacities of the chemical reactors, the mathematical model was taken as the basis to minimize the distances of the materials for each chemical reactor in Table 5.

Material travel costs data have been taken for each of the reactors using the forklift and hydraulic skid to transport sacks costs and costs for each reactor is different for its Table 6 load capacity.

A. Branch and Bound

The equipment used for the experiment was Solver and the point marked are the distances selected with Branch and Bound as the a priori routes as a route a priori is:

Table 4 Distance matrix

	R1	R2	R3	R4	R5	R6	R7	R8	R9	R10	R11	R12	R13	R14
Warehouse	13	13	15	7	19	23	27	12	18	20	23	33	35	15
R1	**100**	4	8	14	23	26	31	14	20	18	20	35	37	21
R2	4	**100**	4	13	20	22	28	11	16	14	16	32	34	18
R3	8	4	**100**	12	17	19	24	9	13	10	12	28	30	16
R4	14	13	12	**100**	12	16	20	6	11	14	17	26	28	8
R5	23	20	17	12	**100**	4	8	9	4	10	12	14	16	4
R6	26	22	19	26	4	**100**	6	12	6	10	11	10	12	9
R7	31	28	24	20	8	6	**100**	17	12	16	16	7	9	12
R8	14	11	9	6	9	12	17	**100**	6	8	11	22	24	7
R9	20	16	13	11	4	6	12	6	**100**	6	8	16	18	6
R10	18	14	10	14	10	10	16	8	6	**100**	3	18	20	12
R11	20	16	12	17	12	11	16	11	8	3	**100**	17	19	15
R12	35	32	28	26	14	10	7	22	16	18	17	**100**	2	18
R13	37	34	30	28	16	12	9	24	18	20	19	2	**100**	20
R14	21	18	16	8	4	9	12	7	6	12	15	18	20	**100**

Table 5 A priori route

	Warehouse														Warehouse
Chemical reactor	**0**	**1**	**2**	**3**	**10**	**11**	**9**	**14**	**5**	**6**	**7**	**13**	**12**	**8**	**4**	**0**
Demand for sacks		**30**	**40**	**28**	**40**	**12**	**29**	**17**	**32**	**15**	**28**	**22**	**18**	**13**	**31**	

Table 6 Cost table

Forklift

Route	Cost
Warehouse	$ -
Reactor 1	$ 21.00
Reactor 2	$ 22.00
Reactor 3	$ 27.00
Reactor 4	$ 15.00
Reactor 5	$ 18.00
Reactor 6	$ 19.00
Reactor 7	$ 25.00
Reactor 8	$ 16.00
Reactor 9	$ 10.00
Reactor 10	$ 13.00
Reactor 11	$ 12.00
Reactor 12	$ 22.00
Reactor 13	$ 22.00
Reactor 14	$ 18.00
Total	**$ 260.00**

Hydraulic skate

Route	Cost
Warehouse	$ -
Reactor 1	$ 11.00
Reactor 2	$ 9.00
Reactor 3	$ 17.00
Reactor 4	$ 8.00
Reactor 5	$ 18.00
Reactor 6	$ 19.00
Reactor 7	$ 18.00
Reactor 8	$ 7.00
Reactor 9	$ 8.00
Reactor 10	$ 7.00
Reactor 11	$ 6.00
Reactor 12	$ 12.00
Reactor 13	$ 13.00
Reactor 14	$ 15.00
Total	**$ 168.00**

Table 7 Distance from route a priori

	R1	R2	R3	R4	R5	R6	R7	R8	R9	R10	R11	R12	R13	R14
Warehouse	13	13	15	7	19	23	27	12	18	20	23	33	35	15
R1	100	4	8	14	23	26	31	14	20	18	20	35	37	21
R2	4	100	4	13	20	22	28	11	16	14	16	32	34	18
R3	8	4	100	12	17	19	24	9	13	10	12	28	30	16
R4	14	13	12	100	12	16	20	6	11	14	17	26	28	8
R5	23	20	17	12	100	4	8	9	4	10	12	14	16	4
R6	26	22	19	26	4	100	6	12	6	10	11	10	12	9
R7	31	28	24	20	8	6	100	17	12	16	16	7	9	12
R8	14	11	9	6	9	12	17	100	6	8	11	22	24	7
R9	20	16	13	11	4	6	12	6	100	6	8	16	18	6
R10	18	14	10	14	10	10	16	8	6	100	3	18	20	12
R11	20	16	12	17	12	11	16	11	8	3	100	17	19	15
R12	35	32	28	26	14	10	7	22	16	18	17	100	2	18
R13	37	34	30	28	16	12	9	24	18	20	19	2	100	20
R14	21	18	16	8	4	9	12	7	6	12	15	18	20	100

0 – 1 – 2 – 3 – 10 – 11 – 9 – 14 – 5 – 6 – 7 – 13 – 12 – 8 – 4 – 0 including coordinates 0 – 1 and 4 – 0 of the start and end of the route where you pass through each of the selected routes starting at the warehouse and ending un in the warehouse with a total 113, Table 7 is presented below.

Once the route has been obtained, a model is worked on to optimize the distribution routes of the aforementioned mathematical model using the TSP strategies to visit all reactors in the same order fixed with the service where the demand of the vehicle's capacity must always be met. The total routes and distances of the sub paths should be taken for all algorithms and the best algorithm should be selected that minimizes the total path of materials in general of the total of the sub paths worked. It Sub routes generated from the a priori route are presented starting and ending in the symmetric matrix in which both vehicles travel through each of the reactors and meet their demand, it is observed in Table 8.

The forklift routes are a total 360 m for the distance generating 4 sub routes and for the hydraulic skid 355 m in distance generating 6 sub routes for a total route of the chemical reactor area of 113 m on an a priori route.

Table 8 Sub routes and distances

Forklift

Sub routes	Route	Distance
1	0,1,2,3,0	98
2	0,10,11,4,9,14,0	98
3	0,5.6.7.13.12.0	115
4	0,8,4,0	49
		360

Hydraulic skate

Sub routes	Route	Distance
1	0,1,2,0	70
2	0,3,10,0	68
3	0,11,9,14,0	58
4	0,5,6,0	47
5	0,7,13,0	50
6	0,12,8,4,0	62
		355

B. Simulated Annealing

The parameters of this algorithm are temperature $ti = 10, 100, 1000, 10000$ and $tf = 0.001$.

A random form improvement is generated according to the Metropolis algorithm with the use of Matlab software. A route is created and divided according to the material load capacity of the transport vehicles to each reactor, i.e. if the vehicles go through the materials to meet the total demand of each reactor, if the route is less than the load of the forklift and hydraulic skid, continue with the route and find a reactor with demand for materials added to the total demand of the load capacity of the vehicles completed on the travel and return to the warehouse with the next reactor where it was last found and generate alternate routes. The data in Table 4 were used and the Simulated Annealing with Matlab software was used using the above temperature listed in Table 9, 10, 11 and 12 showing the numbers of routes, routes, distances in meters and times using different start temperatures and the same final temperature for all tables.

Table 9 Temperature ti=10 and tf=0.001

Number of sub routes	Route	Distance traveled meters	Total time in seconds
1	0 6 3 11 7 14 8 5 1 2 4 13 9 10 12 0	241.2006	11.90
2	0 14 13 4 6 11 8 9 12 3 7 10 1 2 5 0	225.4141	11.86
3	0 9 13 7 6 3 1 11 12 2 4 10 8 5 14 0	202.9653	11.36
4	0 12 8 11 1 6 4 14 9 3 13 10 5 7 2 0	263.2961	11.79
5	0 8 14 7 5 12 10 11 2 1 4 13 3 9 6 0	198.8953	12.42
Total Cost Forklifts	**$ 260**	**Total Cost Hydraulic Skate**	**$ 168**

Table 10 Temperature ti=100 and t f = 0.001

Number of sub routes	Route	Distance traveled meters	Total time in seconds
1	0 12 10 4 6 9 7 8 1 3 13 11 2 5 14 0	241.6034	14.77
2	0 11 14 1 13 10 7 2 12 9 6 3 8 4 5 0	277.8681	14.75
3	0 5 10 14 12 9 4 7 8 6 13 2 11 1 3 0	240.3417	14.15
4	0 12 6 11 2 1 8 9 4 5 3 7 14 10 13 0	235.0186	14.28
5	0 7 10 6 12 11 13 8 5 2 3 1 9 14 4 0	177.3883	15.01
Total Cost Forklifts	**$ 260**	**Total Cost Hydraulic Skate**	**$ 168**

The results of sub paths derived from the solution methods used in the collected data section are validated. This section walked through the routes obtained from the sub paths created from the case study as well as the solution methods to find the route a priori. The forklift route and hydraulic skid were used to chart the route of the routes by meters and other costs present the best sub paths of both methods in Tables 13 and 14.

12 Conclusion

Therefore the reactors have their own demand, if the sum of the demand of the reactors that are visited on the route a priori is greater than the load capacity of the forklift and the hydraulic skid it will return to the warehouse to recharge and continue with the route to attend to the missing reactor and so on[16]. Four sub routes were generated for the forklift with a total travel of 360 m and a total cost of $255 pesos, for the hydraulic skate were generated 6 sub routes with a total travel of 355 m and a

Table 11 Temperature $Ti = 1000$ and $Tf = 0.001$

Number of sub routes	Route	Distance traveled meters	Total time in seconds
1	0 14 12 6 7 5 11 9 13 3 4 8 10 2 1 0	182.7155	17.11
2	0 6 12 4 5 10 13 7 8 1 9 2 11 3 14 0	211.0211	19.81
3	0 12 4 1 6 7 14 5 2 11 8 10 9 3 13 0	251.0376	17.02
4	0 8 14 10 5 13 9 2 3 4 7 1 6 11 12 0	230.0318	17.92
5	0 4 8 3 10 12 11 9 1 14 7 6 2 5 13 0	207.4411	16.71
Total Cost Forklifts	**$ 260**	**Total Cost Hydraulic Skate**	**$ 168**

cost of $165 pesos using Branch and Bound. For the temperatures of the Simulated Annealing, 5 sub routes of each of the initial temperatures are created where the total cost of using the forklift is $260 pesos and the hydraulic skid is $168 pesos where the best sub routes of each of which discovered by temperatures. It was found that during the process of obtaining the sub-routes and distances of the material path of the forklift and the hydraulic skid starting from the warehouse and ending at the warehouse with the use of Branch and Bound and Simulated Annealing a variation in the material path was located with the use of these solution methods.

Both solution methods present different ways of resolving the case study where the costs of using material transport vehicles are similar in terms of distances if there is a small difference, it should be noted that both methods of solution are effective as they look for the optimal local within a target function and minimize the routes in an optimization problem note in Fig. 14.

Figure 15 shows the best result obtained from the two solution methods in terms of cost is better than Simulated Annealing. The advantages of performing this research help in the correct handling of the materials and can also be an improvement for the design of the layout in terms of manufacturing in redesigning the

Table 12 Temperature ti=10,000 and tf=0.001

Number of sub routes	Route	Distance traveled meters	Total time in seconds
1	0 1 5 4 8 3 9 2 13 6 10 14 7 12 11 0	205.0279	20.47
2	0 3 11 9 10 5 8 1 6 12 14 13 2 7 4 0	206.8167	19.61
3	0 13 9 1 4 12 5 2 8 10 14 3 6 7 11 0	240.672	19.84
4	0 1 12 2 5 4 8 6 7 3 14 13 10 9 11 0	249.5516	19.74
5	0 13 2 11 8 9 7 14 5 3 6 1 4 10 12 0	247.1421	20.64
Total Cost Forklifts	**$ 260**	**Total Cost Hydraulic Skate**	**$ 168**

facilities, maintenance, unnecessary movements, downtime, cleaning and process standardization.

It is shown that the Branch and Bound solution method was better than the Simulated Anneal, but this does not mean that it is better as both have their strengths and the Simulated Annealing can vary when working with different start and end temperatures. A simulation was made in the layout design of the chemical company with the use of the forklift and the hydraulic skid. The route was simulated with the Branch and Bound and first we went through the 14 reactors and the materials warehouse with the hydraulic skid, marked with the green color the trajectory through all the layout visiting each reactor creating different paths with the generated sub-routes S1, S2, S3, S4, S5, S6 and with a total of 355 m to meet the demand requested daily and for the freight elevator was marked with the color blue generating blue roads and their corresponding sub-routes were S1,S2,S3,S4 and with a total of 360 m to meet the demand requested. Below is shown in Fig. 16 the layout of the hydraulic skid route using Branch and Bound and in Fig. 17 the layout of the forklift route.

Table 13 Sub routes with branch and bound

Forklifts			
Sub route	**Route**	**Distance**	**Cost**
S1	0 1 2 3 0	98	$ 70
S2	0 10 11 4 9 14 0	98	$ 78
S3	0 5 6 7 13 12 0	115	$ 76
S4	0 8 4 0	49	$ 31
Total Tour:		**360 Meter**	
Total Cost:		**$ 255**	
Hydraulic Skate			
Sub route	**Route**	**Distance**	**Cost**
S1	0 1 2 0	70	$ 19
S2	0 3 10 0	68	$ 24
S3	0 11 9 14 0	58	$ 27
S4	0 5 6 0	47	$ 37
S5	0 7 13 0	50	$ 31
S6	0 12 8 4 0	62	$ 27
Total Tour:		**355 Meter**	
Total Cost:		**$ 165**	

13 Future Work

As future work, it is proposed to apply other methods of solution. Using other metaheuristics such as Ant Colony, BRKGA and GRASP as well as a greater number of reactors, means of transporting materials and the use of simulation software to compare results and include restrictions to the problem to be solved.

Table 14 Best sub routes with simulated annealing

Temperature	Sub routes	Route	Distance
ti=10 y tf=0.001	S1	0 8 14 7 5 12 10 11 2 1 4 13 3 9 6 0	198.8953
ti=100 y tf=0.001	S2	0 7 10 6 12 11 13 8 5 2 3 1 9 14 4 0	177.3883
ti=1000 y tf=0.001	S3	0 14 12 6 7 5 11 9 13 3 4 8 10 2 1 0	182.7155
ti=10000 y tf=0.001	S4	0 1 5 4 8 3 9 2 13 6 10 14 7 12 11 0	205.0279
Total Cost Forklifts	$ 260	Total Cost Hydraulic Skate	$ 168

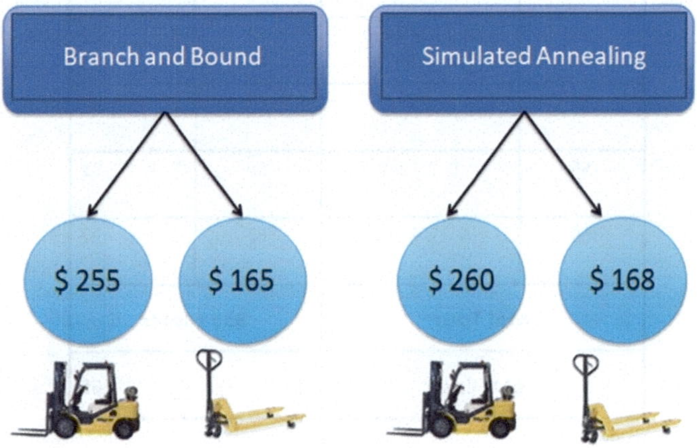

Fig. 14 Branch and Bound result and Simulated Annealing

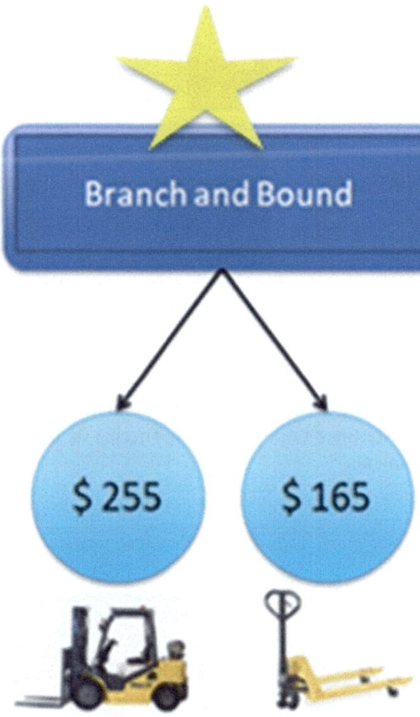

Fig. 15 Best possible solution

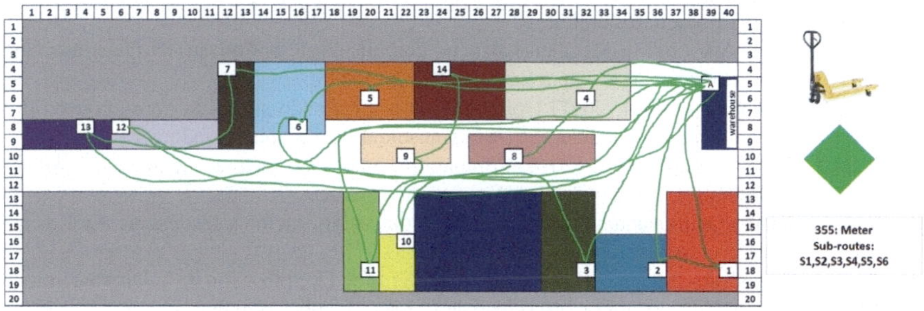

Fig. 16 Material handling using Branch and Bound with hydraulic skid

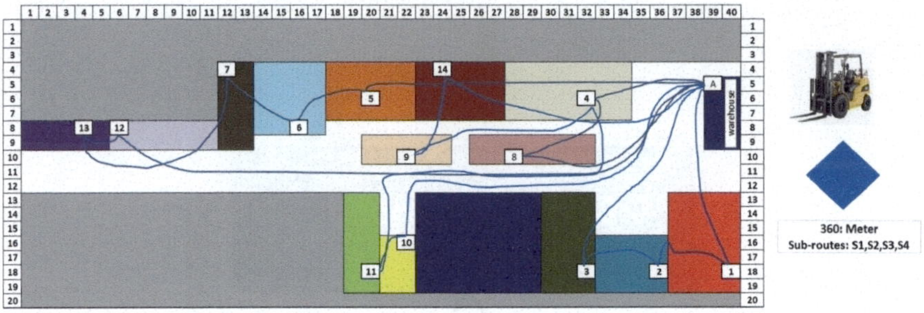

Fig. 17 Material handling using Branch and Bound with a forklift

Acknowledgements We would like to thank the Instituto Tecnológico de Saltillo for providing us with its facilities, Dr Ricardo Martínez López, Dr Alan Tijerina de la Rosa and MII Pedro Henoc Ireta for their advice and comments.

References

1. Tompkins JA, White JA, Bozer YA, Tanchoco JMA (2010) Facilities planning, Fourth ed. John Wiley & Sons, New York
2. Devise O (200) Pierreval. Indicators for measuring performances of morphology and materials handling systems. Int J Product Econ 64(1–3):209
3. Alfredo DM, Aldair AA, Victor MN, Jorge ML (2014) Optimización multiobjetive del problema de distribución de planta: Un nuevo modelo matemático. Colombia, Ingenieria y competitividad
4. Ríos LMR, Ramírez D (2012) Desarrollo de un modelo heurístico para el manejo de material en estibas en una bodega. Colombia
5. Blum C, Roli AA (2003) Metaheuristics in combinatorial optimization: Overview and conceptual comparison. CM Comput Surv 35(3):268–308
6. Araque JA, Rodriguez JLD (2017) Optimización por recocido simulado de un convertidor multinivel monofásico con modulación pwm sinusoidal de múltiple portadora
7. Oncar T, Kuban A, Laporte GA (2016) Comparative analysis of several asymmetric traveling salesman formulations. Computers and Operational Research
8. Robles C (2010) Optimización por colonia de hormigas: aplicación y tendencias. En Revista Ingeniería Solidaria, vol 6, núm. 10, pp 83–89
9. Mauricio G, Rodriguez LA, Ochoa JF (2014) Modelos de Optimización de la Distribución de Planta. Universidad Distrital Francisco José de Caldas, Bogotá Colombia
10. Hosseini-Nasab H, Fereidouni S, Mohammad TFG, Bagher MF (2017) Classification of facility layout problems: a review study. Iran. Department of Industrial Engineering, Yazd University, Yazd 89195–741
11. Heragu SS (1997) Facility design. Boston. BWS
12. Izadinia N, Eshghi KA () robust mathematical model and ACO solution for multi-floor discrete layout problem with uncertain locations and demands. Comput Ind Eng 96:237–248
13. Matsuzaki K, Irohara T, Yoshimoto K (1999) Heuristic algorithm to solve the multi-floor layout problem with the consideration of elevator utilization. Comput Ind Eng 36(2):487–502
14. Amaral ARSA (2013) parallel ordering problem in facilities layout. Comput Oper Res 40:2930–2939
15. Chung J, Tanchoco JMA (2010) The double row layout problem. Prod Res 48(3):709–727

16. Palubeckis G (2015) Fast local search for single row facility layout. Eur J Oper Res 246(3):800–814
17. Drira A, Pierreva H, Hajri-Gabouj S (2007) Facility layout problems: A survey. Tunisia : Institut National des Sciences Appliqueéset de Technologie, Centre Urbain Nord, BP 676, 1080 Tunis
18. Hungerlander P, Anjos MF (2015) A semidefinite optimization based approach for global optimization of multi-row facility layout. Eur J Oper Res 245(1):46–61
19. Alduaji A, Hassan NM (2020) Adopting a circular open-field layoutin designing flexible manufacturing systems. American University of Sharjah, UAE, Department of Industrial Engineering
20. Keller B, Buscher (2015) Single Row Layout Models. European J Operat Res
21. Vázquez EM (2015) Un nuevo algoritmo para la optimización de estructuras: El recocido simulado. Dpto. de Estructuras de Edificación de la UPM, Madrid-España
22. López E, Salar O, Murillo Á (2014) El problema del agente viajero: un algoritmo determinístico usando búsqueda TABÚ. Revista de Matemática: Teoría y Aplicaciones
23. Lopez CA, Mendoza V, Jairo A, Cuartas E (2018) Algoritmo para la exploración de todos los valores posibles en el problema del agente viajero. Colombia, Universidad Tecnologica de Pereira Colombia
24. Querelle (2015) Distancia entre dos puntos
25. Natalia Alejandra G-T, Ricardo Andrés M-M, Henry L-D (2016) Solución del problema de ruteo de vehículos con demanda estocásticas mediante la optimización por espiral. Santander, Colombia, Facultad de Ingenieria
26. Clarke-Wright (1964) Scheduling of vehicles from a central depot to a number of delivery points. Operation Research, No. 12
27. Manzini R (2012) Warehousing in the global supply chain. Springer, Italia
28. Ireta P (2014) Algoritmo Genético para resolver el VRP con Demanda Estocástica. International Congress on Logistics & Supply Chain, Saltillo Coahuila
29. Rocha LB, Gonzalez EC, Orjuela JA (2011) Una revisión al estado del arte del problema de ruteo de vehículos: Evolución histórica y métodos de solución. Universidad Distrital FJC
30. Griffis SE, Bell JE, Closs DJ (2012) Metaheuristics in Logistics and Supply Chain Management. J Buss Logistics 90–106
31. Zhou J (2015) Uncapacitated Facility Layout Problem with Stochastic Demands. China: Department of Mathematical Sciences, Tsinghua University, Beijing 100084
32. Azevedo MM, Crispim JA, Pinho de Sousa JA (2016) Dynamic multi-objective approach for the reconfigurable multi-facility layout problem. Portugal, Faculty of Engineering, University of Porto, Rua Dr. Roberto Frias
33. Zha S, Guo Y, Huang S, Wang F, & Huang X (2017) Robust facility layout design under uncertain product demands. China, College of Mechanical and Electrical Engineering, Nanjing University of Aeronautics, Nanjing, 210016
34. Monge JCM, Juan JF (2015) Modelo de enfriamiento en recocido simulado. Instituto Tecnológico de Costa Rica, Revista digital, Matemática Educación e Internet

Waste Collection of Touristics Services Sector Residues Vehicle Routing Problem with Time Windows to an Industrial Polygon in a Smart City

Diego Hurtado-Olivares, José Alberto Hernández-Aguilar, Alberto Ochoa-Zezzatti, José Crispín Zavala-Díaz, and Guillermo Santamaría-Bonfil

Abstract This chapter analyses the literature review of optimization proposals for waste management, which is concerned with finding cost-optimal routes for solid waste collection trucks. The approach of waste management is support for cities, innovating intelligent solutions in smart cities. This research attempts to demonstrate how optimization techniques can improve collection efficiency, cost-effectiveness, and sustainability. This chapter attempts to serve as a roadmap of research literature produced in the field of waste management, mathematical modeling, and optimization of waste collection processes based on simulated annealing algorithm and a combination of neighborhood operators. More generally, this chapter focused on the mathematical description of the framework of waste management systems, including households as waste sources, treatment sites, collection processes, and a heuristic algorithm that use a greedy algorithm to initialization the route and simulated annealing that uses improvement phases of the route.

Keywords Waste collection · Waste management for smart cities · Vehicle routing problem

---

D. Hurtado-Olivares · J. A. Hernández-Aguilar (✉) · J. C. Zavala-Díaz
Universidad Autónoma del Estado de Morelos, Avenida Universidad 1001, Chamilpa, 62209 Cuernavaca, Mor, México
e-mail: jose_hernandez@uaem.mx

A. Ochoa-Zezzatti
Universidad Autónoma de Ciudad Juárez, Ave. Plutarco Elías Calles y Hermanos Escobar, 32310, Cd Juárez, Chihuahua, México

G. Santamaría-Bonfil
Instituto Nacional de Electricidad y Energías Limpias, Reforma 113, Palmira, 62490 Cuernavaca, Mor, México

© The Author(s), under exclusive license to Springer Nature Switzerland AG 2021
A. Ochoa-Zezzatti et al. (eds.), *Technological and Industrial Applications Associated with Intelligent Logistics*, Lecture Notes in Intelligent Transportation and Infrastructure,
https://doi.org/10.1007/978-3-030-68655-0_6

1 Introduction

In this chapter we consider the Vehicle Routing Problem (VRP) that arises when a set of customers have waste that must be collected by garbage trucks, the VRP deals with the optimization problem at designing optimal routes with minimal traveled distance. Solid waste management is a relevant activity in smart cities; this activity is an interesting area to be analyzed since it contributes to a more efficient logistics system. In Mexico, it is estimated that around 6% of Mexico's gross domestic product (GDP) is reserved for waste transportation [1]. In proportion, 60.97% of the solid waste generated comes from household sources, and the remaining 39.02% from industries, shops, and other sources. Forty-three percent of the average cities in Mexico carry out a design using a technical method to carry out the collection [2]. In this sense, is necessary that the municipality policy-makers propose strategies that allow them to plan and program their actions incorporating new technologies. The current problem is modeled as the vehicle routing problem. The problem is illustrated in Fig. 1 for a single vehicle and multiple disposal sites.

This chapter is organized into six sections. In Sect. 2, we review the relevant literature on the problem. In Sect. 3, we present the problem formulation, the notation for the problem, and define the neighbor sets for each customer. The neighbor set for a customer is those customers that are closest to it, but with a compatible time window. In this section, we also present our procedures: obtaining an initial solution; route evaluation; route improvement; and vehicle reduction. In Sect. 4, we present our metaheuristic algorithms using simulated annealing as well as our metaheuristic based on three stochastic neighborhood searches. In Sect. 5 we present our conclusions, and finally, in Sect. 6 the references.

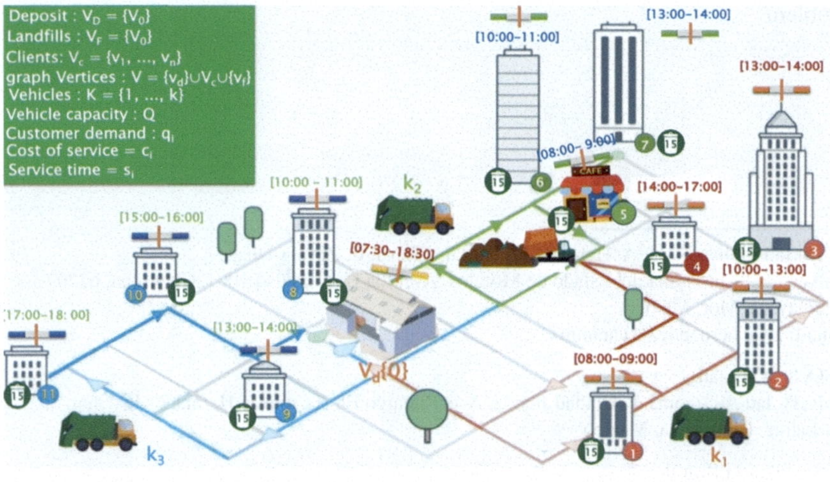

Fig. 1 A model of Waste Collection VRP with Time Windows (WCVRPTW) (Own source)

2 Literature Review

In this section, we survey the literature as well as relevant previous work dealing with the problem of routing vehicles oriented to collect solid waste from customers. Several papers in the literature have a deal with waste management by modeling them as a VRP under different constraints, especially as the VRPTW. The waste collection VRPTW has received attention in recent years. The use of computers to solve waste collection problems has been reported by several researchers over the past 29 years since its first applications with an approach to optimize waste collection as described in [4]. A route using a heuristic was developed with a routing approach using data from one of the districts. VRP requires a considerable amount of time in data collection, and most of the programming effort has been in data management. The actual routing effort constitutes a relatively small part of the system. Tests have shown that with the VRP, constant savings in the distance traveled by vehicles can be achieved, which translates into fuel savings and vehicle maintenance work, and eventually operational costs.

One of the first proposals for the waste collection was proposed by [5] using a methodology that combines a method of operations research with systems engineering. They proposed a model was to consider a set of points for the collection routes instead of the arches that form the streets. This shows how a simple formulation of a problem such as one of the linear programming provides decision-makers with efficient solutions to the complex problem of urban administration. In [6] is considered only one disposal site and is formulated the problem in mixed-integer programming, where they modified the Solomon. In [6] is proposed a heuristic procedure to solve a waste collection VRSP in Hanoi. The problem is characterized by a VRSP with time windows, with loading and unloading, arrival time, and total time restrictions. The proposed heuristic consists of two phases: (1) the construction and (2) improvement phases. Compared to current practice, the proposed heuristic shows good results and in good computational time.

In [7] described a vehicle route planning of recyclable urban waste. They develop an algorithm that assigns routes for each day of the month. Two important aspects that they modeled using a long time and the separate collection of three types of waste: (i) glass containers, (ii) paper containers, (iii) metal plastic containers. The results showed that significant savings. In [8] a VRPTW is modeled for commercial waste collection, taking into account multiple waste treatments and driver rest time. Solomon [9] proposes a benchmark problem set for the vehicle routing problem with time windows and conducts a computational study of several heuristic algorithms using the set. Among the tested heuristics, an insertion heuristic, which is known as Solomon's insertion heuristic, generates the best routing schedules in many cases.

2.1 Algorithms Used for Waste Management

Many methods have been used to solve the VRP TW, because of the complexity of the problem (NP-hard), in some cases exact methods, but usually metaheuristics, classic heuristics such as SA [10], LNS [11], IGA [12], Genetic Algorithm (GA) [13] and

PSO [14], etc. In some cases, exact methods can be used using Branch-and-bound algorithms [15], Branch-and-cut algorithms [16], and dynamic programming [17]. The summary is illustrated in Fig. 2.

Since the point of view of approximate algorithms used to solve VRPTW, we identified Classic Heuristics (sequential improvement, construction algorithms, two-phase, three phases, and two lower bounding), and metaheuristics (population based: Genetic Algorithms, Ant colony, simulated annealing, large neighborhoods, and particle swarm optimization-PSO), and machine learning (neural networks).

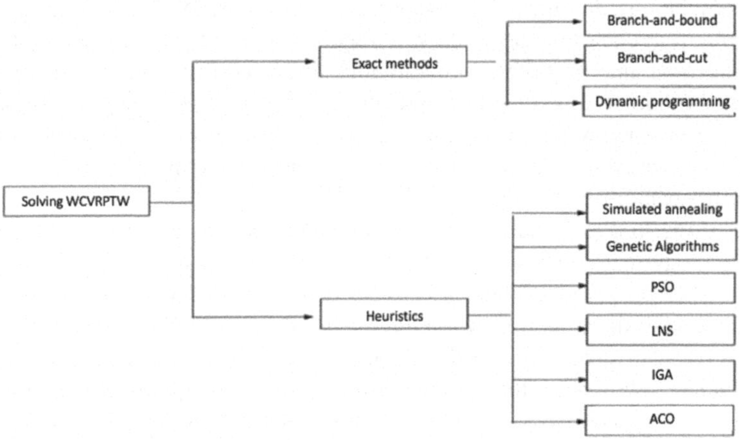

Fig. 2 Summary of methods used to solve the VRPTW

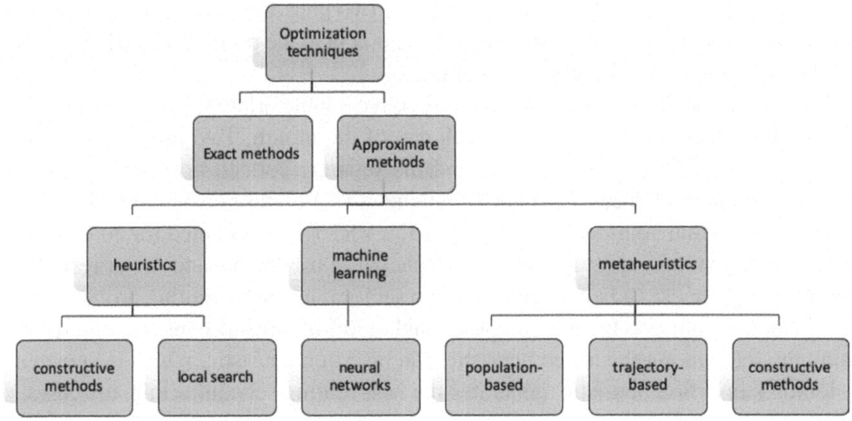

Fig. 3 Solutions methods to solve the VRP

2.2 Complexity

According to [18], VRP and VRPTW belong to the class of NP-hard combinatorial optimization problems. The VRPTW in particular, according to Solomon [9] with time window restrictions remains much more complicated to solve than a simple VRP even for a fixed fleet of vehicles. Therefore, when it is required to solve these types of problems, heuristic procedures are mainly suggested to solve instances of larger problems of the VRPTW. Heuristic methods often produce a near-optimal solution or the optimal solution in some cases; in large problems, they produce a good solution in a reasonable amount of computer time. Therefore, the development of heuristic algorithms that can obtain almost optimal feasible solutions is a common practice. Instances with 100 clients or more are very difficult to solve optimally. However, very few of the benchmarks of Solomon instances involving 100 clients have been optimally resolved [19].

2.3 Solution Methods

The VRP is defined as a type of combinatorial problem. Heuristics and meta-heuristics are used to get realistic and close to the optimal solution. Researchers presently are making use of combinations of different heuristics to shorten the time needed for finding the nearly optimal solutions. Significant progress has been seen in the last 15 years in the development of metaheuristics to solve the VRP. All of these approaches of exploration of the solution space beyond the first local minimum found, and all embedding procedures paddled by classical construction and heuristic improvement (Fig. 3).

3 Model Formulation

In this section we define the problem formulation and the notation used throughout the chapter for the development of the mathematical model focused on the waste collection problem, firstly we have identified the requirements and specifications available to companies in the solid waste collection. Within these constraints like the capacity of the vehicles, the number of clients, the service time of each client, the maximum collection capacity allowed, and other elements that allow the formulation of the model we are considering. Model decisions are related to the allocation of collection routes for demand by a set of customers.

3.1 Problem Formulation

In this section, we formally define the WCVRPTW. The problem is defined on a graph G = (V, E), where the set of nodes V = { $V_d \cup V_f \cup V_c$} consists of a depot $V_d = \{0\}$, m disposal sites $V_f = \{1,\ldots, m\}$, n customers $V_c = \{m+1, 1,\ldots, m+2\}$ with all positive weights; The node $V_d \in$ V represents the depot, that is, from which all k vehicles depart the routes passing through each vertex once, and finally returning to the depot, this is called a tour. The length of a tour is the summation of the lengths of the edges on the path that defines the route. No arc ends at node V_d of the depot and all routes begin and end at depot V_d. The set of vertices represented by customers $V_d \in$ V where V $\in V_d$ and $V_c = \{V_1, \ldots, V_n\}$; and the other vertices are V_c clients; the set of n vertices that denote customers is denoted as N. The set of arcs A denotes all possible connections between the nodes (including the node denoting the depot). E represents the set of pairs (i, j): where i, j belongs $\in$ V to the set of vertices is called the set of edges. E = $\{(V_i, V_j)$: i $\neq$ j} is the set of arcs, each with a non-negative travel cost C_{ij}. We associate a cost C_{ij} associated with the distance or travel time defined for each edge of the set of edges and a time t_{ij} with each (i, j) $\in$ A of the routing network. The travel time t_{ij}, j can include the service time on the client i. Each client $V_c \in$ V / { V_d} has a non-negative demand d_i, i $\in$ C; Each route has a daily limit of maximum garbage collection load allow R for each customer V_c to which it must be satisfied by a vehicle and the capacity C of the vehicle must be respected; A homogeneous fleet with each vehicle with capacity Q, is available in the depot V_d. t_{ij} and C_{ij} represent the travel distance time and the cost associated with the arc (i, j). Respectively each node i $\in V_c$ has an associated service time s_i and a time window $[a_i, b_i]$ and we define the arc *di* as the collection amount of a client $i \in V_c$. A vehicle can arrive before the start of the time window (that is, a_i), which means that it incurs timeout until service is possible. However, no vehicle can reach beyond the closing of a given time interval b_i. Vehicles must also leave the depot within the depot time window $[a_o, b_o]$ and must return before or at the moment $b_n + 1$. Assuming that the waiting time is allowed without cost, we can assume that $a_o = b_o = 0$; that is, all routes start at time 0 (Table 1).

Table 1 Object notation

Notation	Representation	Definition
V	$\{V_d\} \cup V_c$	Vertices of the graph V = $\{V_v$ {v0}, $\{V_1,\ldots, V_n\}$ ιVι = n+2
V_d	$\{V_0\}$	Depot
V_c	$\{V_1,\ldots,V_n\}$	Set of service points vertices
A	{(i, j): i, j $\in$ V}	Set of edges
K	{1, …, k}	Vehicle set
N	$V_i \in$ V, i $\in$	Number of vertices

Table 2 Description of symbols

i, j	Node Indexes, $i, j = 1, \ldots, n$
d_i	Customer demand i (node)
c	Vehicle capacity (maximum weight)
t_{ij}	Travel time between customer i y j
C_{ij}	Associated cost of travel between client i y, j
s_i	Customer Service Time i
x_i	X Coordinate (Length) of the customer i
y_i	Customer latitude i
$[a_i, b_i]$	Customer Service Time Window i
R	Maximum amount of garbage that can be collected
M	Large value positive constant
A	Set made up of the ordered pairs i, j
a	Window start time at a node
b	Time limit of the window in a node
C	Total service cost
K	Vehicle fleet

3.2 Notation

The objective of the WCVRPTW is to assign n customers to the k vehicles between the warehouse, the customers to meet the collection of customer demand to minimize the total distance, complying with the restrictions to build a WCVRPTW problem (appointment). The basic restrictions of the problem consist of each site is visited only once by a single vehicle, each route starts at the same depot, and all demands of all customers must be met. Besides, the vehicle must on time. A vehicle that arrives early will produce waiting time and a vehicle that arrives late will produce delay. All scheduled vehicles must return to the depot within the maximum operating time for each vehicle. Table 2 presents the corresponding notation applied in this research.

3.3 Model Constraints

- Time windows of stops and the depot,
- Vehicle capacity (i.e., volume, weight),
- Route capacity (i.e., the maximum number of lifts, volume, and weight that can be handled per vehicle per day),
- Routing time limit per vehicle,
- Disposal trips (when the capacity of the vehicle is full, the vehicle it must go to a disposal facility),
- Driver's lunch break.

3.4 Model Construction

In this chapter, the mathematical model of the waste collection and vehicle problem is presented, with total travel distance as the objective function.

3.4.1 Analysis of Objectives Function

To solve the problem, we can establish the VRPTW in the following mathematical form as Solomon had used [20], a MILP is a method used to archive an optimal result in a mathematical model in which all the mathematical functions are linear. The mathematical model consists of a linear function to optimize (minimize or maximize), linear constraints, and generally nonnegative variables. The parameters and variables used in the introduced MILP model are presented here followed by the mathematical model and the preprocessing method.

$$\min \sum_{(i,j)\in A}^{n} C_{ij} \sum_{k\in K}^{n} X_{ijk} \tag{1}$$

Equation (1) minimizes the total travel distance C_{ij} to be followed by each vehicle k from a point X_i to X_j (X_{ijk}). The total cost can be interpreted as the total travel route in distance or time units.

$$\sum_{i\in Vc}^{n} \sum_{k\in K}^{n} X_{ijk} = 1 \quad \forall j \in V_c \tag{2}$$

$$\sum_{j\in V}^{n} X_{0jk} = 1 \quad \forall k \in K \tag{3}$$

$$\sum_{i\in V}^{n} X_{i0'k} = 1 \quad \forall k \in K \tag{4}$$

$$\sum_{i\in V_c}^{n} x_{ijk} = \sum_{i\in V_c}^{n} x_{jik} \quad \forall j \in V_c, k \in K \tag{5}$$

$$\sum_{i\in V}^{n} x_{ik}, n+1, k = 1 \quad \forall k \in K \tag{6}$$

$$w_{ik} + s_i + t_{ij} - w_{jk} \leqslant \left(1 - X_{ijk}\right) M_v \quad \forall (i, j) \in A, k \in K \tag{7}$$

$$E \leqslant w_{ik} \leqslant L \quad \forall k \in K, i \in (0, n+1) \tag{8}$$

$$a_i \sum_{i \in V_c}^{n} X_{ijk} \leqslant W_{ik} \leqslant b_i \sum_{i \in V_c}^{n} X_{ijk} \quad \forall i \in V_c, k \in K \tag{9}$$

$$\sum_{i \in V_c}^{n} d_i \sum_{k \in K}^{n} d_{ijk} \leqslant C \quad \forall k \in K \tag{10}$$

$$d_{ik} + d_i \leqslant C + \left(1 - x_{ijk}\right)M \quad \forall i \in V_c \tag{11}$$

$$\sum_{i \in Vc}^{n} d_i \leqslant R \quad \forall i \in V_c \tag{12}$$

$$X_{ijk} \geqslant 0 \quad \forall i \in V_c, k \in K \tag{13}$$

$$X_{ijk} \in \{0, 1\} \qquad \forall (i, j) \in A, k \in K \tag{14}$$

As the mathematical model (see Eqs. (1)–(14)) shows, taken into consideration in the model, they can be described by operation and decision-making strategies. The model was intended to assist operational managers in deciding optimum strategic and tactical plans for operation according to the available real-time information on the generated waste amount and the available technological and logistic capacities.

4 Algorithm

4.1 Algorithm Design

Because of the NP-hard [18] it is hard to solve it using exact integer programming techniques, we chose to apply one of the well-known metaheuristic algorithms; simulated annealing with some neighborhood methods and an initial solution using a greedy algorithm.

4.2 Initial Solution Based on Greedy Algorithm

To effectively solve the VRPTW problem, it is crucial to generate a good initial feasible solution in which all the constraints are satisfied. First, we use a greedy algorithm to obtain a good initial solution. The initial solution is usually constructed with some cheaper insert heuristics. An initial solution is built by attempting to fully

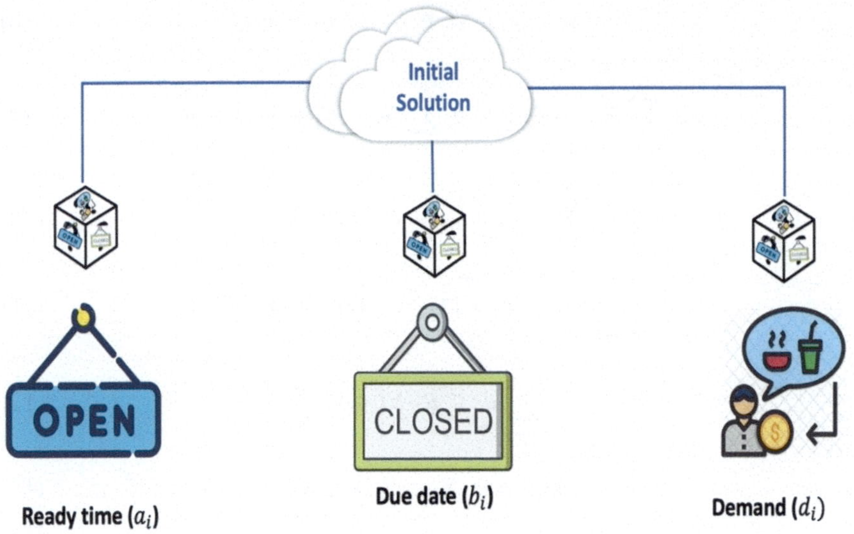

Fig. 4 Selection of the ordering metric of the greedy algorithm

utilize a vehicle during the day in a workday (to minimize the total travel distance). Once a vehicle can no longer be used, then we start a new vehicle route with a new vehicle. The nodes with their closest time window are added for the assignment, as soon as possible according to their time. However, we are adapting a classical greedy algorithm to a random algorithm, that is to say, that each time it generates an assignment of customers for each vehicle, this assignment is carried out randomly according to a metric that is, ordering the customers under a criterion. The first ordering criterion is to order the closest customers to attend by their time window a_i, the second criterion is to order the customers that their time window is the closest to closing b_i and finally order the customers with the highest demand d_i. This allows us to have variations of the initial solution to try to improve the initial solution. In Fig. 4 you can see an illustration of how the selection is made.

Once the ordering criterion of the customers has been selected, the customers are assigned to the vehicle, once the vehicle cannot serve the customer due to restrictions, it goes to the next vehicle and the criteria of order, this procedure is repeated until all clients are visited.

4.3 Simulated Annealing Algorithm

Simulated Annealing (SA) was first used to search the feasible solutions of a combinatorial optimization problem in 1983 [21]. It was inspired by an analogy between the physical annealing of solids and combinatorial problems. In the physical process, a solid is first melted at high temperature and then cooled very slowly to obtain a perfect lattice structure corresponding to a minimum energy state. SA is a local search strategy, but it tries to avoid local minima by accepting worse solutions with some probability. SA includes components during the search procedure, including initial temperature, cooling strategy, objective function, temperature-controlled acceptance rule, and stopping criterion. Simulated annealing is structured as three main methods: the first is to generate a random initial solution; next, permute the current solution by using special neighborhood operators and a probability function to accept a new solution, this avoids the algorithm to scape at a local minimum during the search of the global optimal. Good tuning of these parameters may ensure an optimal search and an efficient process.

The flow diagram is presented in Fig. 5. The temperature of the algorithm t is started with the value of the initial temperature. *Niter* variable determines the number of iterations. This property enables the simulated annealing to archive the best solution in the local area and the exploitation of the search. SA usually use one single operator, but in our SA we permute the solution randomly chosen using three operators:

Relocation: solutions are obtained by relocating a sub-path from one route to another [22].

2-opt: finds an initial random route over the set of clients [23]. Eliminating two arches, reversing one of the two resulting paths, and reconnecting them until there are no more improvements that can be made improve this route.

Swap: This route is generated by changing the position / sequence of a customer with another route [24].

These operators are determined by using several preliminary experiments, by choosing the operators with the best performance to solve the VRPTW.

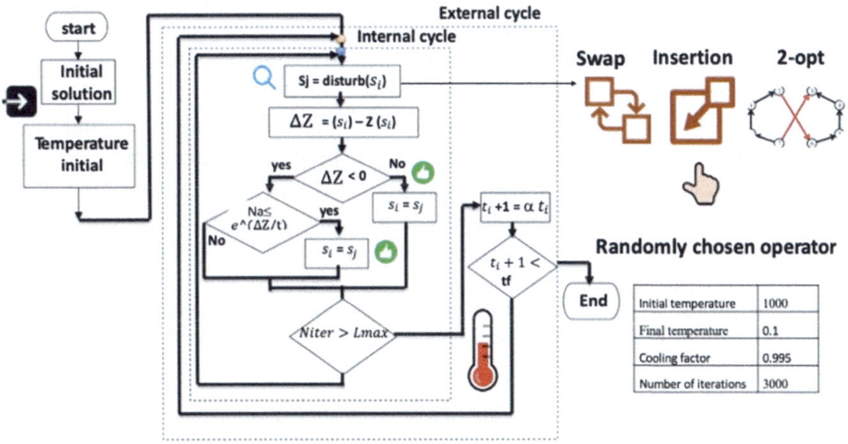

Fig. 5 Simulated annealing algorithm proposed diagram

5 Conclusion

The added value of the chapter is the description of the mathematical model of the WCVRPTW, which makes the description of the impact of waste management, using optimization algorithms to make it possible. The scientific contribution of this chapter for researchers in this field is the mathematical modeling and optimization of waste collection processes based on a simulated annealing algorithm and a combination of neighborhood operators. More generally, this chapter focused on the mathematical description of the framework of waste management systems, including households as waste sources, treatment sites, and collection processes. The role of waste management systems become more and more important because it is a priority to move from reliance on waste dumps that offer no environmental protection, to waste management systems that retain useful resources within the economy and its operation is green and does not endanger the public health through environmental pollution. The results can be generalized because the model can be applied to different fields of waste management systems to manage commercial, industrial, residential, and medical waste.

6 Future Research

However, there are also limitations of the study and the described model that shows directions for further research. Our model takes care of a fleet with different trucks because the algorithm makes it possible to set the loading capacity of each truck considering the time windows of each client. Traditional formal models are generally focusing on cost-based optimization of traditional municipal solid waste collection: the effective service network design using optimization algorithms. However, the determining factors of a successful implementation of waste management system include the use of Industry 4.0 technologies and IoT solutions, but there are other influencing factors like smart bins that can detect the level of waste in the bin and then calculate the volume depending on the bin's type (edge computing) and travel time prediction system based on real-time information on the arrival time of waste collection vehicle. Conventional waste generation prediction models including correlation and regression models generally used demographic and socioeconomic factors. Various independent variables were considered in most of the prediction models.

Acknowledgements We want to thank CONACYT-Mexico for being granted a scholarship during (2018–2020) for the support of this research.

References

1. ONU Hábitat (2015) El 2015 registra el mayor número de días contaminados en el deporte Nacional de Movilidad Urbana en México 2014–2015. https://70.35.196.242/onuhabitatmexico/Reporte-Nacional-de-Movilidad-Urbana-en-Mexico-2014-2015.pdf
2. Sedesol (2017) Manual Técnico Sobre Generación, Recolección y Transferencia de Residuos Sólidos Municipales, Sedesol, México City, México, 2017. Accessed on: Nov. 02, 2018. [Online]. Available: https://www.inapam.gob.mx/work/models/SEDESOL/Resource/1592/1/images/Manual TecnicosobreGeneracionRecoleccion.pdf
3. Buhrkal K, Larsen A, Ropke S (2012) The waste collection vehicle routing problem with time windows in a city logistics context. Procedia-Social Behav Sci 39:241–254
4. Canen AG, Scott LG (1995) Bridging theory and practice in VRP. J Oper Res Soc 46(1):1–8
5. Kulcar T (1996) Optimizing solid waste collection in Brussels. Eur J Oper Res 90(1):71–77
6. Tung DV, Pinnoi A (2000) Vehicle routing–scheduling for waste collection in Hanoi. Eur J Oper Res 125(3):449–468
7. Teixeira J, Antunes AP, de Sousa JP (2004) Recyclable waste collection planning—a case study. Eur J Oper Res 158(3):543–554
8. Kim BI, Kim S, Sahoo S (2006) Waste collection vehicle routing problem with time windows. Comput Oper Res 33:3624–3642
9. Solomon MM (1983) Vehicle routing and scheduling with time windows constraints: Models and algorithms. Ph.D. Dissertation, Department of Decision Sciences
10. Babaee Tirkolaee E, Alinaghian M, Bakhshi Sasi M, Seyyed Esfahani MM (2016) Solving a robust capacitated arc routing problem using a hybrid simulated annealing algorithm: a waste collection application. J Ind Eng Manag Stud 3(1):61–76
11. Mofid-Nakhaee E, Barzinpour F (2019) A multi-compartment capacitated arc routing problem with intermediate facilities for solid waste collection using hybrid adaptive large neighborhood search and whale algorithm. Waste Manage Res 37(1):38–47
12. Delgado-Antequera L, Caballero R, Sánchez-Oro J, Colmenar JM, Martí R (2020) Iterated greedy with variable neighborhood search for a Multiobjective waste collection problem. Expert Syst Appl 145:113101
13. Lacomme P, Prins C, Ramdane-Chérif W (2001) A genetic algorithm for the capacitated arc routing problem and its extensions. In: Workshops on Applications of Evolutionary Computation, Springer, Berlin, Heidelberg, pp 473–483
14. Akhtar M, Hannan MA, Basri H (2015) Particle swarm optimization modeling for solid waste collection problem with constraints. In: 2015 IEEE 3rdInternational Conference on Smart Instrumentation, Measurement and Applications (ICSIMA), IEEE, pp 1–4
15. Toth P, Vigo D (2002) Branch-and-bound algorithms for the capacitated VRP. In: The vehicle routing problem. Society for Industrial and Applied Mathematics, pp 29–51
16. Naddef D, Rinaldi G (2002) Branch-and-cut algorithms for the capacitated VRP. In: The vehicle routing problem, Society for Industrial and Applied Mathematics, pp 53–84
17. Laporte G, Nobert Y (1987) Exact algorithms for the vehicle routing problem. In: North-Holland Mathematics Studies, vol 132, North-Holland, pp 147–184
18. Lageweg BJ, Lawler EL, Lenstra JK, Rinnooy Kan AHG (1981) Computer aided complexity classification of deterministic scheduling problems. Stichting Mathematisch Centrum. Mathematische Besliskunde, (BW 138/81)
19. Desrosiers J, Dumas Y, Solomon MM, Soumis F (1995) Time constrained routing and scheduling. Handbooks Oper Res Manag Sci 8:35–139
20. Hillier FS (2012) Introduction to operations research. Tata McGraw-Hill Education
21. Kirkpatrick S, Gelatt CD, Vecchi MP (1983) Optimization by simulated annealing. Science 220(4598):671–680
22. Balseiro SR, Loiseau I, Ramonet J (2011) An ant colony algorithm hybridized with insertion heuristics for the time dependent vehicle routing problem with time windows. Comput Oper Res 38(6):954–966

23. Suthikarnnarunai N (2008) A sweep algorithm for the mix fleet vehicle routing problem. In: Proceedings of the International Multi Conference of Engineers and Computer Scientists, vol 2, pp 19–21
24. Meihua W, Xuhong T, Shumin W (2011) Hybrid ant colony optimization algorithm for two-echelon vehicle routing problem. Procedia Eng 15:3361–3365

Distribution of Merchandise Through Clarke and Wright Heuristic and Mathematical Model: Case Study

Martha Felicitas Quiroz Flores and Lourdes Loza-Hernandez

Abstract Vehicle routing problems (VRP) have various applications in the industry, they are important for the proper functioning of logistics systems and they are often treated as problems of economic interest because they seek to generate solutions that minimize aspects such as: the size of the routes, the number of vehicles, transfer times, etc., which have a financial impact on companies. In the case study company, there is currently no adequate management of resources, since the way in which transport routes are assigned according to the experience of the workers, who are unaware of the requirements of the delivery points, causing late deliveries or incomplete deliveries. In order to meet the requirements of the delivery points, it is necessary to propose a distribution plan that guarantees the delivery of the requirements at the lowest cost. The Clarke and Wright heuristic and VRP mathematical model are applied to the merchandise distribution network to decrease the company's transportation costs and to propose a solution to vehicle routing problems, in addition to present to the company's personnel a decision-making tool for a better use of the vehicle fleet.

Keywords Vehicle routing problem · Clarke and Wright algorithm · Merchandise distribution · Mathematical model

1 Introduction

Today the industry is made up of a series of logistics networks whose optimization plays an important role in the sustainability of companies. Over the years, various methods have been proposed to solve vehicle routing problems (Vehicle Routing Problem or VRP) which allow the design of delivery routes at a lower cost. This case study aims to find the routes that minimize the cost of transporting goods from a central warehouse to a set of delivery points, while respecting the vehicle capacity restrictions. The model is known as a Capacity Vehicle Routing Problem (CVRP),

---

M. F. Quiroz Flores · L. Loza-Hernandez (✉)
Faculty of Engineering, Autonomous University of the State of Mexico, Ciudad Universitaria, Toluca, Mexico
e-mail: llozahe@gmail.com

© The Author(s), under exclusive license to Springer Nature Switzerland AG 2021
A. Ochoa-Zezzatti et al. (eds.), *Technological and Industrial Applications Associated with Intelligent Logistics*, Lecture Notes in Intelligent Transportation and Infrastructure,
https://doi.org/10.1007/978-3-030-68655-0_7

since it has a vehicle fleet with limited capacity, in addition to the characteristics of a routing problem (Vehicle Routing Problem or VRP). To obtain a solution, the use of the Clarke and Wright heuristic or also known as the Savings Method is implemented, so called because in its execution distance saving matrices are generated to obtain better solutions in each step of its execution. Also a mathematical model is applied to obtain the optimal solution. In this way, routes are generated to improve the use of the company's available resources, minimizing the distance to be traveled by each vehicle and the costs of using the vehicle that is used on each route.

The document contains four sections, the following is section two, which shows the methodology to be used in the development of the work; section three presents the characteristics of the case study and methods applied to solve the problem, and finally section four shows the conclusions according to the results obtained from Clarke and Wright heuristic and mathematical model.

2 Methodology

2.1 Vehicle Routing Problem (VRP)

A. The VRP is one of the most famous combinatorial optimization problems. Dantzig and Ramser introduced the VRP based on the gasoline delivery problem, where they propose a mathematical programming formulation and an algorithm for VRP [1]. Since then, the VRP has spawned a well-known and important field in Operations Research.

Generally, the VRP is applied in the transport of goods between clients and warehouses. All vehicles must start and return to the depot. Each client can only be visited once. The objective of the VRP is to generate a minimum set of routes that satisfy the demand of all customers. The VRP is easy to assert and understand, while it is difficult to solve in the real world, especially in combination with other practical constraints. In the VRP domain, there are numerous problems with specific constraints [2].

The time and computational effort required to solve this problem increases exponentially with respect to the size of the problem, due to the set of constraints built by the number of nodes to be visited by vehicles. For this type of problem, it is often desirable to obtain approximate and good solutions as well as quickly to become useful in decision making [3].

Vehicle routing problems are so varied in the industry due to the wide range of situations, multiple constraints, and the complexity of problems that each industry has. For these problems there are variants that characterize them for their application, however, in general, their main elements are three for a VRP:

- Customers.
- Vehicles.
- Deposits.

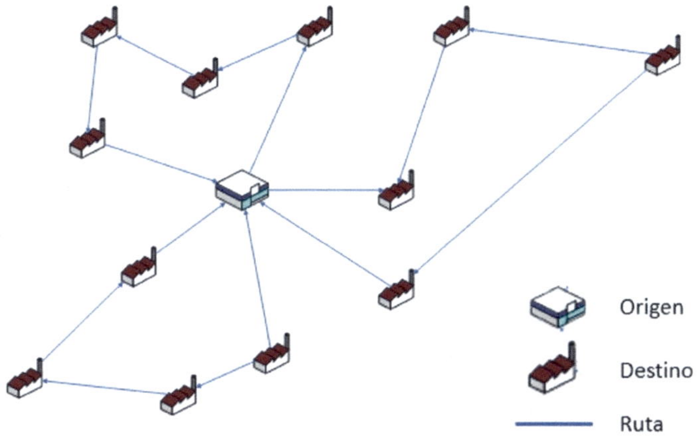

Fig. 1 Distribution network

When there are different delivery points, exist different characteristics of customers, warehouses and vehicles, and different restrictions, then as a result of those differences, there are variants of the VRP model. The assignment of routes is always different because of the VRP is dynamic derived, also when decisions are made.

Figure 1 is an example of the desired result of a distribution network.

The VRP is known to be a complex computational problem where different exact and heuristic algorithms [4] have been proposed to find feasible solutions. The algorithms used to solve the vehicle routing problem are varied in terms of procedures, resolution times and focus with a view to improving the use of resources, as well as the type of algorithm used.

It cannot be guaranteed that an exact algorithm will find optimal paths within a reasonable computation time when the number of destinations is large, this due to the complexity of the problem, for this reason the use of heuristic and meta-heuristic algorithms is recommended for solving this problem. Considering that the real cases of the VRP often exceed hundreds of vertices that the solutions must often be determined quickly, most of the algorithms used in practice are heuristic [5].

The basic version of the VRP is the Capacitated VRP (CVRP) [6], which due to the characteristics of the company, will be used in this case study.

In CVRP all clients are delivered through deterministic demands that are known in advance. The vehicle fleet originates from a central depot, and the restriction is given by the capacity of the vehicles. The goal is to minimize the total cost of serving all customers.

The mathematic model, according to [6], for the CVRP is a two-index vehicle flow formulation that uses $O(n^2)$ binary variables x to indicate if a vehicle traverses an arc in the optimal solution. In other words, variable xij takes value 1 if arc $(i, j) \in A$ belongs to the optimal solution and takes value 0 otherwise.

$$\text{Min} \sum_{i \epsilon V} \sum_{j \epsilon V} c_{ij} x_{ij} \tag{1}$$

Subject to:

$$\sum_{i \epsilon V} x_{ij} = 1 \quad \forall j \epsilon V \setminus \{0\} \tag{2}$$

$$\sum_{j \epsilon V} x_{ij} = 1 \quad \forall i \epsilon V \setminus \{0\} \tag{3}$$

$$\sum_{i \epsilon V} x_{io} = K, \tag{4}$$

$$\sum_{j \epsilon V} x_{oj} = K, \tag{5}$$

$$\sum_{i \epsilon S} \sum_{j \epsilon S} x_{ij} \geq r(S), \quad S \subseteq V \setminus \{0\}, S \neq 0 \tag{6}$$

$$x_{ij} \, \epsilon \{0, 1\} \quad i, j \, \epsilon V \tag{7}$$

Where:

- $V = \{v0, v1, \ldots, vn\}$ is the set of vertex of the graph, where $v0$ is the depot and vertices $i = 1,\ldots,n$ correspond to the customers.
- cij is associated with each arc $(i,j) \subseteq A$, which is the travel between customer vi and vj.
- K is the number of vehicles with C capacity needed for translate the demand.
- S is the subset of vertex where $S \subseteq V \setminus \{0\}$.
- $r(S)$, is the minimum number of vehicle needed for serve all the customers.

The *indegree* and *outdegree* constraints (2) and (3) impose that exactly one arc enters and leaves each vertex associated with a customer, respectively. Analogously, constraints (4) and (5) impose the degree requirements for the depot vertex. Note that one arbitrary constraint among the 2|V| constraints (2)–(5) is actually implied by the remaining 2|V| - 1 ones; hence it can be removed.

The so-called *capacity-cut* constraints (CCCs) of (6) impose both the connectivity of the solution and the vehicle capacity requirements. In fact, they stipulate that each cut $(V \setminus S, S)$ defined by a customer set S is crossed by a number of arcs not smaller than $r(S)$ (minimum number of vehicles needed to serve set S).

2.2 Clarke and Wright Heuristic

Clarke and Wright Heuristic is the most significant classical heuristic for the VRP. This heuristic is a simple procedure that performs a limited scan of the search space and gives an acceptable quality solution in moderate computation time. The heuristic is applied in the distribution network of the case study, in order to find feasible routes and have a minimum total cost.

Next, the procedure of the Clarke and Wright algorithm is described, as well as the variables, constants and indices used during its execution [7].

i: node named as initial customer $\forall\, i = 1, 2, 3, \ldots, n$

j: node named as end customer. $\forall\, j = 1, 2, 3, \ldots, n;\ i \neq j$

n: number of nodes.

O: start node and end node.

Dij: distance matrix between node i and node j.

Sij: savings matrix between node i and node j.

Ri: path to which node i belongs.

Rj: path to which node$j$ belongs.

(1) Steps for creating the Clarke and Wright algorithm

Step 1

- Create the savings matrix S, using the expression:

  $Sij = D0i + D0j\text{-}Dij;\ \forall\, i, j = 1, 2, \ldots, n \wedge i \neq j.$
- Create n routes of the form $(0, i, 0)$, $\forall\, i = 1, 2, \ldots, n$

  Create a matrix $S' = S$.

Step 2

- Choose the maximum value of the Sij matrix;

  Max Sij = Si* j*
- If i^* is the last customer visited on route Ri^*j^*, it is the first customer visited on route Rj^* and all restrictions are met:

  Join the Ri^* route with the Rj^* route.

  Assign S $\leftarrow$S′.

  Remove the Si^*j^* arcs already used in the created routes.

- Else case, eliminate Si^*j^* from matrix S.

Step 3

- If $S \neq 0$, go to step 2; otherwise END.

3 Case Study

This work is developed in a retail company, which has a distribution center from where the merchandise is sent to the points of sale. The allocation of routes for the delivery of merchandise is in accordance with the experience of the personnel who carry out the activity, without following quantitative methods that allow an adequate allocation of resources, and that also consider the capacity of the vehicles, distances and travel times.

The personnel assigned to this activity do not apply a method for planning vehicle routes, so they do not have a decision-making tool that allows them to visualize where to start the delivery of merchandise and where to end the defined route; this activity is carried out empirically, which implies neglecting the points of sale that have a higher demand or causing incomplete deliveries, causing an irregular service level and increasing fuel costs.

In order to meet the demand of the delivery points, it is necessary to propose a distribution plan that guarantees deliveries of orders on time.

The application of quantitative methods as tools that allow the definition of routes for the distribution of merchandise allows to reorganize the current form of operation of companies, generating savings in costs, distribution time, use of labor and transport.

The company requires an analysis of the vehicle routing problem to modify its distribution network and obtain adequate planning of its resources that allows minimizing transport costs, considering the demand of the points of sale and respecting the capacity of the vehicles.

In case study with information obtained from personnel in the area of distribution, 29 destinations are analyzed, considering that there is a leasing fleet of 6 vehicles with a capacity of 2 tons, to satisfy the demand of each of the company's points of sale.

As a result of this analysis, the number of routes and the order of visits to the points of sale are determined, which improves the use of resources.

Therefore, the necessary data to apply the heuristic are:

- Tickets: distance between each point of sale and the distribution center (distance matrix).
- Departures: defined routes with the best savings.

Additional data to identify the best routes and satisfy the constraints of the case study are:

- Demand from each of the points of sale.
- Fuel costs.
- Size of the fleet of vehicles to be used.
- Vehicle capacity.

3.1 Clarke and Wright Algorithm and CVRP Model Parameters

The parameters used to solve the routing problem using the Clarke and Wright algorithm will be described below:

- Objective: minimize transportation costs and design of new delivery routes.
- Known fleet size: 6 vehicles with a capacity of 2 tons each.
- Capacity per vehicle: an average of 125 boxes of merchandise can be transported.
- Performance: 1 L of Diesel approximately 9.5 km.
- A warehouse: distribution center (DC).
- Numbers of delivery points: 29, located in the central area of the country, identified through numbers from 1 to 29.
- Demands: the merchandise is sent by consolidated boxes, according to the type of product, which can be clothing, footwear or accessories. There is an average estimate of daily shipments to each point of sale. The demand is known one day before the merchandise is delivered at the points of sale.
- Average distance: it was obtained with the Google Maps application [8].

Table 1 shows the demand to be delivered at each point:

3.2 Development of the Methodology

Calculation of distances The Google Maps software [9] is used to calculate the distances between each and every pair of points, as well as to check the distance and the estimated time to get from one point to another. These data are part of the basic information that algorithm needs to be able to propose the routes that will later be assigned to the vehicles as solution. The distances between the origin and each destination are shown in Table 2.

Distance matrix: Table 3 shows an example of the way in which the distances between the first seven nodes were calculated. Distances are considered symmetric only as a sample for company staff, as proof that the application of quantitative methods improves the performance of available resources.

To obtain the results of the algorithm, the calculation was made for the 29 destination nodes and the origin. The calculations were performed on a personal computer with an Intel (R) Core (TM) i5-7200U processor @ 2.50 GHz 2.7 GHz 8.00 GB.

The Microsoft Excel version 2007 spreadsheet is used to facilitate the handling of the heuristic by company personnel dedicated to this activity, since the purchase of specialized software in vehicle routing is not considered within the budget of the company.

Fuel consumption to determine travel costs: Based on the characteristics of the fleet of vehicles, their engines use Diesel. The diesel consumption (lt / km: $ 21.03 / lt) of the vehicle was provided by the company according to historical performance

Table 1 Demand per day

Destination	Demand (Boxes per day)
1	11
2	15
3	10
4	10
5	3
6	7
7	6
8	11
9	11
10	6
11	12
12	10
13	20
14	12
15	5
16	19
17	9
18	4
19	9
20	7
21	8
22	7
23	8
24	24
25	5
26	6
27	11
28	9
29	10

data.

$$Fuel\,cost = Fuel\,consumption\,(lt/km) * Price\,of\,fuel\,(lt) \tag{8}$$

It was determined that for a transport unit the performance of a one kilometer distance is equivalent to a consumption of 0.16 L of Diesel. The price of Diesel was determined according to the price suggested by the reference [10], for the month of November 2019, the price per liter was $ 21.03 in Mexico. Then using the data mentioned in the fuel cost expression we have:

Table 2 Distance between origin and destination

Origen	Destination	Distance (Km)
CD	1	73.6
CD	2	60.3
CD	3	66.9
CD	4	75.7
CD	5	60.5
CD	6	14.3
CD	7	75.5
CD	8	10.8
CD	9	91.5
CD	10	58.6
CD	11	65.2
CD	12	11.3
CD	13	58.7
CD	14	81.0
CD	15	68.2
CD	16	68.4
CD	17	55.7
CD	18	43.7
CD	19	83.6
CD	20	109.2
CD	21	59.1
CD	22	77.5
CD	23	69.6
CD	24	61.8
CD	25	60.1
CD	26	61.6
CD	27	39.5
CD	28	66.1
CD	29	67.4

CD: Distribution center of the company

$$Fuel\,cost = 0.16\,(lt/km) * 21.03\,(/lt) = 3.36\,(/km) \tag{9}$$

Table 3 Distance matrix (km)

i,j	0	1	2	3	4	5	6	7	…
0		73.6	60.3	66.9	75.7	60.5	14.3	75.5	
1	73.6		21.4	42.1	23.5	24.2	76.1	33.9	
2	60.3	21.4		23.9	29.7	17.4	74.9	27.5	
3	66.9	42.1	23.9		47.4	30.3	72.5	45.9	
4	75.7	23.5	29.7	47.4		12.2	72.6	10.6	
5	60.5	24.2	17.4	30.3	12.2		71.0	9.2	
6	14.3	76.1	74.9	72.5	72.6	71.0		70.1	
7	75.5	33.9	27.5	45.9	10.6	9.2	70.1		
.									
.									
.									

3.3 Application of the Clarke and Wright Algorithm and Analysis of Results

From the distance matrix, the first 29 routes are read directly, all leaving the Distribution Center and visiting a single destination. Table 4 shows the distances and costs per node.

This first solution has a distance of 3,616 km and requires 16 vehicles. From Table 4 of the initial solution, only the distance traveled is taken into account, and the fuel cost per kilometer traveled is reserved, this last value is multiplied by each distance of the nodes of the initial solution, resulting in the cost for each route whose total impact is $ 12,166.

Symmetric cost matrix

To determine the cost matrix, the fuel cost scalar is used, which has the value of 3.36 ($ / km), this value is multiplied for each of the destinations in the symmetric distance matrix, resulting in the following matrix. Table 5 shows the result of the calculations for the first eight nodes.

Potential cost matrix savings

The way in which the computation for arc of the route is performed is exemplified through the process performed for arc 1–2. The savings calculation is as follows:

$S_{ij} = C_{0i} + C_{0j} - C_{ij}$

$i = 1\, j = 2$

$C_{0i} = C_{01} = 247\, C_(0j) = C_{02} = 203C_{ij} = C_{12} = 72$

$S_{ij} = S_{12} = 247 + 203 - 72$

$S_{ij} = S_{12} = 378$

This calculation is carried out for the entire matrix and the following data is obtained, Table 6 shows the savings for the first nine nodes:

Table 4 Distances and cost per node

Route	Nodes	Distance (Km)	Cost ($)*
1	0-1-0	147	$494.96
2	0-2-0	121	$405.46
3	0-3-0	134	$450.21
4	0-4-0	151	$509.09
5	0-5-0	121	$406.92
6	0-6-0	29	$95.90
7	0-7-0	151	$508.08
8	0-8-0	22	$72.34
9	0-9-0	183	$615.98
10	0-10-0	117	$394.13
11	0-11-0	130	$438.55
12	0-12-0	23	$75.71
13	0-13-0	117	$395.03
14	0-14-0	162	$544.87
15	0-15-0	136	$459.18
16	0-16-0	137	$460.30
17	0-17-0	111	$374.84
18	0-18-0	87	$294.08
19	0-19-0	167	$562.26
20	0-20-0	218	$735.10
21	0-21-0	144	$485.20
22	0-22-0	155	$521.21
23	0-23-0	139	$468.04
24	0-24-0	124	$415.55
25	0-25-0	120	$404.45
26	0-26-0	123	$414.32
27	0-27-0	79	$265.82
28	0-28-0	132	$445.05
29	0-29-0	135	$453.35

*$: Mexican Pesos

Saving Cost

Order the savings from highest to lowest (Table 7):

Routes generated by the Clarke and Wright algorithm

The first arc that is analyzed is 9–22, which generates the route: 0–9-22–0 with an accumulated cost of: 308 + 38 + 261 = $ 607, and with a demand to cover of: 11

Table 5 Symmetrical cost matrix ($ / km)

ij	0	1	2	3	4	5	6	7	8	…
0		247	203	225	255	203	48	254	36	
1	247		72	142	79	81	256	114	255	
2	203	72		80	100	59	252	93	237	
3	225	142	80		159	102	244	154	230	
4	255	79	100	159		41	244	36	243	
5	203	81	59	102	41		239	31	199	
6	48	256	252	244	244	239		236	19	
7	254	114	93	154	36	31	236		195	
8	36	255	237	230	243	199	19	195		
.										
.										
.										

Table 6 Cost matrix

S	1	2	3	4	5	6	7	8	9	…
1	378	331	423	370	39	387	28	461	359	
2	133	51	95	76	94	90	60	121	127	
3	21	37	88	19	87	37	31	110	92	
4	220	159	278	147	178	178	59	126	206	
5	46	46	85	43	95	108	105	70	10	
6	34	420	60	61	90	335	26	0	51	
7	59	166	447	308	131	227	484	446	430	
8	60	36	80	293	31	4	66	18	72	
9	378	331	423	370	39	387	28	461	359	
.										
.										
.										

$+ 7 = 18$ boxes. This route is feasible, as it respects the constraints of the problem and improves the previous solution.

Table 8 contains the routes that were generated using the savings algorithm. From the 29 routes in the first solution only 3 routes were generated with this algorithm, which comply with the capacity restrictions.

New distribution network with Clarke and Wright algorithm

The final solution of the Clarke and Wright method consists of 3 routes that accumulate a total travel cost of $ 2,786, which improves on the first solution that maintained a total travel cost of $ 12,166, with a total traveled distance of 539.6 km, which can be

Table 7 Saving cost

Arc	Saving (MXN)
S9–22	$540
S1–20	$523
S9–20	$522
S9–21	$487
S7–14	$484
.	.
.	.
.	.
S11–16	$1
S12–21	$1
S17–23	$0

Table 8 Routes generated by the algorithm

Route	Nodes	Capacity used	Total cost (MXN)
1	0-9-22-16-24-29-1-20-21-7-14-15-23-19-10-0	100%	$1,503
2	0-6-8-2-29-13-26-4-5-17-25-27-12-0	94%	$744
3	0-11-28-3-18-0	28%	$539

traveled within the working hours of the merchandise distribution staff. In addition, the number of vehicles decreases to 3 units, respecting the restrictions of not carrying more than 125 boxes on all routes.

Based on the benefits obtained, the alternative of implementing the Clarke and Wright algorithm in the merchandise distribution activity is proposed to the company, since it projects the monthly savings for the assignment of routes, in addition, considering the cost savings of leasing, if only 3 vehicles are used instead of the 6 that are in the contract established by the company.

3.4 Application of the Mathematical Model and Analysis of Results

Based on parameters described in Sect. 3.1, the mathematical model is formulated as following:

$$\sum_{i \epsilon V} \sum_{j \epsilon V} c_{ij} x_{ij} \quad \forall i, \{1, 2, \ldots, 29\} \wedge \forall j, \{1, 2, \ldots\} \epsilon\, V \tag{10}$$

Subject to:

$$\sum_{i=1}^{29} x_{ij} = 1 \quad \forall j\{1, 2, \ldots, 29\}\epsilon\ V \setminus \{0\} \tag{11}$$

$$\sum_{j=1}^{29} x_{ij} = 1 \quad \forall i\{1, 2, \ldots, 29\}\epsilon\ V \setminus \{0\} \tag{12}$$

$$\sum_{i=1}^{29} x_{i0} = Q_K \quad \forall K\{1, 2, \ldots, 6\} \tag{13}$$

$$\sum_{j=1}^{29} x_{0j} = Q_K \quad \forall K\{1, 2, \ldots, 6\} \tag{14}$$

$$\sum_{i\epsilon S}\sum_{j\epsilon S} x_{ij} \geq r(S), \quad \forall S \subseteq V \setminus \{0\}, S \neq 0 \tag{15}$$

$$x_{ij}\ \epsilon\{0, 1\} \quad \forall i, j\ \epsilon\ V \tag{16}$$

To solve the CVRP mathematical model, the ORP Tools utility was used in the Python 3.7.2 software, modifying the parameters according to the case study. The calculations were performed on a personal computer with an Intel (R) Core (TM) i5-2520 M processor @ 2.50 GHz 2.5 GHz 4.00 GB.

Figure 2 shows the results obtained in Python:

Table 9 contains the routes that were generated using the mathematic model; 3 routes were generated with this model, which comply with the capacity restrictions.

From the table, total costs show that even the capacity used is not until 100% these are less than the total cost obtained from the Clarke and Wright heuristic.

4 Conclusions

The VRP is a highly combinatorial problem so the search for an optimal solution becomes a challenging problem. To find an optimal solution requires algorithms that consume a high computation time. VRP is one of the problems in which an increase in the number of nodes supposes an exponential increase in execution time, and quickly surpasses the calculation capabilities of the most powerful computers, so using an algorithm that sacrifices the optimal solution for a shorter execution time, as the case of the Clarke and Wright algorithm, reduces the time and gives us an acceptable solution, however the mathematical models obtained the optimal solution to the problem.

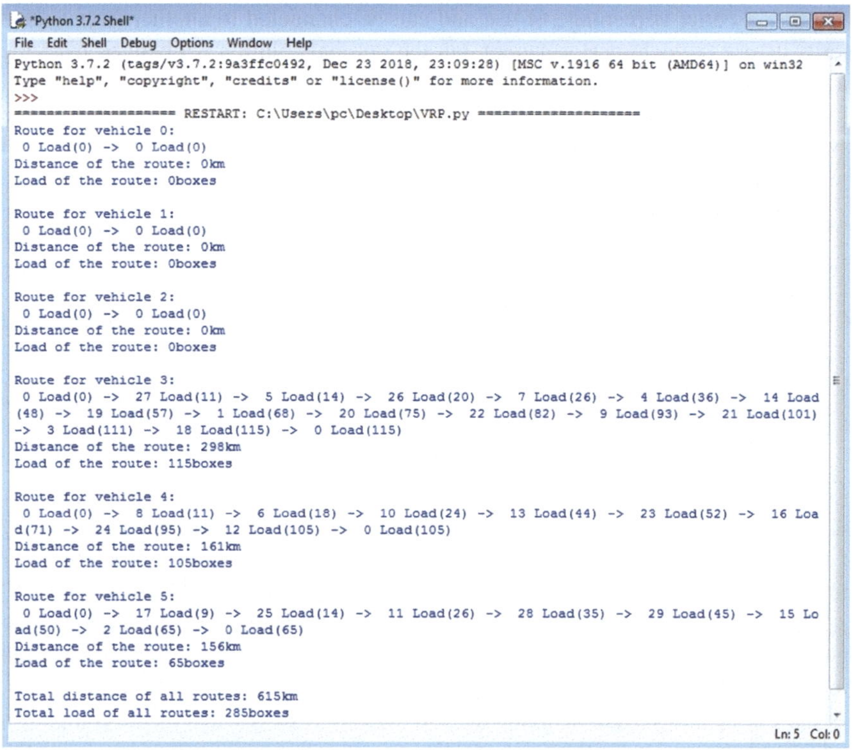

Fig. 2 Results of mathematical model

Table 9 Routes generated by the mathematical model

Route	Nodes	Capacity used	Total cost (MXN)
1	0-27-5-26-7-4-14-19-1-20-22-9-21-3-18-0	92%	$1001
2	0-8-6-10-13-23-16-24-12-0	84%	$541
3	0-17-25-11-28-29-15-2-0	52%	$524

It is important to clarify that the data analyzed belongs to an average daily demand, so the result obtained from the application of the Clarke-Wright heuristic will not always be the same because the demand is not constant. Based on the analysis of results, it is shown that the application of the heuristic allows an improvement in the use of resources and supports the decision-making of the corresponding personnel in the company, since, although the empirical practice prevails in some companies. It is necessary to show through practical applications that the tools offered by the quantitative methods of the Supply Chain allow better decision making.

Applying the Clarke and Wright heuristic improves the use of the company's resources, such as: workforce, vehicles in use and supplies, from now on the vehicles will have a better percentage of use and can be channeled to another activity that merits its use (vehicle performance), but the optimal solution is obtained from the mathematical model.

In relation to the results obtained, it is concluded that: with the application of the savings algorithm and mathematical model presented by the distribution problem of the company, it is achieved:

- Comply with the proposed objective: improve the distribution network,
- Reduce the number of vehicles on the delivery route,
- Make the most of the vehicle fleet,
- Generate 3 feasible routes according to the savings algorithm and the mathematical model,
- Application of quantitative methods as tools that allow the definition of routes for the distribution of goods,
- Reorganize the current way of operating companies, generating cost savings.

The results show that the total costs obtained from mathematical model is better than the total cost got obtained from the Clarke and Wright heuristic. The capacity of the vehicles has better balance with the mathematical model solution, however the Clarke and Wright heuristic is easier to handling for people in the company.

The development of this work serves as a tool for decision makers in the merchandise distribution area of the case study company. The execution of the heuristic and the mathematical model were developed under the use of information technologies. For the Clarke and Wright heuristic the software came be reached by the company's budget and easy to use for the personnel that executes this activity, the above with the aim of motivating the use of new technologies and quantitative methods to facilitate the daily activities and improve the use of resources while meeting customer expectations (points of sale) giving them a better service. For the mathematical model's software is a special quantitative tool that is not cheap for the company, also its handling has to be for qualify people.

References

1. Dantzig GB, Ramser JH (1959) The truck dispatching problem. Manage Sci 6(1):80–91
2. Li X (2015) Capacitated vehicle routing problem with time windows: a case study on pickup of dietary products in nonprofit organization, Proquest Dissertations and Theses, p 87
3. Lüer A et al (2009) El Problema de Rutas de Vehículos: Extensiones y Métodos de Resolución estado del arte, Ceur Workshop Proceedings, 558
4. Nazari M, Oroojlooy A, Takáč M, Snyder LV (2018) Reinforcement Learning for Solving the Vehicle Routing Problem. Adv Neural Inf Proc Syst 9839–9849
5. laporte G (2009) Fifty Years of Vehicle Routing. Transp Sci 43(4):408–416
6. Toth P, Vigo D (2014) Vehicle Routing. Soc Ind Appl Math Philadelphia 1–69
7. Clarke G, Wright JR (1964) Scheduling of Vehicle Routing Problem from a Central Depot to a Number of Delivery Points. Oper Res 12:568–581 https://doi.org/10.1287/OPRE.12.4.568
8. Sanchez F (2015) Aplicación Del Modelo VRP(Vehicle Routing Problem) para la optimización de una red de distribución. Universidad Nacional Autónoma de México

9. Google (2019) Mapa de México en Google Maps. Copyright 2019 INEGI Google. https://www.google.com.mx/maps. Consultado 10 de Noviembre De 2019
10. Intermodal México (2019) Histórico de Precios de Diesel, Copyrights © 2016. https://www.intermodalmexico.com.mx/portal/ajustecombustible/historico. Consultado 10 de Noviembre de 2019

Transport and Movements of Vehicles in a Smart City

Rethinking the Effects of Fatal Falls on an Italian Scooter in a Smart City: An Approach from Ergonomics and an Ideal and Optimal Helmet: Conditioned Helmet for Improved Security on the Road in a Smart City

Alberto Ochoa-Zezzatti, Jose Diaz, José Mejia, Israel Soto-Marrufo, Yesenia Quezada, Liliana Avelar, Juan Hernández, and Emmanuel De León-Evans

Abstract Motorcycle usage varies in all countries for a great variety of factors, while these tend to be a more economic mean of transportation, they also lack the security that cars provide to a certain extent. This means that the necessity of developing a conditioned helmet rises, especially in less developed countries. Traffic accidents are the result of advances in the modern world because as distances have increased between different points, it becomes more necessary to use vehicles. What has led to consolidating this tool as an essential part of life; so much so that, concerning them, a significant fraction of the urban space has been configured. Year after year, the mortality related to this cause takes many lives, also it entails many psychophysical disabilities that produce a significant impact at the level of public health in the population. The damages and sequels generated by the crashes, whatever the form, can be of great importance and the quality of life of people could be affected for a long time. Before the creation of a new design, the necessary tests should be generated to detect the most vulnerable points of the skull, as well as the study of the causes and effects of accidents that occur on motorcycles.

Keywords Italian scooter. humanitarian logistics · Testing falls using an ideal and optimal · Helmet

A. Ochoa-Zezzatti (✉) · J. Diaz · J. Mejia · I. Soto-Marrufo · Y. Quezada · L. Avelar · J. Hernández · E. De León-Evans
Universidad Autónoma de Ciudad Juárez, Av. Hermanos Escobar, Omega, 32410 Cd Juárez, Chihuahua, Mexico
e-mail: alberto.ochoa@uacj.mx

A. Ochoa-Zezzatti
Applied Artificial Intelligence Research Group, Juarez City University, Cd Juárez, Mexico

© The Author(s), under exclusive license to Springer Nature Switzerland AG 2021
A. Ochoa-Zezzatti et al. (eds.), *Technological and Industrial Applications Associated with Intelligent Logistics*, Lecture Notes in Intelligent Transportation and Infrastructure,
https://doi.org/10.1007/978-3-030-68655-0_8

1 Introduction

According to data from the World Health Organization (WHO), road accidents are the second type of violent death in our country; worldwide it has become the leading cause of violent death

(1) The statistics are rising worldwide with about six million deaths due to this event, thus becoming the road traffic in the first way of violent death worldwide. According to the study Load of Disease in Colombia, traffic accidents generate 8.6% of the total healthy life years lost (compared to 24.6% generated by homicides and 7.7% produced by infections and parasitosis)
(2) As personnel of the health area, we have observed that motorcycle traffic accidents generate a great impact on the health system due to the consequences that in morbidity and mortality produce and the respective economic costs that these entails.

Worldwide we can see a steady increment of people who own motorcycles [1], this varies from place to place, mainly true in countries who are seeing an economic development [2].

We all know the motorcycle users are the second most vulnerable on the road, to ensure their safety helmets are introduced to the riders.

Thai [3] specifies in his research that helmets help reduce the likelihood of a head injury in 69% and also a 42% less chance of death just by using a properly adjusted helmet; he also explains that helmet stability is a really important factor, there have been cases where a motorcyclist's helmet rolls-off during an impact.

Researches have been made concerning the most common injuries suffered while on a motorcycle accident, Burns [4] found out in his research that out of 1252 injuries suffered on an accident the most common orthopedic ones were the tibia/fibula, spine and the forearm, while the most common non-orthopedic ones were concussions, skull fractures, facial fractures, hemothorax, and pneumothorax. One of the main causes of death in motorcyclists who are involved in road accidents is the low resistance to the impact of their safety equipment, mostly helmets. Sometimes becoming null the existence of resistance.

Most motorcycle crashes have an impact on the head with an angle, this angle has a considerable difference at the linear and rotational acceleration of the head [5].

Özdöl [6] oversaw a research made in the Antalya Training and Research Hospital situated in Turkey where they studied and collected data about the victims of traffic accidents that were brought to that hospital. A total of 540 victims of motorcycle accidents were brought into the hospital, of those 540, 320 were affected by cranial injuries, "The distribution of cranial injuries was: epidural hemorrhage (12.6%), subdural hemorrhage (15.2%), depressed fracture (10.4%), linear fracture (23%), skull base fracture (5.5%), diffuse axonal injury (9.3%), subarachnoid hemorrhage (25.2%), intracerebral hemorrhage (13.5%), and contusion (26.3)".

In 2017, a total of 34,910 motorcycle collisions happened in Mexico according to data gathered by the National Institute of Statistics and Geography (INEGI), these motorcycle accidents contributed to a total of 367,789 traffic accidents in that year. With a quick calculation, we can determine that roughly 9.5% of the total accidents in Mexico in 2017 were motorcycle accidents, while pedestrians only caused 3.5% of the accidents that year (with a total of 13,153 accidents). Reviewing the data that has been collected so far we can surely state that increased security in regard of motorcycle helmets is not only a good suggestion but more than actual necessity, by keeping this data in mind the ongoing research has the purpose of helping on the design of a more secure helmet by doing a series of impact tests made on a 3D printed head form considering the most common type of impacts suffered on the head while experiencing an accident as well as basing the data obtained on statistics gathered from sources worldwide.

2 Methodology

The Chicago Tribune [7] made an article depicting information of helmets, where they described that the most common materials that these are made of are thermoplastic, fiberglass, composite (usually fiberglass and Kevlar) and carbon/Kevlar, describing the last ones as the "pinnacle of current helmet technology".

In regards to helmet security, oblique impact tests are commonly used in the industry to test the resistance of the helmets that are produced, as the results that this type of test is significantly different than the results of a linear impact test [8].

A possible example is to use a test rig where the helmet and head form are dropped into a steel base with a certain oblique angle [5].

For the collection of information, 3 databases were used, where multiple research papers were considered. Information regarding injuries, impact tests, and accident statistics. For the tests, an accelerometer was used and controlled by Arduino.

The work that this 6-axis sensor serves in the experiment is to give us data concerning the final position versus the starting position of the sensor, which can be calibrated or left untouched, and also records the maximum acceleration reached in the x, y, and z-axis.

Using this information, we can calculate the amount of force that a helmet and head receive on the impacted place. Traffic is not directly predicted by the kriging applied to the map, but the thing with traffic is that is directly related to accidents. Many more vehicles in the streets lead to a higher chance to be hit by a car. Anxiety in drivers as a consequence of the jams will cause some of them to take risky decisions like passing a red light, overtaking the cars in front and changing lanes, might lead to traffic accidents, therefore worse traffic. Also, as mentioned before, motorcycle drivers are in more danger, because in jams, they use to go in between lanes, which makes them vulnerable to other vehicles shifting lanes.

All this information can provide very useful information because knowing the places in which the accidents may happen with more frequency. Thanks to this, projects can arise on researching what causes these places to be more dangerous or prone to accidents, or research on security helmets for motorcycle drivers, looking at the environment surrounding those places and building the equipment accordingly.

Equipment like helmets can prevent fatality from up to 28% of effectiveness [9]. But the effectiveness can be improved by analyzing the scenarios in which these accidents happen with more probability. This kriging applied to maps is the first step in this direction, because it closes down the investigation and study to smaller, denser areas in which the team can work on looking at possible reasons why this happens more often.

3 Proposed Methodology

Conditioned Helmet for Improved Security on the Road is a decisive situation for many people in a Smart City. We must consider that motorcycles are a common transport in the world [10]. In such a way that its sales continue to grow globally, even in the expected growth of 16.1% by 2019, where scooters remain the kings in sales [11].

A noticeable disadvantage in the use of motorcycles and scooters. It's personal safety because no bodywork protects the person from an impact [12]. If we are talking about motorcycles, it is necessary to talk about the necessary safety equipment.

Security team

- Helmet
- Pants
- Long Sleeve Jacket (Leather Jacket)
- Highly visible colored garments.
- Gloves
- Boots

Wearing these garments can help to reduce injuries and even save life, if they are of the right material [13].

The helmet is an indispensable element that in some parts of the world is compulsory to use.

Because of this, our research will be based on the helmet.

4 The Most Common Mistakes When Riding Motorcycles

The reach by not taking the proper distance before braking, as well as driving in the center of the lane without the precaution of being seen (Fig. 1).

When passing by the right of a row of cars people may get out without looking to the sides, especially trucks or taxis (Fig. 2).

Fig. 1 Road accident associated with a pursuit on a high-density road vehicle, and low visibility by the rider coupled with complicated maneuverability of the Italian scooter

Fig. 2 Accident associated with blind spots with a lack of visibility and decreased maneuverability due to the compression of a large vehicle

On the roads, it should not be kept near or between heavy vehicles, in the event of their braking, the capacity to react is almost non-existent (Fig. 3).

When taking bends, it is necessary to anticipate and use the full width of the lane, which will allow a better perspective of the road ahead (Figs. 4 and 5) [9].

Most common injuries:

Bone fracture.

Generated by damping of the fall in more common areas such as limbs.

Brain injuries.

They are recurrent injuries, whether he wears a helmet or not. It can lead to brain trauma and in the worst-case death.

Spinal damage.

Spinal damage can become crippling in some cases.

Burns.

Generated by the friction of the fall at high speed against the ground [14].

Fig. 3 Compression of Italian Scooter due to effects by the movement of a vehicle of long dimensions in a scenario with high vehicular density

Fig. 4 Road accident associated with an Italian Scooter generated by high speed and change of direction combined with low maneuverability by a sharp curve

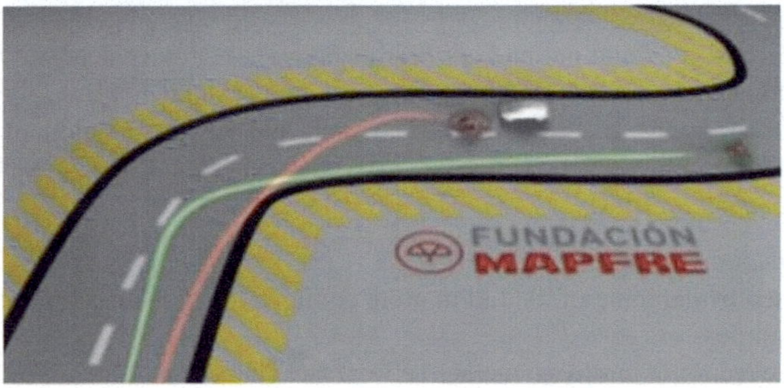

Fig. 5 Road accident due to the reach of a motorized vehicle and low maneuverability due to changes in the direction of two Italian scooters in the same alternate lane

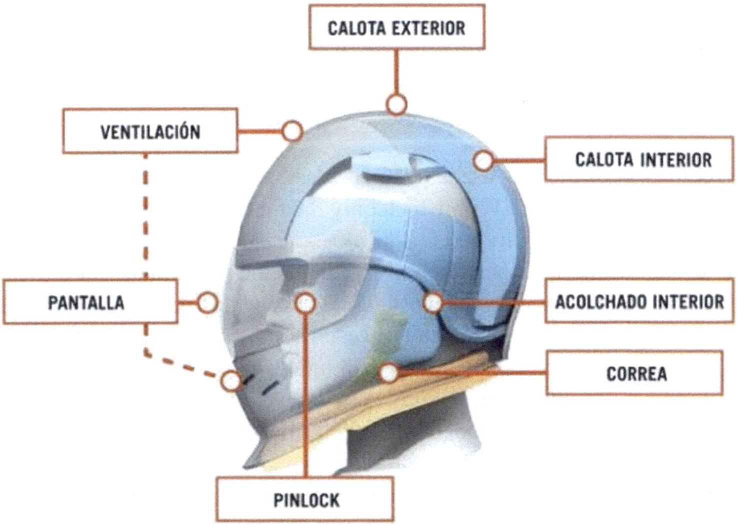

Fig. 6 Desirable printable features in additive manufacturing to optimize the road safety process in a Smart City using a 3D printer Revista moto, "¿conoces tu casco y las piezas que lo componen?", 23-12-17. [Online]: https://revistamoto.com/wp_rm/conoce-tu-casco/ [Consultation: October 20, 2019]

Parts of the helmet were redesigned to improve its functionality with the creation of a new design must start from the basic components of a helmet (Fig. 6).

Internal parts

- Calota interior
- Acolchonado interior
- Correa
- Calota exterior
- Pantalla
- Pin lock

The most common materials in the construction of motorcycle helmets are:

- Thermo-plastic
- Fiberglass
- Carbon fiber
- Kevlar [15].

5 The Function of the Helmet

Protecting the head and decreasing the force of the impact. The above research is used as a basis for finding the ideal size of a helmet. Although it is concluded that a good fit of the helmet is ideal. The helmet should exert a light and uniform pressure on the head without excessive pressure [16]. It should be noted that the skulls of men and women tend to differ in the occipital and frontal areas [17].

Impact recreation

For the recreation of the impact, it was necessary to create a skull in 3D (skull of women aged 23 to 27 years). With which only the first impact that would generate a motorcycle crash will be measured [18].

Considering that now the head will be fixed, and the movement will be by part of the mallet. The mallet will carry a constant velocity within an angle [19]. The skull was coated with moldable foamy to identify each bone in the skull [20]. Subsequently the face was rebuilt most realistically, weighing 4.9 to 5.1 kilos (Average size). To obtain the impact data on each area of the skull covered by the helmet, load cells of 50 kilos are used. That they would amplify their impact through the HX711 module, that it sends the information to our Arduino.

Having the test data will identify which sections are most affected to redesign the helmet using Solidworks.

Some common mistakes

As part of the strategy of the present research to address the above problem is taken as a point of reference a port (52), of 86. The geographical location of this point is Avenida 16 de September (Alcalde) and Avenida Juárez (Avenida Vallarta) specifies the point and in the depot is shown. This port was chosen for being the most demanded of all the SBC according to their number of trips. According to [14], the total number of bicycles for this point is 16, while for other ports the total is between 8 and 10, according to their point of location and demand. Table 1 shows the mean, minimum, maximum and standard deviation of the collected data for men and women.

It is a reality: "all countries should have and enforce a helmet law that makes wearing a helmet obligatory for riders of all powered two-wheelers" (OECD/ITF 2015). To be able to design a helmet there are some things to have in mind: "according to current standards, the performance of a helmet is only tested against radial impact which is not commonly seen in real accidents. Some studies about helmet design have been published but those helmets have been tested for only a few loading conditions" [21]. Also, it's important that "the shell and the inner liner are the key protective components of a helmet. The outer shell is there to resist abrasion and prevent objects from penetrating in the event of an accident. Shells at the lower end of the market are usually injection-molded polycarbonate, although this is sometimes found in some higher-specification helmets. More expensive helmets tend to use laminates of materials such as glass fiber, carbon fiber, and Kevlar". The way we recommend for dealing with these criteria to design the helmet is described by the hierarchical tree shown in Fig. 7.

Data mining and clustering analysis

Table 1 Descriptive statistics

Dimension	Men (n=69)			Women (n=61)		
	MIN	MAX	Mean ± SD	MIN	MAX	Mean ± SD
Eu-Eu	142	170	156.62 ± 6.13	139	166	150 ± 5.65
G-Op	174	219	194.70 ± 9.05	171	201	185.66 ± 7.05
Al-Al	27	47	34.59 ± 3.02	26	36	31.97 ± 2.28
N-Ns	43	59	51.96 ± 3.34	38	57	46.92 ± 3.90
N-Gn	103	140	124.04 ± 6.93	99	128	112.54 ± 6.68
Ft-Ft	75	125	105.35 ± 9.46	77	126	99.2 ± 10.08
Ex-Ex	95	115	104.41 ± 4.34	87	112	99.67 ± 5.20
En-En	26	36	31.25 ± 2.55	24	35	29.34 ± 2.85
Zy-Zy	130	158	142.84 ± 6.85	105	154	133.26 ± 9.57
Perimeter G-Op	534	614	573.52 ± 17.07	508	601	552.57 ± 17.21
G-V-Op	277	383	321.78 ± 20.86	261	525	303.74 ± 34.70
Eu-V-Eu	286	370	320.29 ± 16.53	263	338	301.64 ± 18.01
V-Gn	197	249	224.39 ± 11.68	185	230	204.61 ± 11.08
V-Po	119	158	140.39 ± 9.14	118	160	135.53 ± 9.38
V-N	79	151	103.54 ± 14.30	67	119	92.57 ± 11.95

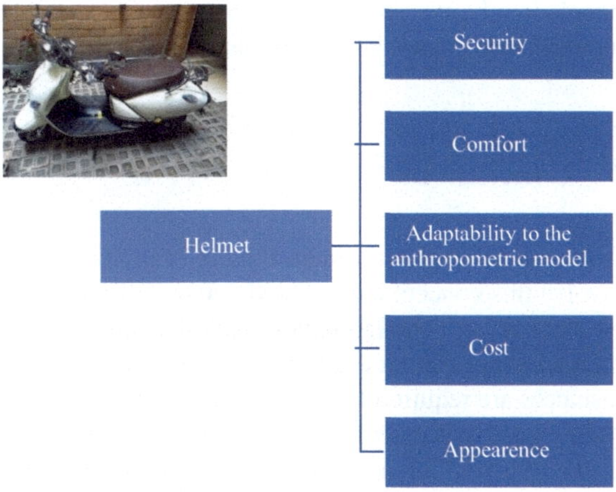

Fig. 7 Hierarchicly tree for helmet design criteria

Clustering was performed on the complete database to find distinctions between people. The algorithm, on a large scale, detected three groups. When doing the comparison with the class gender, there is an error of 35.38% in the classification of the instances, however, the clusters reveal different characteristics of people (see Fig. 8).

The investigators look for situations where the groups of people are producing computer registers (such as documents, Twitter messages, or Web sites and links to

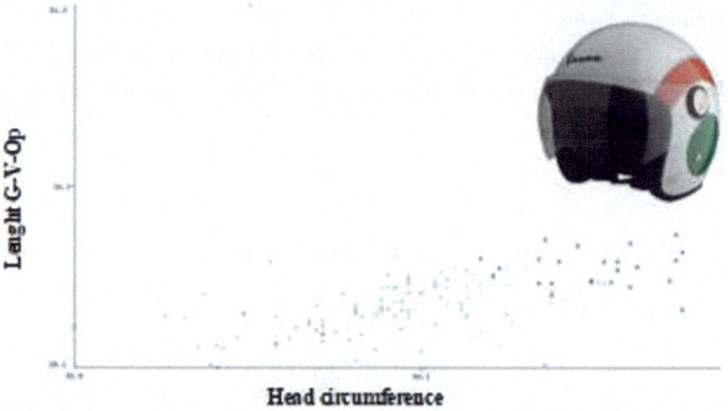

Fig. 8 Data mining analysis to our sample population to propose and ideal and optimal helmet to a shared bike system

groups with a specific profile) as part of its normal activity. The potentially useful information implicit in these files is identified, and the computer techniques to display the results are designed. Thus, the computer discovers and makes explicit the "social tendencies through time" created by a particular type of community as is shown in Fig. 3 with our model to determine the ideal helmet.

In Fig. 9, we organize and reclassify diverse cluster with the sizes of craniofacial people in a border society, we consider a cluster with these features, and modeling an ideal and optimal helmet to this type of Italian Scooter.

6 Conclusions

Although the curse of dimensionality transcends big data scenarios, the ideas that involve this concept cannot be dismissed. That is why, as mentioned in this paper, we did not have enough data to build a complete anthropometric model. We found measures that can be the starting point to design the helmet. A greater number of instances are required to make more forceful affirmations. This follows the comfort need: "good fit is essential. Too small and a helmet will become very uncomfortable, distracting your attention from the task in hand. Too big and it won't give adequate protection. The correct fit should exert light pressure evenly over your head with no excessive pressure at any point". However, we include in Appendix A a table with the centroids that describe every single group that was found.

With these results, it is possible to solve the problem raised in relation to the limited times of users and visitors in journeys and scheduled times, if the entire Smart City is not known if it is possible to know the most significant and symbolic. With the use of this innovative application, using the Djikstra Algorithm associated with a Kriging model can be applied on a mobile device to determine the most important places where it is possible to improve a trip as is analyzed in [9, 13–15].

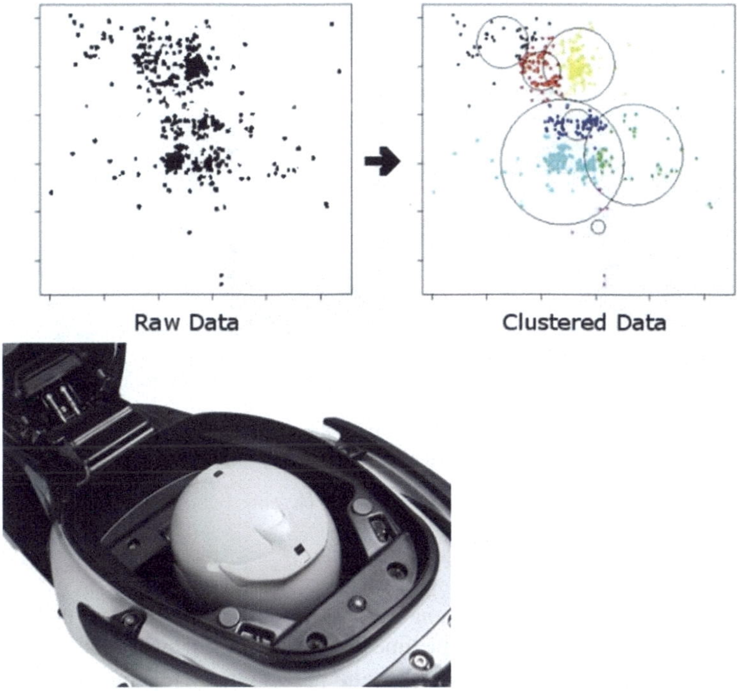

Fig. 9 Data Analysis to determine optimal helmet associated with an Italian Scooter, considering craniofacial sizes

The route of the main attractions is proposed in a smart 3D screen used in social networks, with real-time information in a Mobile device as is proposed in Fig. 7, and the establishment of a trip. Including saving money, time and enjoying with the rest of the community. The most important contribution of the present research is the use of the algorithm in an amusement park where it includes real-time and the symbolic capacity in a route and that allows to avoid the expense of time and money in this type of trip in a Smart City. After the experimentation of the present research, it is observed that through the application of computational techniques it is possible to improve an SBC in terms of distance and time and derived from it is possible to keep the units in good condition. Using an intelligent application, we can determine through IoT the requirements of each user and specify the actions to be carried out in order to use a bicycle or electric bicycle, if the distance is greater or must be transshipped, it also adequately determines the distance between transshipments as is analyzed in [19] and the way to obtain shorter waiting times is efficient, as can be seen in Fig. 10.

Fig. 10 We determine the use of an intelligent application to determine the relevance and ranking of an adequate selection of travel by bicycle un a Smart City

7 Future Research

It is recommended for the application in a mobile device with current photographs and the main recommendations in the time available for the trip. The Intelligent Tool will be used by different types of people who need to travel together with different situations related to displacement. In addition, this application will be used as a recommendation system when traveling to other cities or places in different societies and will explain different scenarios based on time, limited resources and location. Another field topic that will benefit from a more appropriate organization is Logistics of the product or service as it describes the use of Cultural Algorithms to improve a Logistics network. We propose the use of a mobile computing device that can indicate during the tour fun options based on our own tastes and priorities, according to the type of activity you want to perform. We intend to perform more tests on other equipment with different characteristics, either hardware with a graphics card with higher processing cores, a webcam with better resolution or other software versions in order to obtain better results in the quality of the processed video using diverse applied methodology used by kriging as in [1–4]. This platform is also intended to be multiplatform for functions not only in Linux but in other operating systems, whether it is the Windows or Mac environment. One of the main problems in a Smart City are accidents with more than one death associated with reckless driving due to alcohol, we propose that using a mobile device and ubiquitous computing techniques determine through photos of the online status of a group of friends that they will travel together. As a future work it is proposed to submit the SBC instance to solve the CVRP in a parallel environment in order to observe the behavior of both schemes, that is, sequential and parallel. In the near future, making trips of greater distance involves several aspects to be solved, among them mainly the safety of the users through optimal and ideal helmets, a smart grid to help with the energy that the electric bicycles require, a distribution model of space for the smart city and finally an interconnected system that manages to reduce the impacts by scope among the various vehicles, that is why it is proposed to determine how to improve this group of actions in a smart city in the long term. In addition to our future research, it will be specified in a serious game that allows the use of avatars as it should be correctly used helmets proposed in a smart city and the future impact in order to reduce by 2027 the high incidence of deaths by reach in this border society [1]. The excuse in young people that do not use a helmet is that in an Italian scooter they can feel the wind [5, 6, 8]. We are young to do not use a specific helmet as in Wales and Kazakhstan, as is shown in Fig. 11.

Figure 12 shows another future research associated with an innovative Anthropometrics analysis which requires Intelligent Data Analysis as in the determination of a specific helmet to Go-karts in a Fun Park in a Smart City.

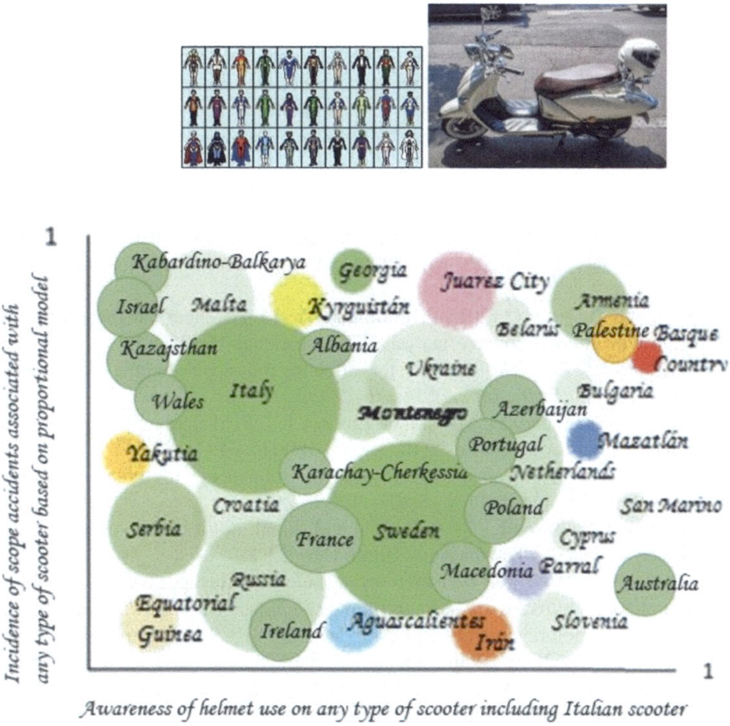

Fig. 11 Comparative of transit accidents–In 39 different societies- relate with any kind of vehicle or Motorbike including Italian Scooter: Green: European and Australian societies; Yellow and Orange: Asian societies; Brown: African Society; Blue and Purple: Mexican Societies and finally in Pink our border society of this study

Fig. 12 The analysis of anthropometries and its posterior analysis with WEKA usage to discover hidden patterns in a specific border society, which practices Go-kart racing

References

1. Xiao Y, Huang H, Peng Y, Wang X (2018) A study on motorcyclist head injuries in car-motorcycle accidents based on real-world data and accident reconstruction. J Mech Med Biol 18(4):19
2. Law TH, Hamid H, Ning C (2015) The motorcycle to passenger car ownership ratio and economic growth: A cross-country analysis. J Transp Geogr 46:122–128
3. Thai KT, McIntosh S, Pang TY (2015) Factors affecting motorcycle helmet use: size, selection, stability, and position. Traffic Inj Prev 16:276–282
4. Burns S, Gugala Z, Jimenez C, Mileski W, Lindsey R (2015) Epidemiology and patterns of musculoskeletal motorcycle injuries in the USA. F1000 Res 4 (114)
5. Ebrahimi I, Golnaraghi F, Wang G (2015) Factors influencing the oblique impact test of motorcycle helmets. Traffic Inj Prev 16:404–408
6. Özdöl Ç, Gediz T, Aghayev K (2019) Cranial and spinal injuries in motorcycle accidents: a hospital-based study. Ulus Travma Acil Cerrahi Derg 25(2)
7. Chicago Tribune (2018) Three best motorcycle helmets, Chicago Tribune (29 de octubre, 2018) [En línea]: https://www.chicagotribune.com/consumer-reviews/sns-bestreviews-automotive-three-best-motorcycle-helmets-bestreviews-20181029-story.html#:~:text=Carbon%2FKevlar%20helmets%20are%20the,command%20a%20particularly%20high%20price.
8. McIntosh A, Lai A (2013) Motorcycle helmets: head and neck dynamics in helmeted and unhelmeted oblique impacts. Traffic Inj Prev 14:835–844
9. Mapfre Foundation (2013) Safe Motorcycle Driving. You Tube (February 4, 2013), 3:59 min. https://www.youtube.com/watch?v=UH_kLst91Ls&t=6s Accessed 30 Oct 2019
10. Emmanuel de León (2019) Helmet conditioned to improve safety on the road. Universidad Autónoma de Ciudad Juárez, Ciudad Juárez, 3 pp [Unpublished]
11. Martin J (2019) Motorcycle passion engine. www.motorpasionmoto.com. Accessed 20 Sept 2019
12. Moto Revista CR (2017) Advantages and disadvantages of motorcycles. Moto Revista CR (September). www.motorevistacr.com Accessed 1 Oct 2019

13. Moto1pro (2018) The biker's equipment, key to your safety. Moto1pro, 106, 17 (2018). www.moto1pro.com Accessed 16 Oct 2019
14. Hospital la paloma, La Paloma hospital. https://www.hospitallapaloma.com/ Accessed 18 Oct 2019
15. We Ride (2017) Parts and materials of motorcycle helmets. Moto Buy Ker's, 3 (July 2017). www.motobuykers.es Accessed 20 Oct 2019
16. Cuevas J (2018) Implementation of a multicriteria analysis model to determine anthropometric characteristics of an ideal and optimal helmet of an Italian Scooter to be used in a Smart City. Universidad Autónoma de Ciudad Juárez, Ciudad Juárez, 18 p [Unpublished]
17. Amores A (2014) Techniques for determining sex from the skull in a current Mediterranean population. University of Granada, Granada, 209 p [Doctoral Thesis]
18. Ocutv (2015) Crash scooter test: results after an accident, You Tube (July 6, 2015), 3:04 min. https://www.youtube.com/watch?v=CjpW-VE70U8 Accessed 1 Nov 2015
19. Secretary of Labor and Social Welfare, Project of Mexican Official Standard PROY-NOM-206, Safety Helmets for the Prevention and Immediate Care of Injuries to the Head of Motorcyclists-Actions to Promote Health-Safety Specifications and Safety Specifications and methods of testing, commercial information and labelling
20. Triglia A (2019) Bones of the head and skull: anatomy, functions and characteristics. Medsalud https://medsalud.com/medicina/huesos-cabeza-craneo. Accessed 29 Oct 2019
21. Fahlstedt M, Halldin P, Kleiven S (2014) Importance of the bicycle helmet design and material for the outcome in bicycle accidents, Proceedings, International cycling safety conference. Diva portal [En línea]: https://www.diva-portal.org/smash/get/diva2:773433/FULLTEXT01.pdf
22. Cuevas J, Ochoa A, Hernandez JL, Mejia J, Avelar L, Mederos B (2020) Implementation of a multicriteria analysis model to determine anthropometric characteristics of an ideal and optimal helmet of an Italian Scooter. In: Castillo O, Melin P, Kacprzyk J (eds) Intuitionistic and type-2 fuzzy logic enhancements in neural and optimization algorithms: theory and applications. Studies in computational intelligence, vol 862. Springer, Cham. https://doi.org/10.1007/978-3-030-35445-9_51

Design of an Urban Transport Network for the Optimal Location of Bus Stops in a Smart City Based on a Big Data Model and Spider Monkey Optimization Algorithm

Marlyn Montalvo-Martel, Alberto Ochoa-Zezzatti, Elías Carrum, and Denise Barzaga

Abstract Taking into account the definition of sustainable development (WCED [132]), a city can be defined as sustainable "if its conditions of production do not destroy over time the conditions of its reproduction" [18]. Hiremath et al. [58] characterize sustainable urban development as "achieving a balance between the development of urban areas and the protection of the environment with a view to equity in income, employment, housing, basic services, social infrastructure, and transport in urban areas". Recently, the concept of Smart City (SC) has become popular in the interest of achieving sustainability goals with the help of "intelligence". It first emerged in the 1990s [1] as an alternative to traditional planning modes, using new technologies to address these problems. According to Lombardi et al. [77], the use of modern technologies in urban life is highlighted in some definitions of intelligent cities, resulting in ecological, efficient, and innovative transport systems, infrastructure, logistics and energy systems.

Keywords Optimal location · Smart cities · Big data · Spider monkey optimization algorithm

1 Introduction

From a logistical point of view, the rapidly growing rates of urbanization and the potential change in people's demands for goods in urban areas justify the need to develop new urban logistics systems. These new systems must ensure efficient urban mobility, not only for people but also for goods [45]. The concept of city logistics is

M. Montalvo-Martel (✉) · E. Carrum · D. Barzaga
Corporación Mexicana de Investigación en Materiales, Calle Ciencia y Tecnología 790, Saltillo 25290, Mexico
e-mail: mmontalvo@alumnos-comimsa.mx

A. Ochoa-Zezzatti
Universidad Autónoma de Ciudad Juárez, Av. Plutarco Elías Calles 1210, Ciudad Juárez 32310, Mexico

© The Author(s), under exclusive license to Springer Nature Switzerland AG 2021

A. Ochoa-Zezzatti et al. (eds.), *Technological and Industrial Applications Associated with Intelligent Logistics*, Lecture Notes in Intelligent Transportation and Infrastructure,
https://doi.org/10.1007/978-3-030-68655-0_9

not exactly defined, however, the most used in the world is that of Taniguchi, who defines city logistics as "a system that totally improves its transport activities, which are generated by private companies in urban areas. This system takes into account the environmental impact of vehicles, vehicle congestion, and energy consumption within the economy. In freight transport, the city's logistics plays a fundamental role in terms of innovation. This is aimed at minimizing costs: environmental, economic, and social, and in the movement of goods; to achieve environmentally sustainable cities [90]. In short, in this context transport is valued as an organizational process that must focus on the customer, for their satisfaction at all times and in an effective manner, with product and information requirements [104].

1.1 Background Relating to Transport

Debnath et al. [35] and Garau et al. [51] propose specific frameworks and indicators for assessing cities according to the intelligence of their transport systems. For example, Garau et al. [50] provide a valid methodology for cities to assess urban mobility intelligence and use this method to compare 17 Italian cities. (2010), Caragliu et al. [19], Correia and Wunstel [32], Dirks et al. [37], Giffinger et al. [52], La Greca et al. [73], Munuzuri et al. [91], Nam and Pardo [92], Steria [113], The Climate Group et al. [121], Think [122], Toppeta [124] and Washburn et al. [131] in their studies address issues related to transport, mobility, and logistics. They optimize logistics and transport in urban areas taking into account traffic conditions and energy consumption. Dahiya [34], Ng and Hills [94], Thynell et al. [123], Tsou and Cheng [126], and Jun et al. [65] state in their research that cities with high population density have traditionally made a greater effort to develop their local public transport systems and, as a consequence, are now in a more favorable position to launch SC initiatives in these environments. On the other hand, Savelsbergh and Van Woensel [110] identify the trends that drove changes in city logistics: the growth of urban populations, the growing importance of e-commerce and rapid supply chains, the rise of the sharing economy, and sustainability. In addition to studying the impacts of the information revolution on city logistics, Taniguchi et al. [120] also describe the big data applications and decision support systems that can be used to improve the design and evaluation of city logistics schemes. Mourad et al. [90] summarize the different types of features and constraints for the shared mobility problem. In addition, they identify variants of problems that consider each type of constraint discussed in order to obtain a clear picture of these variants and their common and different characteristics. Other authors who explore the problem of shared mobility using optimization, simulation, and data analysis are Jain and Hentenryck [64], Heilporn et al. [56], Garaix et al. [49], Herbawi and Weber [57], Berbeglia et al. [12], Kaan and Olinick [67], Parragh and Schmid [100], Huang et al. [61], Kirchler and Wolfler [72], Masson et al. [86], Lehuédé et al. [75], Hosni et al. [60], Atahran et al. [6], Liu et al. [76], Stiglic et al. [114], Lee and Savelsbergh [74], Santos and Xavier [109], Parragh et al. [99], Ma et al. [79], Ritzinger et al. [106], Jung et al. [66],

Masmoudi et al. [84], Braekers and Kovacs [14], Masmoudi et al. [83], Masoud et al. [85], Pimenta et al. [101], Alonso-Mora et al. [2], Stiglic et al. [115], Kalczynski and Miklas-Kalczynska [68] and Wang et al. [130]. Clarens and Hurdle [31], Marwah et al. [82], Ceder and Wilson [22], van Nes et al. [127], Baaj and Mahmassani [8–10], Chien and Schonfeld [27], Furth and Rahbee [47], Chakroborty and Dwivedi [24], Chien and Spasovic [28], Chakroborty [23], Guan et al. [54], Wan and Lo [129], Argwal and Mathew [5], Chen and Yang [25], Fan and Machemehl [40–43], Gao et al. [48], Chiou [29], Yang et al. [135], Fernández et al. [46] and Mauttone and Urquhart [87] take into account the design problems of the transit network, which are summarized by Ibarra-Rojas et al. [63] in his article Planning, Operation and Control of Buses: A Literary Review. Other authors who mention in their research the design of transit networks are Cipriani et al. [30], where they describe a procedure to solve the problem of the bus transit network and its application in the urban area (the city of Rome).

1.2 Description of the Problem

Public transport is considered an important part of sustainable urban development, as it allows more efficient movements throughout the city. By designing and locating well-planned stops it is possible to reduce travel times and thus increase the efficiency of the transit system. One of the most important global trends of the last century was the growth of cities, in terms of surface area and population. So the speed of this phenomenon has prevented well-planned urban development, and with it, long journey times, severe traffic congestion, pollution, traffic accidents, etc. [63]. In many cities around the world, public transport planning becomes more relevant because of the large demographic and economic increase that occurs in many regions, which do not have proper planning, to mitigate the social aspects of the disordered mobility of their urban transport systems, causing them to become a problem [97]. Specifically, in the city of Saltillo, Mexico, the routes for public bus transportation only have some official stops. However, buses make stops each time a user marks the stop no matter where it is, as long as it is within the route being made. This means that the times assigned between stops are not respected and schedules are delayed. At the same time, it causes congestion in the traffic, because when stopping repeatedly it hinders the way of other vehicles. There is also greater fuel consumption each time you stop and restart. For these reasons, it is essential to establish official stops, always taking into account that they do not affect people, that the distance between stops is not too long, the place within the route, among others. This ratifies the importance of having an optimal design of the transport routes used in such a way that it is in accordance with the demand of the clients and that the routes are not too long. vehicle capacities are adequate to ensure the cost-effectiveness of the system [38]. The current problems of establishing transport routes are therefore affected by a large number of restrictions, so that solving them using traditional models is computationally more complex, hence the need to resort to the use of techniques related to artificial intelligence. This ensures the analysis of problems which, according to their typology and characteristics, make it possible to find algorithms that lead to their resolution and make it possible to obtain an approximate solution by using heuristic methods. Bearing in mind the above, the following research problem is raised: The number of the official bus stops established on a route does not guarantee the needs of users, nor the improvement of traffic, nor the number of bus stops established on a route does not guarantee user needs, improved traffic, or compliance with specific time windows for stations.

1.3 Justification

Urban public transport facilitates the movement of people in a city and is, therefore, a fundamental part of it. It reduces pollution, since fewer cars are used to transport people, and it also allows the transfer of individuals who do not have a car and of those who, having a car, sometimes do not use it because of traffic jams or parking difficulties. Cars, trucks, buses, and trains: the world is in the hands of transport in order to favor its economic growth and development. Without transport, people would not arrive at their workplaces, materials would not be delivered to manufacturing companies, and goods would not reach the market. Transport of both people and goods is an indispensable element of global development. The present study consists of designing the location of bus stops to improve traffic and comply with the time windows established for stops. Transit planners and city authorities have made significant efforts to address the problem of traffic and urban road congestion due to rapid population growth and urbanization, especially in developing and emerging countries. This has led to the development of various tools or approaches to the issue of urban transit network design that ideally reflects the interests of users and operators [15]. The main challenge facing planning and operational tasks is between the level of service perceived by the client and the operating costs of the agencies. Users expect to reach their destinations quickly, where travel time breaks down into access and exit to/from stops, waiting, vehicle travel, and transfer. It also expects the ride to be comfortable and hopefully, comfortable. In order for the transit network model to be realistic, customer expectations must be properly assessed in terms of each user's behavior [102, 111]. On the other hand, operators are interested in a cost-effective system, where vehicle usage costs and driver salaries are low. In order to optimize travel frequencies, it is essential to implement logistical tools, such as - simulation, which is relevant in the field of engineering, also allows to know and contextualize the scenario to project and lead to a controlled period of time, - operations research, is a discipline that uses mathematical models, statistics and algorithms in order to make decisions. The latter being the tool selected to provide a solution to the problem posed, as referred to in their research Ibarra-Rojas et al. [63], Szeto and Wu (2010), Yan et al. [134], Amiripour et al. [3], and Buba and Lee [15]. This approach ranges from manual, analytical, heuristic and metaheuristic transport networks both theoretical and realistic. In fact, an exhaustive literature search reveals that no common methodology is used to deal with these types of problems. This is due to the extension of the details incorporated in the formulation of the problem to reflect real-life conditions that differ in many respects [15]. This research has both methodological and practical novelty because the problem of transit network design continues to attract much attention among professionals and planners, especially the bus-based transit system because it is more complex and requires more driving time compared to other modes of transport.

1.4 Research Questions

The following research questions are proposed:

- Why is the location of bus stops important in urban transport?
- What are the different techniques that have been used to optimize transit networks?
- How does demand in a given sector behave?
- How to model demand?
- How to measure the efficiency of the design of the urban transport network in terms of traffic improvement and compliance with the time windows established for stops?

1.5 General Objective

To solve the research problem, the following general objective is formulated: To develop a methodology for the optimal location of bus stops in the city of Saltillo, depending on the demand per sector.

1.5.1 Specific Objectives

The following are proposed as specific objectives:

- To carry out a bibliographical review on the importance of the location of bus stops.
- To Carry out a bibliographic review of transport network problems and the different techniques used to optimize them.
- Determine the behavior of demand in a given sector.
- Select a model for determining demand.
- Validate the results of the application of the proposed urban transport network design in a real environment.

1.6 Hypothesis

With the development of an urban transport network design for the optimal location of bus stops, the needs of users are met, traffic can be improved and the time windows established for stops can be complied with.

1.7 Scope and Delimitations

As delimitations, the procedure will be applied to a route determined from a study in the municipality of Saltillo.

1.8 Contribution (Scientific, Technological and/or Industrial)

As far as the scientific aspect is concerned, this research uses artificial intelligence techniques, which make it possible to deal with unstructured problems in which it is not possible to find algorithms for their solution. Today, artificial intelligence occupies most of the fields of engineering and even the daily life of human beings. It has produced a great variety of products, even at an early stage of development. Although no one can predict the future in detail, it is clear that computers with intelligence at the same level as humans (or higher) will have a great impact on daily life and the future course of civilization [107]. As a technological contribution, a methodology is designed and implemented that allows the optimal location of bus stops in the municipality of Saltillo, Mexico. From the social point of view, the public transportation system of the city of Saltillo is improved, due to the reduction of traffic on highways, accidents, waiting time between stops, among others.

2 Theoretical Framework and State of Art

Next, the proposal of the theoretical frame of reference and the state of the art is developed that contain aspects related to Transport Network, Design of Transport Networks (TDN), methodologies used for transport network problems, location of bus stops, among others; having as support the scientific literature that tackles the problem from the theoretical-practical point of view.

2.1 *Theoretical Framework*

2.1.1 Urban Transport Network

A transport network is infrastructure necessary for the circulation of vehicles transporting people or goods. They are located in an area linking towns and industrial activity [105]. Land transport is the most worked on concerning the optimization of transport networks, as it is where there are the greatest confluence and interactions. Also, it is the one whose networks occupy the terrestrial surface and its axes are visible because they are formed by an infrastructure through which goods and people are transported [112]. Transport networks are classified according to their density into three types: structured networks, unstructured networks, and isolated axes. More technically, a distinction is made between interurban networks and urban networks. Urban networks are one of the most complex aspects faced by public administrations in any city [105]. All cities have a basic road network for the circulation of different vehicles, with streets and sidewalks. In large cities, the basic road network is complemented by ring roads that quickly connect neighborhoods to the most remote cities.

2.1.2 Designing a Transport Network

The Transit Network Design (TND or NDP) consists of the bus network design and frequency configuration. Firstly, the network design resides in establishing a set of viable transit routes on an existing road network and pre-set stopping points (demand points) that make up an efficient transit network, to meet passengers' travel demand with minimum travel time. Secondly, frequencies are assigned to transit routes that establish waiting times and total capacity levels for the routes and makes it possible to calculate the total fleet required for the operation of the network. The TND also defines line designs and operational characteristics, such as the space between stops to optimize specific objective functions, e.g. the minimization of the weighted sum of operators' and users' costs. Due to its complexity, the TND is divided into the following sub-problems that include tactical, strategic and operational decisions [21, 36]:

- Frequency configuration (FS): Characterizes operating periods based on demand patterns (peak and nonpeak schedules) and determines the number of trips per hour needed to meet passenger demand in each planning period.
- Transit Network Scheduling (TNT): Defines bus arrival and departure times at all stops in the transit network to achieve different goals, such as meeting a given frequency, meeting specific demand patterns, maximizing the number of timely passenger transfers, and minimizing waiting times.
- Vehicle Programming Problem (VSP): Determines the allocation of trips and vehicles to cover all planned trips, so that operating costs based on vehicle usage are minimized.

– Driver Scheduling Problem (DSP): Defines the daily tasks that cover all scheduled trips and minimizes the cost of drivers' salaries. A DSP solution must comply with driver-specific work regulations, such as the length of time a driver is required to work, considering minimum/maximum working time, maximum working time without rest and daily rest for all drivers.
– Driver Registration Problem (DRP): Given a set of generic tasks defined in a given time horizon (e.g., one month) for drivers assigned to a particular depot, DRP assigns these tasks to available drivers to their work schedules, called lists, so that labor regulations are met and drivers' wages are minimized. However, each sub-problem can be complicated to solve, so the usual approach to obtaining a solution is to perform sequential methodologies [63].

These methods are iterative procedures that resolve one or more sub-problems in each iteration. Even when a solution is provided for the planning process, the operation of transportation systems is affected by uncertainty of travel times, demand patterns, and disturbances, such as: public protest activities or parades; vehicle breakdowns; weather conditions; and specific day activities. To address these situations, according to Ibarra-Rojas et al. [63], real-time control strategies are implemented to ensure an efficient service during the operation of the system. The most implemented strategies are the following:

– Station control: these are strategies that make decisions, such as stopping vehicles at some stops or omitting specific stops to increase the regularity of the service, reduce travel or waiting times and allow passenger transfers.
– Inter-station control: decisions to be made between stops (also called stations) of transit networks to achieve efficient operation. Some strategies that belong to this category are the control of operating speed and the priority of the traffic signal.

Previous control strategies have benefited from technological advances and the development of monitoring tools such as GPS. With these tools, decision-makers can learn about the actual behavior of the transit network and implement control strategies in real-time. Each of the decision levels: strategic, tactical and operational, as well as real-time decisions, impact the efficiency of the system, which is an important element in achieving a sustainable public transport system in urban areas [63]. Because of these potential benefits, the development of efficient tools for the planning and operation of transport systems remains a challenging area of research that requires careful consideration of different characteristics in the context of dissimilar cities. Therefore, the TND considers elements such as service area, line coverage, stop spacing, line directivity and length, separation time between consecutive trips, load standards and road speeds [40].

Continuing Network Design Problem

According to García-Ródenas [53], the network design problem seeks to modify transport infrastructure by creating new arches or improving the capacity of existing ones, to maximize social benefit (reduction of congestion) and/or minimize the cost of design. The continuous version of this problem appears when the design variables take continuous values and these represent the improvement of the existing capacity in the arcs. At the lower level, there is the traffic allocation model formulated (usually) through the TAP, although Marcotte [81] uses the TAP-VIP. To specify the situation it is considered that the internal problem is a problem with fixed demand, using the DUE principle (deterministic user equilibrium, is the first principle of Wardrop; it assumes that all users perceive the cost in the same way and also know the costs of all routes) to carry out the assignment, formulated in the space Ω, that is, the TAP-VIP model (c, Ω). The variables of the model are:

- lower level: the flow vector in the arcs f $\in \Re^{|\mathcal{A}|}$
- higher level: capacity expansions $\in \Re^{|\mathcal{A}|}$

The capability vector x is a subset of parameters of the travel cost functions in the arcs, i.e. c (x, f). This functional dependence shows how new infrastructures determine new transport costs. The model assumes that there is a function s (x) that provides the investment and operational costs associated with the network design defined by the variable x. The model assumes that there is a function s (x) that provides the investment and operational costs associated with the network design defined by the variable x. The model assumes that there is a function s (x) that provides the investment and operational costs associated with the network design defined by the variable x. On the other hand, a budget is available for the realization of the new infrastructure, which we denote by B and this amount cannot be exceeded. This requirement is taken into account by the budgetary restriction:

$$s(x) \leq B$$

Suppose there is another set of (technical) constraints for the design variables. This set is denoted by X, so the feasible set for design variables is

The problem of network design is formulated

$$\tilde{X} := \{x \in X | s(x) \leq B\}$$

Also, based on the optimal dual variable associated with the budget constraint (denoted *), they give the alternative formulation

$$\begin{aligned} \text{minimize} \quad & Z = \sum\nolimits_{i \in \mathcal{A}} c_l(\mathbf{x}, \mathbf{f}) f_l \\ \text{subject to} \quad & \mathbf{x} \in \tilde{X}, \qquad\qquad [\text{NDP}] \\ & \mathbf{f} \quad \text{solveTAP} - \text{VIP}\big(\mathbf{c}(\mathbf{x}, \cdot), \Omega_f\big) \end{aligned}$$

One result that illustrates the difficulty in solving this problem is the so-called Braess paradox that states that not always adding a new arc to the network, and therefore an increase in the capacity of the transport network, leads to a reduction in travel times used by each user. This is because users choose their route attending exclusively to minimize their travel time.

$$\begin{aligned} \text{minimize} \quad & Z = \sum\nolimits_{l \in \mathcal{A}} c_l(\mathbf{x}, \mathbf{f}) f_l + \mu^* s(\mathbf{x}) \\ \text{subject to} \quad & \mathbf{x} \in \tilde{X}, \qquad\qquad [\text{NDP}] \\ & \mathbf{f} \text{ solve TAP} - \text{VIP}\,(\mathbf{c}(\mathbf{x}, \cdot), \Omega_{\mathbf{f}}) \end{aligned}$$

and do not aim to reduce total travel time on the entire transport network (García-Ródenas 2001).

2.1.3 Operating Time Windows

1. Vehicle route problems (VRP) are a group of alternatives and problem customizations. They are about investigating the routes of a transport fleet to serve customers. This type of problem belongs to the problems of combinatorial optimization. According to Restrepo et al. [103], VRP has different variations.
2. CVRP (capacity VRP): It is the most general VRP and consists of one or more vehicles with limited and constant capacity, responsible for distributing products according to customer demand.
3. MDVRP (Multi-Depot VRP): It is the VRP with multiple tanks. Each depot (with its fleet of vehicles) must serve all customers. PVRP (Period VRP): It contemplates in its approach a horizon of operation of M days, period during which each client must be visited once.
4. SDVRP (Split Delivery VRP): Contemplates split delivery, where a customer can be served by several vehicles if the total cost is reduced, which is important if the size of orders exceeds the capacity of a vehicle.
5. SVRP (Stochastic VRP): Currency one or more random components; customers, demand and times can be stochastic.
6. VRPPD (VRP pickup and Delivery): Allows for delivery and collection. Customers can return certain goods, so keep in mind that the vehicle is capable of both actions. This restriction makes the problem more complex and can cause a misuse of vehicles. If the problem has a restriction to complete all deliveries before starting collections results in a VRPB (VRP Backhauls).
7. MFVRP (Mix Fleet VRP): It contemplates vehicles with different capacities, therefore, it is necessary to consider these capacities in the route that will follow each resource, since a vehicle with greater capacity will be able to cover a wider route (greater demand). VRPTW (VRP with Time Windows): It includes an additional restriction where a time window is associated with each client, that is, each client is only willing to receive the visit of the vehicle during an interval of
8. MFVRP (Mix Fleet VRP): It contemplates vehicles with different capacities, therefore, it is necessary to consider these capacities in the route that will follow

each resource, since a vehicle with greater capacity will be able to cover a wider route (greater demand). VRPTW (VRP with Time Windows): This includes an additional restriction where a time window is associated with each client, i.e. each client is only willing to receive the visit of the vehicle during a given time interval. Due to the problem that the present investigation intends to solve, complying with the time windows established for bus stops is a fundamental step for the optimal location of stops. Therefore, the mathematical formulation of the VRPTW is presented below.

Mathematical Formulation of the VRPTW

According to Restrepo et al. [103] the model indices are: i = starting node i (1, 2, …, n) j = arrival node j (1,2, …, n) n = total nodes k = Vehicle k (1,2, …, K) The variables defined are: , = 1 if the vehicle k is assigned to cross the arc from node i to node j. Zero (0) otherwise. , = 1 if the route from i to j or zero (0) is carried out otherwise. K = Number of vehicles to be used. = start time of the service for the client i The parameters of the problem are: , = Transport cost from node i to node j = Demand at node j = resource capacity k = number of clients H = Cost of the vehicle: = service time H for customer i [,] = time window for customer i = start time of the time window for customer i =closing time of the time window for customer i This is searched for

$$\sum_{j \in A} H y_{0,j} + \sum_{(i,j)e\_A} C_{i,j} y_{i,j} \tag{1}$$

Which is subject to:

$$\sum_{1 \le k \le K} x_{i,j}^{k} = y_{i,j}; \quad \forall \quad \text{i, j} \tag{2}$$

$$\sum_{1 \le j \le n} y_{i,j} = 1; \quad \forall \, \text{i} \tag{3}$$

$$\sum_{1 \le i \le n} y_{i,j} = 1; \quad \forall \, \text{j} \tag{4}$$

$$\sum_{1 \le j \le n} y_{0,j} = k \tag{5}$$

$$\sum y_{1,0} = k \tag{6}$$

$$\sum_{1 \le 3 \le n} \sum_{1 \le j \le n} d_i x_{i,j}^{k} \le u; \forall k \tag{7}$$

$$x_{i,j}^{k} \left(p_i + s_i + t_{i,j} \right) \le p_j \forall k \tag{8}$$

$$a_i \sum x_{i,j}^k \leq p_i \leq b_i \sum x_{i,j}^k \forall \text{ i y k} \tag{9}$$

$$e_i \leq p_1 \leq l_i \quad \forall \text{ k} \tag{10}$$

∀subset Q of (1, 2,…, n)

$$\sum_{k\in} \sum_{j\in\ell} y_{i,j} \leq |Q| \tag{11}$$

$$\text{k} \leq \text{K} \tag{12}$$

$$y_{i,j} \in \{0, 1\};\ \forall(\text{i, j}) \in \text{A} \tag{13}$$

$$\mathcal{X}_{i,j}^k \in \{0, 1\};\ \forall(\text{i, j}) \in \text{A}, \forall\text{k} \tag{14}$$

Set A is defined as A = { (i, j): , = 1} The restriction (2) makes obligatory the assignment of a vehicle to the route (i, j), 1 if it is travelled, and 0 if not. Restrictions (5) and (6) indicate that k is the number of vehicles used in the solution and that all those departing from a place of departure must return to it. Restriction (7) ensures that vehicles do not exceed their capacity. Restriction (8) ensures that vehicle k complies with the time windows. The restriction (9) requires that = 0 each time the vehicle k does not visit the customer i. Restriction (10) ensures that the solution does not contain cycles using nodes 1, 2,…, n. Restriction (11) limits the number of vehicles to be used. Restrictions (12) and (13) indicate that the variables are binary.

2.1.4 Methodologies Used for the Establishment of Transport Routes (Optimization Techniques)

The most representative methodologies used for the establishment of routes, according to Suarez (2009) are grouped into three strands:

- EOQ Models and Continuous Approaches. It analyzes logistics systems based on their most significant variables, assuming that these do not suffer revealing variations in the region of study. It is based on the analogy between the information of each one of the points to some added functions that allow to limit the dimension of the problem. In this way, the size of the problem is significantly reduced to a limited number of variables that provide the analytical resolution of the problem.
- Mathematical programming. It makes it possible to develop mathematical models for the optimization of the logistic system. It formulates an objective function to minimize or maximize from some decision variables to which certain restrictions have been applied. Procura to provide a unique and optimal solution. However, its main problem is the high computational cost.

- Heuristic and metaheuristic algorithms. Due to the computational limitations of the previous methodology another line of research arises which studies the development of heuristic algorithms. These provide a balance between the computational cost and the goodness of the adjustment. The absolute optimal solution ceases to be the absolute objective of the problem and becomes a trend line exclusively, considering as an optimal solution the current solution that corresponds to the lowest cost found so far.

The difficulty of these algorithms lies in that they are inefficient in some problems, so there is a need to improve them. This is how metaheuristic algorithms are developed. These are iterative processes that guide the search through a subordinate heuristic algorithm. They work in an instant of execution with a current state that is varied from a transition of state, disturbance or movement, whose objective is to explore new routes that save on the cost function. The algorithms allow a positive increase in the cost function in exchange for being able to analyze new configurations that lead to new local optimums and thus arrive more efficiently at the optimal global solution.

1. Simulated annealing (SA) is a local search technique considered easy to implement and highly flexible, based on solid annealing in metallurgy. In this process, the temperature of the solid is increased until it melts and then the temperature is lowered, seeking to group the particles that make up the solid in its fundamental state. When extrapolating the above to the resolution of an optimization problem, the cost of the solutions plays the role of the energy of a state and the parameters of temperature control. In summary, this technique seeks to explore the neighborhood of a preliminary solution to "scale up" the quality of the solution currently available (Torres et al. [125]).
2. Genetic algorithms (GA) are adaptive methods that can be used to solve search and optimization problems (Fig. 1). This technique is governed by the principles of the genetic process that takes place in living organisms. In its development, it works with a certain number of individuals that represent feasible solutions to the problem analyzed. These individuals are "crossed" according to their capacity to adapt to the problem, which affects the probability of being chosen for reproduction. From this crossing, new individuals with some of the characteristics of their parents are originated, which form a new population with better characteristics than the previous population. The process is repeated by exploring new spaces in search of possible solutions, which leads to an optimal solution to the problem studied [89].
3. The algorithms known as ant colony optimization (ACO) are models inspired by the behavior of ant colonies (Fig. 2). They are iterative processes wherein each repetition a colony of ants is thrown and each one of them chooses a possible solution to the problem. As a whole, the ants build the solutions probabilistically following the trace of the pheromone previously left by the other ants that chose the same solution, turning the process into one of the best solutions, including feedback by repeating each of the iterations in the same way. Finally,

```
BEGIN /* Algoritmo Genetico Simple */
    Generar una poblacion inicial.
    Computar la funcion de evaluacion de cada individuo.
    WHILE NOT Terminado DO
    BEGIN /* Producir nueva generacion */
        FOR Tamaño poblacion/2 DO
        BEGIN /*Ciclo Reproductivo */
            Seleccionar dos individuos de la anterior generacion,
            para el cruce (probabilidad de seleccion proporcional
            a la funcion de evaluacion del individuo).
            Cruzar con cierta probabilidad los dos
            individuos obteniendo dos descendientes.
            Mutar los dos descendientes con cierta probabilidad.
            Computar la funcion de evaluacion de los dos
            descendientes mutados.
            Insertar los dos descendientes mutados en la nueva generacion.
        END
        IF  la poblacion ha convergido THEN
            Terminado := TRUE
    END
END
```

Fig. 1 Simple genetic algorithm pseudocode. *Source* Vázquez-Solís [128]

Fig. 2 Pseudocode of the algorithm ant colonies. *Source* Vázquez-Solís [128]

```
procedure AS{
    InicializacionDatos
    while (not CriterioFin) {
    ConstruccionSoluciones
    BusquedaLocal
    ActualizacionFeromonas
    }
}
```

an optimal solution is obtained according to the solutions reached in each one of the iterations [11].

4. The tabu search algorithm (TS) is a deterministic method because it does not contemplate the existence of a probabilistic criterion for solution optimization (Fig. 3). It is also characterized by being an iterative process of search and exploration in a field of solutions that evades the local optimal and cyclic behaviors of the algorithm by prohibiting movements of the current solution and structuring of the memory of past solutions. It is based on replacing the solution that is generated without cost condition without taking into account that it is better than the current solution. The main problem is that cyclical processes could be created. To avoid the problem, the algorithm has a list of prohibited or taboo

Fig. 3 Pseudocode of the taboo search algorithm. *Source* Vázquez-Solís [128]

```
procedure BúsquedaTabú{
        s ← SoluciónInicial
        InicializaciónEstructurasMemoria
        while (not CriterioFin) {
        ConstruccionSoluciones
        s' ← MejorSolución
        ActualizaciónEstructurasMemoria
        if ( f(s') < f(s) ) {
                s ← s'
        }
        return s
        }
}
```

operations or solutions. A neighborhood is defined consisting of solutions that do not contain taboo attributes. The taboo list is not static since the problem prevents the use of certain attributes during a certain number of iterations [116].

2.1.5 Location of Bus Stops

A badly located bus stop can cause serious road safety conflicts, often resulting in accidents with victims. Bus stops must be correctly located in sections with good visibility, equipped with vertical signs that warn drivers of their existence, guaranteeing accessibility for all users, disrupting the flow of traffic as little as possible. The following are the characteristics that bus stops must comply with from location-visibility according to the MAPFRE Institute [80]:

- Being located in straight sections.
- Be located far from loading and unloading areas.
- Never at the exit of a curve.
- Never in the vicinity of a change of course.
- Both the stop and the vertical signaling must be visible, without elements that might make detection difficult, such as trees, street furniture, parked cars, etc.

2.2 State of the Art

Over the decades, the improvement of search algorithms along with the advancement of computer technologies has led to optimization and problem-solving techniques, mostly metaheuristic approaches to the problem of urban transit network design. That said, transport planners and practitioners still see the problem as a challenge [4]. An alternative to the discrete optimization problem is to represent TND through analytical models that constitute continuous approximation approaches where passenger demand is represented as a continuous function in geographic space [31, 33, 98]. Studies complementing the reviews of Desaulniers and Hickman [36], Guihaire and Hao [55], Kepaptsoglou and Karlaftis [70], Farahani et al. [44], among others, are presented below. As established by Kepaptsoglou and Karlaftis [70], analytical approaches develop relationships between the components of the public transport network represented by idealized structures. The first studies focus on determining line spacing, stop spacing, and line frequency, assuming that demand is evenly distributed over the service area [17, 59]. Other approaches are developed to consider time-dependent demand [108], space-dependent demand [13, 16], and time- and space-dependent demand [62]. Chien and Schonfeld [27] later define a TND to determine stop locations, lines, and their advances in order to minimize total user and operator costs. To avoid oversimplifying urban areas, the authors propose an analytical model that assumes a rectangular grid and a passenger demand that depends on the area. Daganzo [33] proposes a transit network structure for a square region of idealized demand, combining the network and the concepts of axis and radius. Experimental results suggest the following: a center and radio structure becomes more attractive in cities where infrastructure is expensive; Bus Rapid Transit systems are a good option for cities with medium to high density; user costs dominate operating costs and both decrease with low densities of demand, and the space between bus stops and the distance between lines should be kept above a critical threshold. Estrada et al. [39] then implemented the Daganzo [33] methodology to generate lines in the city of Barcelona. In addition, the authors develop a simulation to map the transit network in the city, obtaining a more accurate estimate of costs and level of service. Subsequently, Medina et al. [88] formulate an optimization problem to determine simultaneously the stop density in a bidirectional corridor and the frequency of the lines during several periods. The proposed approach is implemented to analyze a corridor in the transit network of Santiago, Chile. Numerical results show that the proposed set of stop locations results in a reduction of total costs by approximately 20%, especially in the non-peak period. Kim and Schonfeld [71] propose a probabilistic analytical model for designing and synchronizing a bus transit network that should service from a major terminal to multiple regions. Decisions to be made on this issue include the type of service provided for particular regions, vehicle size, number of zones, frequency, fleet type, and playtimes. A GA is developed to solve examples of the problem and numerical results show that solutions that lead to optimal synchronization have common or integer ratio advances.

2.2.1 Discrete Optimization Approaches for Transit Network Design

To overcome inbreeding passenger demand in discrete optimization versions of the TND, several approaches have been considered. One approach decomposes the problem into sequential sub-problems. Another approach to achieving a user-operator balance is to model the NDT as a two-level optimization problem, where the line design is carried out on a "first-level" and the transit allocation used to determine the flow of passengers on the network is determined on a "lower-level" [29]. A third approach models passenger assignment decisions in the TND using non-linear restrictions or non-linear objective functions, leading to non-convex problems that can be solved by heuristic algorithms. Subsequently, studies are presented that address these three types of approaches.

Sequential Approaches and Simplified Cases for Transit Network Design

Mathematical formulations for transit network design are generally intractable by the exact approaches. However, it is possible to solve them accurately by considering cases such as small issues [129], single-line design cases [54], simplifications that reduce the size of the transit network [78], and determining only some decisions, for example, to stop spacing [47]. Then, advanced meta-heuristics are commonly implemented to address NRT problems, sometimes formulated with two-tier models to consider transit allocation decisions.

Metaheuristic Algorithms for Transit Network Design

While NDT is often broken down to resolve each stage using heuristic algorithms [22, 127], it is important to define feedback between stages that attempt to integrate line design decisions with those of transit cession. This can be naturally integrated into genetic algorithms [23, 119]. Szeto and Wu [119] develop a hybrid solution procedure to simultaneously solve the problem of frequency configuration and transit network design to minimize the number of transfers along with the total travel time of passengers. The methodology consists of hybridizing a GA with a neighborhood search algorithm. Numerical experiments are performed with different travel demand matrices. Compared to the existing network, both the total travel time and the number of transfers are significantly reduced for all races. Yu et al. [136] extend the direct traveler density model of Yang et al. [135] to design a transit network that maximizes the number of direct travelers transported per unit length of a line (passengers/km). The authors implement an ant colony optimization algorithm (ACO) to design the transit network while solving a transit allocation problem based on the minimum travel distance to obtain passenger flows. The proposed approach is tested using data from a Chinese city and the numerical results show that the optimized transit line network is an improvement, in terms of transfers, on the existing transit network. Szeto and Jiang [117] implement three versions of the artificial bee colony algorithm

(ABC) [69] to resolve the NDT, minimizing the weighted sum of the number of transfers and the total travel time of users. ABC algorithms, along with stop sequence improvement heuristics, are used to design lines. The results show that the frequency of each line found by the best-proposed algorithm is reduced by 5.5%, without increasing the size of the fleet, compared to the other two algorithms based on the proposed population. However, the study lacks comparison with other algorithms in the literature. Nikolić and Teodorović [95] implement the ABC to resolve the TND with a static demand that minimizes the number of transfers and total travel time. The authors resolve test instances and compare their results with earlier approaches, such as Baaj and Mahmassani [8] and Chakroborty and Dwivedi [24]. The results show that ABC improves the percentage of demand satisfied without transfers, and demand satisfied with a transfer, in three out of four cases. Improvements are also obtained for average travel time. Later, Nikolić and Teodorović [96] expand on their approach described above. considering the elastic demand and the minimization of the weighted sum of the total number of unsatisfied passengers, the total travel time of all passengers, and the size of the fleet. Nayeem and Rahman [93] develop two versions of a GA with a static demand that minimizes the weighted sum of dissatisfied passengers, the total number of transfers and the total travel time of all passengers served. Also, the authors implemented an alternative algorithm that increases the size of the population after each iteration. The latest study presents a comparison between the proposed algorithms and those of Baaj and Mahmassani [8], Chakroborty and Dwivedi [24], Fan and Machemehl [43], Nikolić and Teodorović [95]. The numerical results show that the AG outperforms the other approaches compared in the study in all three criteria in most of the tested cases. More recently, Canca et al. [18] developed a model to address the problem of integrated rapid transit network design and line planning. A large search metaheuristic is employed to handle network design and line planning problems to solve real-size problems efficiently. The performance of the proposed algorithm is evaluated in a small artificial instance. Computational results indicate that the proposed algorithm outperforms state-of-the-art commercial solvers in all test instances. The algorithm is also tested in a real-life instance using data from the city of Seville.

Two-Level Approaches to Transit Network Design

Gao et al. [48] address NRT where upper-level defined lines minimize the weighted sum of average travel time, average waiting time, and operational cost, while the lower level is a transit allocation problem. The authors propose an algorithm based on sensitivity analysis to solve some numerical examples. Later, Fernandez et al. (2008) present a Transit Network Design problem in the context of Santiago de Chile, with the objective of minimizing the total cost as a function of the travel times of the lines, the operation of the vehicle and the number of passengers. users' travel times. Due to the endogenous characteristics of the demand on the problem, the authors propose a heuristic approach consisting of: (i) determine potential network topologies and then (ii) measure each topology by solving the two-level problem that defines frequency configuration at the top level and traffic allocation decisions at a lower level. Szeto and Jiang [118] formulate the TND where the upper level defines the bus lines minimizing the number of passenger transfers and the problem of the lower level is the allocation of transit with capacity limitations. As in the authors' study for a single-level problem [117], a bee colony algorithm is also implemented to solve the optimization problem. The proposed approach is tested in Hong Kong and Canada.

2.2.2 Robust and Stochastic Transit Network Design

In practice the demand is uncertain. To address this problem, robust, randomly limited and stochastic optimization has been implemented. Chen and Yang [25] address an NRT to minimize total travel time considering the equity issues and uncertainty of demand. The authors propose two different stochastic formulations and to solve the problems developed an iterative solution approach consisting of an AG to solve the NRT and a simulation procedure to determine the uncertainty of demand. The numerical results show that there is a trade-off between the measure of spatial equity and the expected value of total travel time. Stochastic travel times are considered by Yan et al. [133], which addresses a robust optimization version that minimizes the weighted sum of the operator's expected cost value and its variability. To solve the problem, the authors implement a solution methodology that consists of: (i) a shorter K-route algorithm to generate candidate lines; (ii) a simulated annealing algorithm to solve the transit assignment problem; (iii) a Monte Carlo simulation to define the passenger flow load; and (iv) a probit-type discrete choice model included in the simulation to determine passenger options. Amiripour et al. [3] determine a set of lines that are implemented considering seasonal demand patterns with a probability of occurrence. The authors propose a mathematical formulation to minimize the expected value of the weighted sum of total passenger waiting time, unused seating capacity, unsatisfied demand, and fleet size. To solve the proposed formulation, the authors develop a GA algorithm in which the fitness function solves a transit allocation problem.

2.2.3 Multi-target Transit Network Design

Recent studies on NRT problems focus on the analysis of trade-offs between different evaluation criteria. Fan and Machemehl [40] implement a decomposition approach to solving NDT. In their experimental stage, the authors analyze the trade-offs between different targets, as the weighted target functions can be used to approximate Pareto's set of optimal solutions. Mauttone and Urquhart [87] propose a multi-target optimization approach that minimizes user costs and fleet size. The authors propose a General Responsibility Assignment Software Patterns (GRASP) algorithm for the multi-target optimization problem. Chew et al. [26] formulate a bi-objective version that minimizes user costs and operator costs subject to restrictions such as a limited number of lines, a limited number of line nodes, and a limited percentage of passengers using multiple transfers. To resolve the proposed approach, an GA is designed and tested in a Benchmark data set. Most of the research efforts mentioned above focus on the application of GA and its variants to solve the problem of network design and frequency configuration simultaneously. Only a few studies are based on SA, TS, ACO, and ABC. In addition, it is evident that some algorithms produce a good network design based on specific parameters and allocation procedures without paying attention to the comparison of studies with other algorithms in the literature (Table 1).

3 Methodology Proposed

Spider monkey optimization is a recently developed nature-inspired stochastic optimization technique, which particularly finds its place in the category of swarm intelligent techniques. The spider monkey is a species of monkeys which belongs to the class of fission-fusion social structure (FFSS)- based animals (Symington 1990). These monkeys live in groups and exhibit intelligent foraging behavior. They search for their food in different directions by sharing information with other group members. Their intelligent food searching strategy is the source of motivation behind the development of this algorithm. The search space of the optimization problem under consideration can be seen as the food searching area of spider monkeys. Each solution of the optimization problem is represented by the position of spider monkey in this food searching area. The collection of all the solutions is called the swarm. The optimal point is the food source. Fitness of a solution indicates how near is the spider monkey to the food source. Inspired by the food foraging behavior of spider monkeys, this technique follows the rules of information sharing and continuous learning for position updates among the whole swarm like other swarm intelligent techniques. In SMO, each spider monkey strives to update its current position with a better one by using its own experience as well as the experience of other members of the swarm. The detailed mathematical formulation of this technique can be found in Bansal et al. (2014). SMO has four user-defined parameters, namely the perturbation rate, the maximum number of groups, the global leader limit, and the local leader limit. Like all other metaheuristics, this algorithm starts with uniformly generated random initial positions of all the spider monkeys in the swarm. These positions get updated with iterations. The best solution to the whole swarm is called the global leader. If the global leader is not improving for a specified number of iterations, the whole swarm gets divided into groups. The best solution in each group is known as the local leader of that group. Initially, there is only one group. So, initially, the local leader is the same as the global leader. In addition to initialization, SMO has six iterative phases that contribute to improving the position of spider monkeys. These phases are the Local Leader Phase, Global Leader Phase, Global Leader Learning Phase, Local Leader Learning Phase, Local Leader Decision Phase, and Global Leader Decision

Table 1 Summary of the state of the art

Authors (year)	Objective	Constraints	Method of solution	Case
Clarens and Hurdle [31]	Minimize user and operational	Vehicle	Capacity	Analytical Example
Marwah et al. [82]	Minimize bus loading time and operational costs	Satisfaction of demand	Decomposition and Heuristics	Test
Ceder and Wilson [22]	Minimize excess travel, waiting and transfer time	Minimum frequency	Heuristics	Examples
van Nes et al. [127]	Maximise the number of direct journeys	Budget and fleet size	Decomposition and Heuristics	Examples
Baaj and Mahmassani [8–10]	Maximize the number of direct trips	Fleet size and advance restrictions	Decomposition and Heuristics	Examples
Chien and Schonfeld [27]	Minimize user and operational costs	Vehicle Capacity	Analytical	Example
Furth and Rahbee [47]	Minimize road distance, bus boarding delays, and operational costs	The maximum distance between stops	Dynamic Programming	Real
Chakroborty and Dwivedi [24]	Minimize unsatisfied demand, total travel time and number of passengers with more than two transfers	Feasibility of the route	GA	Examples
Chien and Spasovic [28]	Minimize consumer surplus and operating profits	Vehicle Capacity	Analytical	Example
Chakroborty [23]	Minimize passenger waiting time	Fleet size, restrictions policies and maximum transfer time	Decomposition and GA	Example
Guan et al. [54]	Minimize vehicle time and number of transfers	Satisfaction of demand Solver	Market test	Example
Wan and Lo [129]	Minimize cost of operations	Capacity and Frequency Restrictions	CEPLEX	Example
Argwal and Mathew [5]	Minimize costs based on travel times and distances	Frequency restrictions, load factor and demand satisfaction	GA	Real
Chen and Yang [25]	Minimize total travel time	Spatial equity and constraints	GA and Simulation	Example

(continued)

Table 1 (continued)

Authors (year)	Objective	Constraints	Method of solution	Case
Fan and Machemehl [40]	Minimize user weight and operational costs	Fleet size, voyage length and number of lines	Decomposition and Heuristics	Examples
Gao et al. [48]	Minimize user weight and operations	Minimum frequency	Sensitivity analysis	Example
Chiou [29]	Minim ize travel and construction costs	Link capacity	Gradient Method	Examples
Fan and Machemehl [41]	Minimize total costs of operations and users	Restrictions on load factor, line length and fleet size	GA	Examples
Fan and Machemehl [42]	Minimize unmet demand, user and operational costs	Load factor, line length and fleet size	SA	Example
Yang et al. [135]	Maximize the number of direct trips per unit length	Bus capacity, minimum number of direct trips and space between stops	ACO	Examples
Fan and Machemehl [43]	Minimize total cost of operations and users	Restrictions on load factor, line length and fleet size	TS	Example
Fernández et al. [46]	Minimize costs based on travel time, vehicle operations, and user travel time	Allocation of demand and load factor	Heuristics	Real
Mauttone and Urquhart [87]	Minimize VS fleet size. Minimize user cost	Frequency restrictions and fleet size	GRASP	Example
Lownes and Machemehl [78]	Minimize the number of users, operators, and the limit of unsatisfied demand	Fleet capacity and size	TS	Example
Daganzo [33] Minimize total cost of operations and users	Number of stops, spacing between them	Bus stops in a Smart City	Analytical	Examples

(continued)

Table 1 (continued)

Authors (year)	Objective	Constraints	Method of solution	Case
Estrada et al. [39]	Minimize total cost of operations and users	Number of stops, space between them and the capacity system	Analytical	Simulation
Szeto and Wu [119]	Minimize the weight of the number of transfers and total travel time	Fleet size, frequency restrictions, number of intermediate stops and maximum travel time	GA	Real
Cipriani et al. [30]	Minimize user and operational costs and line length	Bus capacity and minimum frequency	GA	Real
Szeto and Jiang [117]	Minimize the weight of the number of transfers and total travel time	Fleet size, frequency restrictions, number of intermediate stops and maximum travel time	ABC	Real
Yu et al. [136]	Maximize travel density	Satisfaction of demand	ACO	Real
Chew et al. [26]	Minimize VS user costs. Minimize operating costs	Number of stops and lines and percentage of passengers using in transfer	GA	Market test
Medina et al. [88]	Minimize total cost of operations and users	Bus capacity, stop capacity and stop density equivalence in each direction	Decomposition and Analytical	Real
Nikolić and Teodorović [95]	Minimize the weight of the number of transfers and total travel time	Satisfaction of demand	ABC	Market test

(continued)

Table 1 (continued)

Authors (year)	Objective	Constraints	Method of solution	Case
Yan et al. [133]	Minimize the expected value of operational costs and their variability	Fleet size, bus capacity, line feasibility and travel time reliability	SA	Examples
Amiripour et al. [3]	Minimize total waiting time, fleet size, total differences of small parts and total empty seating hour	Budget, maximum unsatisfied demand, travel restrictions, number of transfers	GA	Examples
Kim and Schonfeld [71]	Minimize user, operations and transfer costs	Bus size and number of zones	Analytical	Examples
Ouyang et al. [98]	Minimize overall user cost	Fleet size	Analytical	Examples
Nayeem and Rahman [93]	Minimize dissatisfiedpassengers, number of transfers, and travel times	Satisfaction of demand	GA	Real
Nikolić and Teodorović [95]	Minimize the number dissatisfied passengers, travel time, and fleet size	Satisfaction of demand, binding capacity, and vehicle capacity	ABC	Market test
Szeto and Jiang [118]	Minimize transfers	Budget and frequency, fleet capacity and size	ABC	Real
This research (2019)	Optimization of bus stops in a Smart City	Bus capacity and ubiquitous application	Monkey spider optimization Algorithm	Real data

Source Own elaboration

Phase. Each phase has a different purpose to serve in the execution of the entire SMO algorithm. Local Leader Phase and Global Leader Phase are responsible for generating a new trial position for each spider monkey. If this newly generated position is better than the current position, the spider monkey updates its position with the new one; otherwise, it retains its current position. Global Leader Learning Phase and Local Leader Learning Phase are there for selecting the global leader and local leaders, respectively. Local Leader Decision Phase and Global Leader Decision Phase are used to checking and handling stagnation and premature convergence in local groups and the whole swarm, respectively. Figure 4 below shows the methodology followed by this research.

Taking into account the study carried out in the municipality of Saltillo on the urban transport system, the route to be investigated is chosen. Subsequently, a literary revision of the last ten years is carried out, which includes texts related to this subject. A model for determining demand and an optimization method is studied and selected.

Fig. 4 Flowchart of the methodology. *Source* Own elaboration

Then, data collection begins to model the problem situation through the implementation of artificial intelligence techniques. Once this process is concluded, the results are validated to establish the optimal location of bus stops on the selected route.

3.1 Proposed Solution: Implementation of the near neighbor algorithm

When calculating the distance of the equal points of the matrix, for example, m [0] [0] will be placed with a double greater than all the distances shown in Table 2, this modification was made in the distance function this in order to discard these points as a minimum since if they remain as zero they would be taken as small values. The following table shows how it was formed.

In this way it is ensured that the minimum real values are taken, having the minimum of each point two situations necessary for a good result are processed, both situations happen when having the minimum value, one is when finding said value, that value is substituted in the matrix by the greater value this would mark the point as visited and it would be discarded that that point be contemplated again, the next situation is to save the point of the visited matrix in an ArrayList which would be forming the optimal route.

3.2 Validation

A first route was obtained which starts at stop one, followed by two, four, six, seven, five, three, zero and eight ending at the initial stop (1, 2, 4, 6, 7, 5, 3, 0, 8) which threw a total distance of 7.9 km. Each number on the route represents a bus stop, where zero was assigned to the school.

Figure 5 shows how the route would be when following the route.

4 Results

By adding the external cycle to the algorithm and performing the tests with the library, a new route was obtained for the graph, which would start at stop number four, following stop six, seven, five, three, two, one, zero and eight, ending with the initial stop (4,6,7,5,3,2,1,0,8), the total distance of this route is 7.7 km. It is possible to have a better route, since, when starting from a different point, the minimum distances to another bus stop change, or they are not discarded because it is a stop already visited. In Fig. 6, the route is shown following the route that was obtained.

Table 2 Distance matrix with maximum diagonal

	0	1	2	3	4	5	6	7	8
0	100	0.004754	0.008854	0.016618	0.008528	0.013256	0.005438	0.009781	0.008884
1	0.004754	100	0.004237	0.015505	0.007914	0.01495	0.00897	0.013555	0.013046
2	0.008854	0.004237	100	0.014227	0.008267	0.016203	0.01211	0.016556	0.017278
3	0.016618	0.015505	0.014227	100	0.008102	0.009119	0.014198	0.015547	0.024567
4	0.008528	0.007914	0.008267	0.008102	100	0.007995	0.007189	0.010359	0.016796
5	0.013256	0.01495	0.016203	0.009119	0.007995	100	0.008442	0.00738	0.018732
6	0.005438	0.00897	0.01211	0.014198	0.007189	0.008442	100	0.004585	0.010616
7	0.009781	0.013555	0.016556	0.015547	0.010359	0.00738	0.004585	100	0.012128
8	0.008884	0.013046	0.017278	0.024567	0.016796	0.018732	0.010616	0.012128	100

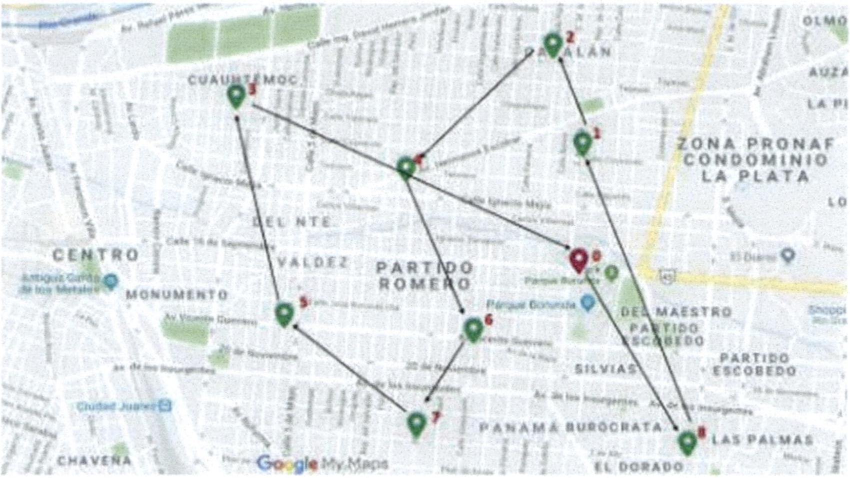

Fig. 5 First route obtained

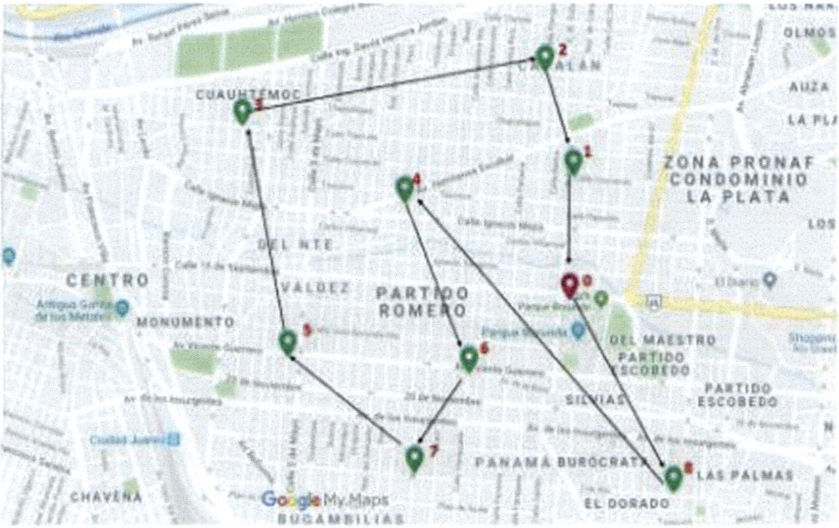

Fig. 6 Optimal route

5 Conclusions and Future Research

When the library was created in the beginning, the Dijkstra algorithm was contemplated. By having the library finished with the said algorithm, the expected results were not obtained, since this algorithm searches for the shortest path from an initial point to an endpoint, this without going through all the points, for the user it would

Fig. 7 The OECD considers that by 2027, seven of the 12 components of a Smart City will be associated with improving school performance, including student mobility

be a bit more complicated to perform the optimization since he would have to add all the adjacent points he needs with their respective distances.

5.1 Future Research

Being a project that improves a route, not only can it be used for schools, it can also be used for parcel companies, delivery companies or transport companies. Using a mobile device associated with a more detailed framework could even determine specific days with more transported objects associated with each student. An aspect of utmost importance will be to adapt in time, the specifications of each individual in childhood that allows it to be competitive, helping it with mobility in a Smart City, as can be seen in Fig. 7.

The route is formed in such a way that the next stop number is the combination with the previous stop number, in this way it is possible to know which is the point of departure and departure, to perform the sum of the distance and corroborate which is the optimal route. An important aspect is to solve with less time the resolution of a new neighborhood with student locations and determine the correct locations of each stop of this route, as you can see in Fig. 8.

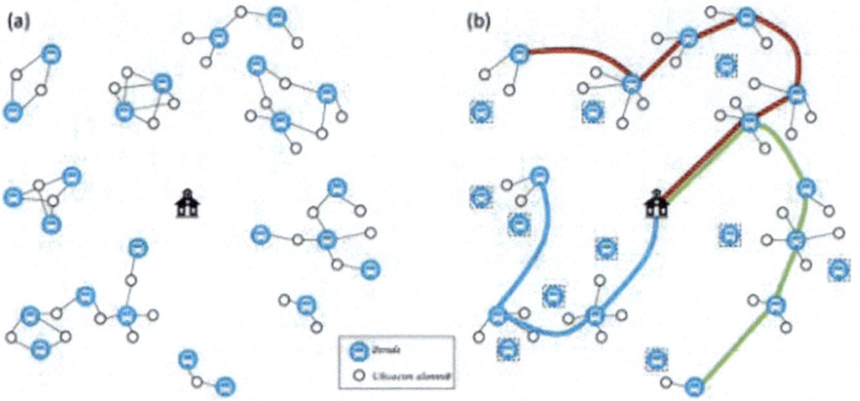

Fig. 8 Description of each group of students and their associated stops (**a**) and then the execution of waiting times (**b**)

References

1. Alawadhi S, Aldama-Nalda A, Chourabi H, Gil-Garica JR, Leung S, Mellouli S (2012) Building understanding of Smart City initiatives. In: Electronic government, 7443 p
2. Alonso-Mora J, Samaranayake S, Wallar A, Frazzoli E, Rus D (2017) Ondemand high-capacity ride-sharing via dynamic trip-vehicle assignment, 201611675. https://doi.org/10.1073/pnas.1611675114
3. Amiripour SMM, Ceder A, Mohaymany AS (2014) Designing large-scale bus network with seasonal variations of demand 48:322–338. https://doi.org/10.1016/j.trc.2014.08.017
4. Amiripour SMM, Mohaymany AS, Ceder A (2015) Optimal modification of urban bus network routes using a genetic algorithm 141(3)
5. Argwal J, Mathew T (2004) Transit route network design using parallel genetic algorithm 18:248–256
6. Atahran A, Lenté C, T'kindt V (2014) A multicriteria dial-a-Ride problem with an ecological measure and heterogeneous vehicles 21 (5–6):279–298. https://doi.org/10.1002/mcda.1518
7. Atzori L, Iera A, Morabito G (2010) The internet of things: a survey. 54(15):2787–2805
8. Baaj M, Mahmassani H (1991) An AI-based approach for transit route system 25:187–209
9. Baaj M, Mahmassani H (1992) AI-based system representation and search procedures for transit route network design:67–70
10. Baaj M, Mahmassani H (1995) A hybrid route generation heuristic algorithm for the design of transit networks 3:31–50
11. Barcos L, Rodríguez VM, Álvarez MJ, Robusté F (2004) Routing Optimization for many-to-many freight transportation problems using Ant Colonies Techniques. In: Presented at the Second World conference on POM and 15th annual POM conference, México
12. Berbeglia G, Cordeau JF, Laporte G (2012) A hybrid tabu search and constraint programming algorithm for the dynamic dial-a-Ride problem 24(3):343–355. https://doi.org/10.1287/ijoc.1110.0454
13. Black A (1979) Optimizing urban mass transit systems: a general model 677, 41–47
14. Braekers K, Kovacs AA (2016) A multi-period dial-a-ride problem with driver consistency 94:355–377. https://doi.org/10.1016/j.trb.2016.09.010
15. Buba AT, Lee LS (2018) A differential evolution for simultaneous transit network design and frequency setting problem 106:277–289. https://doi.org/10.1016/j.eswa.2018.04.011
16. Byrne B (1975) Public transportation line positions and headways for minimum user and system cost in a radial case 9:97–102

17. Byrne BF, Vuchic VR (1972) Public transportation line positions and headways for minimum cost. In: Presented at the Traffic flow and transportation. Elsevier, New York, pp 347–360
18. Canca D, De-Los-Santos A, Laporte G, Mesa JA (2017) An adaptive neigh- bourhood search metaheuristic for the integrated railway rapid transit network design and line planning problem 78:1–14
19. Caragliu A, Del Bo C, Nijkamp P (2009) Smart cities in Europe. In: Presented at the 3rd Central european conference in regional science (CERS 2009). Košice, Slovak Republic
20. Castells M (2000) Urban sustainability in information age 4 (1), 118–122. https://doi.org/10.1080/713656995
21. Ceder A (2007) Public transit planning and operation: theory modeling and practice. Elsevier, Butterworth-Heinemann
22. Ceder A, Wilson N (1986) Bus network design 20:331–344
23. Chakroborty P (2003) Genetic algorithms for optimal urban transit network design 18:184–200
24. Chakroborty P, Dwivedi T (2002) Optimal route network design for transit systems using genetic algorithms 34:83–100
25. Chen A, Yang C (2004) Stochastic transportation network design problem with spatial equity constraint 1882:97–104
26. Chew JSC, Lee LS, Seow HV (2013) Genetic algorithm for biobjective urban transit routing problem
27. Chien S, Schonfeld P (1997) Optimization of grid transit system in heterogeneous urban environment 123:28–35
28. Chien S, Spasovic L (2002) Optimization of grid bus transit systems with elastic demand 36:63–91
29. Chiou SW (2005) Bilevel programming for the continuous transport network design problem 39:361–383
30. Cipriani E, Gori S, Petrelli M (2012) A bus network design procedure with elastic demand for large urban areas 4:57–76
31. Clarens G, Hurdle V (1975) An operating strategy for a commuter bus system 9:1–20
32. Correia LM, Wunstel K (2011) Smart cities applications and requirements. In: White Paper of the Experts Working Group networks european technology platform. Retrieved June 28, 2019, from http://www.scribd.com/doc/87944173/White-PaperSmart-Cities-Applications
33. Daganzo C (2010) Structure of competitive transit networks 44:434–446
34. Dahiya B (2012) Cities in Asia 2012: Demographics, economics, poverty, environment and governance 29(S2)
35. Debnath AK, Chin HC, Haque M, Yuen B (2014) A methodological framework for benchmarking smart transport cities 37:47–56. https://doi.org/10.1016/j.cities.2013.11.004
36. Desaulniers G, Hickman M (2007) Public transit:69–120
37. Dirks S, Keeling M, Dencik J (2009) How smart is your city? IBM Institute for Business Value. Retrieved June 25, 2019, from http://www03.ibm.com/press/attachments/IBV_Smarter_Cities_-_Final.pdf
38. Estrada MA, Robusté F (2008) Análisis de estrategias eficientes en la logística de distribución de paquetería. UPC, Cátedra Abertis
39. Estrada M, Roca-Riu M, Badia H, Robusté F, Daganzo C (2011) Design and implementation of efficient transit networks: procedure, case study and validity test 45, 935–950
40. Fan W, Machemehl R (2004) Optimal transit route network design problem: algorithms, implementations, and numerical results. University of Austin. Texas, USA, Technical Report SWUTC/04/167244-1
41. Fan W, Machemehl R (2006) Optimal transit route network design problem with variable transit demand: a genetic algorithm approach 132, 40–51
42. Fan W, Machemehl R (2006b). Using a simulated annealing algorithm to solve the transit route network design problem 132:122–132
43. Fan W, Machemehl R (2008) Tabu search strategies for the public transportation network optimizations with variable transit demand 23:502–520

44. Farahani R, Miandoabchi E, Szeto W, Rashidi H (2013) A review of urban transportation network design problems 229:281–302
45. Fatnassi E, Chaouachi J, Klibi W (2015) Planning and operating a shared goods and passengers on-demand rapid transit system for sustainable city logistics: 440460. https://doi.org/10.1016/j.trb.2015.07.016
46. Fernández J, de Cea J, Malbran R (2008) Demand responsive urban public transport system design: methodology and application 42:951–972
47. Furth P, Rahbee A (2000) Optimal bus stop spacing through dynamic programming and geographic modeling 1731:15–22
48. Gao Z, Sun H, Shan L (2004) A continuous equilibrium network design model and algorithm for transit systems 38:235–250
49. Garaix T, Artigues C, Feillet D, Josselin D (2011) Optimization of occupancy rate in dial-a-ride problems via linear fractional column generation 38(10):1435–1442. https://doi.org/10.1016/j.cor.2010.12.014
50. Garau C, Masala F, Pinna F (2015) Benchmarking smart urban mobility: a study of italian cities. In: Computational science and its applications. ICCSA, Springer International Publishing, Switzerland
51. Garau C, Masala F, Pinna F (2016) Cagliari and smart urban mobility: analysis and comparison 56:35–46. https://doi.org/10.1016/j.cities.2016.02.012
52. Giffinger R, Fertner C, Kramar H, Kalasek R, Pichler-Milanovic´, N, Meijers E (2007) Smart cities: ranking of european medium-sized cities. Centre of Regional Science (SRF), Vienna University of Technology, Vienna, Austria. Retrieved from http://www.smart-cities.eu/download/smart_cities_final_report.pdf
53. Gracía-Ródenas R (2001) Metodología para el diseño de redes de transporte y para la elaboración de algoritmos en programación matemática convexa diferenciable (Doctoral). Universidad Politécnica de Madrid, Escuela técnica superior de ingenieros aeronáuticos
54. Guan J, Yang H, Wirasinghe SC (2003) Simultaneous optimization of transit line configuration and passenger line assignment 40:885–902
55. Guihaire V, Hao J (2008) Transit network design and scheduling: a global review 42:1251–1273
56. Heilporn G, Cordeau JF, Laporte G (2011) An integer L-shaped algorithm for the dial-a-Ride problem with stochastic customer delays 159(9):883–895. https://doi.org/10.1016/j.dam.2011.01.021
57. Herbawi W, Weber M (2012) The ridematching problem with time windows in dynamic ridesharing: a model and a genetic algorithm. In: Presented at the 2012 IEEE congress on evolutionary computation, pp 1–8. https://doi.org/10.1109/cec.2012.6253001
58. Hiremath RB, Balachandra P, Kumar B, Bansode SS, Murali J (2013) Indicatorbased urban sustainability a review 17:555–563. https://doi.org/10.1016/j.esd.2013.08.004
59. Holryod E (1967). Optimum bus service: a theoretical model for a large uniform urban area. In: Presented at the Third International Symposium on the Theory of Traffic Flow. Elsevier, pp 308–328
60. Hosni H, Naoum-Sawaya J, Artail H (2014) The shared-taxi problem: formulation and solution methods 70:303–318. https://doi.org/10.1016/j.trb.2014.09.011
61. Huang Y, Jin R, Bastani F, Wang X (2013) Large scale real-time ridesharing with service guarantee on road networks 7(14)
62. Hurdle V (1973) Minimum cost locations for parallel public transit lines 7:340–350
63. Ibarra-Rojas OJ, Delgado F, Giesen R, Muñoz JC (2015) Planning, operation, and control of bus transport systems: a literature review 77:38–75. https://doi.org/10.1016/j.trb.2015.03.002
64. Jain S, Hentenryck P (2011) Large neighborhood search for dial-a-ride problems, 6876. https://doi.org/10.1007/978-3-642-237867_31
65. Jun MY, Kwon JH, Jeong JE (2013) The effects of high-density suburban development on commuter mode choices in Seoul Korea 31:230–238
66. Jung J, Jayakrishnan R, Park JY (2016) Dynamic shared-Taxi dispatch algorithm with hybrid-Simulated annealing 31(4):275–291. https://doi.org/10.1111/mice.12157

67. Kaan L, Olinick EV (2013) The vanpool assignment problem: optimization models and solution algorithms 66(1):24–40. https://doi.org/10.1016/j.cie.2013.05.020
68. Kalczynski P, Miklas-Kalczynska M (2018) A decentralized solution to the car pooling problem 8318:1–12. https://doi.org/10.1080/15568318.2018.1440674
69. Karaboga D, Gorkemli B, Ozturk C, Karaboga N (2012) A comprehensive survey: artificial bee colony (abc) algorithm and applications:1–37
70. Kepaptsoglou K, Karlaftis M (2009) Transit route network design problem: review 135(8):491–505
71. Kim M, Schonfeld P (2014) Integration of conventional and flexible bus services with timed transfers 68:76–97
72. Kirchler D, Wolfler R (2013) A granular tabu search algorithm for the dial-a-Ride problem 56:120–135. https://doi.org/10.1016/j.trb.2013.07.014
73. La Greca P, Barbarossa L, Ignaccolo M, Inturri G, Martinico F (2011) The density dilemma. A proposal for introducing smart growth principles in a sprawling settlement within Catania Metropolitan Area 28(6):527–535
74. Lee A, Savelsbergh M (2015) Dynamic ridesharing: is there a role for dedicated drivers? 81:483–497. https://doi.org/10.1016/j.trb.2015.02.013
75. Lehuédé F, Masson R, Parragh SN, Péton O, Tricoire F (2014) A multi-criteria large neighbourhood search for the transportation of disabled people 65 (7), 983–1000. https://doi.org/10.1057/jors.2013.17
76. Liu M, Luo Z, Lim A (2014) A branch-and-cut algorithm for a realistic dial-a-ride problem 81:267–288. https://doi.org/10.1016/j.trb.2015.05.009
77. Lombardi P, Giordano S, Caragliu A, Del Bo C, Deakin M, Nijkamp P, Kourtit K (2011) An advanced triple-helix network model for smart cities performance. In: Vrije Universiteit Amsterdam. Retrieved from http://degree.ubvu.vu.nl/repec/vua/wpaper/pdf/20110045.pdf
78. Lownes N, Machemehl R (2010) Exact and heuristic methods for public transit circulator design 44, 309–318
79. Ma S, Zheng Y, Wolfson O (2015) Real-time city-scale taxi ridesharing 27(7):1782–1795. https://doi.org/10.1109/TKDE.2014.2334313
80. MAPFRE (2004) Guía para la correcta ubicación y señalización de paradas de autobuses escolares en la comunidad foral de navarra. España
81. Marcotte P (1986) Network design problem with congestión effects: a case of bilevel programming 34, 142–162
82. Marwah B, Umrigar F, Patnaik S (1984) Optimal design of bus routes and frequencies for Ahmedabad 994:41–47
83. Masmoudi MA, Braekers K, Masmoudi M, Dammak A (2017) A hybrid genetic algorithm for the heterogeneous dial-A-Ride problem 81:1–13. https://doi.org/10.1016/j.cor.2016.12.008
84. Masmoudi MA, Hosny M, Braekers K, Dammak A (2016) Three effective metaheuristics to solve the multi-depot multi-trip heterogeneous dial-a-ride problem 96:60–80. https://doi.org/10.1016/j.tre.2016.10.002
85. Masoud N, Lloret-Batlle R, Jayakrishnan R (2017) Using bilateral trading to increase ridership and user permanence in ridesharing systems 102:60–77. https://doi.org/10.1016/j.tre.2017.04.007
86. Masson R, Lehuédé F, Péton O (2014) The dial-a-ride problem with transfers 41(1):12–23. https://doi.org/10.1016/j.cor.2013.07.020
87. Mauttone A, Urquhart M (2009) A multiobjective metaheuristic approach for the transit network design problem 1:253–273
88. Medina M, Giesen R, Muñoz J (2013) Model for the optimal location of bus stops and its application to a public transport corridor in Santiago, Chile 2352:84–93
89. Moujahid A, Inza I, Larrañaga P (2008) Algoritmos Genéticos. Departamento de Ciencias de la Computación e Inteligencia Artificial, Universidad del País Vasco
90. Mourad A, Puchinger J, Chu C (2019) A survey of models and algorithms for optimizing shared mobility. https://doi.org/10.1016/j.trb.2019.02.003

91. Munuzuri J, Larraneta J, Onieva L, Cortes P (2005) Solutions applicable by local administrations for urban logistics improvement 22(1):15–28
92. Nam T, Pardo TA (2011) Conceptualizing smart city with dimensions of technology, people, and institutions. In: Presented at the 12th Annual international conference on digital government research, College Park
93. Nayeem M, Rahman M (2014) Transit network design by genetic algorithm with elitism 46:30–45
94. Ng MK, Hills P (2003) World cities or great cities? A comparative study of 5 Asian Metropolises 20(3):151–165
95. Nikolić M, Teodorović D (2013) Transit network design by bee colony optimization 40:5945–5955
96. Nikolić M, Teodorović D (2014) A simultaneous transit network design and frequency setting: computing with bees 41:7200–7209
97. Orea DG, Villarino TG (2013) Salud, ambiente y territorio: una visión entegrada en un mundo globalizado 2:7–19
98. Ouyang Y, Nourbakhsh S, Cassidy M (2014) Continuum approximation approach to bus network design under spatially heterogeneous demand 68:333–344
99. Parragh SN, Pinho de Sousa J, Almada-Lobo B (2015) The dial-a-Ride problem with split requests and profits 49 (2):311–334. https://doi.org/10.1287/trsc.2014.0520
100. Parragh SN, Schmid V (2013) Hybrid column generation and large neighborhood search for the dial-a-ride problem 40(1):490–497. https://doi.org/10.1016/j.cor.2012.08.004
101. Pimenta V, Quilliot A, Toussaint H, Vigo D (2017) Models and algorithms for reliability-oriented dial-a-Ride with autonomous electric vehicles 257(2):601–613. https://doi.org/10.1016/j.ejor.2016.07.037
102. Raveau S, Muñoz J, de Grange L (2011) A topological route choice model for metro 45:138–147
103. Restrepo JH, Medina PD, Cruz EA (2008) Un problema logístico de programación de vehículos con ventanas de tiempo (VRPTW):39
104. Restrepo R (2013) La logística y las soluciones en la gestión de la red de abastecimiento
105. Riquelme P, Gatica G, Orozco E (2015) Diseño de un modelo de operación para ruteo de transporte urbano basado en simulación discreta 2(3):1–12
106. Ritzinger U, Puchinger J, Hartl RF (2016) A survey on dynamic and stochastic vehicle routing problems 54(1):215–231. https://doi.org/10.1080/00207543.2015.1043403
107. Russell SJ, Norvig P (1995) Artificial intelligence: a modern approach. Prentice Hall, Estados Unidos
108. Salzborn F (1972) Optimum bus scheduling 6:137–148
109. Santos DO, Xavier EC (2015) Taxi and ride sharing: a dynamic dial-a-ride problem with money as an incentive 42(19):6728–6737. https://doi.org/10.1016/j.eswa.2015.04.060
110. Savelsbergh M, Van Woensel T (2016) 50th anniversary invited Article'city logistics: challenges and opportunities 50(2),579–590. https://doi.org/10.1287/trsc.2016.0675
111. Schmöcker J, Fonzone A, Shimamoto H, Kurauchi F, Bell M (2011) Frequencybased transit assignment considering seat capacities 45:392–408
112. Serna-Uran CA, García-Castrillón JA, Flórez-Londoño O (2016) Análisis de Rutas de Transporte de Pasajeros Mediante la Herramienta Network Analyst de Arcgis. Caso Aplicado en la Ciudad de Medellín 7(2)
113. Steria (2011) Smart cities will be enabled by smart IT. Retrieved from http://www.steria.com/uk/fileadmin/assets/media/STE3899Smart_Cities_brochure_08_APP.PDF
114. Stiglic M, Agatz N, Savelsbergh M, Gradisar M (2015) The benefits of meeting points in ride-sharing systems 82:36–53. https://doi.org/10.1016/j.trb.2015.07.025
115. Stiglic M, Agatz N, Savelsbergh M, Gradisar M (2018) Enhancing urban mobility: integrating ride-sharing and public transit 90:12–21
116. Suárez MJ (2009) Optimización de la flota de vehículos de servicio entre terminales de consolidación de mercancías con estacionalidad (tesina)

117. Szeto W, Jiang Y (2012) Hybrid artificial bee colony algorithm for transit network design 2284:47–56
118. Szeto W, Jiang Y (2014) Transit route and frequency design: Bi-level modeling and hybrid artificial bee colony algorithm approach 67:235–263
119. Szeto W, Wu Y (2011) A simultaneous bus route desing and frequency setting problem for Tin Shui Wai, Honk Kong 209:141–155
120. Taniguchi E, Thompson RG, Yamada T (2016) New opportunities and challenges for city logistics https://doi.org/10.1016/j.trpro.2016.02.004
121. The Climate Group, ARUP, Accenture & The University of Nottingham (2011) Information marketplaces the new economics of cities. Retrieved from http://www.theclimategroup.org/_assets/files/information_marketplaces_05_12_11.pdf
122. Think (2011) Smart cities initiative: how to foster a quick transition towards local sustainable energy systems. Retrieved from http://www.symple.tm.fr/uploaded/pdf/THINK_smart_cities.pdf
123. Thynell M, Mohan D, Tiwari G (2010) Sustainable transport and the modernization of urban transport in Delhi and Stockholm 27(6):421–429
124. Toppeta D (2010) The smart city vision: how innovation and ICT can build smart, "liveable", sustainable cities. The Innovation Knowledge Foundation. Think!Report
125. Torres MD, Torres A, De la Torre J, Ponce EE, Luna FJ (n.d.) Estudio Comparativo de Algoritmos de Búsqueda Local. Universidad Autónoma de Aguascalientes, México
126. Tsou KW, Cheng HT (2013) The effect of multiple urban network structures on retail patterns–A cease study in Taipei, Taiwan 32:13–23
127. van Nes R, Hamerslag R, Immers B (1988) Design of public transport networks 1202:74–83
128. Vázquez-Solís J (2017) Aplicación del algoritmo de colonia de hormigas al problema de rutas de reparto con destinos móviles. Escuela Técnica Superior de Ingeniería. Universidad de Sevilla
129. Wan Q, Lo H (2003) A mixed integer formulation for multiple-route transit network design 2:299–308
130. Wang X, Agatz N, Erera A (2018) Stable matching for dynamic ride-sharing systems 52(4):850–867. https://doi.org/10.1287/trsc.2017.0768
131. Washburn D, Sindhu U, Balaouras S, Dines RA, Hayes NM, Nelson LE (2010) Helping CIOs understand "'Smart City'" initiatives: Defining the Smart City, its drivers, and the role of the CIO. Forrester Research, Cambridge, MA Inc. Retrieved from http://public.dhe.ibm.com/partnerworld/pub/smb/smarterplanet/forr_help_cios_und_ smart_city_initiatives.pdf
132. WCED (The World Commission on Environment and Development) (1987) Report of the World Commission on Environment and Development: Our common future. (Vol. Research Memorandum 2011–45). United Nations. Retrieved from http://www.undocuments.net/our-common-future.pdf
133. Yan Y, Liu Z, Meng Q, Jiang Y (2013) Robust optimization model of bus transit network design with stochastic travel time 139(6):625–634
134. Yan Y, Meng Q, Wang S, Guo X (2012) Robust optimization model of schedule design for a fixed bus route 25:113–121. https://doi.org/10.1016/j.trc.2012.05.006
135. Yang Z, Yu B, Cheng C (2007) A parallel ant colony algorithm for bus network optimization 22:44–55
136. Yu B, Yang Z, Jin PH, Wu SH (2012) Transit route network design-maximizing direct and transfer demand density 22:58–75

Public Urban Transportation in the Smart City: An Exploratory Study in the Northern México

Arturo Montoya, Aida-Yarira Reyes-Escalante, Diego-Adiel Sandoval-Chávez, and Alberto Ochoa-Zezzatti

Abstract Upon identifying the troubled system of public transportation in the Northern-border City of Juarez, Mexico, this research adopts an exploratory approach to address such problem and provide a plausible path to solve it. Urban mobility has become one of the main challenges for the Smart City. The latter concept is a global tendency to achieve competitive edge via adopting technological development regarding public mobility. Juarez system of public transportation experiences lacks of modernity infrastructure, which results in pollution due to obsolete vehicles, delays, passenger saturation during heavy hours, traffic accidents and users' inconveniences. Although many proposals arose in the past to deal with the prevailing situation, to date the city experiences no improvements whatsoever. This paper examines what the scenario would be if the City embarked in the endeavor of adopting high-tech vehicles for public transportation. Particularly, the paper transits through the sate-of-the-art literature to provide in-sights regarding modern transportation alternatives that not only are environmental-friendly, but also financially viable, promoting productive chaining and local benefits. The proposal centers in buses powered by electricity. The paper discusses financial figures and possible scenarios; it also highlights the need for future urban infrastructure improvement and reconfigurations, especially pavement and transportation routes redesign to allocate charging stations.

Keywords Urban public transportation · Smart city · Urban mobility · High-tech transportation vehicles · Urban infrastructure

---

A. Montoya · A.-Y. Reyes-Escalante (✉) · A. Ochoa-Zezzatti
Universidad Autónoma de Ciudad Juárez (UACJ), Ave. Heroico Colegio Militar y Ave. Universidad S/N, Zona Chamizal. Ciudad Juárez, Chihuahua, Mexico
e-mail: aida.reyes@uacj.mx

D.-A. Sandoval-Chávez
Tecnológico Nacional de México, Campus Ciudad Juárez (TNMCCJ) Ave. Tecnológico 1340 Fracc. Crucero, Ciudad Juárez 32500, Mexico

© The Author(s), under exclusive license to Springer Nature Switzerland AG 2021
A. Ochoa-Zezzatti et al. (eds.), *Technological and Industrial Applications Associated with Intelligent Logistics*, Lecture Notes in Intelligent Transportation and Infrastructure,
https://doi.org/10.1007/978-3-030-68655-0_10

1 Smart City

The term Smart City conveys the idea of a sustainable urban setting that improves citizens "quality of life" at the time that does not threat the environment. To do so, Smart City management promotes the adoption of Information and Communication Technologies (ICTs) for urban development [1]. In the late 2000s, IBM coined the term "smarter planet" with the idea of configuring the Smart City, a division of hardware, software and digital services conceived to help local government authorities to promote sustainability in their cities. Applications include monitoring transportation water and energy consumption and other diverse urban activities.

Smart City concept included a series of indicators that look for the use of technology in diverse realms of the city, such as: public institutions, public space, education infra-structure and many more [2]. These indicators gave place to categories and levels, which drove the assessment and presentation of results.

The Annual Report of Institute of Higher Business Studies (IESE) Business School University of Navarra includes a Cities in Motion Index (CIMI), the advance and progress in intelligent cities. IESE ranked London in first place when it comes to implementation of intelligent strategies and tactics. The runner up was New York, while Mexico City ranked 133th., [3].

A city is intelligent if the appellative "intelligent" accompanies its economy, people, mobility, environment, governance and living. As stated before, ITCs are fundamental to achieve competitive edge and to assure a sustainable future [4].

Expected benefits derived from transforming cities into intelligent entities may go far beyond than originally realized. In 2018 was estimated that the global market value of intelligent cities will be higher than US $ 1 billion by 2020 and US $2.5 billion by 2025. It will also promote a higher interaction and participation among government, private sector and stakeholders, aiming to promote wellbeing, life transformation and safety of citizens [5].

Global cities such as San Francisco, Barcelona, Amsterdam, Montreal, Dublin or Singapur, followed London and New York in approving legislation and documents that seek for the establishment of an integral strategy for the adoption of intelligent guidelines for cities [6]. It is also a tendency to widespread the strategy to reach all levels of citizens. It is now common for government officials of intelligent cities to integrate people's initiatives to benefit common people and small business [7].

A Smart City plan must include the development of a strong bandwidth network to support digital applications so that connectivity is present citywide and available for citizens. The communications infrastructure may be conformed for a mix of technological options, such as data network, cable, optic fiber, state-of-the-art WiFi. Optic fiber is by far the technology that provides whit the fastest connectivity and supports Wi-Fi and other related options, which support a great variety of devices [8].

Mobile intelligent phones technology is a crucial element in the Smart City. These phones serve as a distribution channel of information as well as an intelligent sensor hooked to a global network of information and communication [8].

2 Intelligent Urban Public Transportation

Urban sprawl has affected citizen interaction worldwide. Now people need to commute long distances to arrive to work, school, hospitals, and public offices, or simply to go shopping. The mobility in cities worsen due to an inefficient system of public transportation, which increases the dependency on automobiles, causing traffic congestion and stress.

Besides the negative effect of an inefficient transportation system on mobility, other factors also contribute to affect it. Lack of normative or enforcement on land usage, or even particular interest, which overlooks impacts on mobility, may be worsening the problem. An immediate negative impact of poor public transportation reflects in the environmental conditions, which suffers from massive emissions that contribute to air pollution. By contrast, Smart Cities keep an agenda focused in improving mobility in public transportation by reconverting the powering of vehicles from fossil-based combustibles to clean options, such as gas or electricity. This movement motivated commuters to consider an economic and clean alternative to the use of automobiles, giving place to share rides [9].

On the other side, many solutions have reached the implementation stage. Gassman et al. [10] point out that projects included traffic system guides, sensor-based parking lots (which allowed on-line monitoring), traffic congestions predictors via intelligent traffic lights, and systems for sharing automobiles and bicycles, as well as autonomous transportation, both public and private. Intelligent mobility initiatives pursued the following objectives:

a. To ensure innovative and sustainable transportation systems.
b. To promote access to diverse transportations options.
c. To guarantee citywide availability.
d. To include no-motorized transportation.
e. To integrate ICTs into public transportation system,

According to Gasmman et al. [10] there is no need for a large-scale transformation to improve public transportation; sometimes, big achievements derive from timely, well-implemented small changes using technology. Intelligent solutions in an efficient urban public transportation discourage the use of private vehicles in favor of public commuting. An intelligent transportation option may be innovative, increasing quality service, enhancing accessibility and reducing transportation costs [7].

Public transportation may improve a great deal in the future with the idea of smart systems, to the extent that citizens will not use automobiles, transforming cities into trafficless spaces. Reductions in the use of energy, a byproduct of the process of improving public transportation is a value-added benefit that protects the environment.

Recent evidence reinforces the idea modern public transportation is a topic of interest that has achieved high acceptance among urban population, particularly the adoption of automated vehicles. In their work Bernhard, et al. [11] emphasizes that studies in Germany used IoT to address public transportation considering autonomous vehicles. Other evidence points towards usability and reliability of vehicles. The interest resided on how these factors affected deterioration and performance, providing better framework to establish maintenance policies [12].

Another application study, to optimize public transportation, Panovski and Zaharia proposes a methodology to integrate situational context (including public events, planned interventions and citizen notifications) in the analysis of public transport data, and develop diverse sceneries using data collections [13]. In addition, the research of the new payment technologies for public transport are considered data from smart cards is given as a source of information about transportation movements on public transport [14].

In a very sound approach, Poltavkaya built on the popularity of electronic payment by cards to trace and monitor trip characteristics in public transportation. Panovsky report additional efforts to optimize public transportation, especially to integrate situational contexts, such as public events, planned interventions and citizen notifications [13].

Using open innovation platforms in the public transport, that study develop by Lusikka, Kinnunen and Kostiainen [15] proposed the innovation platforms allow businesses to develop new, better and more personalized services to citizens like an air quality, passenger counting, platform for testing, sensoring and monitoring, and others. Other research has offered alternative improvements for public transport, for example: optimization, vehicle automation, transport sustainability, automatic vehicles, and more. The Table 1 presents a collection of studies related to the development of innovative technologies applied to public transportation.

3 Context of Public Urban Transportation in Juárez

With the exception of Mexico City, in other important cities of the country, there are severe deficiencies in the metrics related to digital infrastructure and connectivity.

The "Global Innovation Index" shows that in Mexico there is a big gap in terms of science, technology and innovation, ranking at the 56th position at global level. The budget as a percent of the GNP destined to science and technology is only 0.5%. An important number a variety of challenges in the areas of economics, social cohesion, transportation and sustainability in general, prevail [27].

Juarez, México, is the largest city of the State of Chihuahua. It is worldwide know that in Juarez prevails a series of negative factors, such as political differences, lack of long term planning, territorial conflicts, violence and insecurity (5th in perception at National level, 72% of the population feels insecure, according to INEGI), corruption, as well as poor coordination among government levels [28]. According to a report from an independent entity, people grade a 15-year old public transportation in Juarez with 5.97 out of 10 and only 23.5% of the people use it. This low grading reflects on the massive use of the automobile, that reaches 46.9% [29]. Public transportation faces an adverse scenario: old vehicles, unpaved roads, high insecurity for passengers

Table 1 Examples of modern public urban transportation research 2019

Piotr Kisielewski, Mariusz Danda, Marek Bauer [16]	Optimization of rosters in public transport
Marcin Koniak, Andrzej Czerepicki [17]	Autonomization of public transport vehicles
Nistha Nakarmi, Sangeeta Singh [18]	Smart infrastructure for sustainable public transportation
Nistha Nakarmi [19]	Smart infrastructure for sustainable public transportation
Manohar Kumar, Neha Kumari, Suryanshu Tomar, Tarun Kumar [20]	Smart public transportation for smart cities
Anatoly A. Zaitcev [21]	Socialization of public transport in the new technological paradigm
Oded Cats, Alex Vermeulen, Martijn Warnier, J.W.C. Van Lint [22]	Modelling growth principles of metropolitan public transport networks
Andreas Lauber, Eric Sax, Markus Wiedemann [23]	Autonomous driving in public transportation depots
Jessica R. Lazarus, Susan A. Shaheen, Stanley E. Young, […], J. Sam Lott [24]	Shared automated mobility and public transport
Andrés Monzón, Sara Hernandez, Andres Garcia-Martinez, Francesco Viti [25]	Public transport in the era of ITS: ITS technologies for public transport
Thomas Holleczek, Liang Yu, Joseph K. Lee, […], Patrick Jaillet [26]	Detecting weak public transport connections from cellphone and public transport data

and other commuters, mainly cyclists and pedestrians [30]. Public transportation in Juarez mobilizes an estimate of 420,000 people daily [31].

Juarez network of public transportation encompasses 26 routes privately operated. It also includes a massive-use, lane-dedicated route named Eco-Bus (see Fig. 1). However, coverage in not complete along the city, there are many zones without formal service. In these zones, private parties provide informal service to transport people to nearby places [31].

The cost for a single one-way trip is $8.00 (50% discount for seniors). Payment options confine almost to cash for all lines, with the exception of Eco-Bus, which only accepts prepaid cards. Service hours vary for each route, but in general they operate from 4:30 AM to 9:00 PM. Not all the bus stops are equipped, some lack benches, signals, boots or sunshades. Some are informal and unequipped.

With the exception of the study of Bouskela, Casseb, Bassi, Cristina De Luca, Marcelo Facchina [8] which provided insights for Chihuahua City to approach the condition of a Smart City in terms of public transportation, no work could be found addressing public transportation from the standpoint of the Smart City.

Fig. 1 Juarez urban area. *Source* Google. (n.d.). [Map of Ciudad Juárez, Chihuahua Mx. on Google maps]. Retrieved on February 5, 2015, from: https://www.google.com/maps/@31.6905492,-106.46087,12.25z?hl=en

4 Method

The methodology was development in two steps:

a. The method consisted in an exploratory studying adopting a data collection instrument (survey) previously designed and validated. The survey included 26 routes; data collection includes the following categories: route, environmental impact, technology. The treatment of data was descriptive and relational, using SPSS v.23. In addition, Weka 3.9.2 and Tableau 15.0 completed the relational analysis (Table 2).
b. Envisioning an intelligent alternative for the public transportation in Juarez. Proposal steps.

5 Results

a. An exploratory studying

Out of 26 routes, 18 (69%) start their way in the downtown zone, the remaining six have a different starting and ending point, but their routes includes some parts of the downtown within a diameter of 1470 m. (see Table 3). The 31% of the routes have a unique path, very well known by users. The remaining have two or more paths, not always known by users due for a lack an of a route map, either printed or digital (Fig. 2).

Table 2 Transportation routes in Juarez, Mexico

Route	Name
1	1-A Morelos
2	1-B Express
3	2-A
4	2-B
5	2-Lazaro
6	3-A
7	3-B
8	4
9	5-A colectivos
10	5-B
11	7
12	8-A
13	8-B
14	10
15	Línea Industria
16	Juárez Aeropuerto
17	Juárez oasis
18	Juárez Zaragoza
19	Mercado de Abastos
20	Oriente Poniente
21	Permisionarios Unidos
22	Poniente Sur
23	Transportes urbano
24	Universitaria
25	Valle de Juárez
26	Ecobus

Results show that up to 850 buses operate daily. Even though more than 1200 concessions are active, many of them do not operate due to diverse reasons, such as obsolesce and poor maintenance. In average, a route has 33 buses circulating, Table 4 show the number or busses by route.

In regards to frequency of routes, 19 (73%) lines have a waiting time between 5–10 min, which route controllers monitor considering bus number and time schedule. 11% fall. In the range of 10–15 min, while 12% do in the range of 16–20 min. Only one line reported a waiting time longer than 20 min.

With the exception of Eco-Bus, which operates buses of models 2017 o newer, school bustype vehicles cover the other routes. The average model is 2005, that is, they have been operating for at least 15 years. These diesel-powered buses (96%) -most of them imported to Mexico already used- lack of adequate maintenance. It is

Table 3 Transport line terminals

#	Route	Final point
1	1-A Morelos	Termoeléctrica
2	1-B Express	Fracc. Sierra vista
3	2-A	Colonia Insurgentes
4	2-B	Colonia Chihuahua
5	2-Lazaro	Montada
6	3-A	Escobedo
7	3-B	Montada
8	4	Hipódromo
9	5-A colectivos	KM 20
10	5-B	Pánfilo Natera
11	7	Parque Ind. gema
12	8-A	colonia zapata
13	8-B	Seguro Nuevo
14	10	Anapra
15	Línea Industria	Smart Ponciano
16	Juárez aeropuerto	sanlorenzo
17	Juárez oasis bu	Fracc. oasis
18	Juárez Zaragoza	Zaragosa
19	Mercado de abastos	Los ojitos
20	Oriente Poniente	las misiones
21	Permisionarios Unidos	las palmas
22	Poniente Sur	KM20
23	Transportes urbano	mariano Escobedo
24	Universitaria	Ciudad Universitaria
25	Valle de Juárez	San Francisco
26	Ecobus	Tierra Nueva

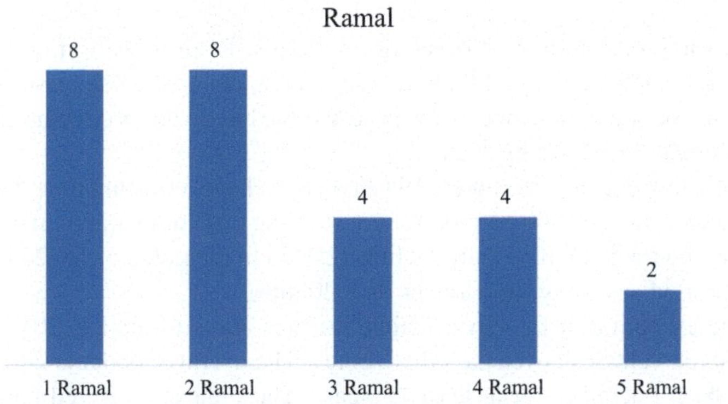

Fig. 2 Total Ramal terminal

Table 4 Busses and routes

Route	Quantity busses
1-A Morelos	50
1-B Express	40
2-A	21
2-B	18
2-Lazaro	43
3-A	45
3-B	25
4	10
5-A colectivos	52
5-B	38
7	8
8-A	38
8-B	31
10	40
Línea Industrial	14
Juárez aeropuerto	73
Juárez Oasis bu	20
Juárez Zaragoza	50
Mercado de abastos	16
Oriente Poniente	50
Permisionarios Unidos	14
Poniente Sur	40
Transportes Urbano	10
Universitaria	57
Valle de Juárez	30
Ecobus	17

very common to operate buses under extreme climate conditions –lacking A/C (in only 8% A/C is available), full of people to overcapacity and over rough-unpaved roads.

By City's ordinance, all drivers must attend a driving course in the facilities of the Department of Public Transportation. It is questionable how and who designed the course and how reliable it is. Moreover, an extra fee may waive the course. Additionally, all drivers must have an adequate license driver to operate public transportation buses. Results showed only two lines provide drivers with extra training as a part of their organizational educational program.

When it comes to unit maintenance, 85% of concessioners provide service in their own mechanic shop, while some do so in informal or improvised shops, or even

Table 5 Factors affecting maintenance frequency

	Option 1	Option 2
Unpaved roads or damaged pavement	19	3
Transport polices	3	2
High cost of spare parts	1	11
Credit restrictions	1	2
Vandalism and incivilities	1	7
Traffic issues	1	1

outdoors, in the public space. Only three reported to maintain units in private-owned certified shops.

The most frequent period of maintenance is weekly (62%), followed by monthly (38%). Maintenance include fluid levels and reposition of worn parts, as well as occasional failures. Very often, these maintenance actions derived from circulating in un-paved roads, damaging suspension and tires. Factors affection the frequency of unit maintenance is the high cost of spare parts, misconduct of users (vandalism and incivilities), see Table 5.

In the case of intelligent options, only one unit reported to have free WiFi (unavailable to users). Technological features concentrate in a GPS system to locate the unit, without keeping a database. Results revealed that, traffic controllers, messaging and phone calls continue in use. Sensing technologies only appeared in four units (15%).

None of the concessioners provide with electronic itineraries. A private initiative created a Facebook page named "Mapas de Ruteras de Ciudad Juárez" (@RuterasJrz · Sistema de transporte público).

b. Envisioning an intelligent alternative for the public transportation in Juarez.

The proposal is to increase the quality of public transportation services via adopting a strategy based upon the implementation of information technologies. That is, to increase public mobility incorporating Smart City's tactics while the environment is not harmed. Three stages compass this proposal: (a) replacing actual unit with modern vehicles; (b) adopting information technologies; (c) developing of an App for the use in the public transportation system of Juarez.

We propose to replace old out-of-times buses with modern state-of-the-art electricity-powered units, equipped with satellite-based location devices, front and rear cameras, users monitoring, passenger on-off and bus stoppage monitoring, WiFi, general use monitoring, emergency alerts and electronic payment devices, among others.

Electrical buses are cleaner and healthier, at the time that bring savings in fuel and maintenance. They are silent and easy to drive, do not emit toxic gases that pollute the environment, but they are nor exempt of disadvantages.

There are advantages and disadvantages in adopting electric vehicles [32]. The advantages include zero emissions, 50–100% reduction in greenhouse effect gases (depending on the way generation process of electricity), reducing life cycle costs of

Fig. 3 Electrical Buss BYD model K9 [32]

batteries, low levels of noise, among others. On the other hand, disadvantages include capital-intensive investment to provide for infrastructure, such as rapid chargers, high price of units, limited options for financing and high operating costs.

The proposal consists in replacing the old fleet with electrical units equipped with energy-storage batteries. The reference is the unit BYD Model K9 [33]. BYD is a China-based company that is the top seller of electrical buses. Recently, BYD expanded operations by opening a manufacturing plant in California. According to BYD's website, the buses operate on a Ferrum-based (Fe) battery, which last for 155 miles at full charge. The recharge time is approximately one hour; the battery has a useful life up to 4000 recharge cycles (see Fig. 3).

To adopt electrical technology implies the need for space to accommodate chargers in the operations base. For this project, we propose the acquisition and installation of at least 60 charge stations located at functional terminals (see Fig. 4).

All intelligent equipment will be integrated into the unit, this refers to tangible services of comfort for the passenger, such as comfortable seats, and intangible services such as monitoring of passengers and driver, WiFi services, emergency signals, cameras, among others, see Fig. 5.

Battery-equipped buses do not have exhaust pipe; instead, electrical cells produce water steam only. Buses' emissions depend upon how previous productive chain links generate electricity and hydrogen fuel. If renewable energies generate the electricity and hydrogen, then there will be no emissions at all [34].

The proposal includes the development of an App. The use of apps goes in line with the adventment of innovations in information technologies, such as intelligent phones and tablets.

The app would interact with the main social networks and user's email. The user will be able to search for the status of routes in real time, as well as for departing and

Fig. 4 Station reload [32]

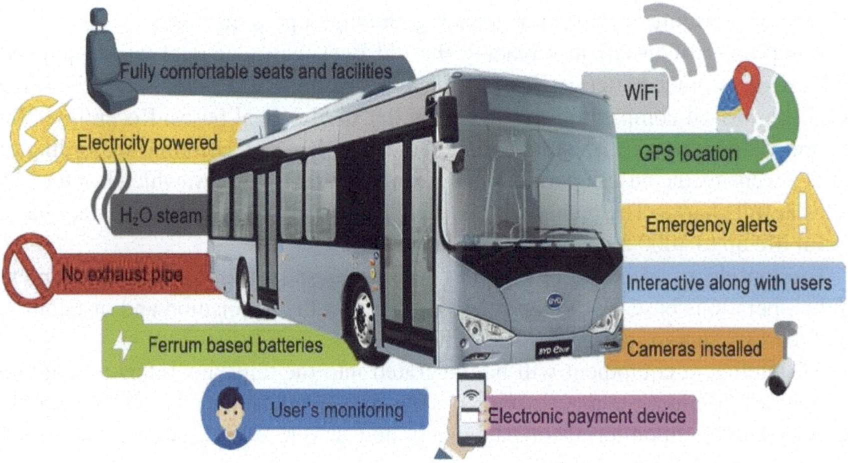

Fig. 5 Intelligent bus

ending points, and transships. The app accepts electronic payments and keep record of the trip history, providing statistics about frequency, distances and favorite routes. The trip planning platform will allow the user to introduce destine details, providing information about nearest bus stop, waiting times, trip times, traffic status, transship options and available transportation lines Users will be able to buy, via credit card or electronic wire transfer, passage for one or more days, up to a month. The app will send to the smart phone o tablet a QR code, which validates when getting on the bus (Fig. 6). The app will also enable the user to report bus conditions and driver behavior, thus controlling unit's condition and driving irregularities.

Proposal pilotage will require an initial investment of MX$280 million, which cover the acquisition of 60 buses at a unit cost of MX$7.5 million. Proper intelligent equipment technology for buses would cost an extra MX$20 million for the whole project (see Table 6).

Even in a pessimistic scenario in which the project requires high investment, the proposal sound because it thrives to overcome the prevalent poor and costly service. The objective aims to keep up with modernity, to avoid extra costs for users and to decrease the use of private automobiles. Then, the proposal suggests a better option for transportation, searching into the buses' market, focusing on sustainability and modernity.

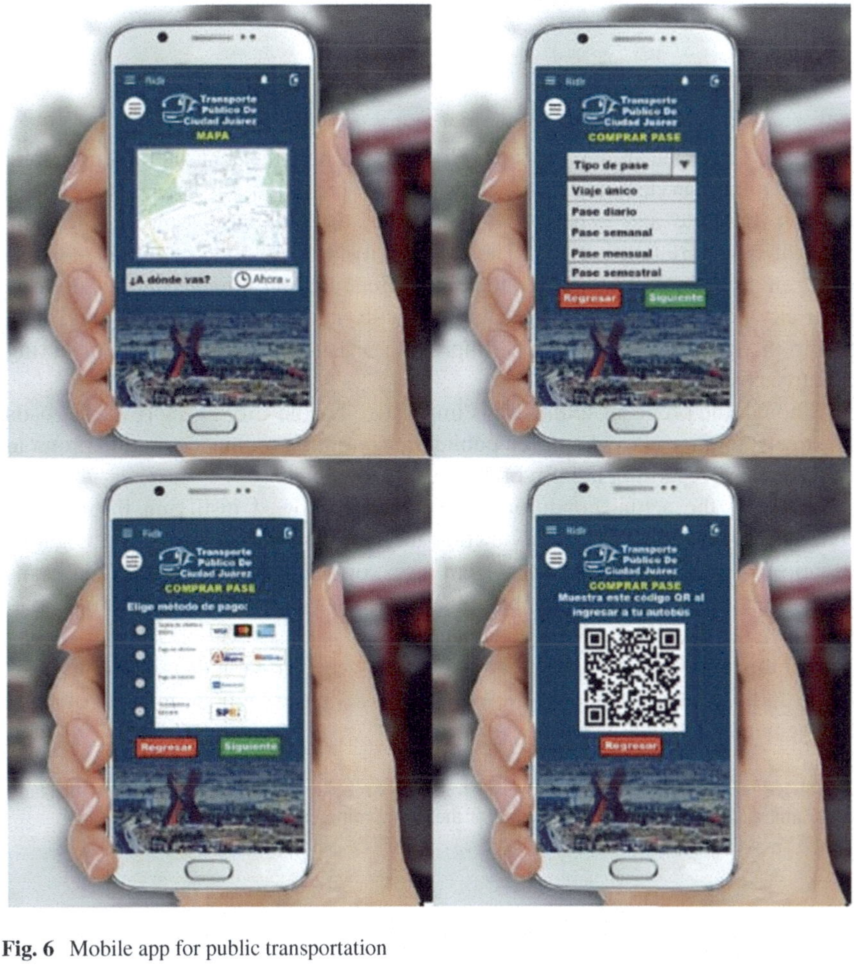

Fig. 6 Mobile app for public transportation

Table 6 Proposal general cost

	Proposal cost
Elements	Cost
Unit cost K9	$7,500,000.00
Number of units	60
Total units cost	$450,000,000.00
Chargers cost	$250,000.00
Number of chargers	60
Total cost of chargers	$15,000,000.00
Pavement route de transp.	$280,000,000.00
Other intelligent equipment	$12,000,000.00
App development	$5,000,000.00
Total	$762,000,000.00

6 Concluding Remarks

This paper envisioned a plausible solution for the prevalent inefficient public transportations service in Juarez, Mexico. By reconfiguring the existing fleet, which is obsolete and expensive to maintain, the city may aspire to have an excellent transportation service. The Smart City promotes the adoption of intelligent, environmental-friendly strategies to offer safe and reliable public services. Electricity-powered buses, fully equipped with state-of-the-art technology, may revamp the transportation services, improving urban sustainability while lowering operative costs.

Although the proposal sounds, it is imperative for the City to improve the roads and streets, to improve and enforce public transportation ordinances and to promote civil education regarding the use of new technologies.

To review the actual conditions of the buses allowed justifying the need to modernize this service. By interviewing, the concessioners revealed the negative scenario the actual transportation system faces: out-of-times, deteriorated buses, as well as archaic management and maintenance practices. This made clear the need to modernize the fleet to provide a better service.

With this proposal, it is easy to visualize the need for an update in the units, as well as in the routes, existent or New-Al this gives place to modern forms of route management.

Acknowledgements we would like to thank to the 26 concessioners who participated in the study for the time dedicated to this research and for the information provided.

References

1. Montava Miró J (2014) Smart Cities. Criterio, análisis y aplicación de la ciudad inteligente. Caso de estudio la ciudad italiana de Matera: Patrimonio de la Humanidad, Nov 2014, Acceded 23 Jun 2020. https://ri-unet.upv.es/handle/10251/44000
2. Giffinger R, Haindl G. Smart cities ranking: an effective instrument for the positioning of cities?, p 12
3. Berrone P, Ricart JE, Duch A, Carrasco C (2019) IESE cities in motion index 2019, Publication Serves of Universidad de Navarra, May 2019. https://doi.org/10.15581/018.st-509
4. Ontiveros Baeza E, López Sabater V, Vizcaíno D, Fundación Telefónica (2017) Las Ciudades del futuro: inteligentes, digitales y sostenibles. Barcelona; Madrid: Ariel: Fundación Telefónica
5. Walcrof D, Chiasson G, Murty N (2018) Creating the smart cities of the future, p 16
6. Fernández DS (2017) Factores de desarrollo de las ciudades inteligentes. Rev Univ Geogr 26(1):135–152
7. Barlow M, Levy-Bencheton C (2018) Smart cities, smart future: showcasing tomorrow. John Wiley & Sons
8. Bouskela M, Casseb M, Bassi S, Luca CD, Facchina M (2016) La ruta hacia las smart cities: Migrando de una gestión tradicional a la ciudad inteligente. Inter-American Development Bank
9. Villa CAP (2020) Imperios contra Estados. La destrucción del orden internacional contemporáneo, Imp. Contra Estados Destrucción Orden Int. Contemp. Accedido: 23 Jun 2020. https://www.academia.edu/34899321/Imperios_contra_Estados._La_destrucci%C3%B3n_del_orden_internacional_contempor%C3%A1neo
10. Gassmann O, Böhm J, Palmié M, Gassmann O, Böhm J, Palmié M (2019) Smart cities. In: Smart cities, pp 25–66. Emerald Publishing Limited,
11. Bernhard C, Oberfeld D, Hoffmann C, Weismüller D, Hecht H (2020) User acceptance of automated public transport: valence of an autonomous minibus experience. Transp Res Part F Traffic Psychol Behav 70: 109–123, abr. 2020. https://doi.org/10.1016/j.trf.2020.02.008
12. Husniah H, Andriana A, Kusmaya K, Gunawan G, Supriatna AK (2020) The economic aspects of internet of things for a public transport system. Proceeding MICEB Mulawarman Int Conf Econ Bus 2, n.º 0, Art. n.º 0, Mar 2020
13. Panovski D, Zaharia T (2020) Real-time public transportation prediction with machine learning algorithms, en 2020 IEEE international conference on consumer electronics (ICCE), ene, pp 1–4. https://doi.org/10.1109/icce46568.2020.9043077
14. Poltavskaya Y (2020) Modeling transport process based on trip generation. https://angtu.editorum.ru/en/storage/viewWindow/53854. Acceded 24 Jun 2020
15. Lusikka T, Kinnunen TK, Kostiainen J (2020) Public transport innovation platform boosting Intelligent Transport System value chains. Util Policy 62, p 100998, Feb 2020. https://doi.org/10.1016/j.jup.2019.100998
16. Kisielewski P, Dańda M, Bauer M (2019) Optimization of rosters in public transport. en 2019 6th international conference on models and technologies for intelligent transportation systems (MT-ITS), Jun 2019, pp 1–8. https://doi.org/10.1109/mtits.2019.8883350
17. Koniak M, Czerepicki A (2017) Selection of the battery pack parameters for an electric vehicle based on performance requirements. IOP Conf Ser Mater Sci Eng 211, p 012005, Jun 2017. https://doi.org/10.1088/1757-899x/211/1/012005
18. Nakarmi N, Singh S (2019) Smart infrastructure for sustainable public transportation. In: IOE Graduate conference 2019
19. Nakarmi N (2019) Smart infrastructure for sustainable public transportation. Project: Sustainable public transportation
20. Kumar M, Kumari N, Tomar S, Kumar T (2019) Smart public transportation for smart cities, Social Science Research Network, Rochester, NY, SSRN Scholarly Paper ID 3404487, Mar 2019. https://doi.org/10.2139/ssrn.3404487
21. Zaitcev AA, З. Александрович А (2019) Socialization of public transport in the new technological paradigm. Transp Syst Technol 5(3), Art. n.º 3, Sep 2019. https://doi.org/10.17816/transsyst2019534-17

22. Cats O, Vermeulen A, Warnier M, van Lint H (2020) Modelling growth principles of metropolitan public transport networks. J Transp Geogr 82: 102567, ene. https://doi.org/10.1016/j.jtrangeo.2019.102567
23. Lauber A, Sax E, Wiedemann M (2018) Autonomous driving in public transportation depots. ATZ Worldw 120(6): 68–71, Jun 2018. https://doi.org/10.1007/s38311-018-0051-6
24. Lazarus J et al (2017) Shared automated mobility and public transport. Jun 2017. https://doi.org/10.7922/g2hq3x3v
25. Monzón A, Hernandez S, Martínez AG, Kaparias I, Viti F (2016) Public Transport in the era of ITS: ITS technologies for public transport. In: Gentile G, Noekel K (eds) Modelling public transport passenger flows in the era of intelligent transport systems: COST Action TU1004 (TransITS), pp 85–128. Springer International Publishing, Cham
26. Holleczek T, Yu L, Lee JK, Senn O, Ratti C, Jaillet P (2014) Detecting weak public transport connections from cellphone and public transport data. In: Proceedings of the 2014 international conference on big data science and computing, Beijing, China, ago, pp 1–2. https://doi.org/10.1145/2640087.2644196
27. W. I. P. Organization, C. University, y INSEAD (2019) The global innovation index 2019: creating healthy lives—The future of medical innovation. Geneva: World Intellectual Property Organization
28. Herrera V «Juárez 2030». Acceded 27 Jun 2020. https://juarez2030.mx/
29. Plan Estratégico de Juárez, «Informe Así Estamos Juárez 2019», Plan Estratégico de Juárez. https://planjuarez.org/documentos/informe-asi-estamos-juarez-2019-2/. Acceded 27 Jun 2020
30. Ortega VMO (2016) El sistema Bus Rapid Transit en Ciudad Juárez y el derecho a la ciudad. Chihuahua Hoy, 14, n.º 0, Art. n.º 0, dic. https://doi.org/10.20983/chihuahuahoy.2016.14.17
31. Cal y Mayo, Análisis carretero en México. https://www.calymayor.com.mx/es/reportes_estadisticos.html (acceded 27 Jun 2020)
32. World Bank (2019) Green Your Bus Ride: Clean Buses in Latin America. World Bank. https://documents.worldbank.org/en/publication/documents-re-ports/documentdetail. Acceded 27 Jun 2020
33. ChinaBuses, BYD Electric Bus K9 - 12 m-13 m–www.chinabuses.org - www.chinabuses.org. http://www.chinabuses.org/product/buses/542.html. Acceded 27 Jun 2020
34. Hamurcu E, Yetik İŞ (2018) Feature selection for optimal weather detection with meteorological radar data. In: 2018 26th signal processing and communications applications conference (SIU), May 2018, pp 1–4. https://doi.org/10.1109/siu.2018.8404610

The Difficulties and Complications of Children When Going to a Zoo and Should Interact with the Colors of the Information in It: An Approach Based on the Use of a Humanoid NAO Robot in an Application for "Smart Cities"

Alberto Ochoa-Zezzatti, Martín Montes Rivera, Julio César Ponce Gallegos, Cesar Velazquez, and Paulo N. M. Sampaio

Abstract One of the decisive aspects in Smart Cities is to provide inclusive spaces, for those who suffer from a condition that limits their interaction with the environment, but in several cases that information is only designed for average people. Zoos visual information is an example of well-designed signals for providing data associated with the species in the exhibition, but all this information is colorful and with different signals, such as the colors of the species, their habitat including their food and geographical distribution, which it means that is not the must inclusive information for people that cannot perceive colors or that has some kind of condition that difficulties metal associations. Thus, any child who has a condition associated with the identification of colors, or certain mental condition will have difficulties to

A. Ochoa-Zezzatti
Doctorado en Tecnología, Universidad Autónoma de Ciudad Juárez, Chihuahua, México
e-mail: alberto.ochoa@uacj.mx

M. M. Rivera (✉)
Universidad Politécnica de Aguascalientes, Aguascalientes, Mexico
e-mail: martin.montes@upa.edu.mx

J. C. Ponce Gallegos
Departamento de Ciencias de La Computación, Universidad Autónoma de Aguascalientes, Aguascalientes, México
e-mail: julk_cpg@hotmail.com

C. Velazquez
Departamento de Sistemas de Información, Universidad Autónoma de Aguascalientes, Aguascalientes, México
e-mail: vace555@hotmail.com

P. N. M. Sampaio
Graduate Program in Computing and Systems – University of Salvador (UNIFACS), Salvador, Brazil
e-mail: pnms.funchal@gmail.com

© The Author(s), under exclusive license to Springer Nature Switzerland AG 2021
A. Ochoa-Zezzatti et al. (eds.), *Technological and Industrial Applications Associated with Intelligent Logistics*, Lecture Notes in Intelligent Transportation and Infrastructure,
https://doi.org/10.1007/978-3-030-68655-0_11

understand the zoo signals. Down syndrome is the most frequent cause of mental disability, presenting similar difficulties among people who suffer from it; a short-term memory capacity, exhausting attention and language delay are some of its most frequent characteristics. In Mexico, only 3% of children with Down Syndrome receive education. However, there are several technological alternatives that has been implemented around the world, and the use of robots is gradually becoming a good alternative for assisting and performing therapy to people that suffers this disability. NAO is an Aldebaran humanoid robot with several sensors, actuators and tools, that can be programmed with python, which has helped to popularize this friendly robot in the assisting of people. In this paper, we propose the use of a humanoid robot for future applications in therapy for children with Down Syndrome. To this purpose, we propose how to improve their ability to work with colors and shapes in a group of children with Down Syndrome.

Keywords Color identification · Down syndrome · Therapy for children with disabilities · NAO humanoid robot

1 Introduction

Currently, the number of related genetic abnormalities is quite high; to mention that there are more than 12,000 described genetic syndromes, one of which is Down Syndrome. In 1866, John Langdon Down identified this syndrome for the first time and, in that year, it was attributed to a delay in the patient's normal development. Later, it was stated that Down's syndrome was a consequence of infectious processes, alcoholism, among others. Only in 1958, Jerome Lejeune and Pat Jacobs discovered the presence of a third chromosome in the pair 21 associated with this condition, since then it has been considered a genetic syndrome. Being the first genetic abnormality described in humans [1, 2].

The Down's syndrome creates deficiencies in the physical and intellectual development of individuals. Children with Down's Syndrome usually have similar physical, neuropsychological, sensory, motor, and cognitive characteristics. The latter is de-scribed within [3–5]:

The attention is unstable, scattered and tiring; to motivate child's learning, it is necessary to apply different rehabilitation activities.

- Short-term memory has no processing of sensory and auditory information, which generally helps to improve with visual efforts. As for long-term
- memory, there is a difficulty in storing and retrieving information. The child retains information by observation because of habits, but has difficulty memorizing, which strengthens the learning of language and vocabulary.
- There is a considerable delay in oral expression compared to other fields of development. There is also a gap between the levels of understanding and oral expression.

- Vocabulary delay is present, possibly because there is a lack of under-standing of the relationship between objects, people, facts, and words that represent them, there is no retention of this relationship or there is no spatio-temporal recognition.
- Slow reaction time. Children with Down Syndrome also demonstrate impulsiveness, low tolerance to frustration, lack of persistence in tasks, low innate motivation, and the need for external supervision to complete a task.

The visual perception of children with down syndrome is significatively lower than the present in average children, and the majority of therapies are more focus on the improvement of the motor capabilities affected by their visual-motor disparity, how-ever the analysis in [5] has shown that grouping all kinds of therapy offers a complementary result improving the motor disabilities in children with this condition.

In that sense a technological solution that helps children with down syndrome must offer not only visual or audio information, but movement assistant too, so that helps improving focus during therapy sessions, and produce enough motivation in the learning process, this reason takes to consider robot assistants in the treatment children with down syndrome or other kinds of intellectual disabilities [6].

There are different situations that must be faced with intelligent technologies and innovation processes, when attending people with intellectual conditions, specially at times when the environment is extremely dynamic being this characteristic very typical within what is now called cognitive era. The above motivates to provide a Conceptual Model of Cognitive-Innovation (CgI-M) as an Archetype that has a formal support that consists of the Systematic Process for Knowledge Management (KMoS-REload), since science cannot be directly applied for producing technological solutions, without find the right conditions for solving a complex problem like the found assisting children with down syndrome [7].

Children with down syndrome are exposed to all kind of difficulties associated with their intellectual, motor, and visual conditions, including education, which together with family support are the main points for improving their lifestyle.

Education in children with down syndrome is complex and is not related with comprehensive material but learning reinforcement, which can be a difficult task for teachers and parents, because it must be very repetitive and patient [8].

Moreover, sometimes the environments are not well designed with inclusive material or technologies, even when they could produce rich and motivative experiences for people with mental conditions.

Zoos are remarkably interesting environments where humans can learn about animals and environments and experiences that are commonly far from human comprehension, in that sense there are maps, guides, signals and colorful materials that help visitants to improve their visit. But these signals are examples of well-designed material for providing better knowledge in the exhibition to average people, such as the colors of the species, their habitat including their food and geographical distribution. On the other hand, the zoo signals are not the must inclusive one for people that cannot perceive colors or who has a condition that difficulties metal associations. Thus, any child who has difficulties with color identification, or cognitive association will have severe difficulties understanding them.

Nevertheless, in the today complex, technological and cognitive society, there are emerging new alternatives for helping children with down syndrome, and ro-bots are a good example of technology that can help with this condition, especially in children because robots are attractive for them and can perform repetitive tasks, use several sensors and tools for perceiving the environment [6].

NAO humanoid robot is an Aldebaran product that has skills related with object recognition, color identification and motor coordination. In addition, it has been used previously for helping children with down syndrome in their education process [6].

In this paper, there is proposed the use of NAO robot for helping children with Down Syndrome. Specifically, with their ability to work with colors and shapes in a group of children with Down Syndrome with difficulties in the perception of a zoo environment.

The remainder of this paper is as follows: Sect. 2 presents the elements found in a zoo related with colors and signals, Sect. 3 shows the general aspects and tools of the NAO robot, Sect. 4 presents the concepts related with color identification as was performed in this work, Sect. 5 presents the results in this work and Sect. 6 the conclusions of this research.

2 Signaling at a Zoo

A necessary aspect in a zoo is the correct signaling in the public space. Therefore, it is important to consider properly where to place the signs and height specifications, as illustrated in Fig. 1. These signals use several colors associated with information related to location and species, but the associations cannot being per-ceived by children with difficulties in the perception of colors or limited by mental disabilities that affect its perception in associations.

The limited perception of color or associations related with it is not a familiar situation for people that have a normal vision or average mental condition, but simulating of that situations could help to perceive the difficulties that children with this affectations could have. The color perception of children with color blindnessor problems associated with a syndrome related to a situation of misunderstanding colors, could be improved by using special signs to understand where they are and where they want to go as is shown in Figs. 2 and 3. Some examples of the difficulties with color signals in zoos are shown in Fig. 4. They areobtained transforming the zoo signals with the technique in [8].

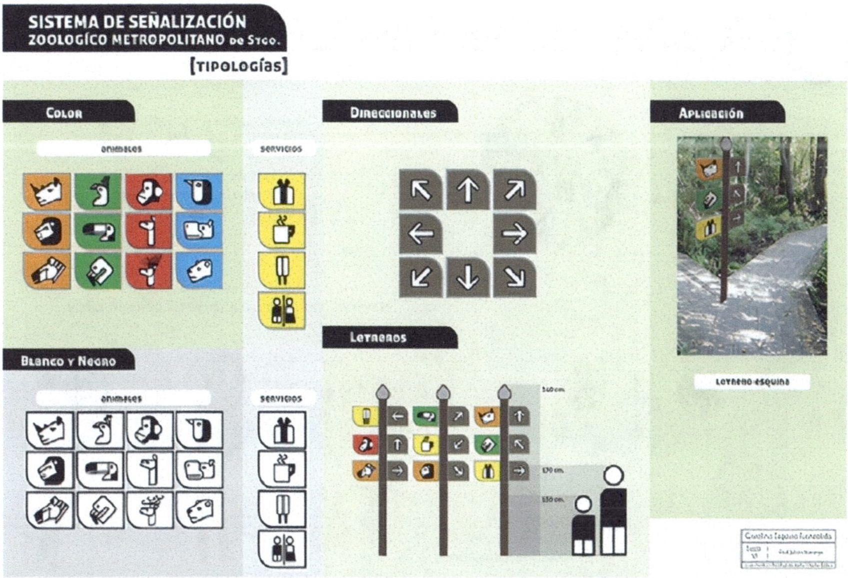

Fig. 1 Graphical description of the distribution of iconography in a zoo and specifications of measurements and height for proper observation

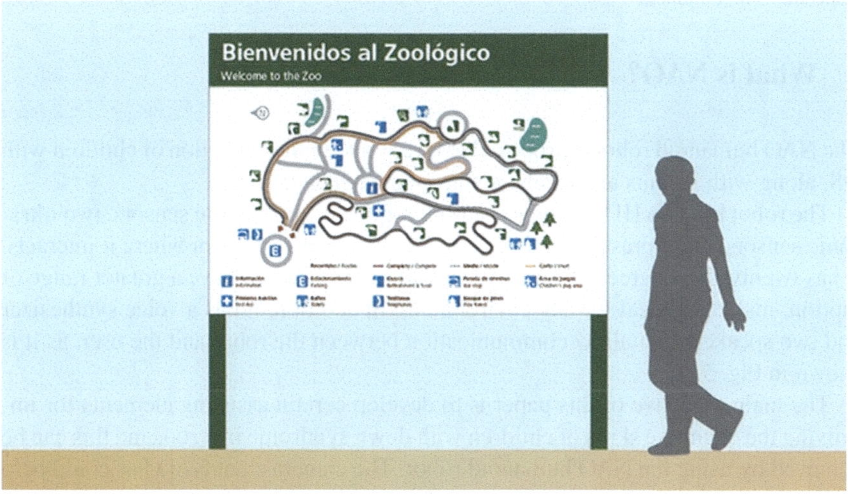

Fig. 2 Use of symbolism for children with a color vision problem

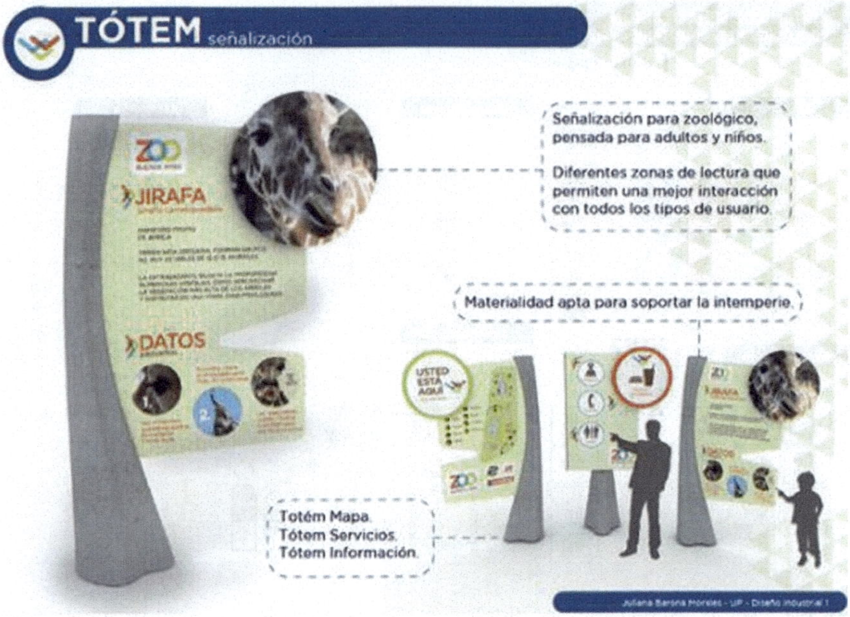

Fig. 3 Specifications of information associated with the data of species and using symbolism associated with blindness

3 What is NAO?

The NAO humanoid robot is proposed to be applied in the education of children with DS, along with parents and teachers who can teach them daily.

The robot has two HD cameras, four microphones, nine tactile sensors, two ultrasonic sensors, eight pressure sensors to perceive the environment where it interacts. It has twenty-five degrees of freedom that allow the robot to have a greater range of motion, making it ideal for any environment. In addition, it has a voice synthesizer and two speakers that allow communication between the robot and the user, as it is shown in Fig. 5.

The main objective of this paper is to develop certain assisting elements for improving the cognitive skills of children with down syndrome in a zoo, and this can be achieved by using the NAO humanoid robot. The cameras that NAO has could help with color recognition, as its actuator and sensor improve motor coordination, that affects children with down syndrome. In addition, it aims to be a tool to moti-vate them to talk and collaborate, since these are common social skills that every child should learn during their development [6].

Through the illustration process, it was possible to recognize all kinds of colors using the NAO camera, so the robot will be able to mention the color of the zoo signals every time the user show a picture to the humanoid with them. In addition, the process was also repeated using toys located by the robot and placed in a container at the

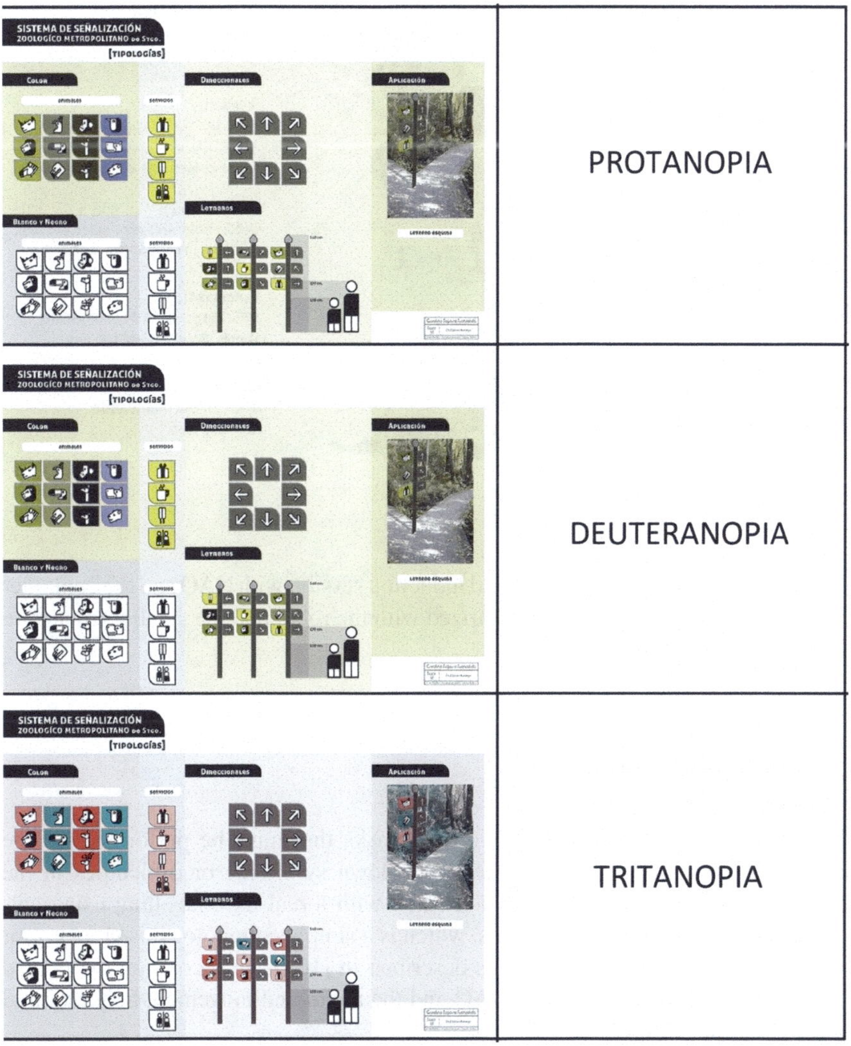

Fig. 4 Difficulties perceived with color disabilities in zoo signals

beginning or end of a hand therapy session improving the motivation and educational activities in the zoo experience.

NAO can assist children with down syndrome in a zoo by identifying the colors in the signals and pointing to different locations and performing different behaviours depending on the signal. Moreover, NAO robot can improve the comprehending during the therapy session by including a computer screen that transform the environment and show different symbols to the children.

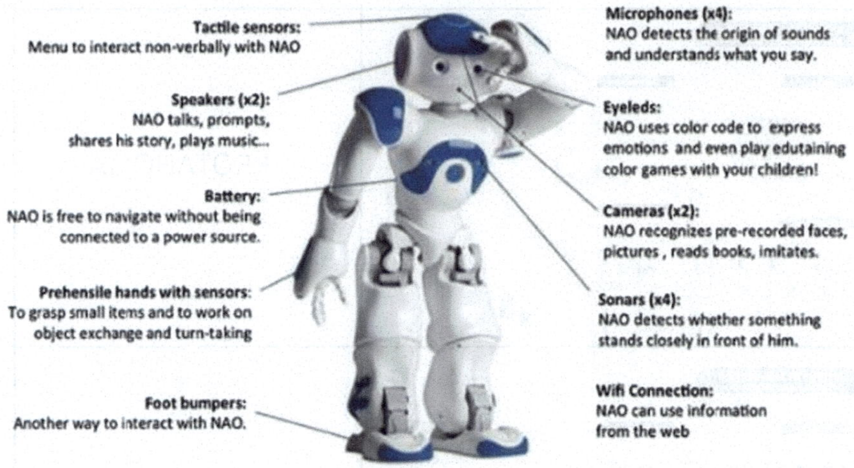

Fig. 5 NAO Sensors and Joints

It is recommendable to perform and induction session with NAO and the computer screen so that the children get familiarized with the robot and the symbols related to the colors or the animals.

4 Color Recognition

The color assisting is divided in two main tasks that must be performed for the complementary assistance of children with down syndrome or difficulties in the perception of colors. The first part is related with a real-time labelling using colorADD symbology in the zoo signals, which is a simpler symbology used for col-or identification recognized by ONU as described in [10]. The second part is related with the recognizing of colors by NAO and the audio and movement help required for assisting the children.

Since the generation of symbols require a more sophisticated technique so that the processing can be performed in real-time, is best to work with the RGB space and the symbology of color ADD, as its described in [11]. The general technique used for this labelling is fully detailed in [9].

In general in this labelling process, the colors perceived by the camera are identified with the equations generated using genetic programming, after that the detected colors are convoluted with images containing the symbology in Table 1 as described in [9].

On the other hand, the NAO processing related with images is simpler because it does not require image streaming with image processing in real time, however the motion processes and audio generation still being required for assisting the children and this is a complicated process too.

Table 1 Symbols of Color ADD for the real-time labeling assistant as described in [9]

Symbol						
Color	Red	Orange	Yellow	Green	Blue	Purple

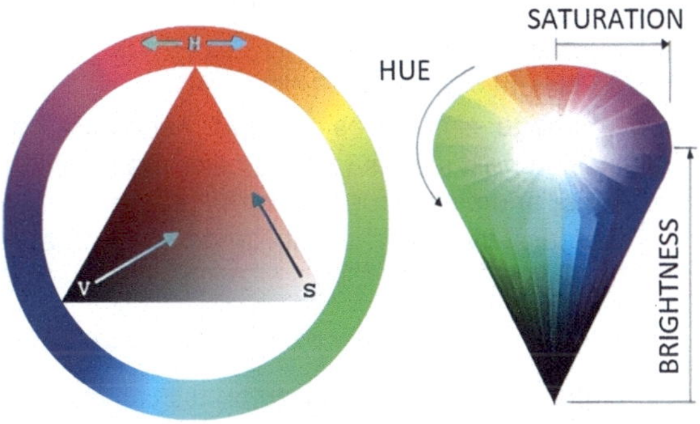

Fig. 6 HSV model as an inverted pyramid

The code illustrated below specifies the use of the top camera, RGB color space and 320 × 240 resolution in the NAO robot. Each pixel is saved in a new ar-ray of the same resolution. Once the array is obtained, the BGR color space must be changed to the HSV color. The HSV model is based on cylindrical coordinates and is derived from the RGB model. In addition, it represents colors that combine three values related to: hue, saturation, and value. The first one lets the color distinguish itself, considering the wavelength ranging from 0° to 360°. However, saturation refers to a sensation that goes from high to low color intensity. Finally, the value refers to the amount of white a color can contain. The previous description is illustrated in Fig. 6.

To carry out color recognition, the humanoid was placed in front of a white table measuring 15 × 41 × 25 cm (L × W × H). As was mentioned in Sect. 3, the humanoid has two HD cameras, in this case, it was decided the deployment of the upper camera due to the position of the robot at the beginning of the action. In fact, we could apply both cameras, however we decided to apply just one of them for simplifying the processing stage. Regarding the objects there were used four red, blue, purple, and orange figures.

For the illustration process, we applied the Numpy module, as it allows work-ing matrices in Python and OpenCV, a visionary computerized library. These modules were imported into Python 2.7 software, since NAO can be programmed using python language as described in Sect. 3.

The first step was to make the connection with the NAO camera through the computer using the following code in Algorithm 1.

```
    Device_Video = ALProxy ('ALVideoDevice', "IP", 9559)
    AL_kTopCamera = 0
    AL_kQVGA = 1
    AL_kBGRColorSpace = 13
    captureDevice = Device_Video.subscribeCamera("Prueba",
 AL_kTopCamera, AL_kQVGA,
    AL_kBGRColorSpace, 10)
    Ancho = 320
    Height = 240
    Image = np. zeros ((Height, Ancho, 3), np. uint8)
    The next step is to build color masks. In this case, orange, red,
 blue, and purple masks were created.
mascara_morado = cv2.inRange(hsv, morado_bajos, morado_altos)
```

Algorithm 1 NAO camera connection algorithm.

After that, each image received by the camera is processed with image morphological operations, each mask was submitted to open and closed morphological transformations. The first transformation consists of converting into zero all the pixels in the illustration that do not completely contain the structured element in its surroundings (erosion). This means that by obtaining matrix B and matrix A, as depicted in Fig. 3, they can overlap each other. If all the pixels in matrix B cross with matrix A, they will be stored in a new matrix, instead they will be stored as zeros. As illustrated in Fig. 7.

The second step of open transformation is dilation. It consists of the superimposition of matrix B and matrix A, in the difference of erosion, in this phase, the first pixel of matrix B that crosses with matrix A, will store all the pixels of matrix B in a new matrix as shown in Fig. 8. The closed transformation is applied first, to the dilation in the hand of the erosion. However, it should be clear that these operations are not contrary.

The combination of open and closed operations can be used for image filtering and segmentation. Furthermore, the OpenCV function "countNonZero" is used to count all the matrix of each color, which are stored in a variable to compare them between the masks. However, the variable with the highest value will indicate the color that is under observation. Finally, "ALTextToSpeech" was used to program the robot to say

Fig. 7 Example of Erosion

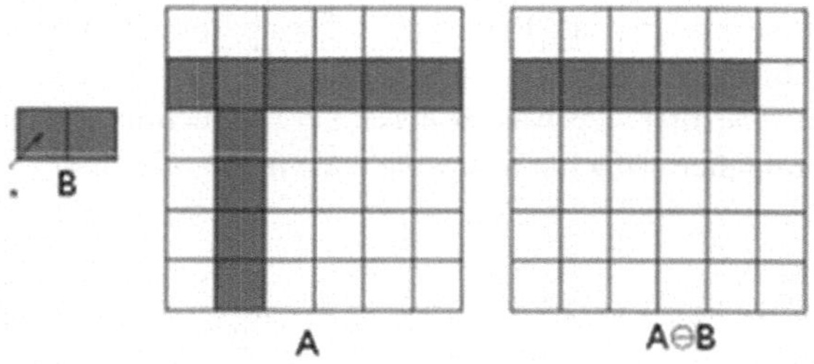

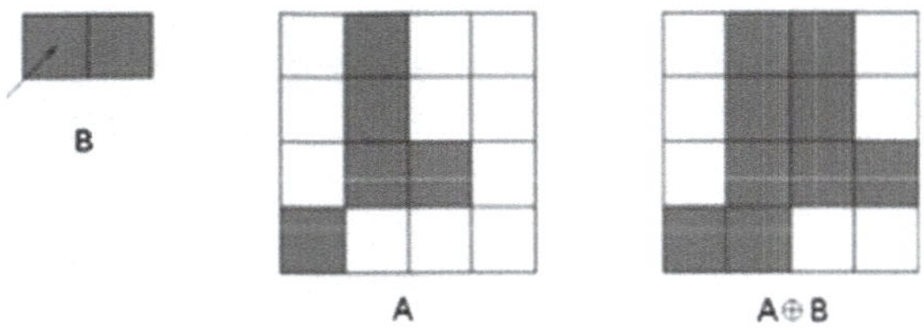

Fig. 8 Example of Expansio

the following sentence: "The object is color-": tts = ALProxy ('ALTextToSpeech', "IP", 9559) tts.say ("The object is blue") 3.

5 Results

The illustration process worked as different color figures were divided into sections and noticed with a small percentage of error, depending on the illumination, as illustrated in Fig. 9.

Once a piece was located on the table, after a few seconds, the robot mentioned the color of the figure. However, there was no need to make screens or restart the software, because the recognition was performed in real time (Fig. 10).

Considering the movement of the figures, the objective is achieved if leases are in the mark or near, at the same time, to the bottle where the figures are placed. This action is done in five main phases: (1) locate the figure, (2) capture the figure, (3) locate the figure and the hand in the bottle, (4) place the figure inside the bottle and

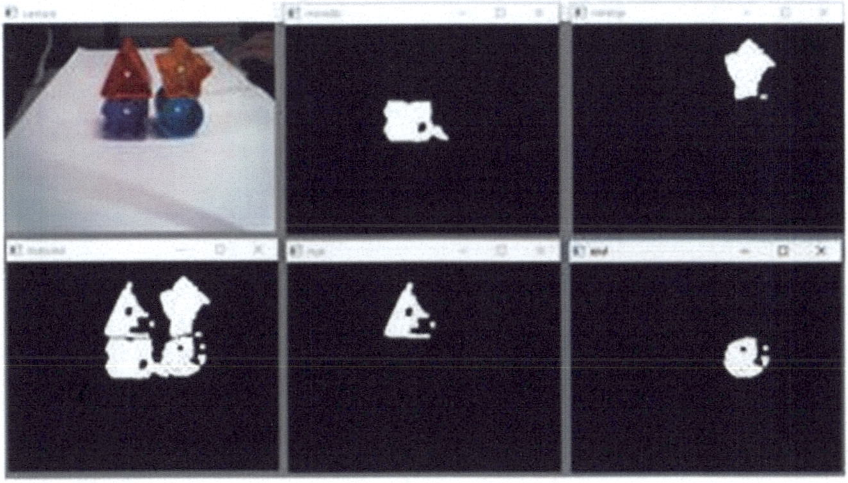

Fig. 9 Color mask with the camera No. **a** Original image, **b** Purple mask, **c** Orange mask, **d** Four-color mask, **e** Red mask, and **f** Blue mask

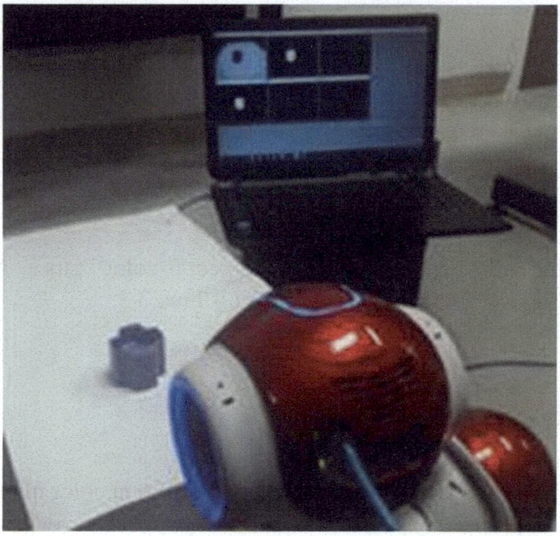

Fig. 10 Image of the Color Recognition of Humanoid Processing Not in Real Time

(5) lift the bottle. It is necessary to locate the pieces one by one in the place of the pointer. Once the humanoid perceives the last movement, he synthesizes the phrase: "We need to save this".

The Algorithm 2 presents the code used to differentiate the colors using the NAO for the above results, and the result may vary slightly.

```
import sys
import cv2
import numpy as np
from naoqi import ALProxy
#In this part you use arrays to define the spectrum of the colors
naranja_bajos = np.array([5,150,50],dtype=np.uint8)
naranja_altos = np.array([15, 255, 255], dtype=np.uint8)

morado_bajos = np.array([110,50,50], dtype=np.uint8)
morado_altos = np.array([150, 255, 255], dtype=np.uint8)

rojo_bajos = np.array([400,100,100], dtype=np.uint8)
rojo_altos = np.array([1000, 500, 500], dtype=np.uint8)

azul_bajos = np.array([80, 50, 50], dtype=np.uint8)
azul_altos = np.array([104, 255, 255], dtype=np.uint8)
```

Algorithm 2 Code used to differentiate the colors using the NAO robot.

In the Algorithm 2, the first four lines define the libraries applied in this pro-gram. After that, we define the color matrices to represent each color. For this purpose, it is possible to search in internet for the colors we want to identify, when the spectrum is identified, it is important also to check the light within the environment. Once this process can affect the way colors are perceived, an additional adjustment is also required. Moreover, it is important to set up the lower and upper spectrum of the color and the larger one. If this adjustment is not carried out, the robot will detect different colors from those required.

The next step is to connect NAO using the proxy associated with the respective NAO API (Application Programming Interface). Therefore, it is important to make sure NAO is applied correctly to connect successfully. The following line activates the NAO's upper or lower camera, depending on the desired use, as de-scribed in the next algorithm.

```
# Use the Proxy model of NAOqi.169.254.17.106
Dispositivo_Video = ALProxy('ALVideoDevice', "169.254.17.106", 9559) # Connect to NAO.
tts = ALProxy('ALTextToSpeech', "169.254.17.106", 9559)
# Use superior camera.
AL_kTopCamera = 0 # Superior camera.
AL_kQVGA = 1 # Resolution 320x240.
AL_kBGRColorSpace = 13 # Color space RGB.
captureDevice = Dispositivo_Video.subscribeCamera( # Use the camera.
"Prueba", AL_kTopCamera, AL_kQVGA, AL_kBGRColorSpace, 10) # 10 fps.width = 320
Ancho = 320
Altura = 240
Imagen = np.zeros((Altura, Ancho, 3), np.uint8) # Creates an image in zeros with the resolution.
```

Algorithm 3 NAO connection using the proxy associated.

Still in the Algorithm 3 above, the image resolution should be set. For this purpose, it is important to use a higher resolution. Nevertheless, the compilation of the program may take longer, and the chances are the system gets blocked.

The space for RGB is then used. After that, the image in the selected camera is captured. The next line (9) makes use of the variables previously defined. The vaustment can also have an impact on the performance of the program as needed, so the resolution is set to 320×240, if a higher resolution is required then the respective values should be adjusted. The latter may turn the program slower, using a larger array and needing to process a larger image. Finally, the image is created with its corresponding resolution, as described in Algorithm 4.

```
while True:
    # Get image.
    Resultado = Dispositivo_Video.getImageRemote(captureDevice); # Take a capture.

    if Resultado == None:
        print 'No se pudo capturar.'
        break
    elif Resultado[6] == None: # In Resultado[6] the image is saved in binary.
        print 'No se obtuvieron datos de imagen.'
        break
    else:

        # Acomodates the binary values in an array for a format of the image.
        Valores = map(ord, list(Resultado[6]))
        i = 0
        for y in range(0, Altura):
            for x in range(0, Ancho):
                Imagen.itemset((y, x, 0), Valores[i + 0])
                Imagen.itemset((y, x, 1), Valores[i + 1])
                Imagen.itemset((y, x, 2), Valores[i + 2])
                i = i + 3
```

Algorithm 4 NAO handling of camera image processed.

In the previous code snippet, the image captured by the camera is handled. In this part, if there is any capture error or if it does not detect the image properly, a message: "It was not possible to capture the image" or "There is not enough data for the image" will be displayed on the screen. The generation of the color masks related with the color classification is shown in Algorithm 5.

If the data is captured correctly, its binary values are stored in an array corresponding to the image format. Starting from 0 within the variable "y" which is re-lated to "Height" (height) and x which is related to "Width" (width).

In Algorithm 5, the captured images are processed, applying the range of colors of the lower and upper spectrum, and defining which color is identified in light or dark tone. In this excerpt, several masks are generated, allowing the detection and separation of this part in several masks, which are applied to separate each color and detect it. To do so, the previously defined variable to establish the color rang-es in the spectrum is applied. Therefore, it is possible to establish the morphology along with the kernel defined for the image. This part shows 8 images, one for each color used, one for the camera which is a "regular image" and one that detects all the previously established colors, showing all the colors that the NAO camera is available to detect.

```
#_,imagen = camara.read()
hsv = cv2.cvtColor(Imagen, cv2.COLOR_BGR2HSV)
mascara_morado = cv2.inRange(hsv, morado_bajos, morado_altos)
mascara_naranja = cv2.inRange(hsv, naranja_bajos, naranja_altos)
mascara_roja = cv2.inRange(hsv, rojo_bajos, rojo_altos)
mascara_azul = cv2.inRange(hsv, azul_bajos, azul_altos)

kernel = np.ones((5,5), np.uint8)

mascara_morado = cv2.morphologyEx(mascara_morado, cv2.MORPH_CLOSE, kernel)
mascara_morado = cv2.morphologyEx(mascara_morado, cv2.MORPH_OPEN, kernel)
Morado = cv2.countNonZero(mascara_morado)

mascara_naranja = cv2.morphologyEx(mascara_naranja, cv2.MORPH_CLOSE, kernel)
mascara_naranja = cv2.morphologyEx(mascara_naranja, cv2.MORPH_OPEN, kernel)
Naranja = cv2.countNonZero(mascara_naranja)

mascara_roja = cv2.morphologyEx(mascara_roja, cv2.MORPH_CLOSE, kernel)
mascara_roja = cv2.morphologyEx(mascara_roja, cv2.MORPH_OPEN, kernel)
Rojo = cv2.countNonZero(mascara_roja)

mascara_azul = cv2.morphologyEx(mascara_azul, cv2.MORPH_OPEN, kernel)
mascara_azul = cv2.morphologyEx(mascara_azul, cv2.MORPH_OPEN, kernel)
Azul = cv2.countNonZero(mascara_azul)

mascara = cv2.add(mascara_morado, mascara_naranja)
mascara = cv2.add(mascara, mascara_roja)
mascara = cv2.add(mascara, mascara_azul)
```

Algorithm 5 Color masks related with the color classification in NAO.

Finally, the applied command is used again in this case to make NAO present and audio with the detected color. To achieve this, it is necessary to recognize the highest spectrum value for the selected color. In the first line it is also possible to see how the color "Purple" needs to be larger than the other colors, such as "red", "blue" and "orange", to allow NAO to be able to differentiate from the other colors and "say" its name, as described in Algorithm 6.

```
if Morado > Rojo & Morado > Naranja & Morado > Azul:
    print 'Morado'
    tts.say("El objeto es morado")

elif Morado < Rojo & Rojo > Naranja & Rojo > Azul:
    print 'Rojo'
    tts.say("El objeto es rojo")

elif Naranja > Rojo & Morado < Naranja & Naranja > Azul:
    print 'naranja'
    tts.say("El objeto es naranja")

elif Azul > Rojo & Azul > Naranja & Morado < Azul:
    print 'Azul'
    tts.say("El objeto es azul")
```

Algorithm 6 Audio assistance with NAO for color classification.

As for movement, the easiest way to implement it is to use the Choregraphe software. Therefore, it is necessary to save the executed moves so that they can be used in conjunction with another program, such as color recognition, to make NAO act in a certain way, depending on the need of the problem. For instance: To perform the movement of the toy parts and store them in an account, the Choregraphe 2.1.4.14 software has been managed. To achieve this goal, the same white table was used to locate the humanoid at 5 cm between the foot and the table, as shown in Fig. 11.

In Fig. 12, the piece is located on the left, and the bottle is placed on the right side. Thus, a cylinder with 14 cm height and a 7.5 cm diameter was used. A mark depicted in was painted to indicating the place where the pieces are located so that

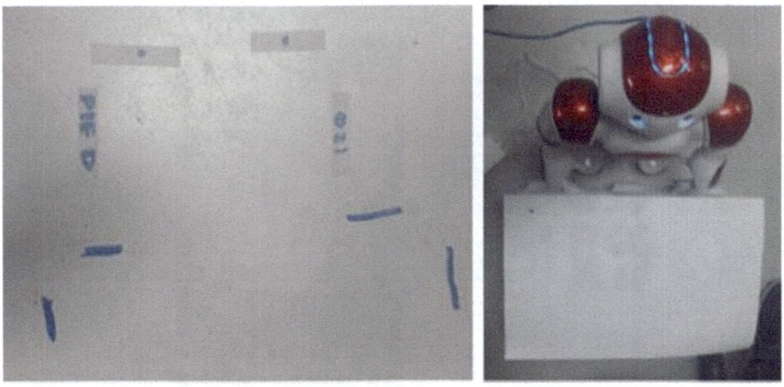

Fig. 11 Localization of NAO, Figures of Figures and Flask marked with red

Fig. 12 Movement of the main moments of the figures: (1) Locate the figure and hand in the bottle, (3) Locate the figure and hand in the bottle, (4) Leave the figure inside the bottle, (5) Lift the bottle

the humanoid can grab each piece and place it inside the container that also has a specific location.

In Fig. 13 you can see our proposal for real-time symbology labelling in the zoo signals, which was obtained by processing the images acquired with an external camera, and then streamed on a computer screen after processing them, so that the

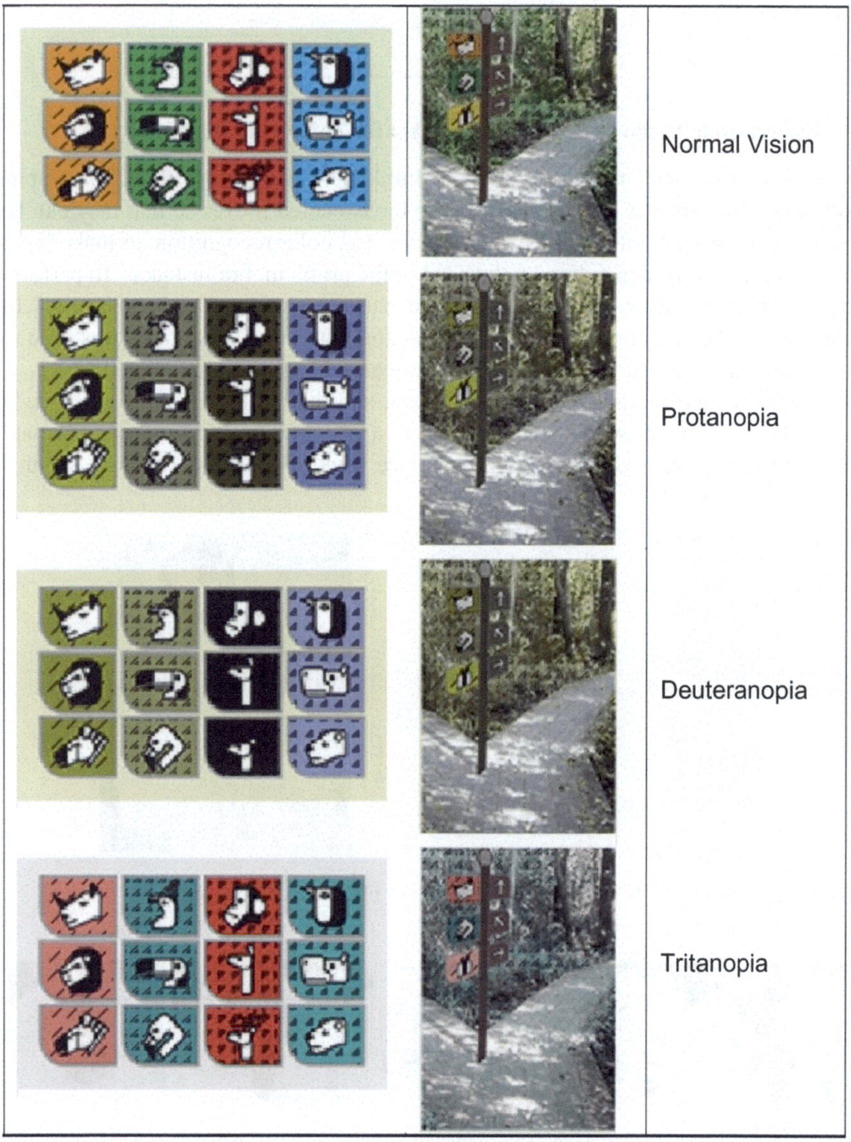

Fig. 13 Visual representation using colorblind symbology, to help children learn more about the park and specify safety criteria

humanoid robot NAO, together with the new symbology can help the children to identify the places of interest in the park.

Once our proposed Color Recognition System has been implemented within the NAO Humanoid Robot, and the control points within the Zoo are fully integrated with the labeling system, NAO program can be improved by connecting it with geolocated indicators, so that the robot can help the children to improve their experience during their visit, as it is shown in Fig. 14.

6 Conclusions and Future Research

Due to the lack of qualified people to educate children with Down syndrome, ninety-seven percent of these children do not receive the special education they need, making difficult to include them in society. For this reason, the use of a NAO humanoid would be a great support for the education of children with Down syndrome. The different resources of this robot are exploited, making use of efficient tools, such as tactile and precision sensors, cameras, microphone, and voice synthesizer. Therefore, all these resources are appealing to attract children´s attention. Considering that children with Down's syndrome have short and long-term disabilities, the activities must be repetitive to make children remember them. For this purpose, NAO promotes three learning modes: kinesthetic, visual, and auditory. Moreover, as shown in this paper NAO can be implemented together with other technologies like augmented reality with color labelling and simplified symbols, enhancing the learning experiences of children with special conditions in not inclusive environments.

This survey lists the first steps in supporting children with Down syndrome to develop certain skills. First, the knowledge of the four colors, which were mentioned earlier, and the location of each piece for the robot to do his task. In addition, motor coordination is expected to be an essential contribution for the children to replicate the movements of the robot. In this context, it is up to the child to place figures inside the container. After that the children are trained in new experiences using the NAO robot, in this research by using new symbols for identifying new knowledge related with colors in the zoo.

Therefore, the main goal it to make children develop knowledge by performing repetitive tasks with objects and the NAO robot, allowing learning procedures with reinforcement that are required by children with down syndrome.

6.1 Future Research

Due to the lack of specialized education personnel for children with Down Syndrome, children are left without education and therefore their inclusion is difficult in new environments. Therefore, the use of the humanoid NAO should be expanded as a support in the education of these children with DS, therefore using the methodology

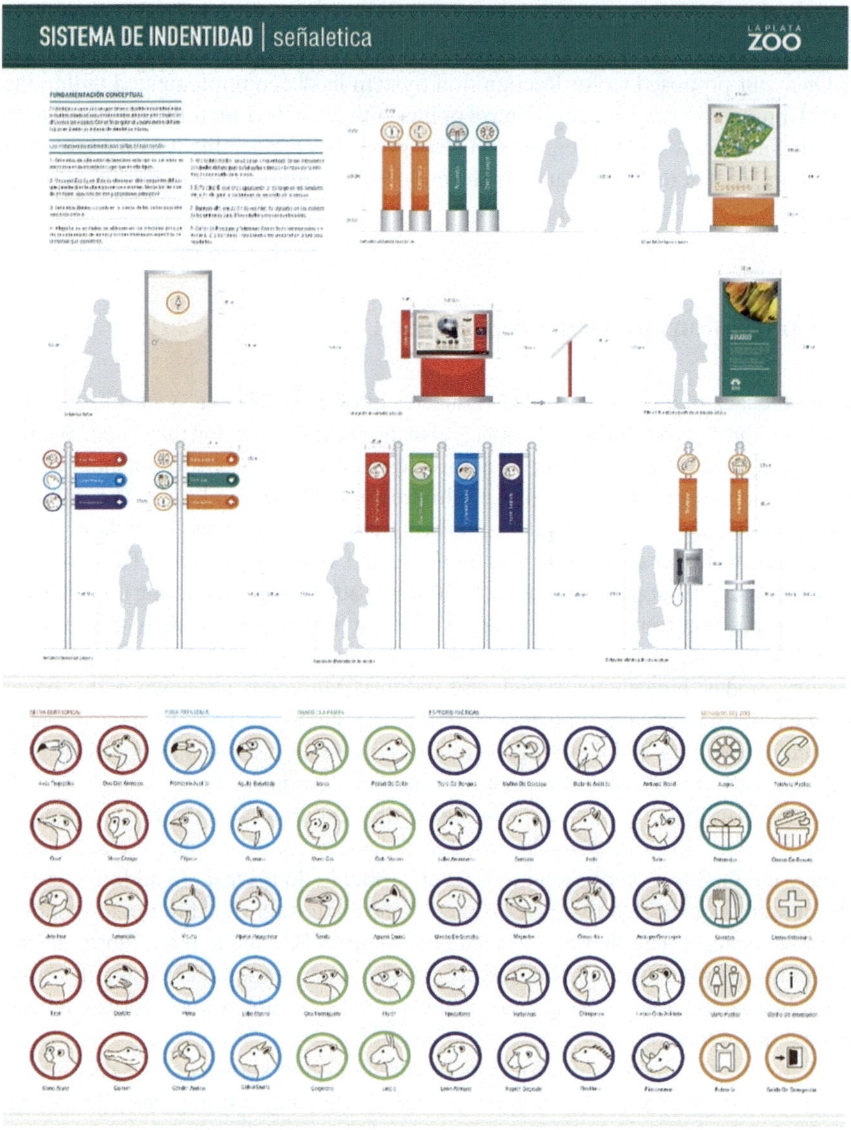

Fig. 14 Geolocation recognition system of species at the zoo and its implementation for a group of species. The Nao Humanoid Robot has a voice system that allows you to easily lo-cate the color labels and transmit them to the symbolism required by the child with color blindness

in this work the research must be implemented for other learning environments to help children to retain knowledge and improve its learning experience in different not inclusive environments, like in the zoo experience presented in this work.

In future smart cities, each child will have a Nao Humanoid Robot that will al-low them to discover features of the world from an early age and where there will be no limit due to a condition such as color blindness or down syndrome. Visionaries, futurists, and scientists specify that this may happen from 2027.

References

1. Smith E, Hokstad S, Næss KAB (2020) Children with Down syndrome can benefit from language interventions; Results from a systematic review and meta-analysis. J Com-Mun Disord 85:105992
2. Strippoli P, Pelleri MC, Caracausi M et al (2013) An integrated route to identifying new pathogenesis-based therapeutic approaches for trisomy 21 (Down Syndrome) following the thought of Jérôme Lejeune. Sci Postprint 1. https://doi.org/10.14340/spp.2013.12r0005
3. Elisa Campana M, Elisa M (2012) An Analysis of Down Syndrome Children and the Importance of Their Cognitive and Communicative Development
4. Fernández JC, Buceta MJ (2003) La educación de personas con síndrome de Down: es-trategias de aprendizaje
5. Uyanik M, Bumin G, Kayihan H (2003) Comparison of different therapy approaches in children with Down syndrome. Pediatr Int 45:68–73. https://doi.org/10.1046/j.1442-200X.2003.01670.x
6. Alemi M, Bahramipour S (2019) An innovative approach of incorporating a humanoid robot into teaching EFL learners with intellectual disabilities. Asian-Pacific J Second Foreign Lang Educ 4:1–22. https://doi.org/10.1186/s40862-019-0075-5
7. Nightingale P (1998) A cognitive model of innovation. Res Policy 27:689–709. https://doi.org/10.1016/S0048-7333(98)00078-X
8. De FM, Lopes Neto D, Llapa-Rodriguez EO (2020) Special education of stu-dents with Down Syndrome for self-care. Esc Anna Nery 24:2020. https://doi.org/10.1590/2177-9465-ean-2019-0129
9. Montes-Rivera M, Padilla-Díaz A, Ponce-Gallegos JC et al (2019) Performance of Human Proposed Equations, Genetic Programming Equations, and Artificial Neural Networks in a Real-Time Color Labeling Assistant for the Colorblind. In: Martínez-Villaseñor L, Batyrshin I, Marín-Hernández A (eds) Advances in Soft Computing. Springer International Publishing, Cham, pp 557–578
10. Neiva M (2018) ColorADD, color identification system. https://www.coloradd.net/imgs/ColorADDAboutUs_2015V1.pdf
11. Rivera MM, Díaz AP, Reich JC et al (2019) Augmented Reality Labels for Security Signs based on Color Segmentation with PSO for Assisting Colorblind People. Int J Comb Optim Probl Informatics 10:7–20

Optimization of Route Planning for the Package Delivery Problem Using Fuzzy Clustering

Paula Hernández–Hernández and Norberto Castillo–García

Abstract Package Delivery Companies face a variety of operational problems every day. One of these problems is to determine the best route to locally deliver the packages to their customers. In this chapter we propose an intelligent logistics methodology based on fuzzy clustering to approach the routing problem of delivery companies. Specifically, we use the well–known fuzzy c–means algorithm to group together those delivery locations whose geographic distances are relatively close. Once we have obtained the clusters, the next step is to optimize the route in each cluster. This is achieved by solving an integer linear programming formulation for the Traveling Salesman Problem. As far as we are aware, this is the first time this methodology is applied in this context. In order to assess the performance of our methodology in practice, we have conducted a computational experiment. The experiment consisted in solving 30 randomly generated instances. The results were compared with a control methodology which optimizes the entire set of delivery locations simultaneously (without the clustering process). The experimental results clearly showed that our methodology outperformed the control methodology in terms of number of optimal solutions found, average computing time, and average relative gap by 68%, 92% and 98%, respectively. The experimental results were statistically validated by the Wilcoxon Rank Sum Test. The test found statistical significance in average computing time and average objective value for a confidence level of 99.99% and 98.86%, respectively.

P. Hernández–Hernández · N. Castillo–García (✉)
Tecnológico Nacional de México/I.T. Altamira, Altamira, Mexico
e-mail: norberto_castillo15@hotmail.com

P. Hernández–Hernández
e-mail: paulahdz314@hotmail.com

© The Author(s), under exclusive license to Springer Nature Switzerland AG 2021
A. Ochoa-Zezzatti et al. (eds.), *Technological and Industrial Applications Associated with Intelligent Logistics*, Lecture Notes in Intelligent Transportation and Infrastructure,
https://doi.org/10.1007/978-3-030-68655-0_12

1 Introduction

Package delivery companies, such as UPS or FedEx, distributes products among its customers worldwide. The products are typically packed in rectangular–shaped boxes of different dimensions. The dimension of each box depends, indeed, on the nature of the product. A customer of these delivery companies might be a person, company or organization. Basically, there are two customers involved in the process: the one who dispatches the product and the other who receives it. The process ends when the final (second) customer gets the package.

Package delivery companies face a variety of operational problems every day. For example, they have to decide how to better pack the products together or how to find the best route in order to efficiently deliver the product. In this chapter, we deal with the operational problems of the last stage of the process, that is, when the packages arrive at the distribution center of the city where they must be delivered.

Locally, the Package Delivery Problem (PDP) can be categorized into two main phases: the packing process and the routing process. The packing process consists in determining the best way to pack the products into the delivery truck. The goal is to minimize the number of delivery trucks required or to minimize the number of times that a single delivery truck must visit the distribution center. For a better explanation of this process we refer the reader to [6]. The route planning process consists in determining the best path that a single delivery truck must follow. The goal is to minimize the total cost of the route. The cost of the route might be expressed in terms of time, distance or fuel, among other resources.

In general, the goal of Logistics is to manage the flow of different kinds of objects between an origin and a destination in order to fulfill the customer requirements [5]. Logistics 4.0 is a relatively new concept. It favors the use of technology and information processing to improve traditional logistics [12]. Thus, in this chapter we approach the route planning process for package delivery companies from an intelligent logistics point of view. More precisely, we propose a methodology based on fuzzy clustering to group destinations that are relatively close to each other. Thus, we expect to significantly reduce the computational effort inherent in the route planning computation when considering the whole number of destinations.

In order to intelligently group the destinations, we use the well–known fuzzy c–means algorithm. The main characteristic of this algorithm is that each destination is assigned to each cluster with a certain level of membership. This gives the fuzzy c–means algorithm the advantage of softening the membership of the destinations to each cluster. This is particularly important when the destinations are very close to each other. Moreover, the number of clusters (groups) is a parameter that delivery companies must determine. For instance, the delivery company could probably set the number of clusters according to the number of delivery trucks available. The fuzzy c–means algorithm has been extensively and successfully used in several domains. We refer the reader to [1, 4, 7, 10, 11, 14] for several examples of the application of this algorithm.

Once the clusters have been obtained, the next step consists in computing the best route for each cluster. This process can be viewed as an application of the Traveling Salesman Problem (TSP). Thus, we determine the best route for each cluster by solving TSP with the destinations of the cluster. In this chapter we use an integer linear programming formulation for TSP and the well–known optimization engine CPLEX to exactly solve TSP. More precisely, we use the mathematical formulation for the asymmetric variant of TSP with the subtour elimination constraints proposed by Miller, Tucker and Zemlin (see pages 142–144 of [2]).

In order to assess the performance of our fuzzy methodology in practice, we conducted a computational experiment with 30 artificial instances. These instances were randomly generated by the automatic tool proposed in [6]. We compare our fuzzy methodology with a control methodology under the same experimental conditions. The control methodology consists in computing TSP for the entire set of destinations (without the clustering process). This gives us a clear idea about the hardness of the route planning process for package delivery companies. Throughout this chapter, we will refer to our fuzzy clustering methodology as FUZZY and to the control methodology as CONTROL.

In order to determine the superiority of one particular methodology, we measure the following four attributes from the experimental data: (i) the number of optimal solutions found, (ii) the average objective value, (iii) the average computing time, and (iv) the average relative gap. The experimental results clearly indicate that FUZZY outperformed CONTROL in effectiveness (number of optimal solutions found), efficiency (average computing time) and average relative gap by 66%, 92% and 98%, respectively. Conversely, CONTROL outperformed FUZZY in solution quality (average objective value) by 0.9%, that is, less than one percent. Thus, the experimental results clearly show that FUZZY considerably reduces the computational effort with respect to CONTROL at the expense of a minuscule loss in solution quality. The experimental results were statistically validated through the well–known Wilcoxon Rank Sum Test [13]. The test found statistical significance in average computing time and average objective value for a confidence level of 99.9% and 98.9%, respectively. Therefore, according to the experimental results, our fuzzy methodology considerably reduce the computational effort related to the route planning process for package delivery companies. Although statistically significant, the differences in average objective values are so small that we consider them negligible.

The remainder of this chapter is organized as follows. Section 2 formally defines the problem approached in this study. In Sect. 3 we describe the solution method proposed. Section 4 reports the numerical experiment conducted to assess the performance of our solution method in practice. Finally, in Sect. 5 we discuss the key findings of this research.

2 Problem Definition

Let $D = \{d_1, d_2, \ldots, d_n\}$ be the finite and discrete set of n destinations for which the company must deliver one or more packages. Each destination can univocally be located by two parameters: the latitude and the longitude on the Earth. Thus, each destination d_i is related to an ordered pair (φ_i, λ_i) which represents its global position. In the ordered pair, φ_i and λ_i are respectively the latitude and longitude of destination d_i, for all $i = 1, \ldots, n$. The geographic distance $\delta(d_i, d_j)$ between destinations d_i and d_j is computed by the haversine formula shown in Eq. (1) [9]:

$$\delta(d_i, d_j) = 2r \tan^{-1}\left(\frac{\sqrt{a}}{\sqrt{1-a}}\right) \quad \forall i, j = 1, \ldots, n, \tag{1}$$

where

- $a = \sin^2\left(\frac{\varphi_j - \varphi_i}{2}\right) + \cos(\varphi_i)\cos(\varphi_j)\sin^2\left(\frac{\lambda_j - \lambda_i}{2}\right)$
- r represents the Earth radius. In this chapter we consider $r = 6{,}371$ km.
- $\tan^{-1}\left(\frac{\sqrt{a}}{\sqrt{1-a}}\right)$ is the angle (in radians) which results from converting the rectangular coordinates $(\sqrt{a}, \sqrt{1-a})$ to polar coordinates.

The goal of the route planning process in the Package Delivery Problem (PDP) is to visit each destination exactly once in such a way that the total cost of the route is minimized. Thus, this problem is equivalent to find the permutation of destinations $\pi^\star$ whose total cost of the route is the minimum among all possible routes. Mathematically:

$$\pi^\star = \min_{\pi \in \Pi}\left\{\sum_{i=0}^{n-1} \delta(d_{\pi(i)}, d_{\pi(i+1)}) + \delta(d_{\pi(n)}, d_{\pi(1)})\right\}, \tag{2}$$

where $\pi(i)$ stands for the i–th destination in the permutation π and Π represents all possible permutations of destinations, that is, the solution space.

3 Solution Method

In this section we describe our proposed method to approach the route planning process for the Package Delivery Problem described in Sect. 2. Our method consists of two phases. In the first phase, we cluster those destinations whose geographic distances are relatively close to each other. Then, in the second phase, we try to find the best route for each cluster. As mentioned previously, in the first phase we use the well–known fuzzy c–means algorithm in order to obtain the clusters [3]. The operation of this algorithm is described in Sect. 3.1. In the second phase we use the integer linear programming formulation for the asymmetric variant of TSP with the Miller, Tucker and Zemlin subtour elimination constraints [2]. The complete mathematical formulation used in this study is presented in Sect. 3.2.

3.1 Fuzzy c–Means Algorithm

The goal of this algorithm is to group together those destinations whose geographic distances are relatively close. The algorithm requires three input parameters to operate, namely, k, m' and ϵ. Parameter k specifies the number of clusters in which the total number of destinations will be divided. Parameter m' controls the level of fuzzification in the process. Although this parameter can take any value in $(1, \infty)$, it is typically used the value of 2 [3]. Parameter ϵ establishes the level of tolerance for the algorithm. This parameter is used to decide whether or not the algorithm should continue its execution. In addition to the previous parameters, the algorithm also requires the set of destinations D described by their latitudes and longitudes (φ_i, λ_i) for all $i = 1, \ldots, n$. Algorithm 1 shows the high level pseudocode of the fuzzy c–means algorithm used in this study.

Algorithm 1: Fuzzy c–means algorithm.

```
Data: k, m', ε and D.
Result: A k–partition of the set D.
Begin U
stop ← false
while not stop do
    computeCentroids()
    computeDistancesToCentroids()
    updateU()
    computeEpsilon()
    stop ← stopExecution()
end
getClusters()
```

At the beginning, the algorithm initializes the matrix **U**, which has k rows and n columns. This matrix is used to store the membership levels of each destination to each cluster. Thus, every entry of **U** is a value in between [0, 1]. We initialize **U** in such a way that the sum of each column is exactly 1 and any row is not the vector **0**, i.e., a vector whose entries are all zero.

Once the matrix **U** has been initialized, the algorithm assigns the value of false to the flag *stop* and enters into the main loop. Then, the algorithm computes the centroids v_i (for all $i = 1, \ldots, k$) based on the current values of **U** and the set of latitudes and longitudes (φ_j, λ_j) (for all $j = 1, \ldots, n$) according to:

$$v_{i,\varphi} = \frac{\sum_{j=1}^{n} (\mathbf{U}_{ij})^{m'} \times \varphi_j}{\sum_{j=1}^{n} (\mathbf{U}_{ij})^{m'}} \quad \forall i = 1, \ldots, k$$

and

$$v_{i,\lambda} = \frac{\sum_{j=1}^{n} (\mathbf{U}_{ij})^{m'} \times \lambda_j}{\sum_{j=1}^{n} (\mathbf{U}_{ij})^{m'}} \quad \forall i = 1, \ldots, k$$

Thus, the centroids are the set of k ordered pairs $(v_{1,\varphi}, v_{1,\lambda}), (v_{2,\varphi}, v_{2,\lambda}), \ldots, (v_{k,\varphi}, v_{k,\lambda})$. After the computation of the centroids, the next step consists in computing the distances from the centroids to the destinations. This is accomplished by computing the euclidean distance d_{ij} between centroid $v_i = (v_{i,\varphi}, v_{i,\lambda})$ and destination $d_j = (\varphi_j, \lambda_j)$ as follows:

$$d_{ij} = \left[(\varphi_j - v_{i,\varphi})^2 + (\lambda_j - v_{i,\lambda})^2 \right]^{1/2}.$$

Once the distances have been calculated, the following step is updating the matrix $\mathbf{U}$. The goal of this process is to recalculate the membership level of each destination to the new centroids. At this point we strongly recommend to store the values of $\mathbf{U}$ (i.e., $\mathbf{U}^{\text{prev}} \leftarrow \mathbf{U}$) since it will be used in the next step (computeEpsilon()). The update of $\mathbf{U}$ is given by:

$$\mathbf{U}_{ij} = \left[\sum_{l=1}^{k} \left(\frac{d_{ij}}{d_{lj}} \right)^{2/(m'-1)} \right]^{-1}.$$

The goal of the computeEpsilon() procedure is to measure the difference between the previous membership levels and the current ones. This computation can be performed by:

$$\mathbf{E}_{ij} = \mathbf{U}_{ij} - \mathbf{U}_{ij}^{\text{prev}}.$$

The last procedure in the main loop (stopExecution()) determines whether or not the algorithm must continue its execution. Specifically, every entry $\mathbf{E}_{ij}$ is compared to the parameter ϵ. If at least one entry is greater than ϵ then the algorithm must continue (the procedure returns false) otherwise the algorithm must stop (the procedure returns true). Finally, in the last procedure the algorithm returns the k clusters found during its execution. It is important to mention that each destination is associated to the cluster for which its membership level is the highest. Figures 1, 2, 3 respectively show the execution of the fuzzy c–means algorithm over three instances with 65, 140 and 260

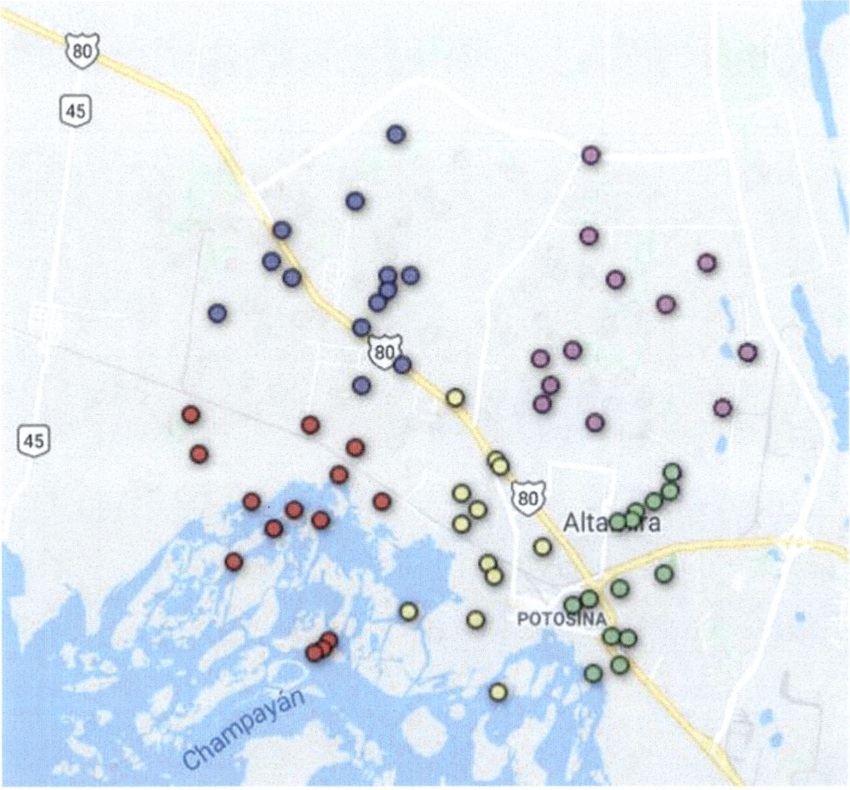

Fig. 1 Example of the execution of the fuzzy c–means algorithm over an instance with 65 delivery locations

destinations and $k = 5$ clusters. These instances were randomly generated by the automatic tool proposed in [6].

3.2 Formulation for TSP

The second phase of our proposed methodology FUZZY consists in solving the route planning for each cluster. More precisely, we propose to solve the integer linear programming (ILP) formulation for TSP with the destinations in each cluster. The general ILP formulation uses the following two decision variables:

x_{ij} is a binary decision variable used to decide whether or not each directed arc $(i, j) \in A$ is in the tour.

u_i is an integer decision variable defined as the sequence number of the destination i in the tour.

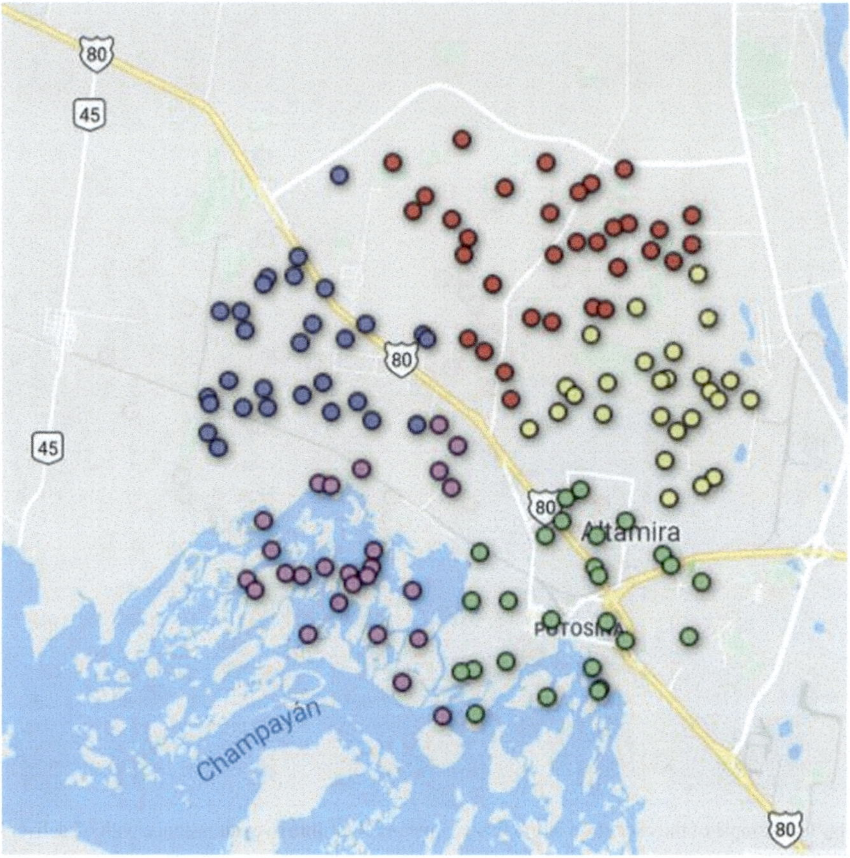

Fig. 2 Example of the execution of the fuzzy c–means algorithm over an instance with 140 delivery locations

The complete mathematical formulation used in our methodology Fuzzy is:

$$\min \sum_{(i,j)\in A} c_{ij}x_{ij} \tag{3}$$

subject to

$$\sum_{\{i:(i,j)\in A\}} x_{ij} = 1 \quad \forall j = 1,\ldots,n \tag{4}$$

$$\sum_{\{j:(i,j)\in A\}} x_{ij} = 1 \quad \forall i = 1,\ldots,n \tag{5}$$

$$u_i - u_j + nx_{ij} \leq n - 1 \quad \forall (i,j) \in A, i, j \neq 1, i \neq j. \tag{6}$$

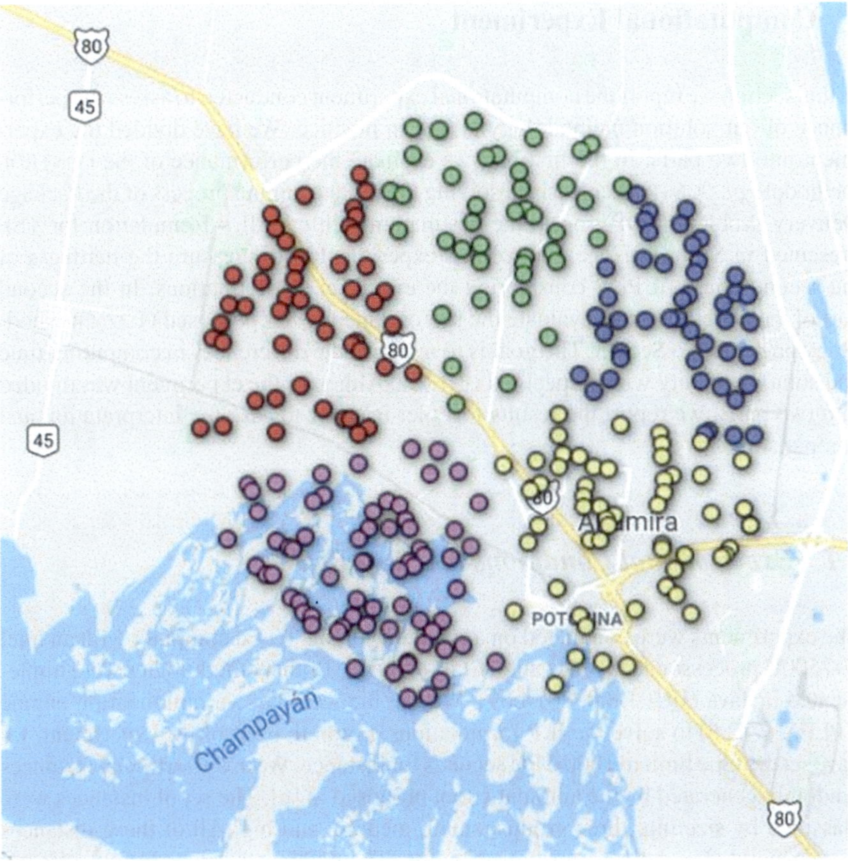

Fig. 3 Example of the execution of the fuzzy c–means algorithm over an instance with 260 delivery locations

Constraints (4) and (5) ensure that each delivery location must be visited exactly once. Constraint (6) corresponds to the Miller, Tucker and Zemlin subtour elimination. Finally, the objective function (3) minimizes the total cost of the route. Let C_i be the set of delivery locations (destinations) in the 8–th cluster ($i = 1, \ldots, k$). In this formulation, the parameter n used in (4) and (5) stands for the number of destinations in the cluster i, i.e., $n = |C_i|$. Moreover, the cost C_{ij} used in the objective function (3) is actually the geographic distance between destinations i and j computed by Eq. (1). Finally, the set A used in the formulation stands for the set of directed arcs computed from the delivery locations in cluster C_i. In this chapter we consider that every pair of destinations in each cluster C_i is mutually adjacent. Therefore, the resulting directed graphs are complete.

4 Computational Experiment

In this section we report the computational experiment conducted to assess the performance of our solution methodology FUZZY in practice. We have divided the experiment into two parts. In the first part we evaluate the performance of the CONTROL methodology. CONTROL consists in solving the route planning process of the Package Delivery Problem (PDP) for all the destinations with the ILP formulation for TSP presented in Sect. 3.2. The goal of this experiment is to measure the hardness of the second phase of PDP considering the entire set of destinations. In the second part of the experiment we evaluate the performance of our proposed FUZZY methodology presented in Sect. 3. The goal is to measure the differences in computing time and solution quality with respect to CONTROL. Although the experiment was divided into two parts, we report the results together in order to ease the interpretation and comparison.

4.1 Experimental Conditions and Test Bed

The experiments were conducted on a standard silicon–based computer with an intel i7-7500U processor at 2.7 GHz and 16 GB of RAM. Both methodologies were implemented in Java (JRE 1.8.0_201-b09). We use the well–known optimization engine CPLEX (v12.8) to solve the ILP formulations. For both parts of the experiment, we have set the time limit to 1800 CPU seconds for instance. We use 30 artificial instances randomly generated by the automatic tool proposed in [6]. The set of instances were classified by size into three groups: small, medium and big. All of these instances were generated under a uniform distribution. The parameter values for our FUZZYy methodology are $k = 5$, $m' = 2$ and $\in = 0.001$. The description of each group of instances is the following:

- **Small**. This dataset contains 10 instances whose number of destinations ranges from 28 to 100.
- **Medium**. This dataset contains 10 instances whose number of destinations ranges from 107 to 199.
- **Big**. This dataset contains 10 instances whose number of destinations ranges from 209 to 287.

Table 1 Experimental results for the CONTROL methodology and our FUZZY methodology over 29 random instances

Methodology	#Opt	O.V.	Gap (%)	Time (s)
CONTROL	7	86.6	1450.8	5.9
FUZZY	27	87.4	114.9	0.1

4.2 Experimental Results and Discussion

Table 1 summarizes the experimental results for the CONTROL methodology and the FUZZY methodology. The table has two rows, one for each methodology. It also has four headings that report important statistics measured from the experiment. Specifically, the table reports the number of optimal solutions found (#Opt); the average objective value (O.V.); the average relative gap (Gap (%)); and the average computing time in CPU seconds (Time (s)). It is important to point out that, in our FUZZY methodology, we consider a solution optimal if and only if CPLEX found the optimal route for each cluster in the time limit. In order to statistically validate the experimental results, we conducted the Wilcoxon rank sum test to both the objective values and the computing times.

The results reported in Table 1 consider 29 instances. This is so because CONTROL could not handle the last instance (the 10th instance) in the dataset Big. From the results we can observe that CONTROL had an effectiveness of 24.14% since it found 7 optimal solutions out of 29. The effectiveness of FUZZY was about 93.10%. Thus, FUZZY outperformed the CONTROL by 68.96% in effectiveness. The solution quality of CONTROL outperformed that of FUZZY by 0.97%. Although this value seems to be insignificant, the Wilcoxon rank sum test determined statistical significance for a confidence level of 98.86%. Figure 4 plots the objective value found by both CONTROL and FUZZY for the 29 artificial instances used in the experiment.

As we can observe from Fig. 4, in most of the instances both methodologies found a similar objective value. Nevertheless,CONTROL outperformed the objective value found by FUZZY in 22 instances whereas FUZZY outperformed CONTROL in

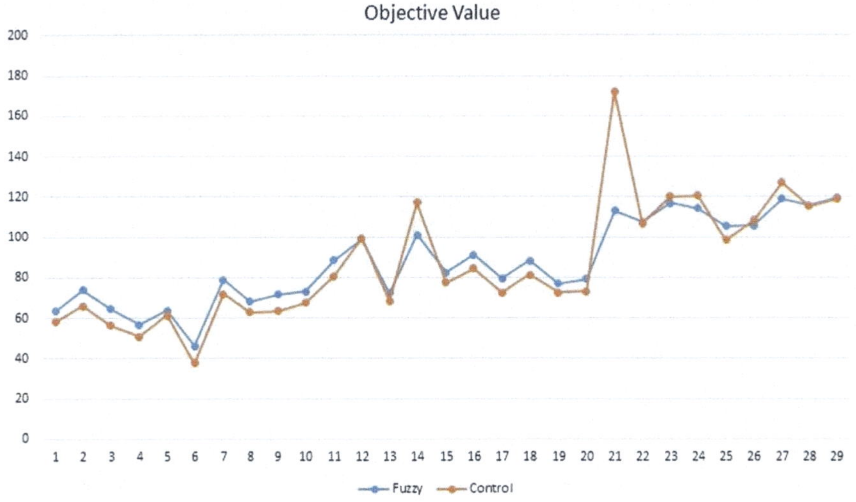

Fig. 4 Comparison of the control methodology and the proposed fuzzy methodology in terms of solution quality, i.e., the objective value found

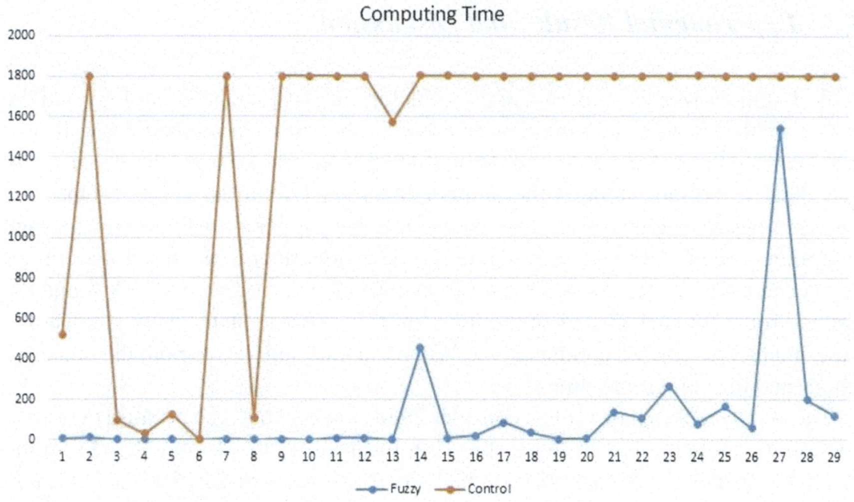

Fig. 5 Comparison of the control methodology and the proposed fuzzy methodology in terms of efficiency, i.e., the computing time

7 instances. The average computing time spent by CONTROL is about 1450.8 CPU seconds. This means that the total time required by CONTROL to carry out the experiment is about 42,072 CPU seconds, that is, 11 h, 41 min and 12 s. Conversely, FUZZY solved the same 29 instances in 3332 CPU seconds (55 min and 32 s). This means that FUZZY saved a considerable amount of time about 92.08% with respect to CONTROL. This represents a saving in uninterrupted computations of 10.76 hours. Figure 5 depicts a plot to compare the computing time of both methodologies. Like the plot shown in Fig. 4, the x-axis of this plot shows the 29 instances. The y-axis of the plot represents the computing time.

As we can see from Fig. 5, the difference in computing time between CONTROL and FUZZY is considerable. In fact, our FUZZY methodology outperformed the CONTROL methodology for all the 29 instances. The Wilcoxon test determined statistical significance with a confidence level of 99.99% for the average computing time. Finally, FUZZY reduced the average relative gap by 5.8% with respect to CONTROL. This means that our FUZZY methodology was much closer to find the optimal solutions for the instances than the CONTROL methodology.

5 Conclusions

In this chapter we have approached the Package Delivery Problem (PDP). This problem consists in locally delivering a set of rectangular–shaped boxes (packages) to a set of destinations (customers). PDP has two main phases: (i) the assignment of the packages to their corresponding delivery truck, and (ii) the route planning process. The first phase can be viewed as an application of the Three–Dimensional Bin Packing Problem with additional constraints [6]. The second phase tries to find the best path for the delivery truck considering the destination of the packages. This phase can be viewed as an application of the Traveling Salesman Problem (TSP). In this chapter we have focused on the second phase. In particular, we proposed to use the well–known Fuzzy c–means algorithm in order to reduce the computational complexity of the problem. The goal is to generate a number of k clusters and to optimize the route in each cluster.

We conducted a computational experiment to assess the performance of our proposed methodology (FUZZY) in practice. We compare our results with a CONTROL methodology. The CONTROL methodology consists in solving the TSP with all the destinations. This gives us an idea about the hardness of the second phase of PDP. The experimental results showed that the FUZZY methodology reduced the computing time of the CONTROL methodology in 92.08% but decreased the solution quality in 0.97%. We have conducted the Wilcoxon rank sum test to the experimental results in order to determine whether or not these results are statistically significant. The test determinates that the differences in computing time and solution quality are both significant with a confidence level of 99.99% and 98.86%, respectively. Therefore, considering all the experimental evidence and the statistical validation, we can conclude that our proposed FUZZY methodology significantly reduces the computational effort in the second phase of the Package Delivery Problem.

As stated previously, as far as we know, this is the first time that this methodology is applied in this specific context. Therefore, as a further work, other solution methodologies can be proposed and evaluated with other experimental designs such as the one proposed in [8].

Acknowledgements The authors would like to thank *Tecnológico Nacional de México*, and especially the authorities of Instituto Tecnológico de Altamira for their support in this research. We also thank the Mexican Council for Science and Technology (CONACYT) for its support through the Mexican National System of Researchers (SNI) (Grant Nos. 72282 and 70157).

References

1. Asha G et al (2018) Energy efficient clustering and routing in a wireless sensor networks. Procedia Comput Sci 134:178–185
2. Chen D, Batson R, Dang Y (2010) Applied integer programming: Modeling and solution
3. Cruz PP (2011) Inteligencia artificial con aplicaciones a la ingeniería. Alfaomega
4. Gao X, Zhou Y, Amir MIH, Rosyidah FA, Lee GM (2017) A hybrid genetic algorithm for multi-emergency medical service center location-allocation problem in disaster response. Int J Ind Eng 24(6)
5. Glistau E, Coello Machado NI (2018) Industry 4.0, logistics 4.0 and materials-chances and solutions. In: Materials science forum, vol. 919, pp. 307–314. Trans Tech Publ
6. Hernández PH, Castillo-García N, Larkins ER, Ruiz JGG, Díaz SVM, Resendiz ES (2019) A fuzzy logic classifier for the three dimensional bin packing problem deriving from package

delivery companies application. In: Handbook of research on metaheuristics for order picking optimization in warehouses to smart cities, pp. 433–442. IGI Global (2019)
7. Miyamoto S, Ichihashi H, Honda K, Ichihashi H (2008) Algorithms for fuzzy clustering. Springer (2008)
8. Montesino-Guerra JA, Puga H, Carpio JM, Ornelas-Rodríguez M, Rojas-Domínguez A, Ortiz-Aguilar L (2020) Combinatorial designs on constraint satisfaction problem (vrp). In: Intuitionistic and Type-2 fuzzy logic enhancements in neural and optimization algorithms: theory and applications, pp 509–526. Springer (2020)
9. Robusto CC (1957) The cosine-haversine formula. The American Mathematical Monthly 64(1):38–40
10. Shokouhifar M, Jalali A (2017) Optimized sugeno fuzzy clustering algorithm for wireless sensor networks. Eng Appl Artif Intell 60:16–25
11. Su S, Zhao S (2018) An optimal clustering mechanism based on fuzzy-c means for wireless sensor networks. Sustainable Comput: Inform Syst 18:127–134
12. Wang K (2016) Logistics 4.0 solution-new challenges and opportunities. In: 6th International workshop of advanced manufacturing and automation. Atlantis Press
13. Wilcoxon F (1954) Individual comparisons by ranking methods. Biom Bull 1:80–83
14. Yamamoto K, Uesugi K, Watanabe T (2008) Adaptive routing of cruising taxis by mutual exchange of pathways. In: International conference on knowledge-based and intelligent information and engineering systems, pp 559–566. Springer

State of the Art for the Creation of a Methodology for the Proper Location of Urban Truck Stops on Route 2A

Marlyn Montalvo Martel, Elías Carrum, and Alberto Ochoa-Zezzatti

Abstract The founding, organization, and growth of large urban agglomerations throughout history have generated significant challenges for transportation and mobility systems. Public transportation is considered an essential piece in sustainable urban development, since it allows to achieve more efficient movements and flows in the city, thus, it would be reduced significantly the environmental pollution because of the car's use decrease. However, making the right decisions to establish efficient urban transportation routes is becoming more complex, thus it is necessary to have innovative and useful methods that are capable of finding the solution to transport problems considering urban growth among other factors. Through, the suitable design of the transportation system and the location of the stops, it is possible to reduce travel times and increase user satisfaction. This document reviews the literature of the last five years on some topics of interest such as network design, public transportation, the demand for public transportation; studies carried out with the distance that users are willing to walk to bus stops and the location of stops were also included so that the design of the transportation network is optimal. The results showed that the heuristics and metaheuristics are the most commonly used methods to solve transportation network problems. Furthermore, demand is an essential factor in locating public transportation stops and the distance that must exist between them is 400–700 m, intending to improve the efficiency of public transportation.

Keywords Transportation network · Demand · Walking distance

M. M. Martel · E. Carrum (✉)
Corporación Mexicana de Investigación en Materiales, Calle Ciencia y Tecnología 790, 25290 Saltillo, Mexico
e-mail: eliascarrum@comimsa.com

M. M. Martel
e-mail: mmontalvo@alumnos-comimsa.mx

A. Ochoa-Zezzatti
Universidad Autónoma de Ciudad Juárez, Av. Plutarco Elías Calles 1210, 32310 Ciudad Juárez, Mexico
e-mail: alberto.ochoa@uacj.mx

© The Author(s), under exclusive license to Springer Nature Switzerland AG 2021

A. Ochoa-Zezzatti et al. (eds.), *Technological and Industrial Applications Associated with Intelligent Logistics*, Lecture Notes in Intelligent Transportation and Infrastructure,
https://doi.org/10.1007/978-3-030-68655-0_13

1 Introduction

The urbanization process that is taking place all over the planet constitutes a global trend that induces profound transformations in cities, which develop strategies and plans to face the great changes and guarantee their sustainability.

In recent years, the smart city (SC) concept has become popular due to the interest in achieving sustainability goals with the help of "intelligence". This concept emerged for the first time in the 1990s [1] as an alternative to traditional planning modes, using new technologies.

Most studies on this subject agree that for a city to acquire the name of smart it must provide mobility, society, quality of life, environment, economy, and government (Fig. 1).

From a logistical point of view, the fast growth of urbanization rates and the potential change in the goods demand in urban areas justify the need to develop new urban logistics systems. These must guarantee efficient urban mobility [2].

Ref. [3] define city logistics (urban mobility) as "a system that improves its transport activities (…), taking into account the environmental impact of vehicles, traffic jam and energy consumption within the economy".

In Mexico, efforts are being made to diversify mobility and implement regulatory measures to reduce the use of cars to improve public transportation [4].

According to a study carried out by the Instituto Mexicano para la Competitividad (IMCO), urban mobility options in the most important cities of the country and their ability to offer accessible public transportation options were compared, only the city

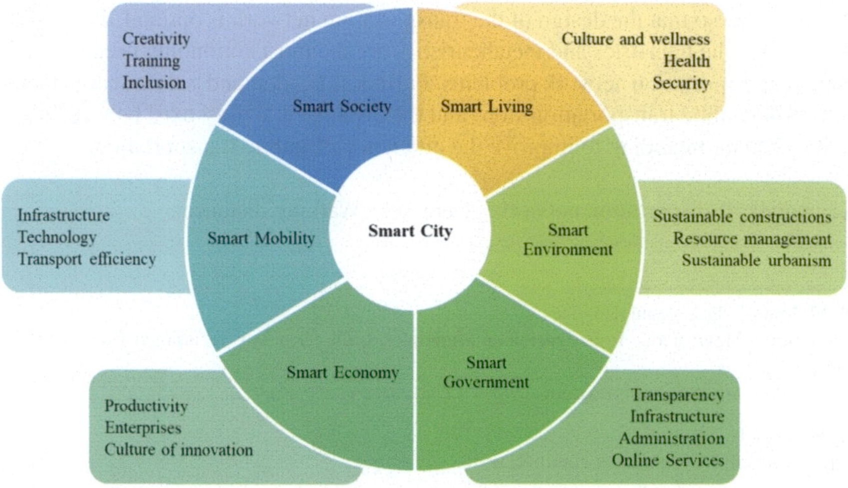

Fig. 1 Pillars of a Smart City

of Saltillo, the Valley of Mexico and Guadalajara met the aforementioned characteristics. These cities, including the capital of Coahuila, are capable of providing adequate public transportation, but they do not have efficient mobility [4].

However, according to [4], Saltillo is working to establish itself as a smart city due to the actions implemented by the municipal authority headed by Eng. Manolo Jiménez Salinas; an example of these actions is the approach with the different factors that make up the ecosystem called Saltillo, the participation of business organizations, citizen councils and committees, the use of new technologies such as LED lighting, solar panels, the use of methane gas. and communication tools to form support groups in which citizens participate in conjunction with the authority to maintain public safety.

The municipality of Saltillo, being the Capital of the State of Coahuila, is one of the main metropolitan areas with the largest population. It is characterized mainly by being an industrial city. According to [5] in Saltillo, there are approximately 58 routes for the public transportation of urban trucks (Fig. 2), in which some official stops are determined.

Despite the establishment of stops, urban trucks stop each time a user stops them regardless of where they are, as long as they are within the route of the journey. This causes the assigned times among stops to not be respected and schedules to be delayed. Also, it causes congestion in traffic, since it obstructs the path of other vehicles when it is repeatedly stopped. There is also a higher cost of fuel and emissions of CO, CO^2, HC, and NOx each time you stop and start again [6].

The present study consists of reviewing the state of the art for the creation of a methodology that allows the proper location of urban truck stops. Besides, it has

Fig. 2 Map of urban truck routes in the municipality of Saltillo [5]

both methodological and practical novelty, because the urban public transportation system needs a lot of analyses and studies in the city. After all, it is more complex and requires more driving time compared to other modes of transportation. Once implemented, this will benefit users who use trucks as an indispensable mobility tool in their daily lives.

2 Literary Review

A transportation network is an infrastructure necessary for the circulation of vehicles that transport people or goods. They are located in an area linking the centers of population or industrial activity [7].

Transportation networks are classified, based on their density, more technically, they differentiate between interurban networks and urban networks. Urban networks are one of the most complex challenges that public administrators face in any city [7].

2.1 Transportation Network Design and Frequency Configuration

The transportation network design is a complex problem that is generally formulated as an optimization problem. The most efficient solution methods are based on heuristic procedures, but their applications are mainly limited to experiments or small networks. In recent years, the evolution of operation researches and the growing power of computer engineering have led to the increase of interest in this problem. Ref. [8–11] propose a global review on the route design and the frequency configuration.

Taking into account the principle of sustainability of smart cities [12] focus their research on the development of an efficient model to solve a variant oriented in a sustainable way of the transportation network design problem. The model incorporates sustainable design goals, considers emission-free (electrical) vehicles, and introduces a direct route design approach with route structure and directivity control.

Among the works carried out over the last five years, there are [7, 13–16], most of which developed an optimization algorithm [13] and used a memetic algorithm (MA) to solve the urban transportation network design problem (UTNDP) with the aim of optimizing both the passenger cost and the unsatisfied demand. The results obtained are more efficient compared to the traditional metaheuristics in the reviewed literature. Ref. [15] carried out a model to address the problem of integrated fast transportation network design and line planning. They use a metaheuristic search. The results indicate that the proposed algorithm outperforms the leading-edge commercial solvers in all test instances.

Regarding the multi-objective nature of the transportation network design problem, [16] formulated a discrete multi-objective mathematical program for the strategic planning of a fast transportation network, which leads to a sensitivity analysis for effectiveness, efficiency, and equity.

Over the past few years, the evolution of operation researches and information technology has brought a great deal of renewed attention to the transportation network design problem. New approaches based on metaheuristic techniques (genetic algorithm, simulated annealing, or taboo search) have been frequently applied to solve optimization problems. Ref. [8] proposed a genetic algorithm (GA) to solve the design of the transportation network and the frequency configuration simultaneously with the objectives of optimizing the costs of passengers and operators. The method uses an algorithm that creates a broad set of transportation routes, a heuristic to generate the initial population, a check of the viability of the set of routes, and a passenger allocation procedure. The results show that the proposed GA with alternate objectives is highly efficient.

Reference [9] propose a two-level solution method that: (i) determines the frequencies in the upper level, maximizing savings in waiting time, considering restrictions such as budget, fleet size, vehicle load and advancement policy; and (ii) running a simulation algorithm at the lower level that models the demand response to the new frequency setting. The proposed approach is tested on the Chicago transportation network. Compared to the baseline scenario, the method achieves a 45.7% reduction in waiting times and a 9% increase in the number of shipments.

Using a different technique [14] applied an analysis methodology to evaluate urban public transportation routes and proposed alternatives that improve their performance. Also, they conducted an analysis aimed at the needs of users and public administrations to find the shortest route for a transportation route, maximizing the number of passengers. This analysis was carried out through a geographic information system (GIS) [17]. The results of the proposed method shown that the Network Analysis allowed to integrate multiple objectives in the solution model such as the number of potential users on the route, speed, travel times, distances.

Several factors that determine the bus system reliability, such as the length of stay, the location of bus stops, and the passengers' number in the loading and unloading areas [18]. The work of [19] was directed at defining the optimal bus schedule for the Pittsburgh bus system to minimize waiting time for passengers.

Besides, the variation in bus demand throughout the different times of the day has a considerable effect on the effectiveness of the system [20]. Ref. [21] proposed a real-time bus monitoring system and passenger information through a mobile application. Using GPS, they created a database with the locations of the buses that belonged to the route, in order to estimate the arrival time at each stop that is part of the route. Additionally, they considered the uncertain conditions as the day progresses to meet the needs of passengers, vehicle drivers, and transportation system administrators.

2.2 Demand

Demand can change unexpectedly or randomly; however, since the demand for transportation depends on economic activities that have a high degree of routine and repetition, there may be a certain tendency to show a more or less stable cyclical behavior [22].

It is possible to establish a pattern for the time variation over the day, that is, a demand that at dawn grows until it reaches a morning maximum (the "peak hour") and can even exceed the transportation capacity offered, then decreases at noon and rises again in the afternoon, to decrease as the population retires to rest during the afternoon and night (Fig. 3). Similarly, there is a variation in demand during the week as shown in (Fig. 4). While demand is normally low on Sundays, it grows on Mondays and it stabilizes on Tuesdays; it can go down on Wednesdays or Thursdays; it shows peaks and congestion on Fridays, to decrease on Saturdays [22].

In general, the researches focused on public transportation make demand estimates to calculate the elasticity in the consumption of the service [23–26].

Based on the estimation of the public transportation demand [23], they expanded the model presented by [27], considering the variation in demand over time and along the route of the line. The variation in demand is modeled assuming the temporal and spatial heterogeneity of the elasticity of the number of users regarding to frequencies. Another interesting research on demand was carried out by [7] where they proposed a simulation model for the Homogeneous Fleet Routing Problem (VRPH) through Arena software in a transportation company in Barranquilla, Colombia. The demand analysis implied the design and execution of a passenger mobility system that allowed to know the arrival rates for each node in the public transportation system.

Fig. 3 Time variation

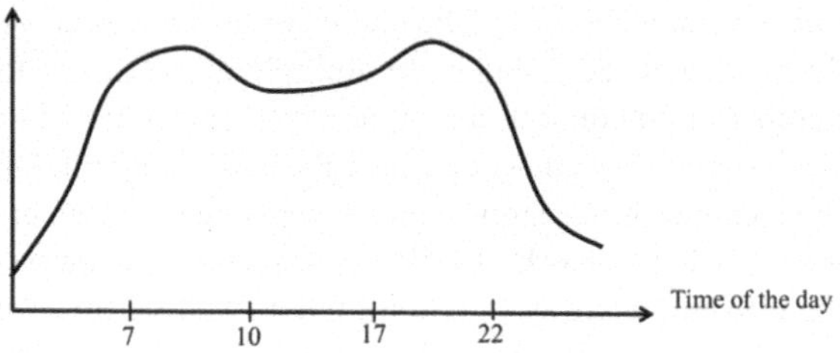

Fig. 4 Weekly variation

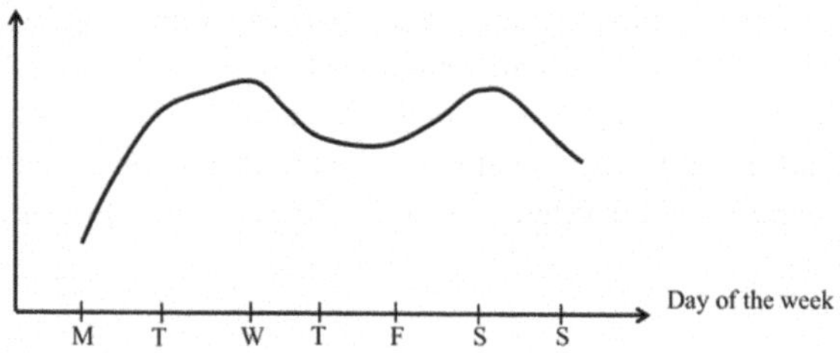

Reference [24] estimated different demand elasticities for urban public transportation, by using two different methodologies as error correction models; dynamic and cointegration models.

Reference [26] presented a discrete choice model to estimate the demand for public transportation in Colombia. The estimation was carried out in the presence of unobservable variables, a methodology initially developed by [28], who proposed a strategy to identify demand parameters in the presence of unobservable variables. From an individual utility model that allows to observe the revealed preferences, the aggregate parameters of demand are identified and the elasticity of demand regarding the cost of the service, the level of income, the frequency, and the speed is calculated.

The problem of determining transportation demand using smart cards was considered in studies such as [29, 30]. Ref. [31] reports the application of various supervised machine learning (SML) techniques to smart card data to forecast demand, and compare these results with traditional linear model estimates. As a result of the analysis of these works, [32] considered a method to determine the transport demand according to the results of the electronic ticket validation: they formulated the problem of determining the traffic demand according to the ticket accounting system electronics and automobile satellite transportation monitoring system (Global Navigation Satellite System).

Taking into account the principle of sustainability of smart cities [12] focus their research on the development of an efficient model to solve a variant oriented in a sustainable way of the transportation network design problem. The model incorporates sustainable design goals, considers emission-free (electrical) vehicles, and introduces a direct route design approach with route structure and directivity control.

2.3 Walking Distance

The researchers seek to know the walking distance that people will accept to access public transportation [33] in order to reduce this distance as much as possible.

The distance that pedestrians are willing to walk to the bus stop is of great importance in planning, modeling, designing, and evaluating public transportation systems. For example, the Public Transportation Infrastructure Manual of Queensland (Australia) states that the reasonable coverage radius for a stop is 400 m [34], this is also consistent with manuals used in the United States and Great Britain. However, the Public Transportation Council [35] in Costa Rica developed a level of service criteria for walking distances to bus stops from A (less than 400 m) to E (more than 700 m).

People walk farther to stopping-places facilities when walking environments are suitable, such as shaded hallways and low traffic impacts [36]. Ref. [37] using data from the Brisbane, Australia Household Travel Survey, they explored the variation in walking time between individuals of different socioeconomic status. Performing a comparative analysis, they reveal that walking time is more sensitive among part-time workers, high-income workers, and elderly travelers and that walking time is less sensitive among higher education students who study and work at the same time.

The distance of access to public transportation is an important metric for planning, modeling, and evaluating public transportation networks, it is common that it is used in objectives and policy statements. However, accurately measuring entry (and exit) distance can be difficult. Because estimates are based on inferences considering census data or small samples of disaggregated data from travel journal surveys. The smart cards used for the payment of fares represent a new data source that can be used to infer access distances for a large sample of users, at a disaggregated level, and at low cost, compared to the daily travel surveys. Ref. [38] estimated the distance of access to public transportation for London using the access card data. Using the access station and the card owner's reported address, the authors calculated a straight-line access distance. The average travel distance of the buses was 0.23 miles (370 m). This methodology of access distance inference allows the analysis of the variation in the access distance through the network.

The demographic characteristics of commuters, such as gender, age, income, and number of vehicles, also determine walking distance [37]. The walk that users make when they reach the final destination of the trip is very important for [39], who using a survey on traffic in the metropolitan area of Minneapolis and St. Paul, developed four models to compare the effects of the surrounding environment of stops within walking distance. Density in work zones is negatively correlated with walking distance, consistent in all four models. Other environmental variables exhibit different impacts depending on the location of the stop. In particular, the combination of land use has positive impacts on the walking distance of users at off-center stops and suburban places of employment, while the density in work zones is more important for suburban centers. Job accessibility and the number of intersections have significant effects on stops within downtown areas but have no significant impacts on stops outside of downtown areas. The number of stopping places has opposite impacts on the walking distance for stops inside and outside the center. Additionally, the environment tends to have a greater impact on walking distance in downtown areas.

In the Latin American context, [40] propose an easy and reliable way to estimate walking distances to bus stops in San José Costa Rica. They determine which statistical distribution fits best the empirical walking distance distribution and compare the actual, perceived, and established preference distances.

A study by [41] clarifies even further the factors affecting user satisfaction, comfort, attitude, and safety of staff were seen. This issue is also addressed by [42], who investigates the "willingness to walk" of pedestrians to the bus stops of Rapid Penang on the island of Penang. They examined parameters, such as sociodemography, walking distance, and time, to determine the factors that influence "willingness to walk." This study also identified the maximum, minimum, and average distances that a pedestrian walked to reach the bus stops (a minimum distance of 50 m and a maximum distance of 2,000 m).

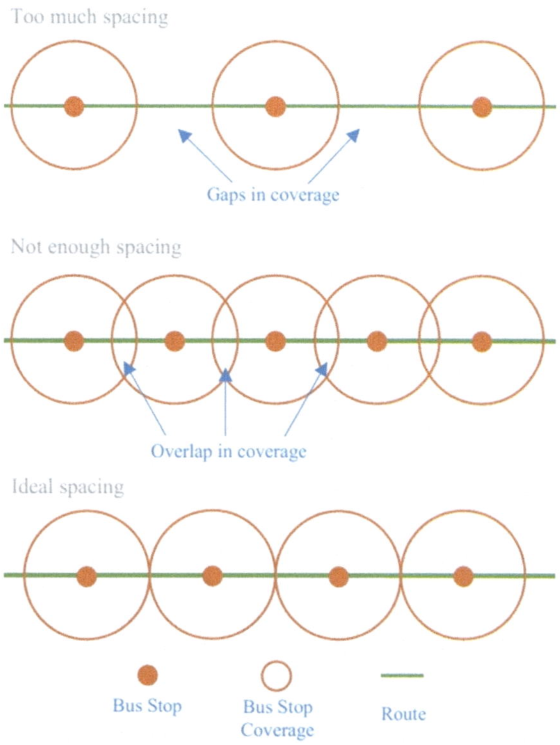

Fig. 5 Spacing impacts on bus stop coverage [52]

2.4 Stops' Location

A review of the available books and documents shows that the spacing between bus stops has attracted the attention of many researchers and urban planners. Most efforts are focused on finding the standard value as an optimal distance between bus stops (Fig. 5).

Traveling to school, work, shops, and other points of interest is a common routine for residents of metropolitan areas [43]. The constant increase in the urban population and the greater difference between the urban and rural population motivate the study of improvements for urban transportation services. These improvements are important to achieve efficiency and accessibility, mainly for residents in peripheral regions. Regarding this problem, recent works use public data sources to diagnose or propose solutions regarding the city's transportation system [44, 45].

Reference [46] analyzed the walking distance to a bus stop in Belo Horizonte, Brazil. They determined the Accessibility Index (AI) for a given region, the shorter the walking distance to a bus stop and the waiting time until the bus arrives, the higher (and better) the index will be. The accessibility of a region is not considered bad, with respect to walking distance, if the distance to a bus stop is, on average, less than or equal to 500 m.

Public transportation services could provide greater accessibility and attract new passengers if there were more embarkation and disembarkation points (also known as pick-up and drop-off points) spatially distributed throughout the city. Genetic algorithms have previously been applied in transportation studies, and boundaries have developed over time. Most of the current studies carried out on the transportation network, obtain the optimal design of the road network, and optimize the distance between stops and their locations.

Researchers who presented an optimization method to determine the space between stops are: [47–51]. A genetic algorithm was proposed by [47] to define the location of ten new bus stops in the city of Nonoichi, Japan. The authors' goal was to minimize the walking distance between the student dormitories and the nearest bus stop connecting to the Kanazawa Institute of Technology. The best solution reduced the walking distance by up to 702 m.

Reference [48] propose an optimization method to define the space between bus stops in the city of Aurangabad, India. Optimization aims to maximize the covered population. The authors use a random walk technique to explore the nodes and edges of the city's road network.

They also [49] propose a bi-objective optimization method to reduce long walks to take a bus, in the city of Dalian, China. The bus stops were chosen based on a set of possible pick-up and drop-off points generated near the passenger destination. The two goals at the job were to minimize travel costs for passengers and operators. The optimization applied the Non-Mastered Classification Genetic Algorithm (NSGA-II).

Reference [53] compare different methods to solve a location routing problem (LRP), testing four artificial intelligence techniques: Maximin, K-means, Fuzzy C-means, and Competitive Learning and two hybrids such as K-means with Competitive Learning and K-means with Maximin to assign the bus stops. They compared the efficiency of the algorithms, in terms of the quality of the solutions. K-means with Maximin provided the best solution, as it minimized the number of bus stop locations and the total travel distance of employees, while K-means with competitive learning, with the same number of stops, provided a higher distance of total trip. The other non-hybrid techniques provided a greater number of bus stop locations. Then they used ant colony optimization (ACO) to determine the optimal route between bus stops, as assigned by K-means with Maximin. The optimal bus route for transporting the factory employees proved very expensive and created monitoring difficulties. To address these concerns, they restricted the number of bus routes; while this increased the overall distance dramatically, it provided a more practical solution for the factory.

Reference [50] propose two integer and metaheuristic programming models to analyze the location of the bus stops and propose new pick-up and drop-off locations to avoid long walks when taking the bus. The proposed computing approaches to optimize the location of bus stops in a scalable way. The optimization method proposed in this work is compared with an adapted version of the algorithm presented by [47]. Having one bus stop for every 800 m of streets implies that, for every address contained in that neighborhood, there will be, on average, one bus stop 400 m away.

This threshold was chosen considering the works of [46, 47]; it is considered reasonable to walk 400 m to reach a bus stop. The experimental results show that many new bus stops are required to improve the quality of service provided to the population.

More recently, [51] propose a CB line planning framework called CB-Planner, which is applicable to multiple sources of trip data. They formulate a mathematical schedule to simultaneously optimize bus stop locations, bus routes, schedules, and passenger probabilities of choosing CB. Then they develop a heuristic solution framework that includes a network density-based clustering method to discover potential travel demands efficiently, a bus stop display algorithm to minimize the number of stops and walking distance, and routing and scheduling algorithms based on dynamic scheduling to maximize profit or estimated performance.

Some other research done on bus stop distance focused on travel and waiting time. Eliminating bus stops also reduces variations in travel times because there are fewer opportunities for delay, for this reason [54] develop a methodology to consolidate stopping places, using a mathematical model and a program that makes decisions to consolidate stops in accordance with the general savings in travel time of users and the desired accessibility. The model iterates until users generalize the travel time savings to the maximum. The results showed that the distance between stops and the probable change in demand are the most influential factors on consolidation. The frequency of the service was also influential.

The location of the bus stops is a complicated issue; by reducing the distances between stops, the walking time decreases, but the total travel time can increase. Ref. [55] study a route in the city of Rasht, in northern Iran. First, they proposed a model for calculating travel time, by which the number of stops and, consequently, travel time can be optimized. Then they used a metaheuristic algorithm to determine the optimal locations and distances of bus stops. They found boarding time along with bus capacity to be the most significant factor affecting travel time. Consequently, they recommend focusing on these factors to improve the efficiency of the transportation system.

Reference [56] conduct a study to optimize and eliminate some of the bus stops in Kuwait in order to improve bus mobility, minimize the negative impact on the environment, decrease processing cost, and mitigate traffic jam. They carried out a place research and survey to indicate the problems affecting bus users and considered different elements such as the spacing between bus stops, rush hours and length of stay; which can be defined as "the average amount of time that a bus stops on the sidewalk to attend the movements of passengers, including the time required to open and close the doors" [57]. Field observation showed that as the number of passengers increases, the length of stay increases. Zero passenger stations were eliminated and the distance between stops was optimized, whereby a standard distance of 1–1.4 km should be adopted between every two stops and the stay time should be increased to more than 6 s per passenger.

3 Conclusions

Due to population growth and the increase in the rate of urbanization, transportation represents a challenge for the development of countries. Therefore, analyzing and studying problems related to mobility and transportation are so complex that traditional algorithms cannot provide a solution. In this sense, heuristic and metaheuristic algorithms are capable of finding a suitable solution for the urban transportation system.

In the demand analysis for public transportation, it is essential to consider at least three key elements such as supply points (urban bus stops), potential demand (number of people using public transportation trucks) and the distance that it separates the potential demand from the supply points (distance that users must walk to reach the closest stop). Therefore, in order to establish the location of the stops, it is necessary to know the areas of greatest demand to expand coverage and accessibility to users. However, this would affect the travel time. In addition, the demand is directly related to the area, this factor significantly affects the number of stops that must be established to satisfy user demand.

The distance that pedestrians are willing to walk to the bus stop is of great importance for modeling public transportation systems, in this sense the studies reviewed above propose that the ideal distance is 400 m, however, some studies suggest a greater distance (700 m) due to the characteristics of the cities and the demand for public transportation in the area.

References

1. Alawadhi S, Aldama-Nalda A, Chourabi H, Gil-Garcia JR, Leung S, Mellouli S, Nam T, Pardo TA, Scholl HJ, Walker S (2012) Building understanding of smart city initiatives. In: Scholl HJ, Janssen M, Wimmer MA, Moe CE, Flak LS (eds) Electronic government, Springer Berlin Heidelberg, Berlin, Heidelberg, pp 40–53. https://doi.org/10.1007/978-3-642-33489-4_4
2. Fatnassi E, Chebbi O, Chaouachi J (2015) Viability of implementing smart mobility tool in the case of Tunis City. In: Saeed K, Homenda W (eds) Computer information systems and industrial management. Springer International Publishing, Cham (2015), pp 339–350. https://doi.org/10.1007/978-3-319-24369-6_28
3. Taniguchi E (2002) The institute for city logistics
4. Duque-García A (2019) Trasporte y movilidad eficiente en el ámbito nacional e internacional ¿Por qué la ciudad inteligente es necesaria?
5. Vázquez, A. Rutas de Camiones de Saltillo y su Área Metropolitana - rutadirecta.com, https://saltillo.rutadirecta.com/
6. Pan Y, Chen S, Li T, Niu S, Tang K (2019) Exploring spatial variation of the bus stop influence zone with multi-source data: a case study in Zhenjiang. China. J Transp Geogr 76:166–177. https://doi.org/10.1016/j.jtrangeo.2019.03.012
7. Riquelme P, Gatica G, Orozco E (2015) Diseño de un Modelo de Operación para Ruteo de Transporte Urbano Basado en Simulación Discreta. Investigacion e Innovación en Ingenierias. 3 https://doi.org/10.17081/invinno.3.2.2026

8. Arbex RO, da Cunha CB (2015) Efficient transit network design and frequencies setting multi-objective optimization by alternating objective genetic algorithm. Transp Res Part B: Methodological 81:355–376. https://doi.org/10.1016/j.trb.2015.06.014
9. Verbas O, Mahmassani H (2015) Integrated frequency allocation and user assignment in multi-modal transit networks: methodology and application to large-scale urban systems. Transp Res Rec J Transp Res Board. 2498:37–45. https://doi.org/10.3141/2498-5
10. López-Ramos F, Codina E, Marín Á, Guarnaschelli A (2017) Integrated approach to network design and frequency setting problem in railway rapid transit systems. Comput Oper Res 80:128–146. https://doi.org/10.1016/j.cor.2016.12.006
11. Buba AT, Lee LS (2018) A differential evolution for simultaneous transit network design and frequency setting problem. Expert Syst Appl 106:277–289. https://doi.org/10.1016/j.eswa.2018.04.011
12. Pternea M, Kepaptsoglou K, Karlaftis MG (2015) Sustainable urban transit network design. Transp Res Part A: Policy Pract 77:276–291. https://doi.org/10.1016/j.tra.2015.04.024
13. Zhao H, Xu W (Ato), Jiang R (2015) The Memetic algorithm for the optimization of urban transit network. Expert Syst Appl 42:3760–3773. https://doi.org/10.1016/j.eswa.2014.11.056
14. Serna-Uran, García-Castrillón J, Flórez-Londoño O (2016) Análisis de Rutas de Transporte de Pasajeros Mediante la Herramienta Network Analyst de Arcgis. Caso Aplicado en la Ciudad de Medellín Analysis of Passenger Transportation Routes through Network Analyst Tool of Arcgis. Case Applied in Medellin City. Ingenierias USBMed. 7. https://doi.org/10.21500/20275846.2631
15. Canca D, De-Los-Santos A, Laporte G, Mesa JA (2017) An adaptive neighborhood search metaheuristic for the integrated railway rapid transit network design and line planning problem. Comput Oper Res 78:1–14. https://doi.org/10.1016/j.cor.2016.08.008
16. Gutiérrez-Jarpa G, Laporte G, Marianov V, Moccia L (2017) Multi-objective rapid transit network design with modal competition: the case of Concepción. Chile. Computers & Operations Research. 78:27–43. https://doi.org/10.1016/j.cor.2016.08.009
17. Durango-Vanegas CE, Zapata-Jaramillo CM (2015) Una representación basada en Semat y RUP para el Método de Desarrollo SIG del Instituto Geográfico Agustín Codazzi. Revista Ingenierías USBMed. 6:24–37
18. Arhin S, Noel E, Anderson MF, Williams L, Ribisso A, Stinson R (2016) Optimization of transit total bus stop time models. J Traffic Transp Eng (English Edition). 3:146–153. https://doi.org/10.1016/j.jtte.2015.07.001
19. Kornfeld S, Ma W, Resnikoff A (2018) Optimizing bus schedules to minimize waiting time. 14
20. Zhang K, Xu Y, Sun DJ (2017) A mixed frontier model for urban bus performance evaluation. Proceedings of the Institution of Civil Engineers - Transport. 171:1–10. https://doi.org/10.1680/jtran.15.00123
21. Chandurkar S, Mugade S, Sinha S, Misal M, Borekar P (2018) Implementation of real time bus monitoring and passenger information system
22. Islas-Rivera VM, Rivera-Trujillo C, Torres Vargas G (2002) Estudio de la demanda del transporte. Instituto Mexicano del Transporte
23. Verbas O, Frei C, Mahmassani H, Chan R (2015) Stretching resources: Sensitivity of optimal bus frequency allocation to stop-level demand elasticities. Public Transport 7:1–20. https://doi.org/10.1007/s12469-013-0084-6
24. Montalvo J (2016) Elasticidad-precio de la demanda del transporte público urbano: un análisis para los servicios de ómnibus y subterráneo de la Ciudad Autónoma de Buenos Aires. Actualidad Económica. 26:7–19
25. Dike DN, Ejem E, Erumaka O, Chukwu O, Ibe C (2018) Estimation of inter-city travel demand for public road transport in Nigeria. J Sustain Dev Transport Logistics 3:88–98. https://doi.org/10.14254/jsdtl.2018.3-3.7
26. Toro D, Cantillo VM, Espinosa A (2018) La demanda de transporte público en Colombia, 2000–2010. Economía & Región. 12:181–204

27. Furth PG, Wilson NHM (1981) Setting frequences on bus routes: theory and practice. In: Transportation Research Record
28. Berry ST (1994) Estimating Discrete-Choice Models of Product Differentiation. Rand J Econ 25:242–262. https://doi.org/10.2307/2555829
29. El Mahrsi MK, Côme E, Oukhellou L, Verleysen M (2017) Clustering smart card data for urban mobility analysis. IEEE Trans Intell Transp Syst 18:712–728. https://doi.org/10.1109/TITS.2016.2600515
30. Fadeev, A., Alhusseini, S., Belova, E.N.: Monitoring Public Transport Demand Using Data From Automated Fare Collection System. In: Engineering Research (2018). https://doi.org/https://doi.org/10.2991/avent-18.2018.2.
31. Palacio S (2018) Machine learning forecasts of public transport demand: a comparative analysis of supervised algorithms using smart card data. SSRN Electron J https://doi.org/10.2139/ssrn.3165303
32. Fadeev A, Alhusseini S (2020) Determining the public transport demand by validation data of the electronic tickets. IOP Conf Ser Mater Sci Eng 734, 012148. https://doi.org/10.1088/1757-899X/734/1/012148
33. Park S, Deakin E, Jang K (2015) Can good walkability expand the size of Transit-Oriented Developments? transportation research record. J Transp Res Board. 2519:157–164. https://doi.org/10.3141/2519-17
34. Department of Transport and Main Roads (2015) TransLink Division Public Transport Infrastructure Manual. State of Queensland, Australia
35. Consejo de Transporte Público de Costa Rica (2015) Manual para la Evaluación y Calificación de la Calidad del Servicio Público de Transporte Remunerado de Personas. , Costa Rica
36. Park S, Choi K, Lee JS (2015) To walk or not to walk: testing the effect of path walkability on transit users' access mode choices to the station. Int J Sustain Transp 9:529–541. https://doi.org/10.1080/15568318.2013.825036
37. Chia J, Lee J (Brian), Kamruzzaman M (2016) Walking to public transit: exploring variations by socioeconomic status. Int J Sustain Transp 10, 805–814. https://doi.org/10.1080/15568318.2016.1156792
38. Viggiano C, Koutsopoulos H, Attanucci J, Wilson N (2016) Inferring public transport access distance from smart card registration and transaction data. Transp Res Rec J Transp Res Board. 2544:55–62. https://doi.org/10.3141/2544-07
39. Wang J, Cao X (2017) Exploring built environment correlates of walking distance of transit egress in the Twin Cities. J Transp Geogr 64:132–138. https://doi.org/10.1016/j.jtrangeo.2017.08.013
40. Rodriguez-Gonzalez MB, Aguero-Valverde J (2017) Walking distances from home to bus stops in San José, Costa Rica: Real, perceived, and stated-preference distances. Adv Transp Stud 43:63–74. https://doi.org/10.4399/97888255077375
41. Efthymiou D, Antoniou C (2017) Understanding the effects of economic crisis on public transport users' satisfaction and demand. Transp Policy 53:89–97. https://doi.org/10.1016/j.tranpol.2016.09.007
42. Sukor NSA, Fisal SM (2018) Factors influencing the willingness to walk to the bus stops in Penang Island. Plann Malays J 16. https://doi.org/10.21837/pmjournal.v16.i5.423
43. Logiodice P, Arbex R, Tomasiello DB, Giannotti MA (2015) Spatial visualization of job inaccessibility to identify transport related social exclusion. In: GeoInfo
44. Santos S, Davis C Jr, Smarzaro R (2017) Analyzing traffic accidents based on the integration of official and crowdsourced data. J Inf Data Manag 8:67–82
45. Jerônimo CLM, Campelo CEC, de Souza Baptista C (2017) Using open data to analyze urban mobility from social networks. J Inf Data Manag 8, 83–83
46. Veras DFG, Pinto GFN, Lobo CFF, Cardoso L, Garcia RA (2016) Acessibilidade urbana em Belo Horizonte: apontamentos sobre a acessibilidade aos serviços de tranporte colectivo municipal. Presented at the 7o. Congresso Luso Brasileiro para o Planejamento, Urbano, Regional, Integrado e Sustentável (Pluris) , Maceió, Brazil

47. Takakura M, Furuta T, Tanaka MS (2015) Urban bus network design using genetic algorithm and map information. In: Proceedings of the Eastern Asia society for transportation studies
48. Nalawade DB, Nagne AD, Dhumal RK, Kale KV (2016) Multilevel framework for optimizing bus stop spacing. IJRET: Int J Res Eng Technol 05, 298–304
49. Yao B, Cao Q, Jin L, Zhang M, Zhao Y (2017) Circle line optimization of shuttle bus in central business district without transit Hub. PROMET - Traffic & Transportation. 29:45. https://doi.org/10.7307/ptt.v29i1.2015
50. Monteiro C, Martins F, Davis Jr C (2017) Optimization of new Pick-up and Drop-off points for Public Transportation. In: Proceedings of the Brazilian Symposium on GeoInformatics
51. Lyu Y, Chow C-Y, Lee VCS, Ng JKY, Li Y, Zeng J (2019) CB-Planner: A bus line planning framework for customized bus systems. Transp Res Part C: Emerg Technol 101:233–253. https://doi.org/10.1016/j.trc.2019.02.006
52. Maryland Transit Administration (2014) Bus stop optimization policy (Pilot)
53. Leksakul K, Smutkupt U, Jintawiwat R, Phongmoo S (2017) Heuristic approach for solving employee bus routes in a large-scale industrial factory. Adv Eng Inform 32:176–187. https://doi.org/10.1016/j.aei.2017.02.006
54. Hassan MN, Hawas YE (2017) A methodology for rearranging transit stops for enhancing transit users generalized travel time. J Traffic Transp Eng (English Edition). 4:14–30. https://doi.org/10.1016/j.jtte.2016.09.012
55. Bargegol I, Ghorbanzadeh M, Ghasedi M, Rastbod M (2017) Evaluation of effective factors on travel time in optimization of bus stops placement using genetic algorithm. IOP Conf Ser Mater Sci Eng 245:042002. https://doi.org/10.1088/1757-899X/245/4/042002.
56. AlRukaibi F, AlKheder S (2019) Optimization of bus stop stations in Kuwait. 44:726–738. https://doi.org/10.1016/j.scs.2018.10.037
57. Al-Mudhaffar A, Nissan A, Bang K-L (2016) Bus stop and bus terminal capacity. Transp Res Procedia 14:1762–1771. https://doi.org/10.1016/j.trpro.2016.05.142

Humanitarian Logistics

Financial Analysis Over the Smartest Companies

Sergio Ignacio Villalba, Esther Guadalupe Carmona, Blanca Lidia Márquez, and Juan Mascareñas Perez-Iñigo

Abstract The companies that have better expectations of being successful are those that have higher skills to innovate; the Massachusetts Institute of Technology (MIT), every year publishes the smartest companies ranking, it's a list that includes disruptive enterprises, these companies are transforming the world and the way of doing business, MIT classifies them in 5 sectors biotechnology, transports, energy, computer and telecommunications, and finally Internet and media; so this paper is focused to analyze the profitability of those companies, moreover of evaluating their risk and generate a perspective for the future to know how attractive to invest in them. We found that the Internet and media sector has the best perspectives, the energy sector is having a lower growth and the biotechnology sector always should be tracked.

Keywords Innovation · Risk · Financial ratios

S. I. Villalba (✉) · E. G. Carmona · B. L. Márquez
Universidad Autónoma de Ciudad Juárez, Ave. Heroico Colegio Militar Y Ave. Universidad S/N, Zona Chamizal C.P. 32300 Cd. Juárez, Chihuahua, México
e-mail: sergio.villalba@uacj.mx

Business Sciences Department and Finance Program, UACJ, Juárez, México

E. G. Carmona
e-mail: esther.carmona@uacj.mx

B. L. Márquez
e-mail: bmarquez@uacj.mx

J. M. Perez-Iñigo
Universidad Complutense de Madrid, Campus Somosaguas, Pabellón de sexton—despacho 53, 28223 Pozuelo de Alarcón, Madrid, Spain
e-mail: jmascare@ucm.es

Financial Economics and Accounting III Department, Universidad Complutense de Madrid, Madrid, Spain

© The Author(s), under exclusive license to Springer Nature Switzerland AG 2021
A. Ochoa-Zezzatti et al. (eds.), *Technological and Industrial Applications Associated with Intelligent Logistics*, Lecture Notes in Intelligent Transportation and Infrastructure,
https://doi.org/10.1007/978-3-030-68655-0_14

1 Introduction

In the last century, the economy was based on mass production, most of the strategies from the managers were focused on taking advantages over fixed costs; financial managers commonly should decide if they dare to obtain high yields buying fixed assets to reach higher operative leverage, or they might decide work with variable cost avoiding the risk, that decision depends mainly on the macroeconomic situation if the economy would have a good perspective of growing, or the central bank follows an expansive policy and Chief Financial Officers (CFO) were motivated to get high profits from the operative leverage; when the economy is not in the best moment or when the central bank wants to keep a restrictive policy, CFO avoid to work with fixed assets, and they look for a more flexible strategy to cover the production [1]. Now in the new century, being part of a global world, sometimes is more important the knowledge than the money [2], the money has a constraint, it's finite and the central banks cannot be printing with no limits, the knowledge is limitless and it can be being used in the whole world, it has no restriction of place or time, a constraint that the knowledge has is that sometimes it can be expensive to generate it, but in the same way it compensates perfectly because it improves quality's life of the people. In the beginning, the cost of developing a new technology might be onerous, but in the future, it will give considerable retribution to the investors [3], in fact, for the research and development (R&D) departments, spend a good part of their budget when the project is beginning, if they do not assign enough money, the project could be more expensive when the products are in the market [4].

Investing in R&D would not be seen as a heavy charge for the finance department, these areas are becoming more than indispensables in the companies, especially because the markets and the costumers are changing in their preferences, currently, the people are less able to accept a standard product, costumers are looking for flexible products, they are buying goods that satisfy their needs no more, no less; the mass market has gone and now the enterprises must-do marketing adapted to the costumers [5]. So, what about the companies that are doing more innovations, every year the Massachusetts Institute of Technology (MIT), publishes a list of the smartest companies in the world, that list refers all the enterprises that are developing disruptive technologies and in the same way, they will be creating value for the shareholders.

2 Theoretical Framework

So, for innovation it can be defined something that is beyond of redesigning or improving a product, innovation means to create a new concept [6], an innovator must have the skill to solve a problem, or develop a product that satisfies entirely a needed creatively. Companies that are generating substantial value to the shareholders are, all those that are generating disruptive innovations, it means that generates new markets and value networks that displaces to the common leader market [7]. The disruptive business initially has lower gross margins, or often they might have losses, this kind of companies begins in small markets that are not attractive to the biggest companies. Investing in small enterprises is accepting by shareholders that they can have lower returns [8], at this point we should ask ourselves, what is better, a good brand or a good innovation? Having a constant cash flows is like a law of conservation, it will keep the company in a reasonable state of peace and a comfortable brand position, therefore, what is better a good brand or a good innovation? Innovation makes the brand, not in the other way, in the middle and the long term the innovation will do the difference between being a leader or a follower.

To analyze the situation of the companies the CFO does a diagnose of the business through the financial ratios or just observing general information market; by one hand, an investor quickly can see a general performance of the company, if he sees the enterprise value, it is necessary to say that there are two kinds of values, the book value (VB) means the shares' price initially [9], and market value, it represents the price of the shares in the open market, the market value includes the book value of the company, plus the net cash flows that the company is expecting to earn, that cash flows must be discounted to present value [10], it can be calculated as:

$$V_M = \sum_{j=1}^{N} CF_j(1 + WACC)^{-j} \tag{1}$$

where V_M is the market value and CF_j represents the cash flow in j, it must be discounted at weight average cost of capital (WACC) at time j; therefore, the differential between VM y VB with respect to the time will be the value creation to the company as a consequence of a good administration. Another index to monitoring the company's performance is the market capitalization, it is the result of multiply the stock's price times the number of shares outstanding [11].

In the other hand, a deeper analysis can be done through financial ratios specially if we see all of them that evaluate the profitability of the company. The Price-to-Book value ratio ($P \rightarrow V_B$) measures the proportion in what the market value has exceeded to book's value; in that case as higher the result as much better. It is calculated as the quotient of dividing the market value (V_M) between the book value (V_B):

$$P \rightarrow V_B = \frac{V_M}{V_B} \tag{2}$$

If the quotient obtained is higher than 1, we assume that the enterprise's value is rising, but if it is lower than 1, the company has more value if it considers to itself in bankruptcy than if it continues operating.

Another profitability ratio is the Price-to-Earnings ratio ($P_\theta \rightarrow E$), it is the result of dividing the price of the share in the market (P_θ), between the earning per share (EPS). It gives us a perspective of the expected potential of the company in the market. The higher the ratio, the higher the investors' expectations regarding earnings growth, it can be obtained as:

$$P_\theta \rightarrow E = \frac{P_\theta}{EPS} \tag{3}$$

The PEG ratio (Price/earnings to growth), is a valuation metric for determining the relative trade-off between the price of a stock, the earnings generated per share (EPS), and the company's expected growth.

In general, the P/E ratio is higher for a company with a higher growth rate. Thus, using just the P/E ratio would make high-growth companies appear overvalued relative to others. It is assumed that by dividing the P/E ratio by the earnings growth rate, the resulting ratio is better for comparing companies with different growth rates [12], PEG ratios can be defined as:

$$PEG\ Ratio = \frac{P_\theta}{EPS_g(E)} \tag{4}$$

where (P_θ) is the stock price in the open market, E represents the earnings in the period (usually computed annually) and $\left(EPS_g\right)$ means the annual earnings per share growth. Another financial ratio, one of the most important is the Return on Assets (ROA), it shows the percentage of how profitable a company's assets are in generating revenue, ROA is obtained as a quotient of dividing net profit between total assets, therefore:

$$ROA = \frac{P_N}{A_T} \tag{5}$$

where (P_N) is the net profit and (A_T) means the total assets. ROA is as much important as ROE, the Return on Equity (ROE) is fundamental to take assertive decisions by the CFO; ROE is a measure of the profitability of a business in relation to the book value of shareholder equity, also known as net assets or assets minus liabilities, in the same way is a measure of how well a company uses investments to generate earnings growth, ROE is computed as net profit (P_N) between the equity from the shareholders (E_S), so, it can be written:

$$ROE = \frac{P_N}{E_S} \tag{6}$$

The earning per share (EPS) means the profit that a shareholder earns per each share that he has bought, EPS is the quotient of dividing the net profit (P_N) between the number of shares outstanding (SO), in that case EPS is:

$$EPS = \frac{P_N}{SO} \tag{7}$$

A non-financial ratio is the Beta index (β), it is a regression slope from the hold period return on a specific stock, with respect with the market returns, beta index might be seen as risk measure over the market, since it gives the sensitivity of stock with respect the market; beta index is useful to compute the cost of capital with the Capital Assets Pricing Model [13–16],the Beta index is computed as:

$$\beta = \frac{\frac{1}{n}\sum_{j=1}^{N}\left(R_j - \overline{R}\right)\left(Rm_j - \overline{Rm}\right)}{\frac{1}{n}\sum_{j=1}^{N}\left(Rm_j - \overline{Rm}\right)^2} = \frac{\sigma_{jm}}{\sigma_m^2} \tag{8}$$

If $\left(R_j\right)$ is the return form asset in$j$ and $(\overline{R})$ is the expected value for the return of an asset in j; $\left(Rm_j\right)$ is the return from the market in j, and $(\overline{Rm})$ is return expected in the market; in other words, if we apply the formula and then we rewrite it, we get that (β), is the quotient from the covariance $\left(\sigma_{jm}\right)$ from the returns of an asset in j and the market returns (Rm), between the variance $\left(\sigma_m^2\right)$ from the market returns [17].

3 Methodology

Initially were contemplated the 50 smartest companies published by the Massachusetts Institute of Technology in its index [18], but unfortunately, 20 companies remain as private enterprises, it means, they have not done an initial public offer (IPO) to list in the stock exchange, so they keep the information confidential even though and the investors have good expectations,so the final sample was with 30 companies, they were analyzed by country and sector, to posteriorly to be compared among them. The information was retrieved from the Reuters Finance database in the currency of the country where the company belongs, then it was converted to U.S. Dollars to standardize.

3.1 Companies to Analyze

Unfortunately, 40% of companies are not listing into a stock exchange, so the information was not available, in Table 1 these companies are listed and sorted by sector and country.

As we see, most of the companies are from the United States (US) and the sectors that are having more innovations are computer and telecommunications, energy and biotechnology. In Table 2 are listed the rest of the companies, all of them are public and are listed in different stock exchanges around the world.

As it is showed more than half of the companies are from the United States, almost one third belongs to biotechnology and another third is from computer and telecommunications.

Table 1 Shows the name of the companies that are not listed in a stock exchange and maintain a private status, the country of origin from the company and the sector of which it belongs

Company	Sector	Country
Enlitic	Biotechnology	United States
Oxford Nanopore	Biotechnology	United Kingdom
RetroSense Therapeutics	Biotechnology	United States
Veritas Genetics	Biotechnology	United States
23andMe	Biotechnology	United States
Aquion Energy	Energy	United States
IDE Technologies	Energy	Israel
24 M	Energy	United States
Sonnen	Energy	Germany
Improbable	Energy	United Kingdom
Carbon	Energy	United States
Seven Bridges	Computer and telecommunications	United States
LittleBits	Computer and telecommunications	United States
Bosch	Computer and telecommunications	Germany
Slack	Computer and telecommunications	United States
Transferwise	Computer and telecommunications	United Kingdom
Didi Chuxing	Transports	China
SpaceX	Transports	United States
Coupang	Internet and digital media	South Korea

Source Own source

Table 2 Shows the companies that are listing on the different stock exchanges around the world classified per sector

Company	Sector	Country
Bristol-Myers Squibb	Biotechnology	United States
Cellectis	Biotechnology	France
Editas Medicine	Biotechnology	United States
Illumina	Biotechnology	United States
Intrexon	Biotechnology	United States
Monsato	Biotechnology	United States
Nestlé	Biotechnology	Switzerland
Spark Therapeutics	Biotechnology	United States

(continued)

Table 2 (continued)

Company	Sector	Country
T2 Biosystems	Biotechnology	United States
Airware	Computer and telecommunications	United States
Fanuc	Computer and telecommunications	Japan
FireEye	Computer and telecommunications	United States
Huawei	Computer and telecommunications	China
IBM	Computer and telecommunications	United States
Intel	Computer and telecommunications	United States
Microsoft	Computer and telecommunications	United States
Mobileye	Computer and telecommunications	Israel
Nvidia	Computer and telecommunications	United States
First Solar	Energy	United States
Alibaba	Internet and digital media	China
Alphabet	Internet and digital media	United States
Amazon	Internet and digital media	United States
Baidu	Internet and digital media	China
Facebook	Internet and digital media	United States
Naver	Internet and digital media	Japan
Tencent	Transports	China
Tesla Motors	Transports	United States
Toyota	Transports	Japan
Snapchat	Internet and digital media	United States

Source Own source

4 Data Analysis

We examined the companies' data from the Yahoo Finance, Reuters Finance, and EMIS, after that we computed the financial ratios per company and were written in the Table 3, it shows the financial ratios and beta index, the enterprise value and market capitalization are in billions of dollars.

To do a better data analysis the companies were classified by sector, to correct the skewness generated by the missing data, they were replaced for the average of the industry. Graph 1 shows the market value of the companies and the capitalization of the enterprise.

The Internet and media sector followed by transport includes the companies that are more valuables; the energy sector is the most lagged, so it could represent an excellent deal since the shares are not so expensive right now. Graph 2 shows the financial ratios per sector.

As we can appreciate in Graph 2 the P/E ratio in the sector of internet and media is extremely high, it was because of Amazon and Alphabet, the expectations of the value

Table 3 Shows the different ratios and relevant financial information per company

Company	Enterprise value	Market cap	$P \rightarrow V_B$	P/E ratio	PEG ratio	ROA (%)	ROE(%)	β	EPS
Alphabet	470.14	540.82	4.23	30.49	1.27	9.13	15.01	0.98	26.30
Microsoft	388.47	447.52	6.23	27.35	2.32	7.23	22.09	1.36	2.10
Amazon	383.30	381.99	23.09	200.3	2.74	4.12	13.64	1.64	N/A
Facebook	344.17	367.46	7.27	61.17	0.92	11.15	13.41	N/A	2.09
Editas Medicine	316.97	517.67	2.73	−4.03	N/A	−23.73	−56.42	N/A	(1.91)
Toyota	286.31	177.16	1.12	8.47	0.32	3.67	13.33	0.72	14.55
Alibaba	264.43	266.37	7.89	38.15	52.74	6.00	21.51	N/A	2.82
Nestlé	262.64	239.01	4.08	27.53	N/A	6.21	13.75	0.73	2.79
Intel	186.98	175.95	2.87	18.04	1.44	8.85	16.96	1.15	2.07
IBM	182.04	148.14	9.42	12.60	4.02	7.54	80.35	0.79	12.31
T2 Biosystems	177.06	192.48	8.49	−3.48	N/A	−47.00	−180.38	N/A	(2.27)
Bristol-Myers Squibb	96.46	94.37	6.33	33.03	1.10	9.27	19.34	1.01	1.71
Baidu	60.95	66.45	5.22	14.11	−167.13	4.94	41.02	1.90	13.91
Monsato	54.66	45.29	8.86	45.55	3.47	8.56	14.93	1.24	2.38
Nvidia	31.32	34.77	7.76	42.70	1.48	10.24	40.48	1.37	1.52
Tesla Motors	31.26	30.71	12.18	−24.55	−6.21	−6.34	−69.57	0.94	(8.45)
Illumina	25.90	26.29	12.79	61.41	3.18	9.35	25.28	0.75	2.91
Airware	9.73	6.45	N/A	−1.33	N/A	−216	N/A	−0.18	(0.06)

(continued)

Table 3 (continued)

Company	Enterprise value	Market cap	$P \rightarrow V_B$	P/E ratio	PEG ratio	ROA (%)	ROE(%)	β	EPS
Mobileye	9.11	9.35	15.77	109.9	1.25	10.88	17.68	N/A	0.34
Cellectis	5.98	8.99	3.35	−19.74	N/A	−8.51	−16.12	1.28	(1.28)
Intrexon	2.98	3.24	5.23	−17.24	0.49	−6.86	−33.66	1.53	(1.59)
First Solar	2.40	3.79	0.65	5.46	−0.50	6.88	12.86	1.79	6.82
FireEye	2.25	2.39	2.56	−4.00	−26.13	−13.42	−53.58	0.51	(3.59)
Spark Therapeutics	1.51	1.84	4.77	−20.07	1.22	−14.74	−26.06	N/A	(2.95)
Huawei	N/A	1.82	0.59	11.01	N/A	6.00	6.12	0.32	0.03
Tencent	N/A	260.85	12.40	6.66	1.22	11.05	32.12	1.09	32.26
Fanuc	N/A	3.40	0.02	0.24	N/A	10.26	11.80	0.78	7.05
Naver	N/A	26.21	7.92	39.38	N/A	24.35	28.87	1.02	20.21
Snapchat	20.44	21.35	9.15	N/A	−1.5	−40.94	−29.62	1.09	−0.72

Source Own source

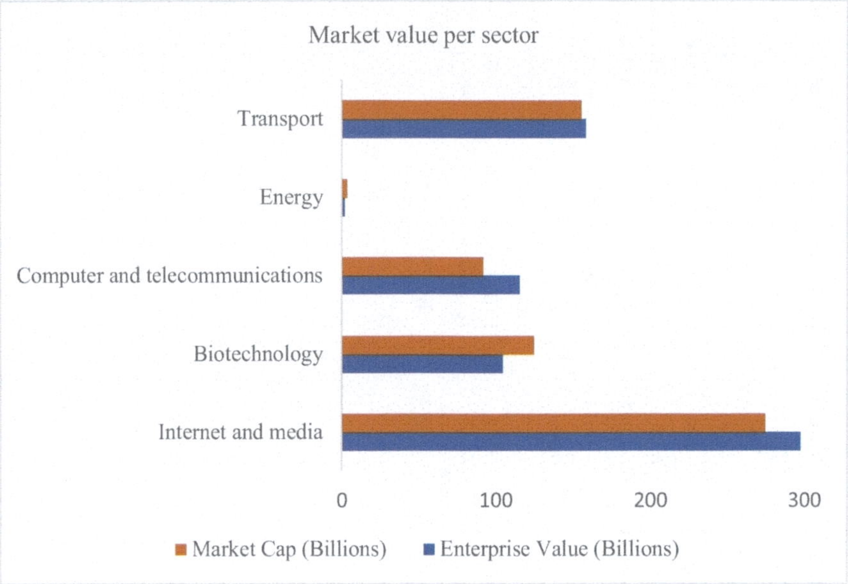

Graph 1 Shows the market value and the capitalization market per sector. *Source* Own source

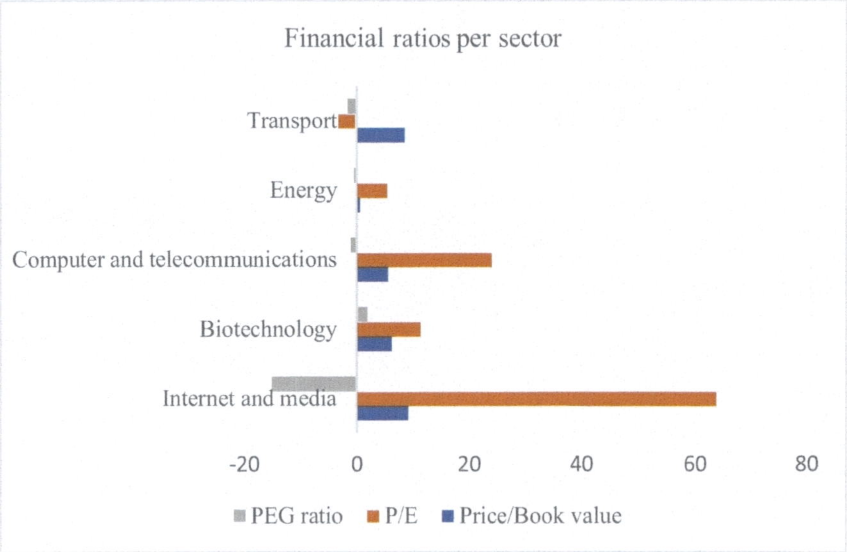

Graph 2 Shows the most relevant information from financial ratios. *Source* Own source

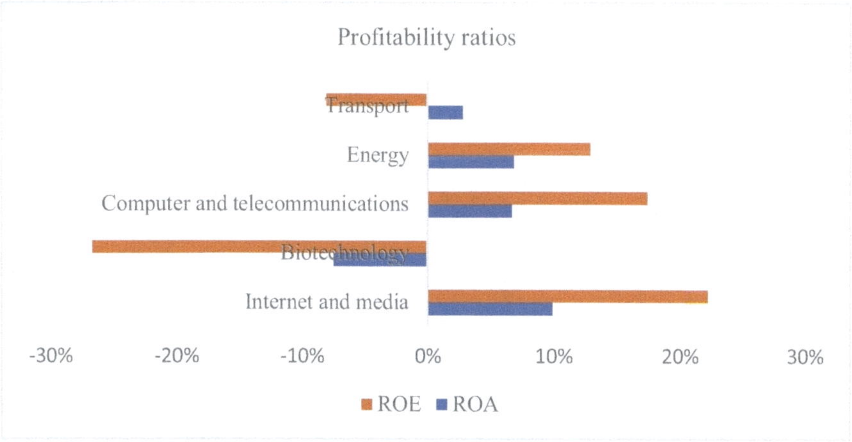

Graph 3 Shows the ROA and ROE per sector. *Source* Own source

of that companies if because they have diversified their products and are investing high amounts of money in their research and development departments (R&D); biotech sector as computer and telecommunications have excellent expectations in the future. Graph 3 shows the profitability of the smartest companies.

Investing in biotechnology implies accepting the probability of high losses, it is caused by the high rates of failure in R&D departments, on the other hand, the Internet and media are giving very attractive returns to the shareholders. Graph 4 shows the Beta index; it helps as a risk market indicator in the function of the CAPM.

As we see, the energy sector and internet and media have higher risks than biotechnology or computer and telecommunications, it is caused because of the goods and services they sell, they have no elastic, so companies have a better margin to operate with the price. Graph 5 shows the earning per share per sector.

Internet and media as the transport sector are giving excellent yields, as much as the energy sector does at least for the last year (2015–2016), the biotech sector had serious problems for generating profits, but is something common in this sector.

5 Conclusion and Future Research

This the first part of the research, and we have found that the yields and returns from smartest companies are very attractive, and it compensates perfectly the operative and the market risks, is deeply recommended to include into a portfolio share like Alphabet or Amazon; in the same way, would be a good idea to include biotech companies, by their essence with a small amount of money, a portfolio could be

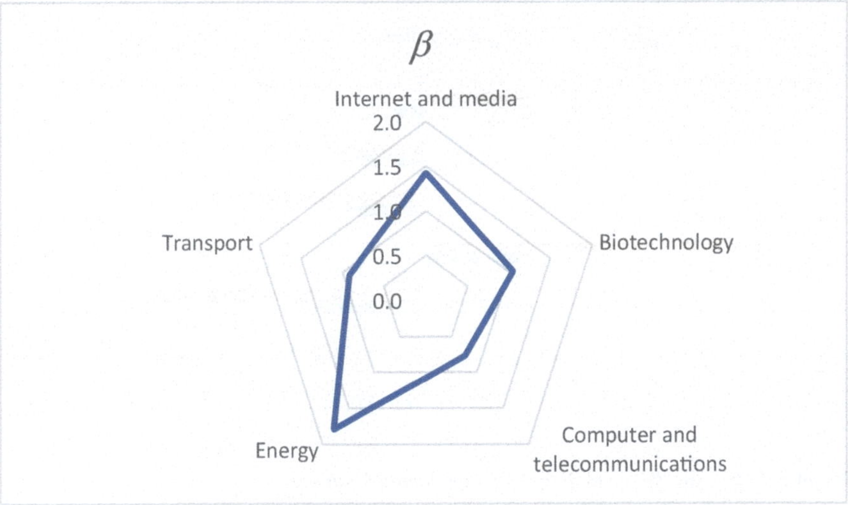

Graph 4 In a radial graph, is showed how the beta index changes in function of the sector. *Source* Own source

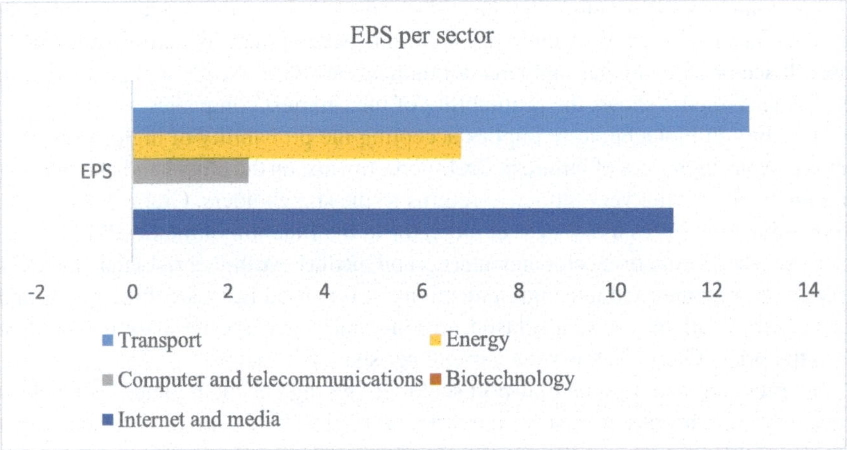

Graph 5 Shows the Earning per share per sector. *Source* Own source

leveraged, having an acceptable range of opportunities to get profits. In the future is shaped to do the same analysis for the past years and then we would be able to do a forecast about the indicators.

References

1. Hirt GA, Block SB, Danielsen BR (2014) Corporate finance foundations—global edition. McGraw-Hill Education
2. Toffler A, Toffler H (1993) War and anti-war: survival at the dawn of the 21st century: Little, Brown
3. Dawes SS, Pardo TA, Simon S, Cresswell AM, LaVigne MF, Andersen DF, Bloniarz PA (2004) Making smart IT choices: understanding value and risk in government it investments. Center for Technology in Government, University at Albany, State University of New York
4. Cresswell AM (2004) Return on investment in information technology. Center for Technology in Government; University at Albany SUNY, Albany, NY
5. Kotler P, Armstrong GM, Adam S, Denize S (2011) Principles of marketing. Pearson, Australia
6. Henderson RM, Clark KB (1990) Architectural innovation: the reconfiguration of existing product technologies and the failure of established firms. Adm Sci Q 35(1):9–30. https://doi.org/10.2307/2393549
7. Christensen CM, Bower JL (1996) Customer power, strategic investment, and the failure of leading firms. Strateg Manage J 17(3):197–218
8. Christensen CM (2000) The innovator's dilemma: the revolutionary national bestseller that changed the way we do business. Harper Business.
9. Harrison WT, Horngren CT (2004) Financial accounting. Prentice Hall
10. Gitman LJ, Joehnk MD, Franco JCA (2005) Fundamentos de inversions. Pearson Educación
11. Maginn JL, Tuttle DL, McLeavey DW, Pinto JE (2007) Managing investment portfolios: a dynamic process. John Wiley & Sons
12. Easton PD (2002) Does the PEG ratio rank stocks according to the market's expected rate of return on equity capital?
13. Lintner J (1965) The valuation of risk assets and the selection of risky investments in stock portfolios and capital budgets. Rev Econ Stat 47(1):13–37. https://doi.org/10.2307/1924119
14. Mossin J (1966) Equilibrium in a capital asset market. Econometrica 34(4):768–783. https://doi.org/10.2307/1910098
15. Sharpe WF (1964) Capital asset prices: a theory of market equilibrium under conditions of risk. J Financ 19(3):425–442
16. Treynor JL (1961) Toward a theory of market value of risky assets. Unpublished manuscript, 6
17. Ross SA, Westerfield R, Jordan BD (2008) Fundamentals of corporate finance: McGraw-Hill, Irwin
18. MIT Technology Review (2016) 50 Smartest companies. https://www.technologyreview.com/lists/companies/2016/

Simulating Crowd Movements During Emergency Fire Situations: Mexico City Airport Simulation Case

Roberto Contreras-Masse, Alberto Ochoa-Zezzatti, Vicente García, and Ana Moheno

Abstract Flying is the first choice when traveling long distances. With several options available for passengers, it is noticed an increase of passengers every year, then airports become more crowded, but, how well prepared are in case of a fire emergency? In this paper, we research a fire situation forcing to evacuate Gate 75 at Mexico City airport. Using Menge crowd simulator movement engine, we use four classes of agents to simulate different passenger movements as well as different moving speeds based on sex and complexion. In our results bottlenecks were detected and improvements to gate layout were suggested. Then, another simulation set was run with the proposed changes to determine how affects the evacuation times. In order to have a successful change, it most improves at least 10% evacuation time, without changing agents' classes, cardinality or speed.

Keywords Crowd simulation · Airport · Fire emergency · Improvement

1 Introduction

Mexico City International Airport (often referenced and abbreviated as AICM in Spanish, or MEX due to IATA airport code) is the busiest and largest airport in Mexico. In 2018, more than 47 million passengers landed or depart from this airport, and 2019 suggest a raising number [1]. Comparing June 2019 and 2018, there is a 6.3% increase of passengers. AICM has two terminals. Terminal 1 is the oldest and original one and Terminal 2 was built by 2006, currently handling near 1.7 million passengers per month. Terminal 2 is divided in two large connected aisles, north zone is for international gates (Gate 52 to 62), and south zone (gate 63 to gate 74) to serve domestic flights. Both zones provide 23 gates and one common gate (gate 75) for remote terminals (Fig. 1).

Gate 75 is divided in 4 areas in two floors. Gate 75A and 75B are in upper level, Gate 75C and 75D are in lower level. The access among floors is conveyed by single

R. Contreras-Masse (✉) · A. Ochoa-Zezzatti · V. García · A. Moheno
Universidad Autónoma de Ciudad Juárez, 32410 Ciudad Juárez, Chihuahua, Mexico
e-mail: rcontreras@itcj.edu.mx

© The Author(s), under exclusive license to Springer Nature Switzerland AG 2021

A. Ochoa-Zezzatti et al. (eds.), *Technological and Industrial Applications Associated with Intelligent Logistics*, Lecture Notes in Intelligent Transportation and Infrastructure,
https://doi.org/10.1007/978-3-030-68655-0_15

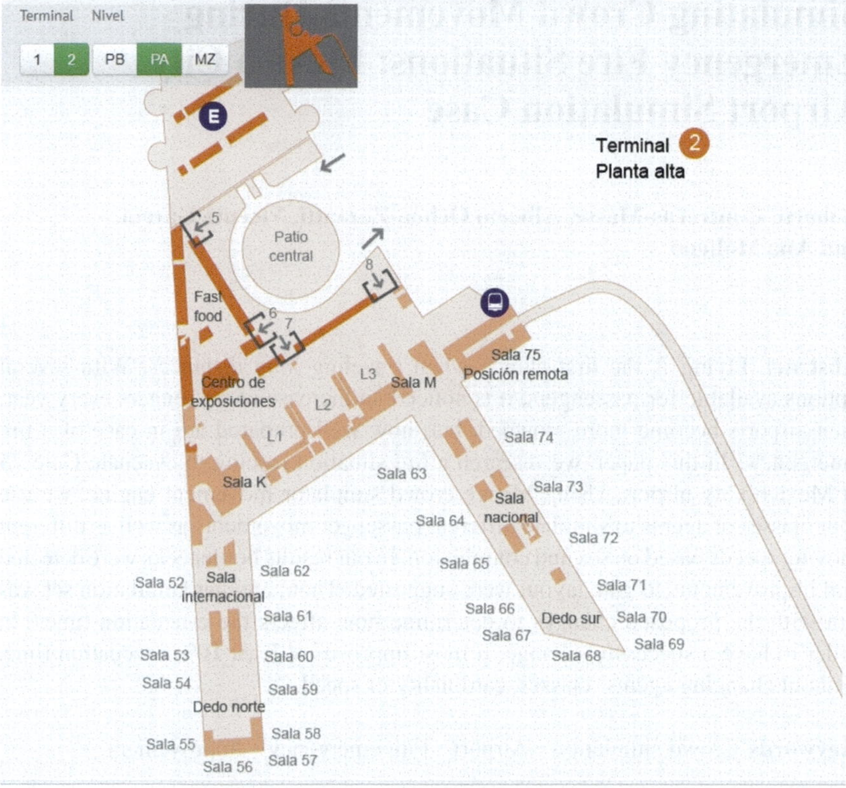

Fig. 1 Terminal 2 departures of AICM. *Source* AICM

narrow ramp of about 2 meters wide. Ramp walls are tempered glasses, providing a sense of amplitude and open space while going down. Most of the human flow goes from upper level to lower level. At each gate (75A-D) there are 3 counters, meaning, in the best-case scenario, airport can server 12 flights boarding at the same time, plus any delay that may happen due by weather or airline fault.

1.1 *Specific Crowd at Gate 75*

The number of daily departures from Gate 75 is numerous. There is no precise statistics about traffic at Gate 75, however, it can be estimated by just being familiar with that place. Flight departures served by this gate are usually for small planes, Embraer 170, 190, ATR42, ATR72. Also, Terminal two host few airlines (Aeromexico, Delta, Copa, Aeromar y Wingo) national and international. Fleet served at Gate 75 corresponds to Aeromexico Connect and Aeromar equipment (Aircraft E170 carries 76 passengers, E190 has 99 seats, while ATR-72 has 72 seats, ATR-42 has 48 seats). Based on public flight status, it is possible to gather the number of flights departed by day from this specific gate, as shown in Table 1. Using an average number of seats per airline and flights per period, it is possible to estimate number of passengers around.

It is important to notice Gate 75 was remodeled and enhanced in 2016 going from 2,000 m^2 to 9,000 m^2 in passenger space. But this space is not only for standing people; it has chairs, glass walls, few steps, restrooms, ramps and counters. Although this effort made, gate 75 is always crowded and busy under normal circumstances, worsen by flight delays. As there are two floors, each one has 4,500 m^2. In Figs. 2 and 3 can be seen the facilities as was thought in contrast with real and current conditions. Also, by personal experience being a regular user of this airport, it is noticed there are not evacuation paths, there is not enough space to walk to upper level through single ramp when there is people going down floor, and there are not emergency exits. Adding to this panorama, there are restaurants located at the beginning of gate (where is the passage to main building), kitchen equipped meaning gas and fire.

Table 1 Estimated passengers at Gate 75

Airline	Avg. Pax	Number of flights per shift		
		6-11:59a	12-5:59p	6-11:59p
AM Connect	87	52	50	28
Aeromar	55	13	8	4
Total Pax		5239	47900	2656
Grand Total				12685

Fig. 2 Gate 75 upper floor panoramic view. *Source* Notimex

Fig. 3 Gate 75 upper floor normal crowd 2019. *Source* Personal footage

The purpose of this manuscript is to study the human crowd movement in this particular gate in the case of an emergency event such as fire and propose an improvement in the layout to do evacuations faster. Earthquakes are not taken in consideration because in case of this event the right thing to do is stay in safe places instead of run. Fire, in the other hand, is a situation that people is not familiar, making people run.

This manuscript is organized in sections. Section 2 present the literature review and provides context for the study. Section 3 models the venue and establish the conditions, variables and restrictions of our simulation and test improvements options. Section 4 discusses the results, proposes a new layout and compare simulation results. Section 5 provides conclusions and future work.

2 Literature Review

From all emergency events, fire is the center of life safety prevention and reaction projects. Studies in human behavior includes awareness, beliefs, attitudes, motivations and decisions, taking in consideration also mimic others [2]. There is a typical timeline of people response when there is a fire emergency. This timeline is divided in pre-evacuation period and movement period. During pre-evacuation period starting by ignition, alarms may or not may go off. People will spend time seeking information, such as source of fire, how big it is, processing the information to enter evacuation decision step. If a decision to evacuate is made, protection of self and others will play in conjunction of seek an escape route, starting the movement period.

The protective action decision model (PADM) developed by Lindell et al. provides a general framework to explain decision making flow in emergencies. It takes different sources of inputs, have a stage of pre-decisional process, adds a layer of perceptions and finally decision making based on those variables [3].

In PADM, inputs are from a vast array of channels, including environmental cues, information channels, information sources. Also, there are behavioral sources and personal beliefs involved, like social context, person personal training, personality, or even physical characteristics as age, body type, agility, weight, or height. Group behavior is other important observation. The affiliative behavior (how humans tend to form groups during evacuation), capacity to help others, and convergence clusters are few of behaviors observed during a fire emergency. Michael Bratman provides a methodology for decision according to beliefs, desires and intentions (BDI) based on cognitive model of humans, in artificial intelligence field, useful for simulations [4] for agent based simulation (ABS). There is also social force models (SFM) as the proposed by Helbing and Molnar to describe pedestrian motion, arguing that a limited number of parameters can simulate individuals and their displacements better when considering SFM [5, 6] and applied in studies for airport passengers movements simulation [7, 8].

The Distinct Element Method is used in Crowd Behavior Simulator for Disaster Evacuation, Proposed by Rahman considers person density in perception domain of an individual, function switching actions and physical plus psychological field perception [9]. Standard field of vision is about 124 degrees, using both eyes, while focusing on objects and symbols narrows it to 60 degrees in binocular vision as shown in Fig. 4.

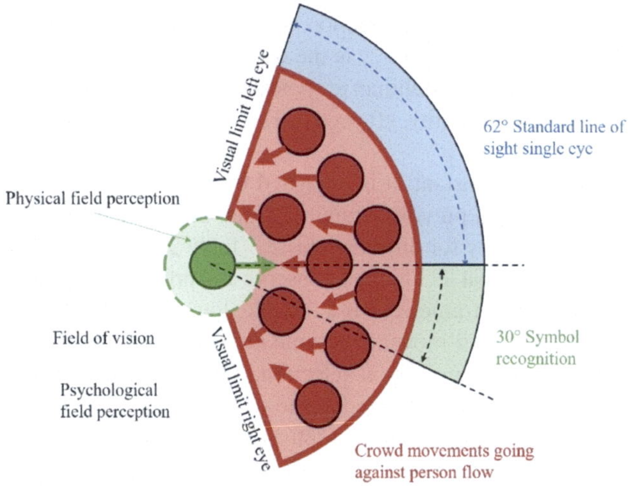

Fig. 4 Person perception in crowd movements with field of vision. *Source* Adapted from [9]

The estimated evacuation time can be calculated using Togawa equation [10], which indicates the elapsed time in seconds to evacuate equals number of people (N), output width in meters (A), the experimental constant (K) 1.3 people/meter-second, total distance to travel in meters (D) and horizontal speed of 0.6 m/s, stairs 0.4 m/s.

$$T = \frac{N}{AK} + \frac{D}{V} \tag{1}$$

However, as it can be observed in Fig. 6, we need to consider two additional evacuation times, one coming out from the ramp and evacuation of Gate 75B. Thus, our Eq. (1) changes to

$$T = \sum_{t=0}^{n} \left(\frac{N_i}{A_i K} + \frac{D_i}{V} \right) \tag{2}$$

Togawa's observations works well during an evacuation under normal circumstances; the process is conveyed in organized form and most likely, practiced by employees. When there is a threat to life like fire in a public building, non-trained persons will try to escape in accelerated way and collisions may occur. This behavior is modeled by the Switching Action Function (SAF) proposed by Rahman [9] and studied in airports as:

$$\int_{t-\alpha_{sw}}^{t} f_{sw}(\tau) g_{sw}(t-\tau)\, d\tau \tag{3}$$

Where f_{sw} is SAF, f_{sw} is the input function with population density number in the perception domain of person *i*, and g_{sw} is the unit response function reflecting the delay to switch course of action. Rahman proposed to use the exponentially decaying function for computational stability, $g(t) = e^{-t}$. This change of direction depicts how a person avoids collision.

Human velocity is not constant, it is impacted by factors such as age, ability to perform, body type, or even previous training. Datta et al. provides the mean and standard deviation of velocity for different classes. Table 2 shows those values [11].

Lastly, simulation is a great technique to observe and predict results with low cost and low investment. Menge framework is a flexible, low cost and extendable option for research. It realizes crowd movement simulation with a particular abstraction, decomposing the main problem in goal selection, plan computing, plan adaptation and spatial queries. The novelty is how adapts to obstacles and therefore, change the plan to reach the goal. Also, it transforms preferred velocity into a feasible velocity, considering other crowd agents during simulation. These agents' trajectories are computed by an initial value problem of type

$$x_i(t) = v_i(t) = V_i(t, \mathbb{C}(t)) \tag{4}$$

Table 2 2 Persons' evacuation velocity by class. *Source* [11]

Evacuee Class	Mean (m/s)	Std. Deviation (m/s)
Male child able (MCA)	1.08	0.26
Male child disable (MCD)	0.92	0.34
Male adult able (MAA)	1.24	0.45
Male adult disable (MAD)	1.06	0.26
Male elderly able (MEA)	1.05	0.15
Male elderly disable (MED)	0.91	0.13
Female child able (FCA)	1.08	0.26
Female child disable (FCD)	0.92	0.34
Female adult able (FAA)	1.30	0.38
Female adult disable (FAD)	1.06	0.26
Female elderly able (FEA)	1.04	0.16
Female elderly disable (FED)	0.89	0.14

where v_i is the instantaneous velocity of agent *i* at time *t*. Goal, Plan and Adaptation can be abstracted with the following mappings:

$$G_i : S_i \times \mathbb{R}^2 \rightarrow \mathbb{R}^2$$
$$P_i : t \times \mathbb{R}^2 \rightarrow \mathbb{R}^2$$
$$A_i : S_i \times \mathbb{R}^2 \rightarrow \mathbb{R}^2$$

These mappings are simplified to: $G_i : t \rightarrow \mathbb{R}^2$, $P_i : \mathbb{R}^2 \rightarrow \mathbb{R}^2$, and $A_i : \mathbb{R}^2 \rightarrow \mathbb{R}^2$. The velocity v_i can be substituted in (4) as:

$$v_i(t) = A_i(P_i(G_i(t))) \tag{5}$$

this simplification allows for orthogonal elements, specifying particular elements and relationships in a Behavioral Finite State Machine (BFMS), materialized in XML configuration files. Curtis presents Menge framework in a very good detail [12].

3 Modeling and Simulation

Gate 75 A to D has a two-level layout referred as upper layer (Gates 75A and B) and lower level (Gates 75 C and D). Both layers are above ground floor where is the access to airport tarmac and buses for remote positions. All access between levels and ground floor is by narrow ramps about 1.5 m width. To provide the feeling of open spaces, ramp walls are made of tempered glass. There are also elevators at the end of the building and elevators. It's important to notice in order to access ground

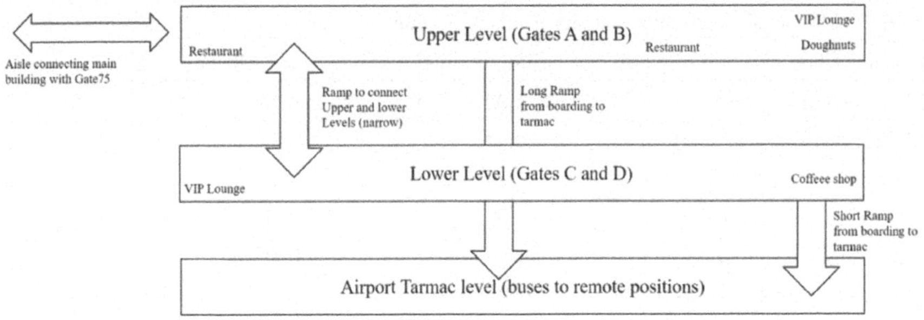

Fig. 5 75 lateral view schematics. *Source* own creation

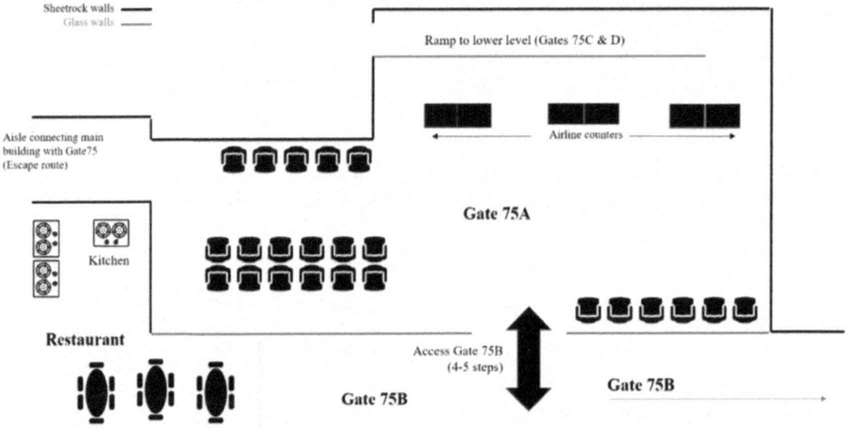

Fig. 6 Gate 75A floor plan. *Source* own creation

floor from upper level, people need to use ramp to lower level and stairs to ground floor. Our modeling of evacuation in case of fire emergency is based on upper level, Gate 75A where is the passage to main building. Figures 5 and 6 shows scenario schematic views. Figure 7 shows the natural evacuation plan and goal exit marker (goal flag).

3.1 Modeling Scene

Floor plan is modeled as Mange scene specifications, as stated by instructions and literature found [12]. We defined walls and waiting chairs as obstacles, using the coordinate system in tag Vertex. We need to define coordinates using center of screen as the origin (0,0). Each unit represents one meter. As an example, obstacle definition can be defined as:

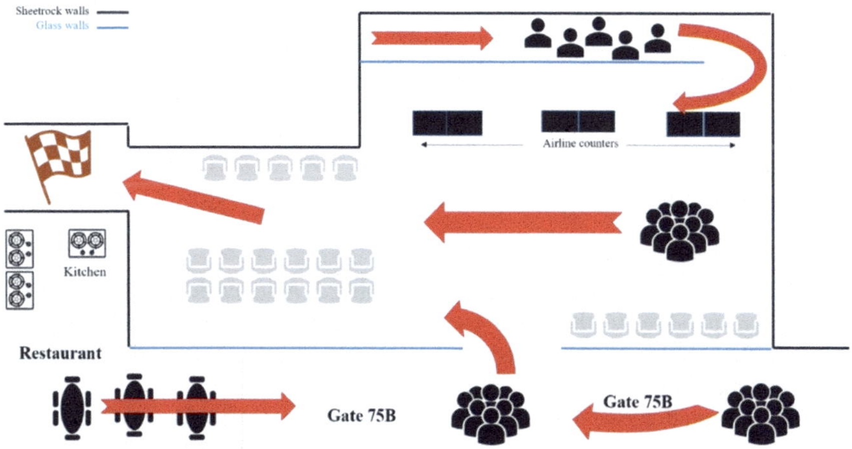

Fig. 7 Gate 75A evacuation. *Source* own creation

```
<ObstacleSet type="explicit" class="1">
    <!-- East wall definition -->
    <Obstacle closed="1">
        <Vertex p_x = "5" p_y = "-10"/>
            <Vertex p_x = "19" p_y = "-10"/>
            <Vertex p_x = "19" p_y = "6.5"/>
            <Vertex p_x = "1" p_y = "6.5"/>
            <Vertex p_x = "1" p_y = "4.5"/>
            <Vertex p_x = "17" p_y = "4.5"/>
            <Vertex p_x = "17" p_y = "-7"/>
            <Vertex p_x = "5" p_y = "-7"/>
    </Obstacle>
</ObstacleSet>
```

3.2 *Modeling Persons*

In crowd simulation, a person is often referred as agent, therefore, Menge uses same concept. We define four classes of agents in tag *AgentProfile*. All agents share the same algorithms for simulation, but each class has unique preferred speed, maximum

Table 3 Agents configuration

Evacuee Class	Speed (m/s)	Radius
Male adult big (MAB)	1.24	0.75
Male adult thin (MAT)	1.24	0.55
Female adult big (FAB)	1.30	0.65
Female adult thin (FAT)	1.06	0.45

speed and maximum acceleration, leveraging Table 2. Also, each class has different radius, corresponding to the space used. Maximum number of neighbors are 9, following SAF equation [9] solved as 9 persons in the vantage field of a person at a given time. Table 3 contains agents' configuration.

For our simulation, we define three groups of agents, (1) 40 Agents at the ramp arranged in single line, coming from lower level; (2) 70 Agents at Gate75A arranged in a grid; and (3) 100 Agents at Gate 75B arranged in hex lattice. Each group contains random agent classes generated by normal distributions.

3.3 *Materials*

For this research it was used AMD PRO A12-9800B R7, 12 Compute cores 4C + 8G laptop with 8.00 GB RAM (7.43 usable), 64-bit Windows 10 operating system, Menge v0.1 (Menge obtained from: http://gamma.cs.unc.edu/Menge/) and Matlab R2017b(9.3.0.713579).

3.4 *Scenario Creation*

Using the schema on Fig. 6, the model of Gate 75A is detailed as walls and obstacles. Our agents have to follow the escape route in stages. For agents coming out of the ramp we adjusted the maximum acceleration to 3.5 and reduce the speed in 14.4%, assuming a direct relation of one percent per slope degree. Ramp path follow two goals, first, reach main area of Gate 75A, and finally go to exit.

For agents escaping from Gate 75B, they don't need any adjustment, as the conditions are exactly the same as if they were in main Gate 75A. However, all agents need to pass through a narrow space that simulates the connection between 75A and 75B. Therefore, therefore, the escape goals are reach Gate 75A, then exit. Our Menge initial scenario is displayed in Fig. 8.

Agent classes are also differentiated by color. These are the default colors available in Menge. It can be observed three crowds in Fig. 8. At top is the ramp where basically agents are displayed in a single row. This was intended as many people travel with bags, carry-ons, laptops, and are hesitant to leave their belongings behind. Same can

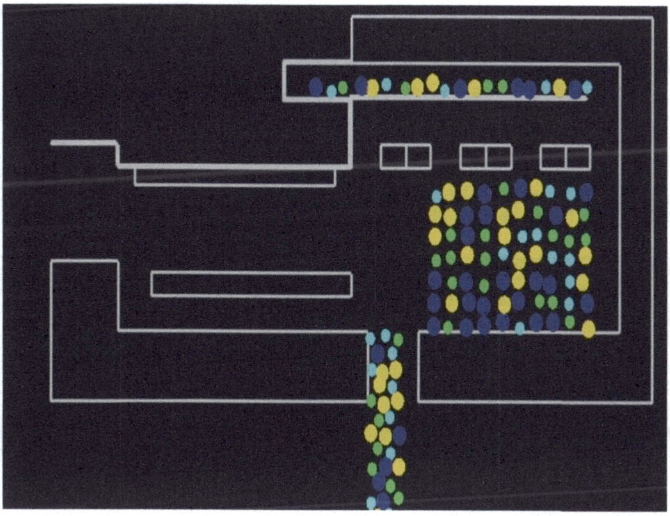

Fig. 8 Simulation 1. *Source* Menge Screenshot

be observed in the lower part of simulator, as there is space available at the right of passage from Gate 75B to 75A. Crowd in the center-right corresponds to agents in Gate 75A.

4 Results

Our simulation was executed until the point where 98% or more agents reached the goal. This limit is considered because when crowds dissipate, people tend to calm and start walking slowly as the danger perception lowers. Figure 9 has four screenshots from simulation. Step (a) shows the initial state. State (b) was taken at the moment where three agent groups collide. We observed few agents opted to go between wall and chairs. Also, agents at ramp had to turn around the glass wall and selected to move behind boarding counters, as there is room to reach the goal. Step (c) shows how Gate 75A is being cleared, but due to crowded space in the middle, there are agents stuck at first boarding counter. Step (d) shows a clear aisle but the corner between chairs and wall suggest being a spot where agents got slowed down.

That behavior was verified visually every time the simulation run, leading to suggest improvements. Our goal is to improve at least a 10% evacuation time. We propose as improvement with lowest cost, to remove chairs in the middle of the aisle. While this could be uncomfortable for passengers, it may improve evacuation time. Now, in order to test our suggestion, our new simulation removed the chairs and keep the same variables and goals as constants.

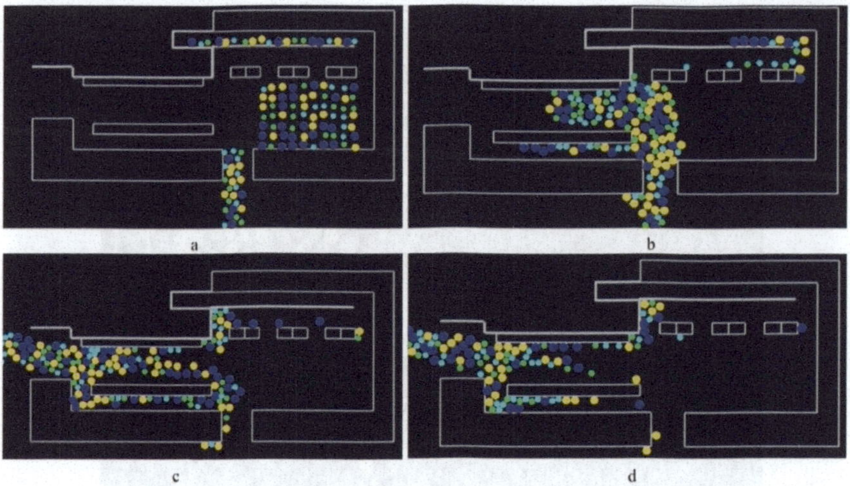

Fig. 9 Simulation 1 run. *Source* Menge Screenshot. **a** initital state; **b** three agent group collide; **c** clearing gate 75A; **d** clear aisle but congestion in corners

4.1 Improvements Simulation

Our simulation initial state is the same as original simulation, with clear aisle. Figure 10 has the screenshots from simulation run, this time with perspective (this is just to avoid confusion with simulation 1). Step (a) confirms the same state as

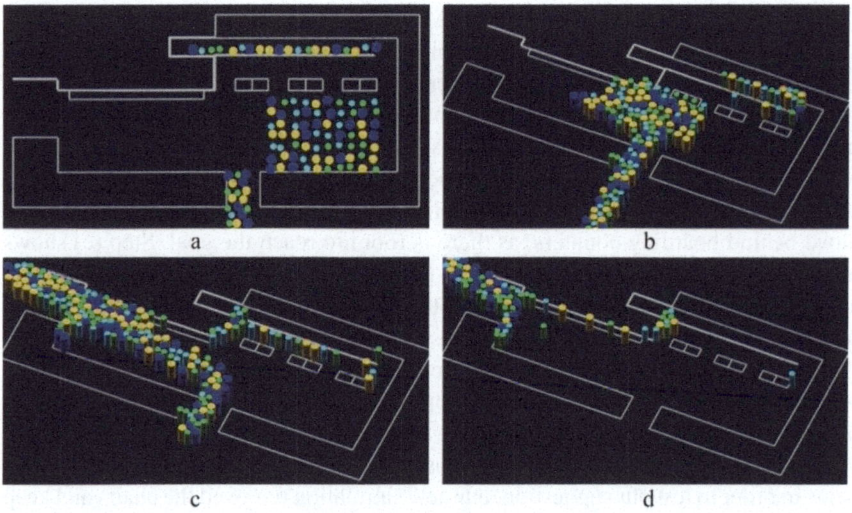

Fig. 10 Simulation 2 run. *Source* Menge Screenshot. **a** initial state, similar to simulation 1; **b** faster clearing; **c** fewer agents stuck; **d** faster clearing

Table 4 Simulation statistics

Feature	Simulation 1	Simulation 2	Difference	Improvement %
Maximum	61.7941	51.9729	16.2074	26.34
Minimum	55.0833	45.0324	3.5167	6.36
Average	58.6960	48.2722	10.4238	17.63
Std. Dev.	2.0791	2.0277	3.1514	4.9366

simulation 1. Step (b) showed a faster and smoother movement of agents toward the exit. However, agents at ramp presented almost same behavior as simulation 1. The difference was appreciated at step (c) where we can visually confirm there is few agents stuck in the problematic areas observed in simulation 1. this last part is observed in step (d).

Both simulations were running 100 times and the main results are in Table 4. Comparing simulation 1 vs simulation 2, we obtained a maximum performance improvement of 26.34% and a minimum improvement of 6.36%, while in average, the improvement was 17.63%. Figure 11 shows the comparison of simulated evacuation time (as simulation 1 and 2) and includes the difference in each run.

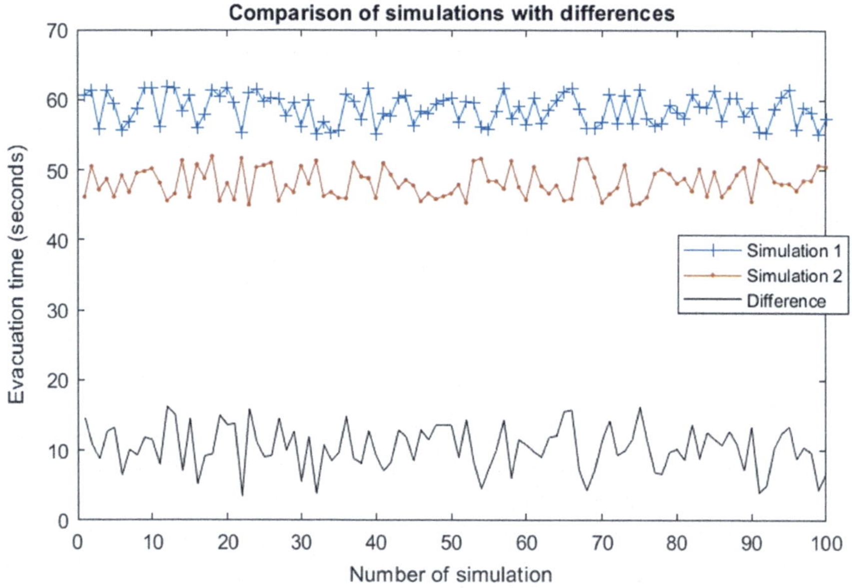

Fig. 11 Simulation comparison. *Source* Own creation

5 Conclusions and Future Research

Comparison of simulation for current layout and proposed changes to layout provides interesting findings that lead to the following conclusions. First, it was observed the benefit of simulation to measure and observe how an evacuation plan could result. In this research, we found Gate 75 has opportunities for improvement. The results observed show a bottleneck that is provoked by having chairs in the middle of an aisle. Also, the lack of escape routes in Gate 75B makes a natural evacuation path to the main building, forcing passengers to move to Gate 75A in order to escape.

Second, passengers coming from lower level through the ramp will have trouble to escape. This ramp in particular could introduce other problems during normal operations. As an example, it would be a problem when a gate change occurs, having to move a flight from Gate 75C or 75D to a boarding gate in the main building, while having other passengers move towards their right gate in the lower level. The ramp is so narrow that a collision will occur.

Third, using a crowd simulation software is convenient, as it allows to play with different hypothesis before deciding what to build. AICM is currently immerse in remodeling as it will add a new branch of gates in Terminal 2. Therefore, as construction company is changing layouts to accommodate the construction efforts, it would be convenient to do a quick simulation to find the best way to apply those changes while keeping the passengers safe and easy flow.

Last, we could observe how a small change in layout can improve evacuation times by more than 10%, testing this change is adequate and is valid to be implemented.

5.1 Future Research

This research has inspired us to continue to explore human crowd movements. As next steps we will use Menge in conjunction of better 3D-graphics engine such as Unity 3D to provide a more accurate representation of scenes. This is feasible due to Menge's architecture; the engine libraries can be called from other software packages.

Also, as a community service, this research is intended to be disclosed with authorities of AICM, so it can be used as an alternative for them to model different scenarios, not only for evacuation, but for passengers moving within terminals, simulation of flow to use free train between terminal buildings, or crowd attempting to check-in at airlines counters. The simulation scenarios are virtually endless. It is our best intention to inspire the main airport of Mexico to use this kind of technology and tools.

References

1. AICM (2019) Estadísticas Aeropuerto Internacional de la ciudad de México
2. Kuligowski ED (2016) Human behavior in fire. In: SFPE Handbook of fire protection engineering, Springer, pp 2070–2114
3. Lindell MK, Perry RW (2012) The Protective action decision model: theoretical modifications and additional evidence. Risk Anal 32(4):616–632
4. Bratman ME, Israel DJ, Pollack ME (1988) Plans and resource-bounded practical reasoning. Comput Intell 4(3):349–355
5. Helbing D, Molnár P (1995) Social force model for pedestrian dynamics. Phys Rev E 51(5):4282–4286
6. Bakar NAA, Majid MA, Adam K (2019) Simulation and modelling the human crowd evacuation. In: IOP Conference series: materials science and engineering, vol. 551, no. 1, 12038 p
7. G. Tesoriere, T. Campisi, A. Canale, A. Severino, and F. Arena, "Modelling and simulation of passenger flow distribution at terminal of Catania airport," in AIP Conference Proceedings, 2018, vol. 2040, no. 1, p. 140006
8. Schultz M, Lehmann S, Fricke H (2007) A discrete microscopic model for pedestrian dynamics to manage emergency situations in airport terminals. In: Pedestrian and evacuation dynamics 2005. Springer, pp 369–375
9. Rahman NA (2019) Switching action behavior model in crowd behavior simulator, pp 21–37
10. Zhang Q, Liu M, Wu C, Zhao G (2007) A stranded-crowd model (SCM) for performance-based design of stadium egress. Build Environ 42(7):2630–2636
11. Dattala S, Behzadan AH (2019) Modeling and simulation of large crowd evacuation in hazard-impacted environments
12. Curtis S, Best A, Manocha D Menge: A modular framework for simulating crowd

Modular Framework for Crowd Simulation "Menge" from a Production Warehouse Simulation Perspective

Irving Bruno López, Rafael Saldaña, and Gilberto Rivera

Abstract The topic of warehouse optimization is a critical and popular topic within the scientific literature [1, 4, 5, 7]. Particularly, order-picking optimization presents several profound difficulties: modeling, analysis, and data representation. Althoug these are necessary and obvious requirements for any research project, the visualization is an important aspect that has its challenges and complications. Menge is an open-source platform for crowd simulation. It provides an extensible common for research and development. In the present study, it is shown how the simulation platform known as Menge was adapted as a possible complement to the optimization process for the simulated visualization of its results. This chapter presents a solution using Menge to separate the modeling for the routing algorithm from visualization and simulation programming. The purpose of this work is to establish a starting point of industrial simulation in collaboration with algorithms for order-picking optimization in warehouses.

Keywords Production Warehouses · Menge simulation · Order-Picking Policies

1 Introduction

There are publications in the scientific literature that deal with the problem of Order picking; however, a few of them include the display of the simulation of the routes suggested by optimization algorithms. Some examples to model production warehouses for simulation are reported by [1], as shown in Fig. 1.

Because creating a simulation model is a separate and exhausting job, [6] proposes a method to create effective simulation models. As shown in Fig. 2.

I. B. López · G. Rivera (✉)
Universidad Autónoma de Ciudad Juárez, Av. Hermanos Escobar Omega, 32410 Cd Juárez, Chihuahua, Mexico
e-mail: gilberto.rivera@uacj.mx

R. Saldaña
Instituto Tecnológico de Ciudad Juárez, Chihuahua, Mexico

© The Author(s), under exclusive license to Springer Nature Switzerland AG 2021

A. Ochoa-Zezzatti et al. (eds.), *Technological and Industrial Applications Associated with Intelligent Logistics*, Lecture Notes in Intelligent Transportation and Infrastructure,
https://doi.org/10.1007/978-3-030-68655-0_16

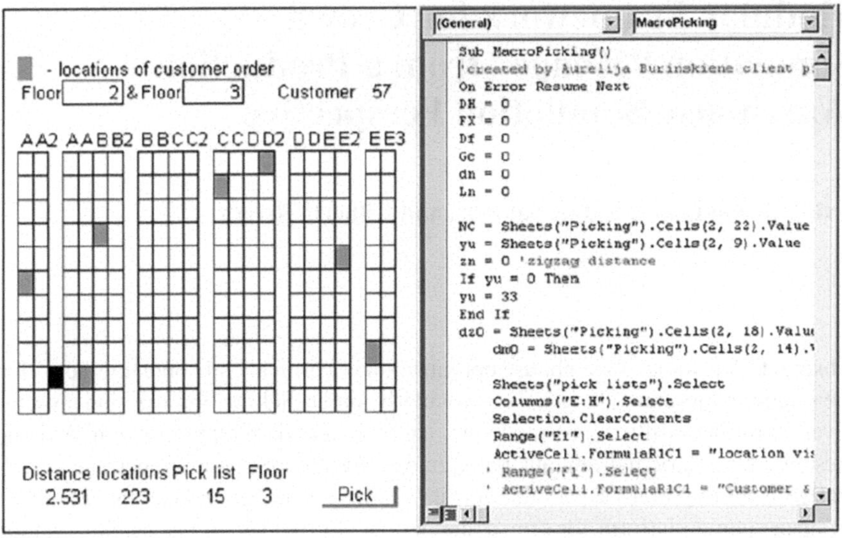

Fig. 1 Simulation model:simulation of order and routing algorithm [1]

A micro-simulation model for an intelligent logistics platform is published by [4] to optimize the transportation tasks and, consequently, reduce costs. These authors establish a flowchart for common logistics of a receiving area and, from this flowchart, a mathematically and visually characterized model is created to display different optimized routes for the order-picking problem. The model obtained is shown in Fig. 3.

Similarly, other authors have reported the creation of tools to create layout models for measurable and verifiable results. Some examples of the latter are the studies published by Sharp and Krueger [5]. A model using Petri nets is shown in Fig. 4.

Defining the layout of production warehouses is critical because warehouse productivity depends on the item locations within the warehouse. The simulation of the picking routes helps to determine the design and control of the warehouse. This is achieved by interrelating three aspects: the type of systems, the layout, and the policies to control all operational processes [7].

Menge is a cross-platform, extensible, modular framework for simulating pedestrian movement in a crowd. Menge creates abstractions for related problems and provides a plug-in architecture so that a novel simulator can be dynamically configured by connecting built-in and bespoke implementations of solutions [3]. The use of Menge is intended to simplify the work required to visualize the routes resultant from optimization algorithms in production warehouses.

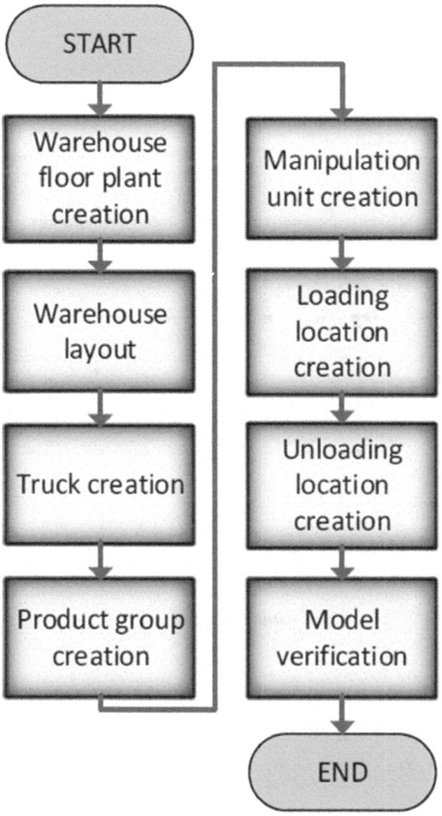

Fig. 2 Simulation model creation procedure [6]

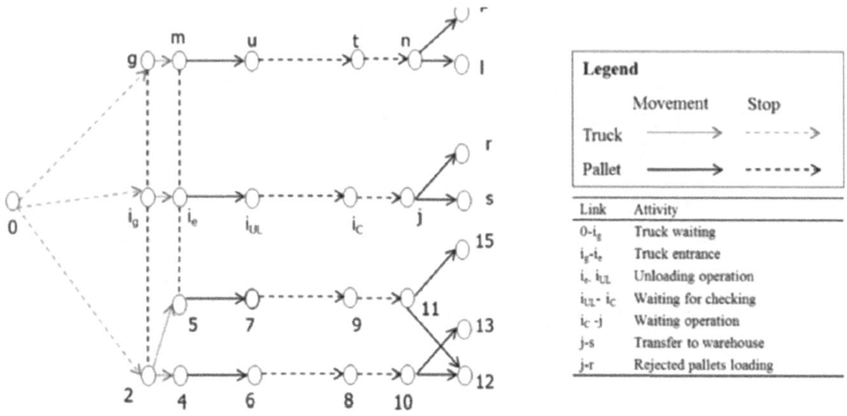

Fig. 3 Receiving Area model [4]

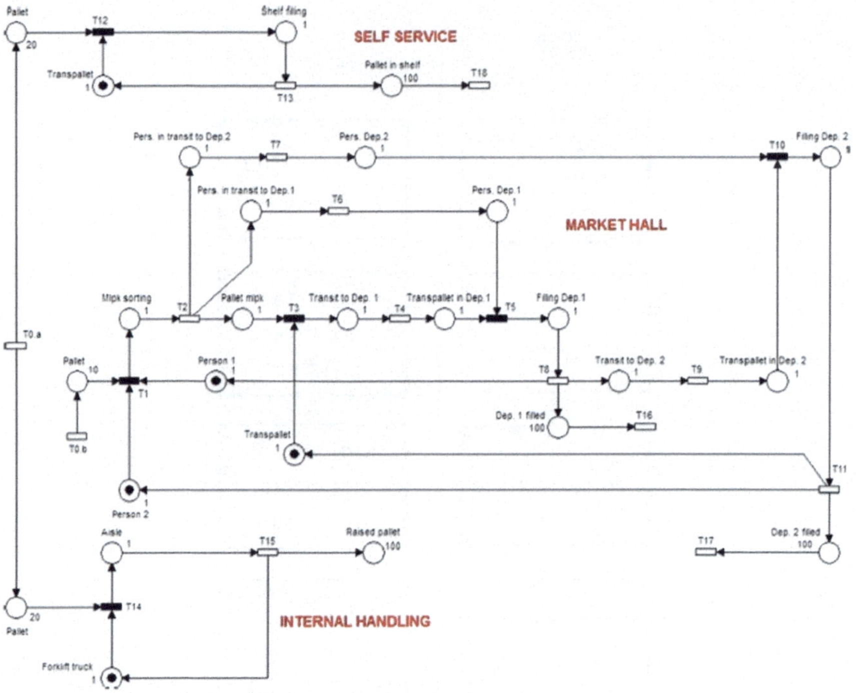

Fig. 4 The Petri Net model of a logistic system [5]

2 Materials

The equipment (software and materials) used for implementing the simulations are the following: Computer equipment to run the simulation, Operating System: Windows 10 Home Single Language 64-bit x64-based processor, System Manufacturer: Lenovo, Processor: AMD Ryzen 3 2200U with Radeon Vega Mobile Gfx 2.50 GHz, Memory: 8 GB RAM and OS Memory: 7.78 GB RAM.

2.1 Software

Visual Studio 2017 Community Edition. Menge: A framework for modular pedestrian simulation for research and development, free code [3].

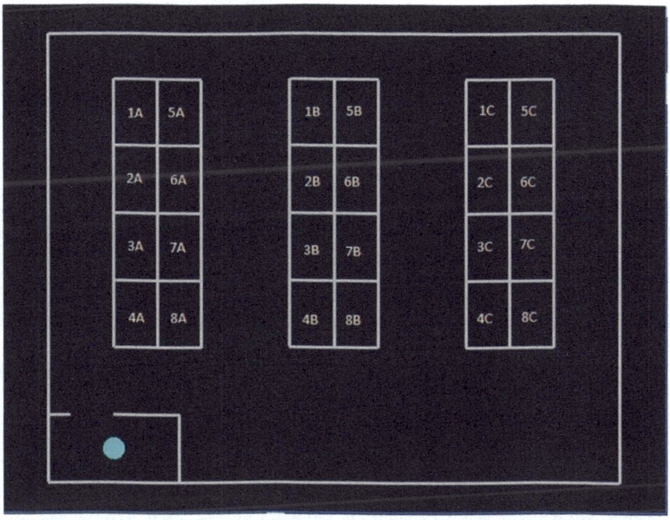

Fig. 5 Menge Scenario for a small production warehouse

2.2 *Scenarios*

The scenario is represented by a warehouse with three racks and one block. Each rack has eight available spaces or locations, four per aisle. A single picker is represented as an agent, which will explore the picking aisles of the warehouse following a specific list of collection points or predetermined goals (Fig. 5).

Some orders were generated with products located at the points 4A, 5A, 4B, 7B, 1C, and 6C, each of which will create a picking node. In this case, in the layout, there are eight crossing nodes (nodes CN1 to CN8) and a single node (DN1) of departure and arrival for the picking agent. Figure 6 shows the representative sequence of the warehouse modeled in Menge with the picking, crossing and destination nodes. The picking nodes are represented in green, the crossing nodes in red and the destination with a blue node, respectively.

2.3 *Simulation*

The simulation was performed using Menge software by modifying an open-source example called steerbench¥curves.xml for setting multiple and consecutive goals and another open-source example called globalNavSwap-roadmap.xml [2] to determine a roadmap as a guide for the simulation agents. These examples were studied and adapted to the conditions of the scenario as well as the objectives to be achieved during the simulation of an order-picking activity in production. The way of use Menge is like a black box. The Menge documentation shows how to adapt the setting

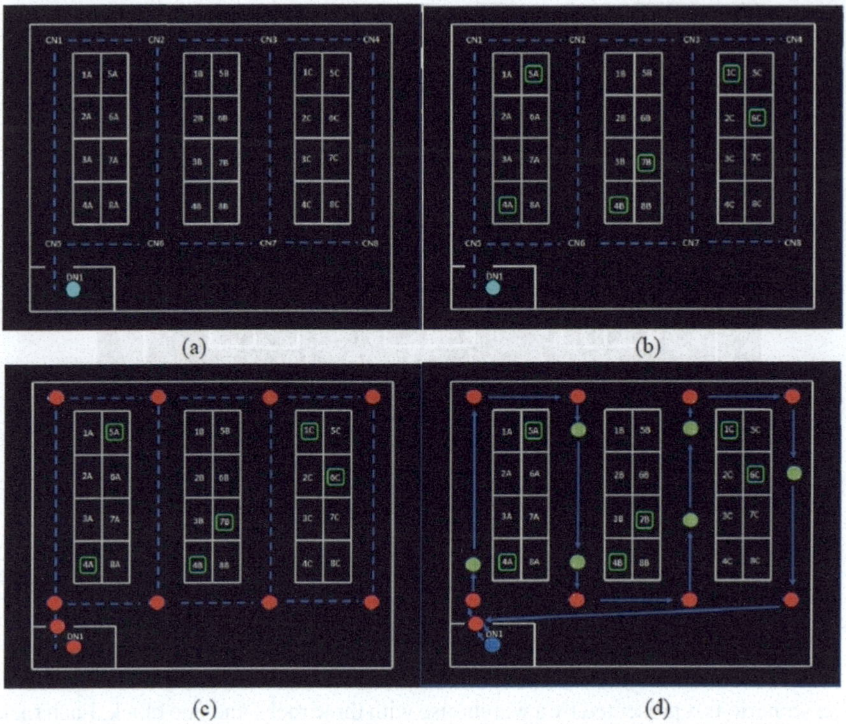

Fig. 6 Menge goals determination case based. **a** Cross nodes labeling. **b** Picking nodes locations. **c** Cross nodes model map. **d** Cross, picking and depot nodes for an order instance

files: the most relevant are the Scene Specification, Behavior Specification, and View Specification.

The obstacles of the scenario and the agents are defined in the Scene Specification file. The goals, as well as the trajectories, are established in the Behavior Specification file. The View Specification was not modified for this project. The mid points serve as a guide for the agent; avoiding having direct contact with the obstacles (following a specific notation in the file graph.txt). This grid of mid points is called roadmap. For this simulation, the S-shape policy was used. The final scenario was set as shown in Fig. 7.

For the agent to follow a pre-established route, it was established that the agent would always leave the depot towards the nearest crossing node (CN5 crossing node), then it will go to the nearest picking point until it reaches the next crossing point, when it arrives at the next picking aisle it will repeat the operation until it has visited all the picking points and, finally, go back to the delivery point. As observed, this simulation considers multiple objectives, which must be reached by the agent. These objectives are declared as a group of points in the behavior configuration file following this pattern: <Goal type="point" id="0" x="-10" y="-8.5"/> .

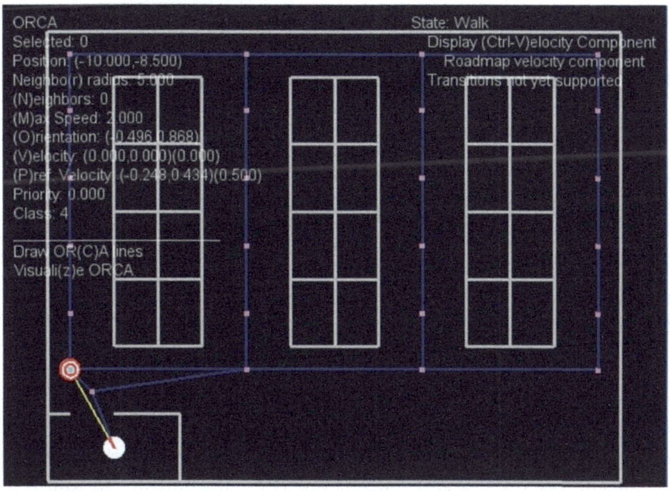

Fig. 7 The agent is heading towards the first goal (simulation parameters are shown)

A group of goals determined by the picking order used to validate the simulation is as follows:

```
<Goal type="point" id="0" x="-10" y="-8.5"/> <!--INICIO¥FIN-->
<Goal type="point" id="1" x="-11" y="-6"/> <!--INTERMEDIO-->
<Goal type="point" id="2" x="-12" y="-5"/> <!--PUNTO INTERMEDIO ABAJO A-->
<Goal type="point" id="3" x="-12" y="-2.5"/> <!--ESTANTE 4A-->
<Goal type="point" id="4" x="-12" y=".5"/> <!--ESTANTE 3A-->
<Goal type="point" id="5" x="-12" y="3.5"/> <!--ESTANTE 2A--> 5
<Goal type="point" id="6" x="-12" y="6.5"/> <!--ESTANTE 1A-->
<Goal type="point" id="7" x="-12" y="9"/> <!--PUNTO INTERMEDIO ARRIBA A-->
```

There is a goal for each point of interest (picking node, crossing node or destination node); however, the route will be defined by the content of the orders. Only the picking points sorted by number and letter are selected and inserted into the Behavior.xml configuration file, relating the goals to the trajectories. In Menge, these trajectories are known as walks. Below, an example of the first walks created for this order-picking case is presented:

```
<State name="Walk" final="0" >
<GoalSelector type="explicit" goal_set="0" goal="2" />
<VelComponent type="road_map" weight="1.0" file_name="ruta.txt"/>
</State>
<State name="Walk2" final="0" >
<GoalSelector type="explicit" goal_set="0" goal="3" />
<VelComponent type="road_map" weight="1.0" file_name="ruta.txt"/>
</State>
<State name="Walk3" final="0" >
<GoalSelector type="explicit" goal_set="0" goal="7" />
<VelComponent type="road_map" weight="1.0" file_name="ruta.txt"/>
</State>
```

The results obtained for this agent's walk for an order of six items located in different locations were 3.6s consistently after 100 runs. The roadmap was defined using the file graph.txt for the simulation. For this simulation, 26 guide points and 29 interconnections were defined. The guide points were defined by specifying the number of points they would connect to and their location within the layout plane. The interconnections indicate which guide point connects to another. An example of this configuration is shown below:

```
26
1 -10 -8.5
3 -11 -6
3 -12 -5
...
29
0 1
1 2
1 13
...
```

The agent identifies its immediate target and orients its direction vector as directly as possible. If the visibility towards its assigned target does not allow a direct connection, the agent goes to the mid point furthest from the Roadmap and closest to the goal. The following set of images shows a simulation run. In Fig. 8, the agent goes to the node AN1; in Fig. 9, the agent is heading towards CN7 (Cross node 7) from PN4B (Picking node 4B).

In Fig. 10, the direction vector assigned is the one indicated by the yellow line. This line is automatically calculated by Menge as the fastest way between the agent and its destination considering the guide points of the route map. The agent uses the points of the route map only as a reference, does not strictly follow the blue lines of the route map unless they coincide with the direction between the current position and the next goal.

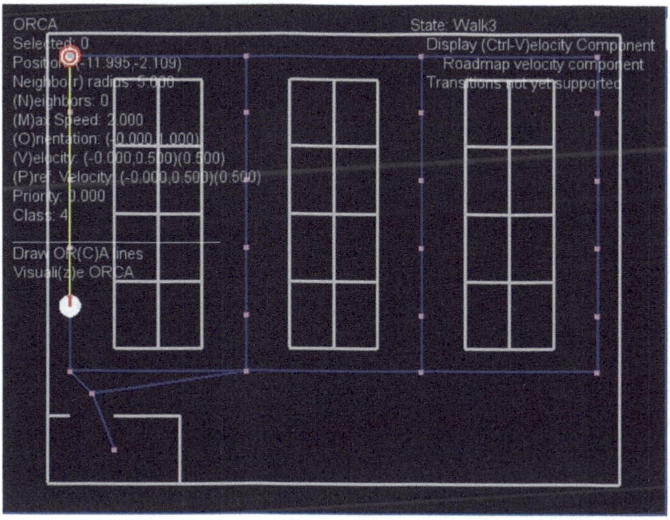

Fig. 8 The agent is heading towards AN1 (Cross node 1) from PN4A (Picking node 4A)

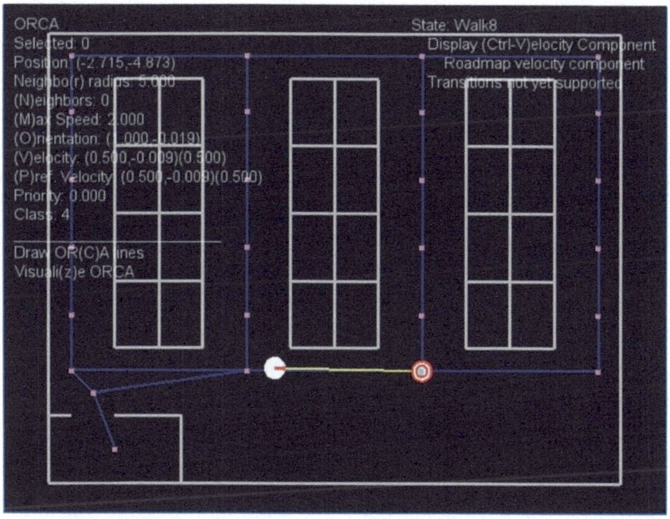

Fig. 9 The agent is heading towards CN7 (Cross node 7) from PN4B (Picking node 4B)

3 Conclusions and Future Work

Here, a notable difference with the most common approaches for the Menge simulation platform is presented. For a specific layout a few points are commonly determined for many people or agents in the simulation; but, in this case, a sequence

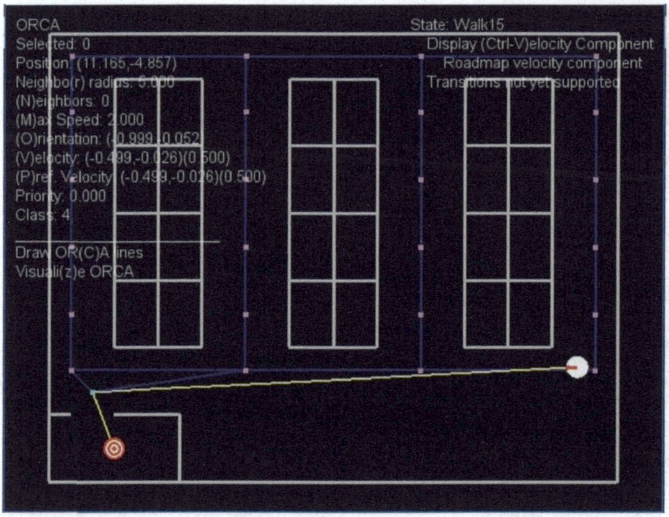

Fig. 10 The agent is heading towards destination DN1 (Depot node 1)

of points are determined for a single agent. This application shows the flexibility of parameter configuration that the Menge tool allows for a wide variety of applications [7].

By using Menge, it was possible to visualize a complete route for a randomly assigned picking order. Menge allows the user to see clearly and fast routes for different orders without the need to understand the optimization model. Menge simulations also do a visualization of other algorithms. The visualization of the simulated route coincided with the assigned route without abrupt changes of direction and traffic blockages. Once the terms used in the specification files are understood, Menge's flexibility allowed to quickly define the layout of the production warehouse, agent and route map. An important aspect that was not yet considered in this work is the lack of picking time for each product.

As future work, it would be proposed the creation of a visual interface that allows the following features:

(1) Visual definition of the warehouse layout.
(2) The definition of a folder where the picking orders to be simulated are automatically read.
(3) The inclusion of more picking agents.
(4) The definition of different capacities for different groups of agents.

We conclude that Menge is a tool that facilitates the visualization of routes and trajectories of agents, making an adequate complement for academics and researchers that are focused also on the optimization of routes and not only on the difficulties of visualization.

References

1. Burinskienė A, Davidavičienė V, Raudeliūnienė J (2018) Simulation and order picking in a very-narrow-aisle warehouse aisle warehouse. Econo Res-Ekonomska Istraživanja, 1–16. https://doi.org/10.1080/1331677X.2018.1505532
2. Curtis S, Best A, Manocha D (2013) MENGE
3. Curtis S, Best A, Manocha D (2016) Menge: a modular framework for simulating crowd movement 40: 1–40. https://doi.org/10.17815/CD.2016.1
4. Gattuso D, Cassone GC, Pellicano DS, Gattuso D, Cassone GC, A, D. S. P., Gattuso D (2017) A micro-simulation model for an intelligent logistics platform: specification and calibration results, 8312(April). https://doi.org/10.1080/16258312.2014.11517358
5. Gerini C, Sciomachen A (2018) Evaluation of the flow of goods at a warehouse logistic department by Petri Nets. Flexible Serv Manuf J. https://doi.org/10.1007/s10696-018-9312-3
6. Šaderová J, Marasová D, Galliková J (2018). Simulation as logistic support to handling in the warehouse: case study 7(1): 112–117. https://doi.org/10.18421/TEM71-13
7. Taylor P, Roodbergen KJ, Vis IFA, Jr, GDT (2014) Simultaneous determination of warehouse layout and control policies, (April 2015), 37–41. https://doi.org/10.1080/00207543.2014.978029

Mobile Application for the Detection of COVID19 Suspicious Cases in Mexico Using an Intelligent Model of Virtual Patients

Alejandro Acosta, Alberto Ochoa-Zezzatti, Alberto Hernandez, and José Mejia

Abstract Nowadays, the medical practice is having complications derived from the fact that it is not possible to have the most adequate logistics to interact with patients physically, that is why we propose an intelligent model of virtual patients that will help to support the identification of different diseases and considering three decisive aspects that are: the diagnosis, prognosis and monitoring to see the evolution of each individual and their immune response for each disease. The determination of suspected cases of COVID19 is the first step in the detection of persons infected with the virus. Making it quicker and simpler for people to know if they are a potential suspect case is something that should be easy these days thanks to all the technology around us. Thanks to the internet and connectivity from mobile devices it has been possible to create applications for any industry and need. For Covid19 it is possible in the same way.

Keywords Covid19 · Virtual patients · Apps for mobile devices · Applied artificial intelligence · Environmental intelligence

1 Introduction

It is very important for any medical personnel in training to be able to interact with real patients to improve learning and therefore the performance of medical knowledge. This research will try to identify the potential of the use of virtual patients as an intelligent resource to help in the midst of the pandemic. Several authors have identified how e-learning can improve the learning skills required to make appropriate

A. Acosta
Universidad Cuauhtémoc, San Luis Potosí, Mexico

A. Ochoa-Zezzatti (✉) · J. Mejia
Juarez City University, Ciudad Juárez, Mexico
e-mail: alberto.ochoa@uacj.mx

A. Hernandez
Universidad Autónoma del Estado de Morelos, Cuernavaca, Mexico

© The Author(s), under exclusive license to Springer Nature Switzerland AG 2021
A. Ochoa-Zezzatti et al. (eds.), *Technological and Industrial Applications Associated with Intelligent Logistics*, Lecture Notes in Intelligent Transportation and Infrastructure,
https://doi.org/10.1007/978-3-030-68655-0_17

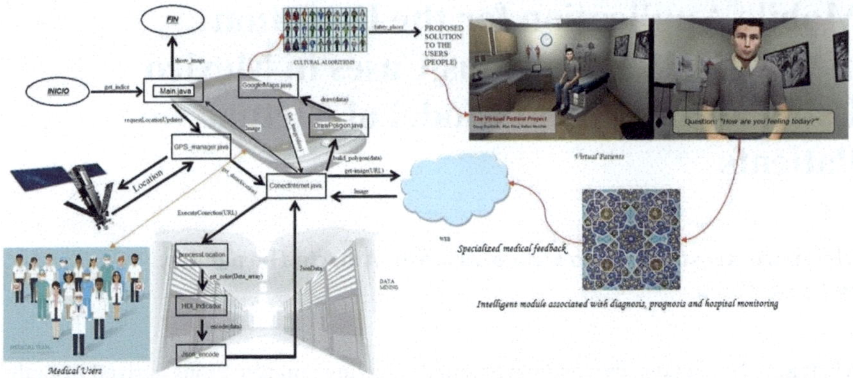

Fig. 1 Model associated with the use of virtual patients in a medical learning environment under the paradigm of ambient intelligence

diagnoses, prognoses and follow-ups, as proposed in [1] According to Kaptchuk et al. [2], there are several mobile applications that facilitate the monitoring of infected people and the flow of these people throughout their lives in order to determine whether they have infected others as they did. One of the main problems is the adoption of these apps. It is impossible to achieve follow-up if people are not willing to use these technological apps. Awareness initiatives must be created so that people are willing to adopt the technology in order to fight the pandemic. In Fig. 1 is possible see a specific concptual diargam associated with our intelligent mobile application related with Virtual Patients.

2 Detecting Suspicious Cases

This paper describes the start of what may be an app that is capable of performing the registration of people along with a series of answers to questions that will allow to detect if a person is suspicious, as is possible see in Fig. 2.

The mobile application is based on a record of basic demographic data such as name, age, state (Mexico) and gender.

The choice of this basic data will make it possible to create profiles of people and manage to segment vulnerable groups in the future or, failing that, strong groups and thus create strategies for isolation, as is possible see in Fig. 3.

It is a fact that the more surveys that are conducted about the pandemic, the better and more accurate the data will be for making predictions. It is extremely important that these surveys ask the right questions and that the answers are 100% honest.

The following diagram shows in Fig. 4, describe how survey questions can be categorized and ordered for better results.

Finally, at the end of the survey the application will be able to send the data to a micro service that can create the prediction and determine whether the user can

Fig. 2 Basic demographic data about use

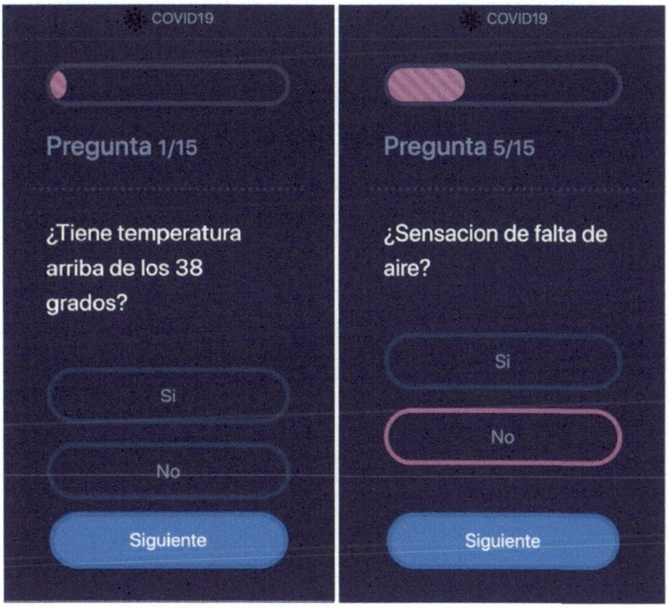

Fig. 3 Survey questions that seek to record data

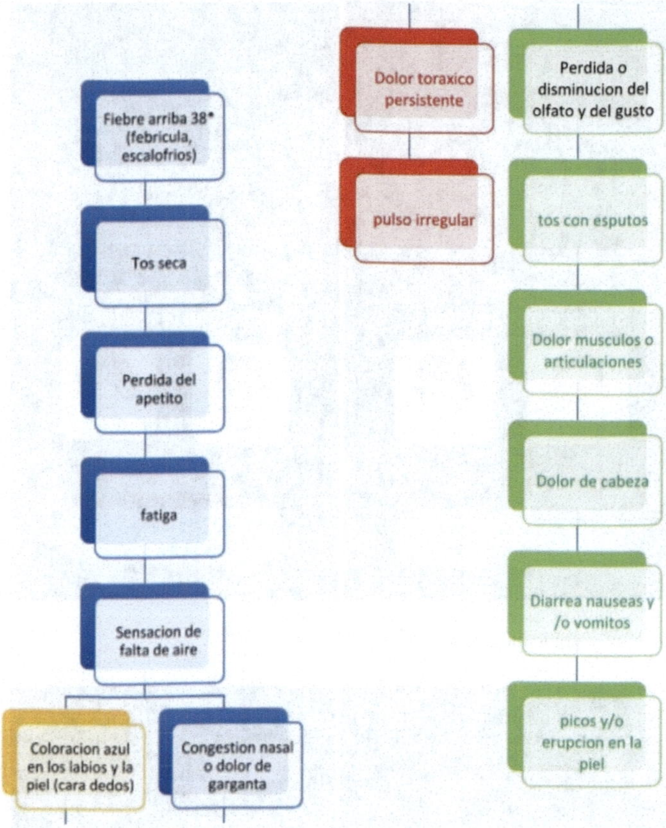

Fig. 4 Basic survey questions (Blue, Yellow, Green and Red)

be considered a suspicious case or not. Based on that detection, recommend the treatment that best suits the diagnosis made by a predictive algorithm [3, 4].

In an intelligent mobile application, this practice results in having the player perform the activities in a way that is fun, and this is achieved by having a balance between the skills and the challenges that we face the player, we will achieve That the game is entertaining and motivating.

According to Csikszentmihalyi, the components of a flow experience are as follows [3] (Table 1).

Table 1 The components of a flow experience

1. Clear objectives (expectations and norms can be perceived, and goals are attainable appropriately with the set of skills and abilities)
2. Concentration and focus, a high degree of concentration in a limited field of attention (a person related to a single activity will have the opportunity to focus and delve into the matter)
3. Direct and immediate feedback (successes and failures in the course of the activity are obvious, so the behavior can be adjusted as needed)
4. Balance between skill level and challenge (the activity is neither too easy nor too complicated)
5. The activity is intrinsically rewarding, so no effort is noticeable when running

3 Conclusions

Deficit Mind Theory mentions that people with Asperger show serious difficulties to take the place of the other and intuit their mental world. People with Asperger's are unable to intuit the mental world of others, so that the consequences of this inability or difficulty very serious [2].

Some of the consequences of this limitation for understanding the mental world of others would be the following [2] (Table 2).

To analyze the existence of statistically significant differences in the pre-test and post-tests measurements of the two groups, the patients was utilized for independent samples. This was complemented by an effect size with a statistic that corresponded, which in this case was Cohen's d. In a second analysis, the posttest and pre-test measurements of each group were compared using the Student's t for related samples. The third analysis involved the use of Cohen's d to assess the magnitude of the change produced during the experience. Finally, the descriptive statistics were utilized which were gathered from the participants 'evaluations and answers. The analysis was carried out using the statistics program Weka. As can be observed in Table 3, the average difference test between the pre-test measurements of the experimental group and the control group did not reveal the existence of statistically significant differences between the two with the variables analyzed. However, there were statistically significant differences between the two groups for all the variables evaluated following the intervention. By using Cohen's d, it was con- firmed that the differences between the groups after the activity were moderate. The means and

Table 2 The limitations for understanding the mental world Difficulty predicting the behavior of others

Difficulty in realizing the intentions of others and knowing the true reasons that guide their behavior
Difficulty understanding emotions, both own and others, which leads them to show few empathic reactions
Difficulty to consider the degree of interest of the interlocutor on the topic of conversation
Difficulty anticipating what others may think about their behavior Difficulty to lie and to understand deceit

Table 3 Average difference test between the pre-test measurements of the experimental group and the control group

	Pre-test			Post-test		
	t	p	d	t	p	d
Facial recognition	0.833	0.405	0.127	−3.477	0.001	−0.481
Social disabilities	−1.146	0.252	−0.280	−3.815	0.000	−0.533
Total	−1.069	0.286	−0.165	−3.894	0.000	−0.597

standard deviations of the variables in the study that correspond to the experimental and control groups for each study phase are displayed in Table 4. The analysis of the post-test–pre-test scores of the control group revealed no statistically significant differences with respect to any of the variables evaluated, as can be observed in Table 4. However, significant differences were found in the same analysis for the scores of the experimental group, both in the total of the questionnaire score in relation to serious game and its two factors, namely, facial recognition and social dis- abilities. Consequently, it can be seen how the Asperger's problems decreased among the people who participated in the serious game program. About the scope of the effect, it was observed that the program had a strong impact on reducing Asperger problems in facial recognition but was weak in terms of affecting other social disabilities. Regarding the assessment carried out by the participants, the4program was given a high score for usefulness (7.2 average) and a slightly lower average score for interest (6.1). 75% of the participants said they would recommend playing serious game to a relative or friend. With regards to the characteristics of the different stories in the game, it can be observed that participants easily identified that said stories were not emotionally well.

4 Future Research

The next steps in this work are the creation of the predictive algorithm based on a framework as a web platform which allows to perform the desired action, as is possible see in Fig. 5.

After generating the appropriate predictions, the next step is to capture that information and visualize it in a dashboard that offers a clear picture where the decision making is fast and effective. This information can be delivered to the health instances that require this type of processed data.

This was not the case of the participants who used other video games not related to Asperger's. As a result, the game's effectiveness in reducing misconceptions is clearly observed, particularly in relation to a friendly therapy. It must be taken into account that the preconceived notion of violence or danger associated with people with severe mental.

Table 4 Means and standard deviations of the variables in the study that correspond to the experimental and control groups for each study phase

	Experimental					Control				
	Pre-test	Post-test	Pre-test			Pre-test	Post-test	Pre-test		
	M (SD)	M (SD)	t	p	d	M (SD)	M (SD)	t	p	d
Facial recognition	5.11 (2.13)	3.55 (2.56)	16.899	0.000	0.662	4.86 (1.79)	4.66 (2.02)	1.244	0.22	0.104
Social disabilities	4.41 (3.29)	3.96 (3.33)	3.301	0.001	0.136	5.48 (4.27)	6.00 (4.27)	−1.748	0.088	−0.121
Total	9.54 (4.78)	7.54 (5.24)	10.406	0.000	0.399	10.35 (5.03)	10.64(5.15)	−0.768	0.447	−0.056

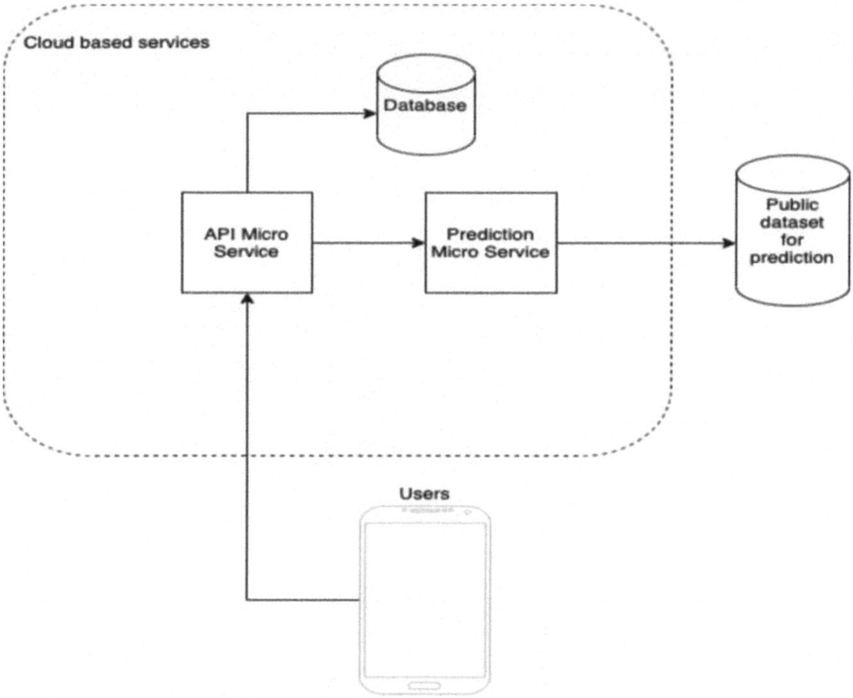

Fig. 5 Infrastructure to be followed for full system interconnectivity

The general assessment of serious game by the patients revealed interesting results as well. The participants scored close to eight points (7.2) for the game's usefulness and a slightly lower score for interest (average score of 6.1). Some of the patients' comments made about the game were the following: "everyday examples help to under- stand this disorder a little better," "it is fundamental to help and accept any individual, and to be patient," "people with Asperger as not necessarily an alien", and "I have learnt that I should treat people with same problems naturally." Additionally, 75% stated they would recommend trying this game to a relative or a friend. In light of the results obtained, serious game has proved to be an effective tool for raising awareness among people about Asperger's and for providing information about these disorders, which makes it possible to dispel myths.

References

1. Penteado BE, Ochs M, Bertrand R, Blache P (2019) Evaluating temporal predictive features for virtual patients feedbacks. In: IVA 2019, pp 88–90
2. Kaptchuk G, Goldstein DG, Hargittai E, Hofman JM, Redmiles EM (2020) How good is good enough for COVID19 apps? The influence of benefits, accuracy, and privacy on willingness to adopt. CoRR abs/2005.04343
3. Olsen JK, Oertel C (2020) Supporting empathy training through virtual patients. In: AIED 2020, pp 234–239
4. Yao H, Alappattu M, Robinson ME, Lok B (2020) Investigating trainees' nonverbal behaviors in virtual patients communication in virtual reality. In: VR workshops 2020, pp 165–166

Humanitarian Logistics for the Optimal and Timely Evacuation in High Buildings Within a Smart City Using an Adaptive Metaheuristic Context

Peter Savier Oropeza-Martínez, José Alberto Hernández-Aguilar, Alberto Ochoa-Zezzatti, and Diego Hurtado-Olivares

Abstract Nowadays, a growing middle class is conquering vertical spaces in almost all societies of Latin America, where it is believed that by 2027, 17% of the population will live in buildings of more than seven floors, Brazil leads this population group with an average of 11 floors in their buildings of "Alto padrão" as are known in social culture in Brazil, in Monterrey, Mexico are currently building the seven tallest buildings in Latin America exceeding 87 floors, however this boom is taking vulnerability to accidents in skyscrapers, especially fires. In this research, we evaluate different metaheuristics to develop an adaptive model associated with variations in it, such as number of people, obstacles, and other aspects associated with the modelling of the stampede as also as a heterogeneous crowd simulation based on personality traits and emotional state to study how these agents can react in dynamic scenarios. This application is oriented to all previous construction with buildings similar to those of the project, to contemplate the different types of behavior that people may have and to study the maximum number of people per floor and the corresponding areas to be able to evacuate.

Keywords Buildings on fire · Multiagent system · IOT · Smart City · Metaheuristics · OCEAN Model

1 Introduction

One of the biggest challenges in computer graphics is the fact of simulating a multiagent system with crowd behavior [1]. To achieve this goal, it is necessary to conduct a study of real mass behaviors. The branch of psychology studies human nature and the possible responses that we can generate from an event, and thus determine certain

P. S. Oropeza-Martínez · J. A. Hernández-Aguilar (✉) · D. Hurtado-Olivares
Universidad Autónoma del Estado de Morelos, 62209 Cuernavaca,, Morelos, México
e-mail: jose_hernandez@uaem.mx

A. Ochoa-Zezzatti
Universidad Autónoma de Ciudad Juárez, Av. Hermanos Escobar, 32410 Ciudad Juárez, Chihuahua, México

© The Author(s), under exclusive license to Springer Nature Switzerland AG 2021
A. Ochoa-Zezzatti et al. (eds.), *Technological and Industrial Applications Associated with Intelligent Logistics*, Lecture Notes in Intelligent Transportation and Infrastructure,
https://doi.org/10.1007/978-3-030-68655-0_18

patterns of behavior. There is a lot of scientific material where they incorporate models of psychology in a multiagent system. In this research we focus on incorporating personality behavior in large groups of agents. Also, we use Ant Colony Optimization Algorithm (ACO) to determine an optimum evacuation route for each agent. This gives us a lower bound of evacuation time for each agent, due to is an estimated time to evacuate. To determine an upper bound, we can change behavior parameters accordingly to bad behavior. This gives us a possible non-collaborative reaction by the agent to achieve their goal, which is the exit of the building. Once having the lower and upper bound we can appreciate the results of the simulations and calculate the average of the evacuation time and see how the estimated time is influenced due to certain behaviors in the crowd. The personality of a person determines patterns in their behavior, their temperamental and emotional state and mental characteristics. In psychology there is a big controversy of determining how many personality traits exist to define a behavior model. Although, among the most popular models is the OCEAN (openness, conscientiousness, extroversion, agreeableness, and neuroticism) model [2]. The opening describes the imaginative and creative aspect of human character, which leads to accept ideals that the person does not know. Consciousness determines at what level a person is organized in their routines and how they take care of things. Extroversion is related to the extroverted or little shy and sociable that a person is. Kindness is the ease and the tendency to get along with other people. Finally, neuroticism refers to emotional instability such as panic and the tendency to experience negative emotions. Each factor is bipolar and is used to determine or describe how can a person is by certain adjectives linked to those personality traits. We've used a personality trait mapped to HiDAC (High-Density Autonomous Crowds) crowd simulation system proposed by Pelechano [3] and incorporate an emotional factor that can affect the agent's behavior.

2 Theory

In a review of the literature, different researches associated with human stampedes were considered, which occur mainly in stadiums such as those in France; night music concerts like what happened in Sweden, Montenegro and Karelya; in holy places of Islam, especially affecting Iranians as in Mecca; in Catholic processions like in Goa and Macau; or provoked by natural actions such as floods, forest fires, capsizing of riverboats, shipwrecks in lakes and earthquakes, as can be seen in [4]. In most of the contingency situations associated with a human stampede, people were not prepared for such a contingency, so the increase in deaths is associated with the lack of fire safety measures in the premises that host a concert of music as happened in 2013 in Brazil, as is shown in Fig. 1. This event left a total of 239 people dead and 124 injured.

In many dangerous situations, the crowds do not pay attention to the existing security measures, due to the high amount of adrenaline their brains are receiving because they are enjoying a concert at high decibels. The factors such as delayed reaction

Fig. 1 Fire in music concert building occurred in 2013 in the center of Sao Paulo

to alarms, the walking speed depending of the type of person (age and locomotion speed), the fact that usually every passenger try to find his family members before being evacuated and problems with instructions from the crew, lacks of signals, the language of the signals or even panic, these human's factors problems should be addressed as part of the evaluation of efficiency of evacuation [5]. Other related dangerous situation is terrorism. In October 1st of 2017, occurred a massive shooting from a 32° floor in the city of Las Vegas, Nevada, US during a country music festival and with a balance of 59 deaths and 851 injuries, which is the worst massacre in US after the 9/11 attack to the twin towers in New York city. Both attacks weren't logistics issue, but once the attack begins, truly it is. Scenarios like terrorism, logistics fall in hands of authorities like police department, military, between others.

To analyze how to organize the best way crowds, we need first to study people's reaction in different situations and how can they interact between each other to accomplish certain objective or on the other hand, not cooperate to achieve the goal.

After an exhaustive related research, we can determine in Table 1 what authors accomplished in each one of their previous works, aiming if they use BDI model, multi agent system, ontologies, labelling, emotions model and the art of aesthetics.

2.1 High Concentration in Different Building Levels

Evacuation systems often are complicated to logistics problems due to different behaviors which people can react, based on current environments. Scenarios with 10 people in only one structure level maybe is possible to control the situation, but having a crowd for instance of 500 people in a building of more than four floors, logistics can be more complicated, not based on evacuation rules or the structure safety normative, but in the crowd's behavior.

Taking this issue to high level buildings like skyscrapers among other inferior buildings, the main issue are the latter's, space where people are often guided to

Table 1 Table of related research on multiagent system

Significant feature of this study and relevant hazardous multiagent system								
Authors	BDI	MAS	Ontologes	Labelling	Model motion	Art and Aesthetics	Year	Title
Hearst M			✓	✓			1992	Automatic Acquisition of Hyponyms from Large Text Corpora
Georgeff M	✓	✓					1997	The Belief-Desire-Intention Model ofagency
Sanderson et al.			✓	✓			1999	Deriving concept hierarchies from text
Hearst M			✓	✓			1999	User Interfaces and Visualization
Scherer K			✓	✓			2000	Psychological models of emotion
Parrott W					✓		2001	Emotions in Social Psychology
Yee K-P				✓			2003	Faceted metadata for image search and browsing
Vicente JJ		✓	✓	✓			2003	Estudio de metodos de desarrollo de sistemas multiagente
Dumais S et al.			✓	✓			2003	Stuff I've seen: A system for personal information retrieval and re-use
Naaman M		✓	✓	✓			2004	Context Data in Geo-Referenced Digital photo Collections
Mani Let AJ		✓	✓	✓			2004	Automatically Inducing Ontologies from Corpora
Guerra-Hernandez A	✓	✓					2004	Learning in BDI Multi-agent Systems

(continued)

Table 1 (continued)

Significant feature of this study and relevant hazardous multiagent system								
Authors	BDI	MAS	Ontologes	Labelling	Model motion	Art and Aesthetics	Year	Title
Guerra-Hernandez A							2005	Automatically Organizing Images using Concept Hierarchies
Phung T	✓	✓					2005	Learning Within the BDI Framework: An Empirical Analysis
Cambria E					✓		2012	The Hourglass of Emotions
Yo W	✓	✓					2012	An Extension Dynamic Model Based on BDI Agent
Henriquez C			✓	✓			2016	Ontologies for aspects automatic detection in sentiment analysis
Edmund T					✓	✓	2017	Neurobiologlcal foundations of aesthetics and art
Pentti M					✓	✓	2017	Emotions, values, and aesthetic perception
Righ S						J	2017	Aesthetic shapes our perception of every-day objects: An ERP study
Baitiche H	✓	✓					2017	Towards A Generic Predictive-Based Plan Selection Approach for BDI Agents
Alex MG				✓	✓		2018	Applying multi-label techniques in emotion identification of short texts

(continued)

Table 1 (continued)

Significant feature of this study and relevant hazardous multiagent system

Authors	BDI	MAS	Ontologes	Labelling	Model motion	Art and Aesthetics	Year	Title
Fernández M				✓			2018	Labelled Port Graph—A Formal Structure for Models and Computations
Tinio PP				✓			2018	Characterizing the emotional response to art beyond pleasure: Correspondence between the emotional characteristics of artworks and viewers' emotional responses
Cela-Conde CJ					✓	✓	2018	Art and brain coevolution
Siri F						✓	2018	Behavioral and autonomic responses to real and digital reproductions of works of art
Christensen JF						✓	2018	Introduction: Art and the brain: From pleasure to well-being
Che J						✓	2018	Cross-cultural empirical aesthetics
Zaidel DW						✓	2018	Culture and art: Importance of art practice, not aesthetes, to early human culture
Ochoa A	✓	✓	✓	✓	✓	✓	2018	Specifying Ontologies in Metroflog's labels for a sample obtained with a Multiagent System
This study	✓	✓	✓	✓	✓			

evacuate the current building, and where gets more density of crowd congestion to find a safety exit. When people collaborate to help each other to accomplish evacuate, exists less people with injuries in the process, deaths inclusive based on human stampedes. Where panic is an important variable and can lead to behaviors where an individual wouldn't know until that moment. Leading to push, scream, tread, etc. a person next by without considering possible damage to others.

Once performed the literature review, we determined to use flocking algorithm and OCEAN model mapped to HiDAC system to simulate crowd behavior during evacuation sinister in high buildings due to the characteristics mentioned in Table 2, where one model includes collective or chaotic behavior in crowd masses, as also as the OCEAN model mapped to HiDAC system due to its psychological inclusion in computer science based on personality traits involving certain behavior in individuals and have an approximation of real life scenarios. BDI was not included in this research due to its characteristics mentioned in Table 2.

2.2 Modelling Crowds

In this research, the simulation of human behaviors is explored in front of high buildings space using different models, which are the following Herd Algorithm and OCEAN model mapped to HiDAC model.

2.2.1 Herd Algorithm

A herd algorithm is used to model the behavior of crowds, virtual environment where the agents interact. To this end, the conceptual and generic framework of the simulation of human behavior in front of an accident is introduced. This research has mainly focused on types of danger associated with crowds in situations of contingency of nature; in this case it is a fire that causes the spread of a structure built by man. One of the most popular and simple ways to simulate a crowd is with the Reynolds flock algorithm [6] that has been used to simulate herds of animals. It is a general alignment repulsion algorithm for the grouping of multiagent systems. In the natural environment, organisms exhibit certain behaviors when traveling in groups. This phenomenon, also known as "flocking" (Pack), occurs at microscopic scales (bacteria) and macroscopic scales (fish). Using computers, these patterns can be simulated by creating simple rules and combining them. This is known as emergent behavior and can be used in games to simulate a group movement chaotic or similar to life.

The representation of birds in the Reynolds flock simulation model [5], each agent must have at least the following attributes to describe the state it is in:

Location: The x, y coordinates of the agent's current position.

Course: The angle of the current course of the agent.

Speed: Speed of movement of the agent.

Table 2 Comparative table of behavior models of multi agent system

Characteristics	BDI	OCEAN	Flocking algorithm	HiDAC
Objectives	Simulate mental attitudes, Beliefs, Desires and Intentions, representing their informational, motivational and deliberative states	Simulate personality as a basis for agent psychology and classifies a personality based on its level of Openness, Conscientiousness, Extraversion, Agreeableness and Neuroticism	Simulate collective or chaotic behaviour movement of large number of interacting agents by creating simple rules and combining them	Simulate the local motion behaviors of crowds moving in a natural manner within dynamically changing virtual environments
Main components	• Beliefs • Desires • Intentions • Events	• Openness • Conscientiousness • Extroversion • Agreeableness • Neuroticism	• Alignment • Cohesion • Separation	• Leadership • Trained or untrained • Communication • Panic • Impatience • Pushing • Right preference • Avoidance or personal space • Waiting radius • Waiting timer • Exploring environment • Walking speed • Gesturing

(continued)

Table 2 (continued)

Characteristics	BDI	OCEAN	Flocking algorithm	HiDAC
Solutions/Limitations	• Agents lack any specific mechanisms to learn from past behaviour and adapt to new situations • Do not have complete axiomatizations and are not efficiently computable • Does not explicitly describe mechanisms for interaction with other agents and integration into a multi agent system	• The metric is not in variant to changes in time step and measurement units • Closely matches the desired personality trait	• Simple to implement • Powerful results	• Closely matches the desired personality behavior • Delivers a wide variety of emergent behaviors driven by the psychological and physiological capabilities of each agent

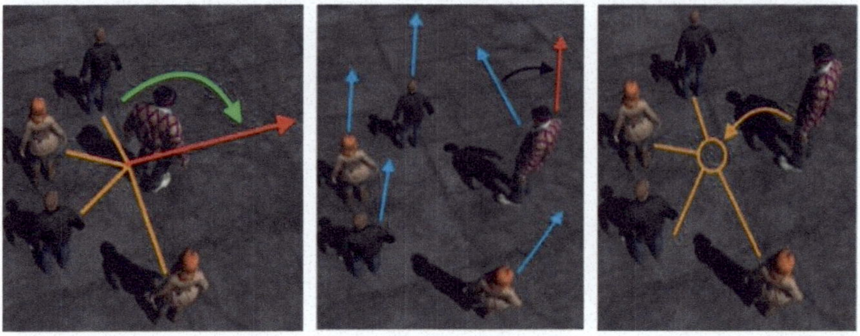

Fig. 2 An overview of basic rules of the flocking algorithm (Separation, Alignment and Cohesion)

The course and speed could be represented by a velocity equivalent to the magnitude of the distance vector. But often more attributes are needed to make the simulation more convincing.

Setting an upper limit to the speed with which Agents can move and turn is a common improvement, as is shown in Fig. 2.

The study of the crowds demonstrates the basic rules of the flock algorithm - Alignment, separation and Reynolds Cohesion [5]. The rules of the algorithm are:

1. **Separation:** All agents must avoid collisions with their neighbors.
2. **Alignment:** Each agent must match the address of their neighbors.
3. **Cohesion:** Each agent tries to go towards the center of the crowd.

Figure 2 presents the flocking rules which collectively lead to flocking behavior. In the flock algorithm each agent has direct access to the geometric description of the whole scene, but reacts only to flock mates within a certain small radius. The basic flocking model consists of three kind of simple steering behavior.

Separation. Gives an agent the ability to maintain a certain distance from others nearby. This prevents agents from crowding too closely together, allowing them to scan a wider area.

Cohesion. Supplies an agent with the ability to cohere (approach and form a group) with other nearby agents. Steering for cohesion can be computed by finding all agents in the local neighborhood and computing the average position of the nearby agents. The steering force is then applied in the direction of that average position.

Alignment. Gives an agent the ability to align with other nearby characters. Steering for alignment can be computed by finding all agents in the local neighborhood and averaging together the'heading' vectors of the nearby agents.

In algorithm 1, we summarized the pseudo code of the overall algorithm. In this section, we present a minimal algorithm that is capable of driving collective multiagent systems towards a stable, collisionless, self-organized correlated flocking state. This algorithm is based on the early models of animal swarms. The desired velocity of the agents is now a sum of interaction terms and some extra terms that define the self-propelling behavior and interactions with a bounded arena.

Algorithm 1. Flocking Algorithm.

```
For i until m do
For each agent
do
    Separation (agent);
    Alignment (agent);
    Cohesion (agent);
end
For each agent
do
    agent. X ← cos (agent. direction) * b. speed * dT time;
    agent. Y ← sin (agent. direction) * b. speed * dT time;
    draw (agent);
end
end
```

2.2.2 OCEAN Model

Personality is considered a group pattern of behavioral, temperamental, emotional, and mental traits for a person. Nowadays, there is still considerable controversy in personality research over how many personality traits exist, but the Five Factor or OCEAN model is popular; it is one of the models we have chosen for this study [2]. The OCEAN model, which are orthogonal dimensions of the personality space, are openness, conscientiousness, extroversion, agreeableness and neuroticism, these dimensions can be appreciated in Fig. 3.

Openness describes a dimension of personality that portrays the imaginative, creative aspect of human character. Conscientiousness determines how much an individual is organized and careful. Extroversion is related to how outgoing and sociable a person is. Agreeableness is friendliness, generosity and the tendency to get along with other people. Finally, neuroticism refers to emotional instability and the tendency to experience negative emotions. Each factor is bipolar (positive and negative values) and composed of several traits, which are essentially the adjectives that are used to describe people [2]. Some of the relevant adjectives describing each of the personality factors for each pole are given in Table 2. We have mapped these

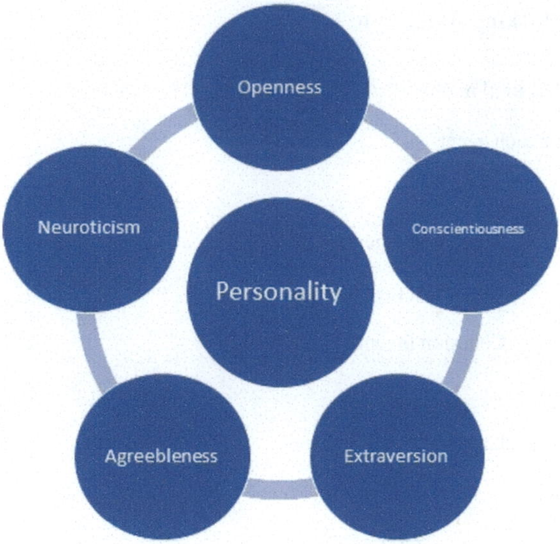

Fig. 3 OCEAN model

trait terms to the set of behaviors in an existing crowd simulation system, HiDAC (High-Density Autonomous Crowds) [].

HiDAC models individual differences by assigning each individual different psychological traits such as impatience, panic, and leadership behaviors and physiological traits such as energy level, speed, etc. The user normally sets these parameters to model the non-uniformity and diversity of the crowd. In this extended work, we free the user of the tedious task of low-level parameter tuning and combine all these behaviors in distinct personality factors. Here we combine a standard personality model with a high-density crowd simulation to create plausible variations in the crowd and permit a novice user to dictate these variations. Furthermore, we examine how these parameters affect the global behavior of a crowd. We will show that novel, emergent formations are realized, and behavior timings are also affected, as is shown in Table 3.

2.2.3 HiDAC

HiDAC [7] is a high-density crowd simulation system, which addresses the simulation of local behaviors and global wayfinding of crowds in a dynamically changing environment. The behaviors of autonomous agents in HIDAC are governed by the combination of geometrical and psychological rules. Psychological attributes include impatience, panic, and leadership behaviors. Physiological attributes are determined by traits such as locomotion, energy levels, maximum speed, etc. Agents are provided with skills such as navigation in complex environments, communication, learning

Table 3 Trait-descriptive adjectives

Trait	Adjective
O^+	Curious, alert, informed, perceptive
O^-	Simple, narrow, ignorant
C^+	Persistent, orderly, predictable, dependable, prompt
C^-	Messy, careless, rude, changeable
E^+	Social, active, assertive, dominant, energetic
E^-	Distant, unsocial, lethargic, vigorless, shy
A^+	Cooperative, tolerant, patient, kind
A^-	Bossy, negative, contrary, stubborn, harsh
N^+	Oversensitive, fearful, dependent, submissive, unconfident
N^-	Calm, independent, confident

and certain kinds of decision-making. Furthermore, they have perception so that they can react to obstacles, other agents, and dynamic changes in the environment.

In order to achieve realistic behavior, collisions are handled both by avoidance and response forces. Over long distances, collision avoidance is applied so that agents can steer around obstacles. On the other hand, collision response is utilized over shorter distances to prevent overlapping of agents with each other and with the environment. In addition to usual crowd behavior, agents might show pushing behavior or can wait for other agents to pass first depending on their politeness and patience. Pushing behavior arises from varying the personal space threshold of each individual. Impatient agents do not respect others' personal space and they appear to push their way through the crowd. In addition, relaxed agents temporarily stop when another agent moves into their path, while impatient agents do not respond to this feedback and tend to 'push'.

2.2.4 Integrating the OCEAN Model into HiDAC

The crowd is composed of subgroups with different personalities. Variations in the characteristics of the subgroups influence the emergent crowd behavior. The user can add any number of groups with shared personality traits and can edit these characteristics during the course of the animation. An agent's personality π is a fivedimensional vector, where each dimension is represented by a personality factor, Ψi. The distribution of the personality factors in a group of individuals is modeled by a Gaussian distribution function N with mean μi and standard deviation σi:

$$\pi =< \Psi O, \Psi C, \Psi E, \Psi A, \Psi N >,$$
$$\Psi_i = N(\mu i, \sigma i2), \text{ for } i \in \{O, C, E, A, N\} \text{ where } \mu \in [01], \sigma \in [-0.10.1]$$

Table 4 Low-level parameters versus trait-descriptive adjectives

Agent behavior	Personality factor	Ocean factor
Leadership	Assertive, social, unsocial, calm, fearful	E,N
Trained or untrained	Informed, ignorant	O
Communication	Social, unsocial	E
Panic	Oversensitive, fearful, calm, orderly, predictable	N,C+
Impatience	Rude, assertive, patient, stubborn, tolerant, orderly	E+, C,A
Pushing	Rude, kind, harsh, assertive, shy	A,E
Right preference	Cooperative, predictable, negative, contrary, changeable	A,C
Avoidance or personal space	Social, distant	E
Waiting radius	Tolerant, patient, negative	A
Waiting timer	Kind, patient, negative	A
Exploring environment	Curious, narrow	O
Walking speed	Energetic, lethargic, vigorless	E
Gesturing	Social, unsocial, shy, energetic, lethargic	E

The overall behavior β for an individual is a combination of different behaviors. Each behavior is a function of personality as:

$$\beta = (\beta 1, \beta 2, \ldots, \beta \mathrm{n})\beta j$$
$$= \mathrm{f}(\pi),\ \mathrm{for}\, \mathrm{j} = 1 \ldots \mathrm{n}$$

Since each factor is bipolar, Ψ can take both positive and negative values, as is shown in Table 4.

2.2.5 Personality-To-Behavior Mapping

The agents' personality factors (adjectives) are mapped into low level parameters and some of the built-in behaviors in the HiDAC model are shown in Table 2. A positive factor takes values in the range [0.5, 1], whereas a negative factor takes values in the range [0, 0.5). A factor given without any sign indicates that both poles apply to that behavior. For instance, E^+ for a behavior means that only extroversion is related to that behavior; introversion is not applicable. As indicated in Table 4, a behavior can be defined by more than one personality dimension. The more adjectives of a certain factor define a behavior, the stronger is the impact of that factor on that behavior. Thus, we assign a weight to the factor's impact on a specific behavior. For instance, ωEL is the weight of extroversion on leadership and it takes a value in the range [0 1]. The sum of the weights for a specific type of behavior is 1. Now we can see how the mapping from a personality dimensions to a specific type of behavior for an agent i as follows:

Leadership. Leaders tend to have more confidence in themselves. They remain calm during emergencies. Each agent has a leadership percentage determined by extroversion and stability. We compute leadership behavior as:

$$\beta_i^{Leadership} = \omega_{EL}\Psi_i^E + \left(1 - \Psi_i^N\right) \tag{1}$$

where

$$\beta_i^{Leadership} \infty E,\ \beta_i^{Leadership} \infty^{-1} N,\ and\ \beta_i^{Leadership} \in [0, 1] \tag{2}$$

Trained. Trained agents have complete knowledge about the environment. Because being trained requires curiosity and because trained people are informed, we associate this parameter with openness. Being trained is a Boolean parameter, so we use a probability function to represent it, as is possible see in Eqs. 3 to 5. As openness increases, the probability that the agent is trained increases:

$$P_i(Trained) = \Psi_i^0 \tag{3}$$

$$\beta_i^{Trained} = \left\{ \begin{array}{l} 1\,if\ P_i\ (Trained)\ \geq 0.5 \\ 0\,otherwise \end{array} \right\| \tag{4}$$

where

$$P_i\ (Trained) \infty\ 0\,and\ \beta_i^{Trained} \in \{0, 1\} \tag{5}$$

Comunication. This parameter determines whether agents communicate. Similar to being trained, communication depends on the probability of agent behavior. As extroversion increases, the probability that the agent communicates increases:

$$P_i(\,Communication\,) = \Psi_i^E \tag{6}$$

$$\beta_i^{Communication} = \left\{ \begin{array}{l} 1\ if\ P_i(\,Communication\,) \geq 0.5 \\ 0\ \ otherwise \end{array} \right\| \tag{7}$$

where

$$P_i(\,Communication\,) \infty\ E\,and\ \beta_i^{Communication} \in \{0, 1\} \tag{8}$$

Panic. In emergency situations, agents display panic behavior depending on their stability and conscientiousness. When they panic, their walking speed increases and they don't wait. We compute panic as:

$$\beta_i^{Panic} = \omega N P \Psi_i^N + \omega C P f\left(\Psi_i^C\right) \tag{6}$$

$$f\left(\Psi_i^c\right) = \left\{ \begin{array}{l} -2\Psi_i^c + 2\, if\ \Psi_i^c \geq 0 \\ 0\ \ otherwise \end{array} \right\| \tag{7}$$

where

$$\beta_i^{Panic} \infty N,\ \beta_i^{Panic} \infty^{-1} C+,\ and\ \beta_i^{Panic} \in [0, 1] \tag{10}$$

Impatience. We implement this parameter by modifying the route selection dynamically on the basis of environmental changes. This parameter depends on an agent's politeness and assertiveness. We compute impatience as:

$$\beta_i^{Impatience} = \omega_{EI} f\left(\Psi_i^E\right) + \omega_{AI}\left(1 - \Psi_i^A\right) + \omega_{CI}\left(1 - \Psi_i^C\right) \tag{11}$$

$$f\left(\Psi_i^E\right) = \left\{ \begin{array}{l} 2\Psi_i^E - 1\, if\ \Psi_i^E \geq 0 \\ 0\ \ otherwise \end{array} \right\| \tag{12}$$

where

$$\beta_i^{Impatience} \infty E+,\ \beta_i^{Impatience} \infty^{-1} A, C,\ and\ \beta_i^{Impatience} \in [0, 1] \tag{13}$$

Pushing. HiDAC can realistically simulate a person's respect for others. Agents can try to force their way through a crowd by pushing others, exhibit more respectful behavior when desired, make decisions about letting others walk first, and queue when necessary. Disagreeable agents tend to push others more because they're harsh and impolite. Similarly, extroverted agents display pushing behavior because they tend to be assertive. We compute pushing as:

$$P_i^{Pushing} = \omega_{EP} f\left(\Psi_i^E\right) + \omega_{AP}\left(1 - \Psi_i^A\right) \tag{8}$$

$$\beta_i^{Pushing} = \left\{ \begin{array}{l} 1\, if\ P_i\ (Pushing) \geq 0.5 \\ 0\ \ otherwise \end{array} \right\| \tag{9}$$

where

$$P_i(Pushing) \infty E,\ P_i\ (Pushing) \infty^{-1} A,\ and\ \beta_i^{Pushing} \in [0, 1] \tag{10}$$

Right preference. People prefer to move toward the right side of the obstacle that they're about to encounter. This behavior shows the person's level of conformity to the rules. The right-preference behavior is a probability function. If an agent is disagreeable or non-conscientious, that agent can make a right or left preference with equal probability. On the other hand, an agent that prefers the right side increases the probability proportionally to the agent's agreeableness and conscientiousness values, if these values are positive. We compute right preference as:

$$P(Right) = \left\{ \begin{array}{l} 0.5\, if\ \Psi_i^A < 0\, or\ \Psi_i^c < 0 \\ \omega_{AR}\Psi_i^A + \omega_{CR}\Psi_i^C\ otherwise \end{array} \right\| \Bigg\} \tag{11}$$

$$\beta_i^{Right} = \left\{ \begin{array}{l} 1\, if\ P_i(\,Right\,) \geq 0.5 \\ 0\ otherwise \end{array} \right\| \Bigg\} \tag{12}$$

where

$$P_i\,(Right) \infty A, C\ and\ \beta_i^{Right} \in \{0, 1\} \tag{13}$$

Personal space. Personal space determines the territory in which a person feels comfortable. Agents try to preserve their personal space when they approach other agents and when other agents approach from behind. However, the values for these two situations aren't the same. According to research, the average personal space in Western cultures is 0.7 m in front and 0.4 m behind. The personal space of an agent i with respect to an agent j is:

$$\beta_{i,j}^{Personalspace} = \left\{ \begin{array}{l} 0.8 * f(i, j)\ if\ \Psi_i^E \in \left[0, \frac{1}{3}\right] \\ 0.7 * f(i, j)\ if\ \Psi_i^E \in \left[\frac{1}{3}, \frac{2}{3}\right] \\ 0.8 * f(i, j)\ if\ \Psi_i^E \in \left(\frac{2}{3}, 1\right] \end{array} \right\| \Bigg\} \tag{14}$$

$$f(i, j) = \left\{ \begin{array}{l} 1\, if\ i\ is\ behind\ j \\ \frac{0.4}{0.7}\ otherwise \end{array} \right\| \Bigg\} \tag{15}$$

where

$$\beta_{i,j}^{Personalspace} \infty^{-1} E\ and\ \beta_{i,j}^{Personalspace} \in \{0.5, 0.7, 0.8\} \tag{16}$$

Waiting radius. In an organized situation, people tend to wait for available space before moving. We call this space the waiting radius; it depends on a person's kindness and consideration—that is, the agreeableness dimension. We compute the waiting radius as.

where:

$$\beta_i^{WaitingRadius} = \left\{ \begin{array}{l} 0.25\ if\ \Psi_i^A \in \left[0, \frac{1}{3}\right] \\ 0.45\ if\ \Psi_i^A \in \left[\frac{1}{3}, \frac{2}{3}\right] \\ 0.65\ if\ \Psi_i^A \in \left(\frac{2}{3}, 1\right] \end{array} \right\| \Bigg\} \tag{17}$$

$$\beta_i^{WaitingRadius} \infty A \, and \, \beta_i^{WaitingRadius} \in \{0.25, 0.45, 0.65\} \tag{18}$$

Waiting timer. If two people are heading in the same direction, each waits for the other to move first. The time they wait—that is, the duration during which they display patience toward each other—depends on their agreeableness. We compute the waiting time as

$$\beta_i^{Waiting\,Timer} = \left\{ \begin{array}{l} 1 \;\; if \; \Psi_i^A \in \left[0, \frac{1}{3}\right] \\ 5 \;\; if \; \Psi_i^A \in \left[\frac{1}{3}, \frac{2}{3}\right] \\ 50 \; if \; \Psi_i^A \in \left(\frac{2}{3}, 1\right] \end{array} \right\| \right\} \tag{19}$$

where

$$\beta_i^{Waitingtimer} \infty A \, and \, \beta_i^{WaitingTimer} \in \{1, 5, 50\} \tag{20}$$

Exploring the environment. HiDAC assigns people specific behaviors. The number of actions they complete depends on their curiosity. Open people will more likely explore different experiences and perform more actions. The openness factor determines the time cap where a person explores the environment. The number of actions that a person completes increases by the degree of openness. We compute the exploring parameter as

$$\beta_i^{Exploring} = 10\Psi_i^0 \tag{21}$$

where

$$\beta_i^{Exploring} \infty \, 0 \, and \, \beta_i^{Exploring} \in [0, 10] \tag{22}$$

Walking speed. A person's energy level determines that person's maximum walking speed. Because extroverts tend to be more energetic and introverts more lethargic, the extroversion trait controls this parameter. We compute the walking speed using Eqs. 29 to 32 as:

$$\beta_i^{Walkingspeed} = \Psi_i^E + 1 \tag{23}$$

where

$$\beta_i^{Walkingspeed} \infty E \, and \, \beta_i^{Walkingspeed} \in [1, 2] \tag{24}$$

Gesturing. The amount of gesturing during a conversation indicates a person's sociability. Outgoing people use more gestures than shy people, which indicates extroversion. We compute the gesturing parameter as

$$\beta_i^{Gesturing} = 10\Psi_i^E \tag{25}$$

where

$$\beta_i^{Gesturing} \infty E \; and \; \beta_i^{Gesturing} \in [0, 10] \tag{26}$$

2.3 Simulating Crowds in Fire Middle-Upper Levels of Buildings

The evacuation process is very complicated considering the variety of fire middle–upper levels of buildings, this may cause polytrophic evacuation processes. These types of levels influence the evacuation direction and cause delays in the evacuation time, especially in the case of a fire due to path obstruction and difficulty to navigate while exists smoke and other factors. People movement can also be restricted due to middle–upper levels by the spatial geometries. These constraints can be inherent in the design or can be caused by improper usage of the space. A middle–upper levels of building may have aisles and stairs too narrow to accommodate easy exit by a large crowd, sometimes, queuing behavior emerges spontaneously when a crowd gathers at an exit, permitting the crowd to "stream" out of the exit in an orderly manner and inadequate number of exterior exits, obstructed passageways, locked exterior doors, stairs or doors obscured by dim lighting or confusing signs, etc. As more people attempt to exit at once, the less of them are able to get out successfully because of the congested and jammed routes. For example, the rules derived at the building level can be utilized to build the decision making module of a building, and the rules extracted at the social interaction and group levels can be incorporated to model the interactions among agents in a virtual environment. To accomplish an evacuation in 3D model building scenario, it's necessary to build a NavMesh, which is the plane where agents can travel through space. In Fig. 4 we can appreciate the NavMesh of each floor of the building with a blue color.

Once the scene is ready, we can start designing the experiments to accomplish. In Fig. 5 we can see an example of a sinister occurring in the 4th floor and how agents travel in space trough each floor to reach their safety.

2.4 Humanitarian Logistics in Fire Buildings

The problem of tall fires in tall buildings over 12 floors.

The problem of the fire at height is a relatively new phenomenon in the world and in large part, is the result of new building concepts, which occurred after the end of the 2nd. World War. Until the tragic fire of the ANDRAUS building in Sao Paulo,

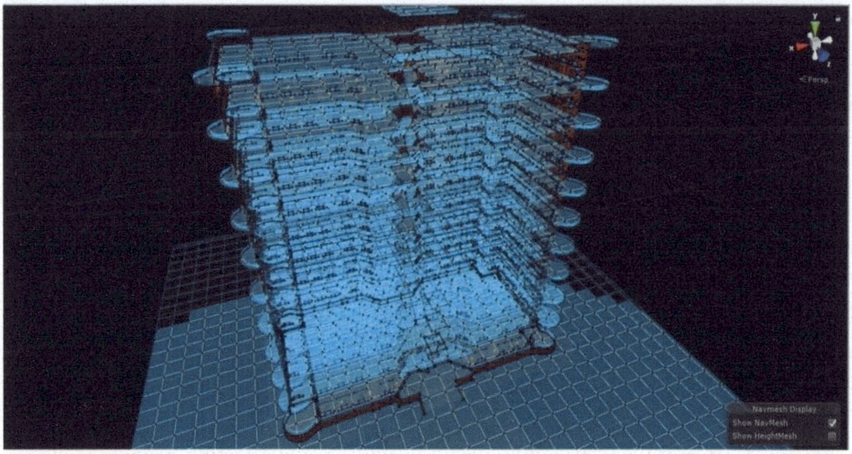

Fig. 4 High building model with NavMesh and graph

Fig. 5 Fire in a 27-story building occurred in 2018 in the center of Sao Paulo

Brazil, its eventual and catastrophic consequences had not been fully evaluated in the literature and never considering modelling using a Multiagent system. In the US, which has this type of construction since the end of the last century, the problem was in permanent evaluation and analysis for a long time. Suffice it to recall that the first high altitude fire occurred in New York in 1898 affected the 16-story building of the "HOME LIFE INSURANCE", and today, it is estimated that the Fire Department of that city attends thousands of alarms annually and that an important percentage of them, by the characteristics of New York, occurs in tall buildings. Relevant situations makes them specialists in such a complex technique and their experience has been very useful to those who are interested in the study of this exciting subject. The former commander and Fire Commissioner of the City, Mr. John T. O'Hagan, wrote in 1977 the basic guide on the subject "HIGHRISE FIRE & LIFE SAFETY" (Building of fire and safety) in which, through experiences and analysis of real cases, determines different conclusions. Of special interest to us is the analysis about the causes of the catastrophic fires in Latin America. Commander O'Hagan attributes them basically to the fact that the norms and codes related to this matter are not strictly applied by the competent authority, either at the stage of the design or construction of the building. Non-compliance with these norms is attributed to three basic reasons:

1. To the average of this type of fires in our continent is infrequent, in comparison to the USA or Europe and, therefore, there is a tendency to give little importance to the respective legislation, to security practices and, in conclusion, to the true dimension of the risk of a fire of this type.
2. The problem of tall buildings is relatively new in Latin America and on many occasions, the authority lacks the necessary knowledge or interest regarding possible design and construction errors that may cause risk to human health and life.
3. Finally, many areas of the continent are poor and developing, and in order to overcome it, the authority omits or does not make compulsory control and security systems, which necessarily increase the cost of works, become a limitation from the same. Added to the construction of buildings, there exist low margins of security that leads to potential threats to the life and safety of users.

In U.S.A. and Europe, with the strictest legislation, it is unlikely, but not impossible, that this will happen. What happens is that in front of a Latin American reality of buildings designed with open stairs, combustible coatings on the floor and walls, and lacking in practice of self-protection system, in the most advanced countries the stairs are closed and pressurized, materials are used non-combustible in coatings and floors and have a complete system of protection elements. And even, in front of the unique staircase, common in Latin America, in Germany and the USA for example, in each inhabited or working place "there must be 2 independent escape routes". Conventional Fire Brigades are not effective in tall buildings unless they are self-sufficient. From the outside they cannot: evacuate people, ventilate the areas, introduce jets of water or enter the building if it is out of reach of the largest available equipment. We are talking about minimum heights of twenty meters, about seven floors in residential buildings, five or six floors in commercials. These limitations are further reduced when the buildings have several levels or are located behind in lots, there are parked vehicles, poles and cables and architectural accessories such as ramps, fountains, water mirrors, recreational are as, among others.

Being a new problem there are different concepts of the definition of tall building. We could say that it is "one that lacks viable external access to reach the upper floors in case of fire and in which firefighters must be based to extinguish it in the systems and elements with which it has been endowed and built." On the other hand, firefighters have a practical definition and states: "It is one in which, due to the excess height of the same Firefighter, he is deprived of the possibility of applying a conventional external attack strategy, while his internal attack is limited by the ways of Access that the building design allows.

3 A Sociocultural Perspective for Simulating Human Stampedes

Comparison between high and Monterrey buildings (m & m) and high rise buildings in Brazil.

The simulation of a pedestrian system model, which is implemented for the analysis of human behavior during the evacuation of the seventh or more—level structure belonging to an living spaces building, is described by the unexpected event of a massive vertical fire as an unpredictable phenomenon, the analysis and simulation are performed considering the behavior in a state of paranoia and the various sociocultural factors of the people, in order to evaluate the amount of possible human losses during this scenario and with that to establish proposals for structural improvements as well How to optimize evacuation strategies and stratagems, taking into account that the factors that determine human losses can be those as: total or partial collapse of the structure of the building or suffocation caused by the pressure generated between people in a state of paranoia, including the effects of smoke associated with the massive fire, as is possible see in Fig. 5. In order to carry out the study and simulation of people' behavior under paranoia and their decision-making in the catastrophic event, factors such as: population density in the temporal space, level of paranoia associated with stress, factor as Cultural partner between the people taking into account the level of altruism, perception of the environment and maturity, the magnitude of the massive fire and the level of the structure in which the people is located, these factors are used in the analysis and experimental design. The factors mentioned above are assessed using the proposed model to generate the corresponding analysis and thus suggest a pattern of evacuation behavior, structural improvements or even perform safety assessments of existing buildings under these conditions associated with fire, an important aspect to consider is the lifestyle of many people in these building, which is of kind "groutier" and determine that a nocturnal fire will be decisive to loss lives in this situation.

3.1 Architecture of System Using CUDA & FlameGPU

Agent Based Modelling is a technique for the computational simulation of complex interacting systems through the specification of the behavior of several autonomous individuals acting simultaneously. This is a bottom up approach, in contrast with the top down one of modelling the behavior of the whole system through dynamic mathematical equations. The focus on individuals is considerably more computationally demanding but provides a natural and flexible environment for studying systems demonstrating emergent behavior. Despite the obvious parallelism, traditionally frameworks for ABM fail to exploit this and are often based on highly serialized algorithms for manipulating mobile discrete agents. Such an approach has serious implications, placing stringent limitations on both the scale of models and the speed at which they may be simulated. The purpose of the FLAME GPU framework is to address the limitations of previous agent modelling software by targeting the high-performance GPU architecture. The framework is designed with parallelism in mind and as such allows agent models to scale to massive sizes and ensures simulations run within reasonable time constraints. In addition to this visualization is easily achievable as simulation data is held entirely within GPU memory where it can be rendered directly.

FLAME GPU framework is a template-based simulation environment that maps formal descriptions of agents into simulation code [8]. The representation of an agent is based on the concept of a communicating X-Machine (which is an extension to the Finite State Machine which includes memory). Whilst the X-Machine has very formal definition X-Machine agents can be thought of state machines which are able to communicate via messages which are stored in a globally accessible message list. Agent functionality is exposed as a set of state transition functions which move agents from one internal state to another. Upon changing state, agents update their internal memory through the influence of messages which may be either used as input (by iterating message lists) or as output (where information may be passed to the message lists for other agents to read). FLAME GPU uses agent function scripting for this purpose where script is defined in a number of Agent Function Files. Simulation models are specified using a format called X-Machine Markup Language (XMML) which is XML syntax with Schemas governing the content. A typically XMML model file consists of a definition of a number of X-Machine agents (including state and memory information as well as a set of agent transition functions), a number of message types (each of which has a globally accessible message list) and a set of simulation layers which define the execution order of agent functions (which constitutes a single simulation iteration). Throughout a simulation, agent data is persistent however message information (and message lists) is persistent only over the lifecycle of a single iteration. This allows a mechanism for agents to iteratively interact in a way which allows emergent global group behavior. The process of generating a FLAME GPU simulation is described by the 1. The use of XML schemas forms a large part of the process where polymorphic like extension allows a base schema specification to be extended with a number of GPU specific elements. Given an XMML model definition, template driven code generation is achieved through Extensible Stylesheet Transformations (XSLT). XSLT is a flexible functional language based on XML (validated itself using a W3C specified Schema) and is suitable for the translation of XML documents into other document formats using a number of compliant processors (although the FLAME GPU SDK provides its own). Through the specification of a number of XSLT Simulation Templates a

Dynamic Simulation API is generated which links with the Agent Function Files to generate a simulation program, as is shown in Fig. 6.

Figure 7 shows how agents in FLAME GPU can have a big concentration density over a bottleneck, due to a specific exit route which does not contemplate large scale crowds' trough that exit, as is shown in Fig. 7.

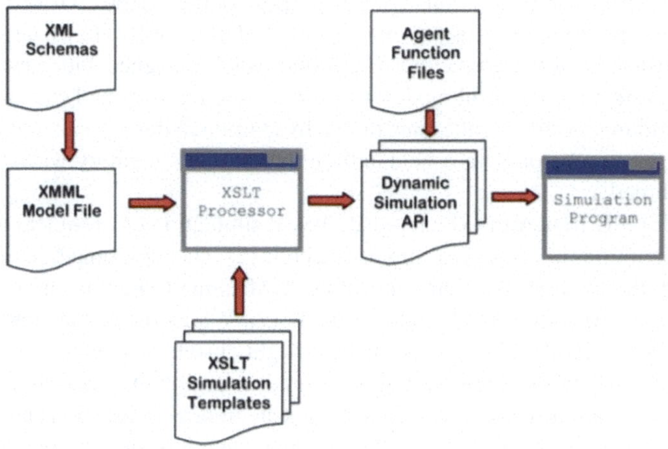

Fig. 6 FLAME GPU Modelling and Simulation Processes

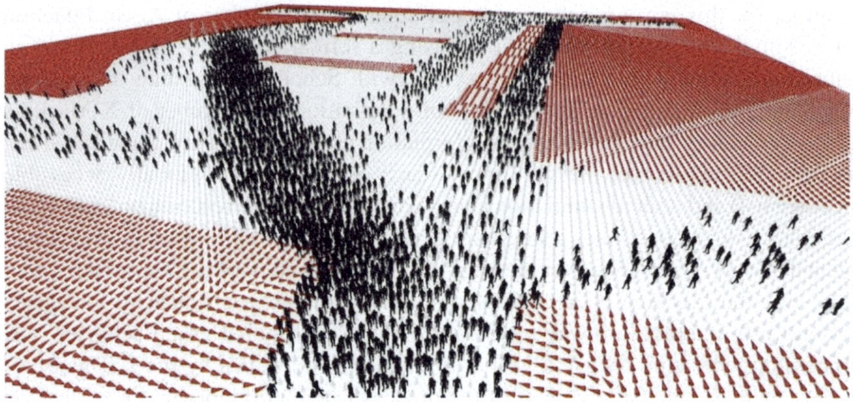

Fig. 7 High crowd concentration bottleneck

3.2 Experiment Design of Evacuation Proposed Model

In order to obtain the most efficient arrangement of a Pedestrian speed model associated with a fire in a tall building, we developed a cluster for storing the data of each of the representative individuals. The narrative guide is made with the purpose of distributing an optimal form for each the evaluated Pedestrian speed model associated with a fire in a building [9]. The main experiment consisted in implementing a Pedestrian speed model associated with a fire in a building, with 500 agents. The stop condition is reached after 75 runs; this allowed generating the best selection of each kind and their possible location in a specific Model. A location is obtained after comparing the different cultural and economic similarities of each Pedestrian speed model associated with a fire in a building and the evaluation of the Multiple Matching Model as is proposed in [10]. The vector of weights employed for the fitness function is Wi = [0.6, 0.7, 0.8, 0.5, 0.6, 0.7], which respectively represents the importance of the particular attributes.

Pedestrian speed model associated with a fire in a building is associated with an increase index. Cost–Benefit and Equipment's on the Pedestrian speed model associated with a fire in a building. Then, the flocking algorithm will select the specific location of each Pedestrian speed model associated with a fire in a building based on the attributes similarity. Each attribute is represented by a discrete value from 0 to 7, where 0 means absence and 7 the highest value of the attribute. The experiment design consists of an orthogonal array test with interactions amongst the attribute variables; these variables are studied within a location range (1 to 400) specific to a coordinates x and y. The orthogonal array is L-N (2**5), in other words, 7 times the N executions. The value of N is defined by the combination of the 7 possible values of the variables, also the values in the location range. In Table 5, we list some possible scenarios as the result of combining the values of the attributes and the specific location to represent a specific issue (Pedestrian speed model associated with a fire in a building). The results permit us to analyze the effect of the variables in the location selection of all the possible combinations of values. As is shown in Table 5.

The use of the orthogonal array test facilitates the reorganization of the different attributes. Also the array aids to specify the best possibilities to adequate correct solutions (locations) for each Pedestrian speed model associated with a fire in a building.

Table 5 The orthogonal array test

Increase index	Light and entropy	Pedestrian velocity	Panic in the people	Cost/Benefit security	Alert equipment	Scape conflict level	Propagation sequence
4	1	2	2	3	4	4	3
3	1	2	2	3	3	3	5
2	1	3	2	4	4	1	1
5	2	3	2	5	2	2	2

Fig. 8 Rectangular building, Cuernavaca Morelos

Different attributes were used to identify the real possibilities of improving a Pedestrian speed model associated with a fire in a building set in a particular environment, and to specify the correlations with other Pedestrian speed model associated with a fire in a building with similarity necessities. The locations will be choosing based on the orthogonal test array.

4 Experimentation of Diverse Scenarios

4.1 Use Case of a Building of More Than 27 Levels High

We modelling three different buildings with more of 27 levels to try justify our experimental model as is possible see in Figs. 8, 9 and 10.

4.2 Simulation Results

We have checked the validity of the behavior models by implementing the algorithms presented in Sect. 2.2.1. Our primary goal was to test the stability of the algorithms under realistic environmental conditions, including for example fire and its randomly changing direction. The first algorithm to be described is the flocking's algorithm. The algorithm starts with a fixed number of agents that occupy a randomly generated

Fig. 9 Spiral structure building

Fig. 10 Convex structure building

position in this space. Each agent moves around the spatial data, testing the neighborhood of each location in order to verify whether a point can have some desired properties. Each agent follows the rules of movement described in Reynolds' model. In addition, this model considers four different kinds of agents, classified on the basis of some properties of data in their neighborhood. Different agents are characterized by a different taxonomy. In Fig. 11, the flocking algorithm is illustrated.

Once the component accelerations have been calculated, the weighted sum can be taken, and the final acceleration can be applied to the boid's velocity. We characterize the flocking behavior of a group of agents without influencing agents in the large and

Fig. 11 Simulation of flocking algorithm behavior

herd settings. We measure the number of clusters of agents that are path connected and facing the same direction; each of these clusters forms a small flock, as is shown in Fig. 12.

In Fig. 13, the result of an experiment performed can be visualized, and it allows observing that the group of agents to be evacuated was able to leave unharmed due to the establishment of a correct escape model validated by our research.

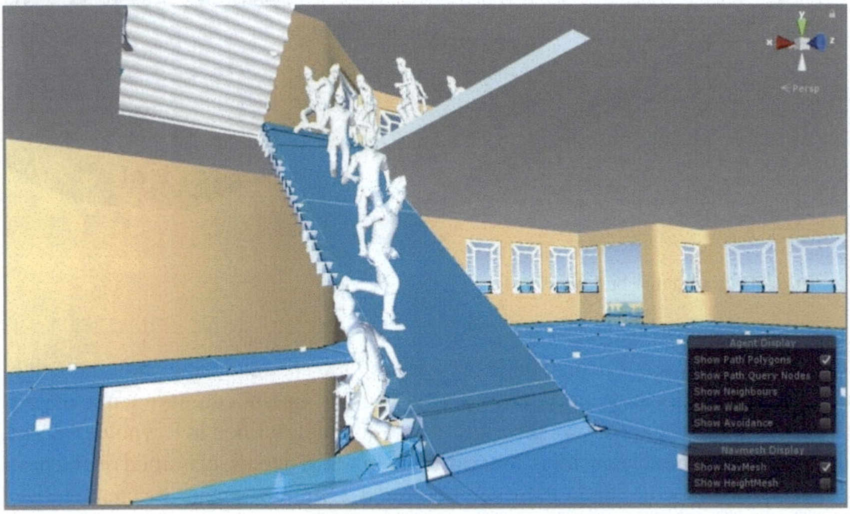

Fig. 12 Simulation of OCEAN model behavior evacuating the building

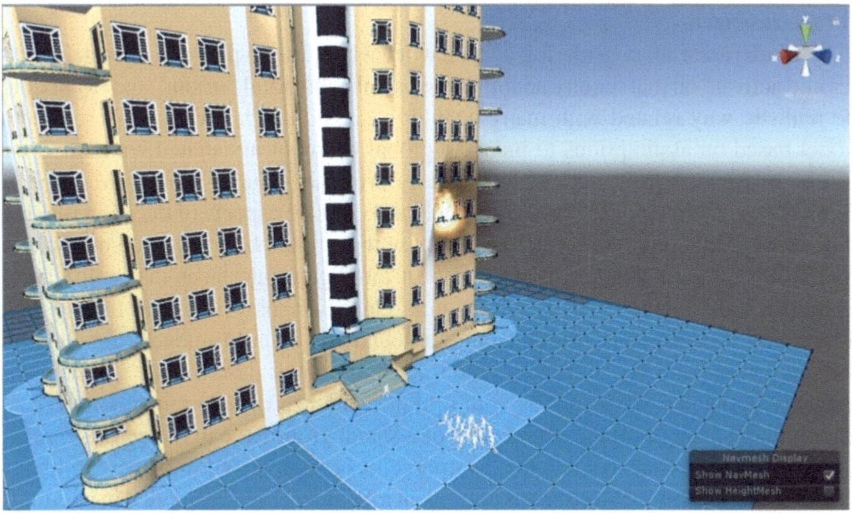

Fig. 13 High building model fire in 4th floor and agents evacuating the structure

5 Conclusions

This research uses a multi-agent system based on approach for modelling and simulating dynamic human crowds, but it introduces the use of specific anthropometric and sociocultural aspects that have an impact on individual and group behavior. Different to existing models that focus on the flow dynamics of individuals in particular spaces and therefore provide a partial abstraction of individuals, our approach includes anthropometric and sociocultural aspects because:

(i) We aim to address human logistics in which modelling both the urban space and the individuals of the crowds is important to predict and prevent catastrophes;
(ii) We use virtual worlds to visualize the simulations of dynamic crowds in public spaces. In order to make virtual worlds realistic in big and fine grain it is important to reproduce avatar (i.e., agent) diversity considering both their physical characteristics and their behavior.

The challenge is to reproduce realistic behaviors of the individuals interacting among each other and evolving in the space; but also, of the social groups identified within the dynamic crowd, and the dynamic crowd as a whole entity. The relevance of this research is that it derives a behavioral model simulation of the dynamics of the crowds from empirical observation so it will help to prevent crisis scenarios. Using and supporting simulations and visualization by efficient and ad-hoc data management, makes it possible to address real time dynamic crowd observation and organization.

Future research.

A future research, considering a framework which uses a plethora of metaheuristics and activate on that require adapting the social behavioral "in situ" and expanding the realistic way related with many people in fire buildings. We have described a model based on assumptions to induce a global behavior associated with emotions in a high risk scenario, this model produces promising initial results. We hope to improve the accuracy of the model, as well as to induce aspects of decision making under uncertainty with the emotional response that should be considered to improve the existing model. The results will support intelligent interfaces that will lead to more efficient searches and the moderators can reasonably integrate the existing community models. A BDI architecture associated with a Multiagent System and with an incremental vocabulary of emotions could better describe the situations of danger associated with scenarios with a high incidence of determinants related to paradigmatic changes in the perspective of a social group, as can be seen in Fig. 14.

Also, It is important to stress that as smart cities continue to grow, smart buildings will not be left behind. The issues of intelligent logistics should be applied in buildings through the Internet of Things (IoT), we want to propose as future work a flame sensor that detects fire and automatically send a signal to a microcontroller to activate an antifire system, and save the event data in a database for future analysis of information and perform an intelligent prediction system from data. As well as, send an alert to the administrator or person in charge of operations of the building, warning that fire is happening on a certain floor. As shown in Fig. 15.

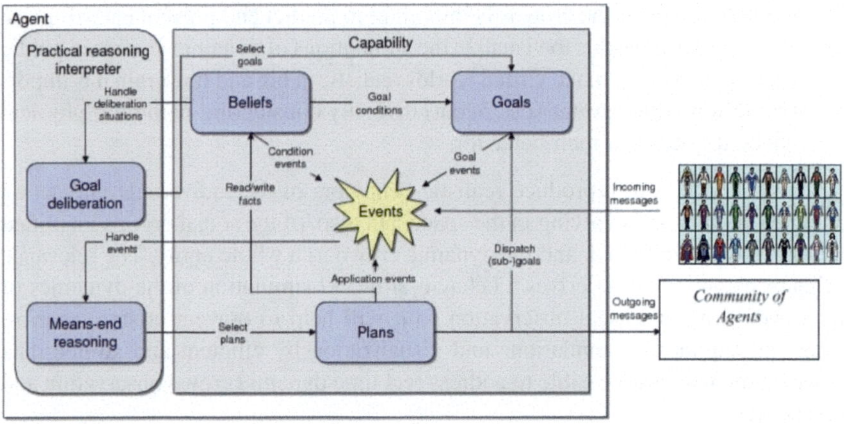

Fig. 14 Diversity of opinions in a BDI Architecture associates with a model of Community perspective

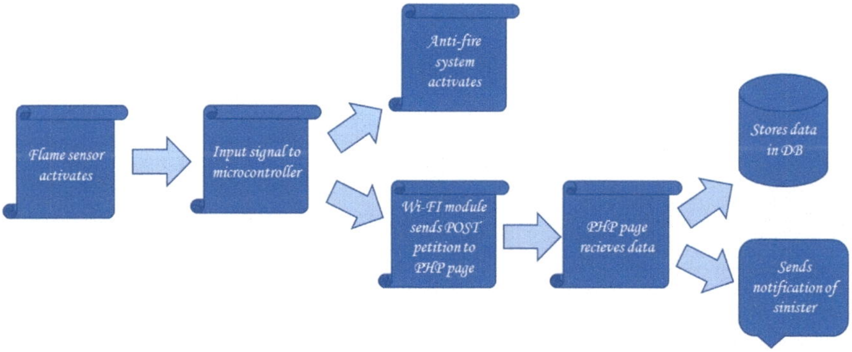

Fig. 15 IoT process for smart buildings

Acknowledgements We want to thank to organization CONACYT for the granted scholarship, and to organization of Protección Civil de Morelos for its comments.

References

1. Pan X, Han CS, Dauber K, Law KH (2007) A multiagent based framework for the simulation of human and social behaviors during emergency evacuations. AI Soc 22(2):113–132
2. Durupinar F, Allbeck J, Pelechano N, Badler N (2008) Creating crowd variation with the ocean personality model. In Proceedings of the 7th international joint conference on Autonomous agents and multiagent systems-Volume 3 (pp. 1217–1220). International Foundation for Autonomous Agents and Multiagent Systems
3. Durupinar F, Pelechano N, Allbeck J, Gudukbay U, Badler NI (2011) How the ocean personality model affects the perception of crowds. IEEE Comput Graphics Appl 31(3):22–31
4. Ochoa A, Vargas-Solar G, Espinosa-Oviedo JA, Rudomín I (2017) humanitarian logistics based on multiagent systems for the prevention of a plethora of disasters related with crowds
5. Oropeza P, Hurtado D, Gil E, Maldonado P, Ruiz M (2016) Characterization of a Passenger Vessel Sinking and Finally Foundering in Lacustrine Environment. Res Comput Sci 122:37–47
6. Su H, Wang X, Chen G (2009) A connectivity preserving flocking algorithm for multi-agent systems based only on position measurements. Int J Control 82(7):1334–1343
7. Pelechano N, Allbeck JM, Badler NI (2007) Controlling individual agents in high-density crowd simulation. In Proceedings of the 2007 ACM SIGGRAPH/Eurographics symposium on computer animation (pp. 99–108). Eurographics Association
8. Richmond P, Chimeh MK (2017) "FLAME GPU: Complex system simulation framework," In International conference on high performance computing & simulation (HPCS), Genoa, 2017, pp 11–17 https://doi.org/10.1109/HPCS.2017.12
9. Liao C, Zhu K, Guo H, Tang J (2019) Simulation research on safe flow rate of bidirectional crowds using Bayesian–Nash equilibrium. Complexity 7942483:1–7942483:15
10. Kielar PM, Borrmann A (2018) Spice: a cognitive agent framework for computational crowd simulations in complex environments. Auton Agent Multi-Agent Syst 32(3):387–416

E-commerce, Marketing and Mobile Application of Logistics Including Human Factor

Using Machine Learning to Predict Online Buying Behaviour, Wholesale and Fashion Marketing at Zara, an Analysis Including Z Generation

Stefania Piedrahita Orozco, Alberto Ochoa-Zezzatti, and Gustavo Delgado Lechuga

Abstract Zara aims to satisfy the needs of its customers at all times. Being a company located in the main cities of the world is not easy since each country is governed by different rules and customs, Zara in 39 years of life has become the number one in the retail distribution of textiles worldwide; in the first years that Zara was making incursions into fashion, it carried out an expansion activity extending to a national level with a scheme called oil stain [1], this technique consists of the expansion of a company in the market at a national level, first by opening establishments in cities near the headquarters and then by progressively opening establishments in more distant points; In this way, the Inditex group with its company Zara opened stores in different Galician cities and then expanded to the rest of Spain [2].

Keywords Machine learning · Fashion business analytics · Big data applied to textile organizations

1 Problem Statement

In the years 1975 and 1980 it opened in different Galician cities, and so they gradually reached more distant points, in 1983 and 1986 the first stores were opened in major Spanish cities such as Valladolid, Zaragoza, Barcelona, Seville, Valencia, Madrid, Bilbao, and Malaga. In 1988 began its internationalization in a country very close to the Portuguese city of Porto where the culture is very similar to that of Spain and the risks of failure were lower. In 1989 and 1990, Zara was situated in two of the world's fashion capitals, New York and Paris. Zara's use of new trends in data analysis through the information obtained by its clients' purchases online has been due to the strong and accelerated changes in the textile sector in recent years that have triggered

S. Piedrahita Orozco · G. Delgado Lechuga
Maestria en Big Data, Universidad Cuauhtemoc, Aguascalientes, Mexico

A. Ochoa-Zezzatti (✉)
Universidad Autónoma de Ciudad Juárez, Av. Hermanos Escobar Omega, 32410 Cd Juárez, Chihuahua, Mexico
e-mail: alberto.ochoa@uacj.mx

© The Author(s), under exclusive license to Springer Nature Switzerland AG 2021
A. Ochoa-Zezzatti et al. (eds.), *Technological and Industrial Applications Associated with Intelligent Logistics*, Lecture Notes in Intelligent Transportation and Infrastructure,
https://doi.org/10.1007/978-3-030-68655-0_19

alterations in the traditional policies of large companies and led it to implement the new strategies with competitiveness [3], in mitigation the economic crisis is presented that forces the company to remain attentive and propose communication strategies, marketing, image, quality, and customer service with greater dynamism, focusing on the optimization of its resources to keep the company positioned against future competition in the market. It also has a department of professionals who are looking for all the social trends that customers are looking for; these collaborators frequent discotheques, shopping areas, London Ups, chic districts of Milan [3], among other places, where fashion can be detected in order to quickly and easily include loyal customers within the stores. It is not then the designers who set the fashion trends within Zara, it is really the customers who are using different styles of clothing that set the trend within the design of the same company.

Meanwhile, for 2007 the company Zara carried out the first online projections of the group with the presentation of the first Zara Home store through which it collected the first online data of its buyers in relation to sizes, colors and garments in each of the stores [1], having such information allowed the company to store millions of data around the purchases generated by its customers, information that would have to be processed and analyzed in order to generate new trends and from 2015 onwards, large volumes of data are generated, so it is possible to think about the predictions of e-commerce sales in ZARA company according to the areas of location of each of the stores using the data set containing products, color, size, sales and stock provided on the official website of the company [4]. The research will be developed in a time frame of 4 months that will allow to execute the Machine learning technique to predict sales data and fashion trends for the Zara company using the WEKA platform an integrated development environment, being a software platform for machine learning and data mining. The main source of information within the research revolves around the data generated by the Zara company in its official pages which record the evidence of the work carried out around sales, marketing and fashion trends carried out in other ways within the same company. This is how it becomes the main source of study where the data downloaded from the same page of the company officially available on the web for studies that decide to carry out as it is. Predict Zara's e-commerce sales according to the location areas of each of the stores using data sets containing products, color, size, quantity of online purchases, stock, quality of perception, fashion style, average age, symbolic capital, store country, difficulty of online purchase and Zara's company continent, provided on the official website of the company, applying the information through machine learning, to determine future behavior in relation to products, color, size, quantity of online purchases, stock, quality of perception, fashion style, average age, symbolic capital, country of the store, difficulty of online shopping and continent the company Zara; In this way, the information obtained through machine learning is analyzed, and the correlation between the products, color, size, sales and stock variables of the Zara company is determined, establishing the sales and fashion trend.

1.1 Research Questions

Does the e-commerce model project a sales trend in the Zara company?

Is it possible to predict Zara's e-commerce sales according to the location areas of each of the stores using data sets containing products, color, size, quantity of online purchases, stock, quality of perception, fashion style, average age, symbolic capital, store country, difficulty of online purchase and Zara's company continent provided on the company's official website?

Applying the information through machine learning, to determine future behavior in relation to products, color, size, quantity of online purchases, stock, quality of perception, fashion style, average age, symbolic capital, country of the store, difficulty of online purchase and continent of the company Zara?

Once applied Machine Learning is it possible to analyze the information obtained to predict sales, and future trend behavior in relation to products, color, size, quantity of online purchases, stock, quality of perception, fashion style, average age, symbolic capital, country of the store, difficulty of online purchase and continent of the company Zara?

Is there a correlation between the variables of products, color, size, quantity of online purchases, stock, quality of perception, fashion style, average age, symbolic capital, country of the store, difficulty of online purchase and continent of the company establishing the sales trend?

1.2 Justification

This research sets fashion trends at Zara thanks to the prediction models generated with the data from the purchases of different customers. This fashion trend will be based on people's tastes but not on social impositions. Likewise, the company avoids consumer travel by optimizing the time spent shopping at online stores since the products advertised are focused on the customer's taste. In addition, the company will generate products that are related to the customer. On a theoretical level, the research uses the Machine learning tool to carry out practices through the use and study of Zara's company, generating a methodological structure of production and sale according to the data analysed, causing a reorganisation in the company's production system that is feasible in terms of the consumer's needs. Zara, being a company with presence in all continents, manufactures men's, women's and children's clothing at reasonable prices to sell in mass with an acceptable quality [5]. In addition, Zara's store managers report via internet to the corporate headquarters exactly what has been sold so that they can make decisions on the new production lines they are trying to introduce around the world and according to the needs of the consumer [1], in line with the needs to generate good clothing around the world and in line with fashion trends, react faster than their competitors to capture the trends that are being generated in the world according to climate and fashion changes. Therefore, it is necessary a constant analysis to the sales reports generated in each one of the stores to lead the company to the manufacture of clothes coherent with the needs of the consumer and to fidelize their purchases.

Zara's success in the countries where it is located is focused on promotion and distribution, they do not invest in advertising what they do is invest in the stores, in the location, decoration using large windows, with wide and wide doors facilitating the entrance to the store so that the income is more fluid and they invite people to enter giving a better image of the brand. Zara attends the new collections of the most important designers in the world so they imitate their clothes before they are launched to the market, handling more economical materials and making modifications to the garments.

1.3 Hypothesis

H0. The sale of the product depends on the colour and size determining the e-commerce sales trend of the Zara company.

H1. The sale of the product does not depend on color and size and does not determine the e-commerce sales trend of the Zara company.

1.4 Description of the Thesis Organization

The research will be developed in the framework of four chapters, thus initiating Chapter 1 Formulation of the problem, understanding the approach of the problem as the possibility to delimit the objectives of research, research questions, justification of the research and evaluation of the evidence in the knowledge of the problem, this according to [6] in chapter II theoretical framework being a development of the theoretical perspective the literature is reviewed to detect, obtain and consult pertinently carrying out extraction of the information of interest [6], in Chapter III Methodology establishes the scope of the research through the choice and design of the instrument to be applied within the population sample in Chapter IV Discussion of results, finally in Chapter V, Conclusions and Future Work tends to the evaluation of reliability, validity of measurement instruments to analyze and interpret through machine learning algorithms the hypotheses raised [6]. The following is a general outline of the concepts inherent to the case study:

Conceptual map of the theoretical framework.

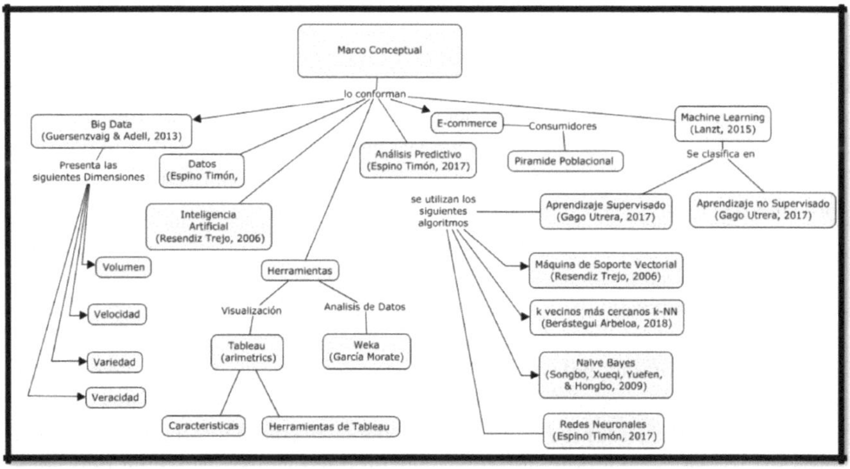

1.4.1 Big Data

The term Big Data is used to refer to "large digital data sets that require computer systems to capture, store, search, manipulate and display them" [7].

Dimensions of Big Data

When we speak of Big Data, the term is usually associated with the processing of large volumes of data, however, Big Data is an approach that encompasses other dimensions. They are briefly named and described below. For some authors, Big Data has become the 4Vs approach [8].

- Volume: accumulation and constant increase of data.
- Speed: necessary to generate, obtain and process the data.
- Variety: of origin, source and format of the data.
- Veracity: reliability of the data.

1.4.2 Data

The data is the source from which the variables are obtained, the relations between them, the knowledge induced or the behaviour patterns identified, becoming a vital element of all predictive analysis [9].

1.4.3 Artificial Intelligence (AI)

Artificial Intelligence (AI) is the branch of computer science that develops processes that imitate the intelligence of living beings. The main application of this science is the creation of machines for the automation of tasks that require intelligent behavior. This is one of the simplest concepts of AI framed fundamentally to the computer approach [10].

1.4.4 E-Commerce

E-commerce is a commercial transaction where the parties involved make purchases and sales of different products through the Internet. This allows merchants to reduce intermediaries and costs for the sale of their products. E-commerce has become a sales alternative for physical stores, offering their products online, allowing consumers ease of purchase and reduced time, travel, thus increasing the use of technology and encouraging buyers with different promotions of their products such as cyberday, cyber weekend, black Friday among others. People buy online because online stores have a 24 h a day, seven days a week, allowing purchases at any time of day, in this case saving time on mobility, with online shopping at different stores consumers have the ease of comparing prices on different websites contributing to economic savings and convenience of choosing a product to another, in the online stores they offer exclusive products for online purchase, this means that in the physical store the product is not displayed, in the different platforms you can see that it is much easier to visualize the catalog, this way the buyer has more options to choose the product to his liking, today people can also make purchases worldwide (Fig. 1; Table 1).

Population Pyramid Colombia 2019

1.4.5 Predictive Analysis

Predictive analysis is an area of data mining that consists of the extraction of existing information from the data and its use to predict trends and behaviour patterns, which can be applied to any unknown event, whether in the past, present or future. Predictive analysis is based on the identification of relationships between variables in past events, and then exploiting those relationships to predict possible outcomes in future situations [9]. In order to carry out predictive analysis, a large amount of data must be available, data from the present and the past, in order to establish behaviour patterns and generate knowledge from the analyzed data.

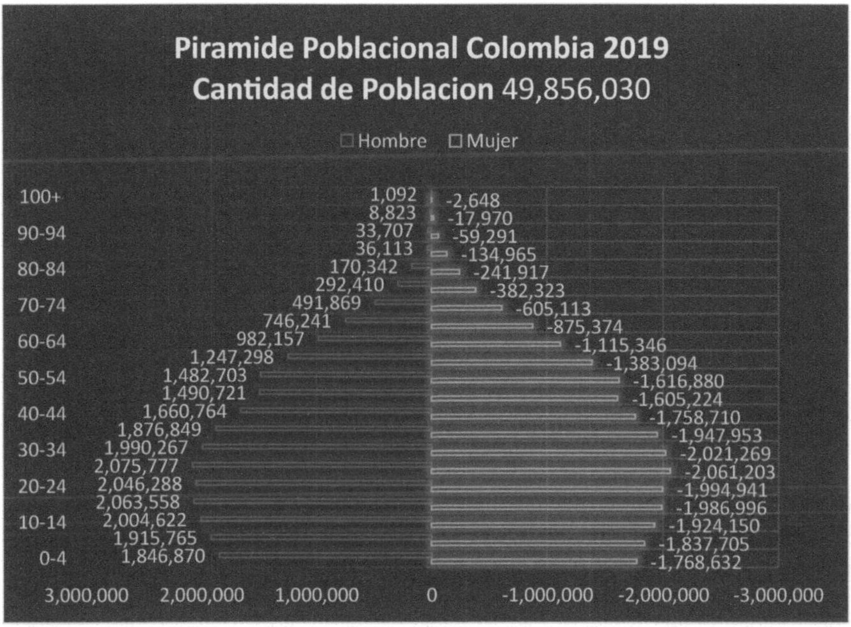

Fig. 1 Population pyramid of Colombia 2019. *Source* Own elaboration

Table 1 Population data of Colombia 2019

Age	Male	Female
100+	1,092	−2,648
95–99	8,823	−17,970
90–94	33,707	−59,291
85–89	36,113	−134,965
80–84	170,342	−241,917
75–79	292,410	−382,323
70–74	491,869	−605,113
65–69	746,241	−875,374
60–64	982,157	−1,115,346
55–59	1,247,298	−1,383,094
50–54	1,482,703	−1,616,880
45–49	1,490,721	−1,605,224
40–44	1,660,764	−1,758,710

(continued)

Table 1 (continued)

Age	Male	Female
35–39	1,876,849	−1,947,953
30–34	1,990,267	−2,021,269
25–29	2,075,777	−2,061,203
20–24	2,046,288	−1,94,941
15–19	2,063,558	−1,986,996
10–14	2,004,622	−1,924,150
5–9	1,915,765	−1,837,705
0–4	1,846,870	−1,768,632

Source Own elaboration

1.4.6 Machine Learning (ML)

According to [11] defines Machine Learning as a subfield of computer science and a branch of Artificial Intelligence whose objective is to develop techniques that allow computers to learn, i.e. to generalise behaviour and knowledge from information provided in the form of an example. According to [12] Machine Learning is a data analysis method that automates the construction of analytical models. It is a branch of artificial intelligence based on the idea that systems can learn from data, identify patterns and make decisions with minimal human intervention. These methods were born from its application to pattern recognition and from the idea that machines can learn without being programmed to perform specific tasks. One of the most important aspects of the ML is the fact that they are interactive methods, i.e. as models are exposed to new data they can adapt independently. They learn from previous computations to produce reliable and repeatable decisions and results.

Machine learning algorithms are classified into two classes as mentioned [12] in your article:

Supervised Learning

This type of algorithm makes predictions based on a set of examples. With these techniques we have an input variable that consists of a set of labeled training data and a desired output variable to analyze the training data uses an algorithm to learn the function that assigns the input to the output. This inferred function maps new and unknown examples generalizing from the training data to anticipate results in unseen situations. In the supervised learning [13] indicate that the data have additional attributes that are tried to predict within this category highlights the qualification algorithms in which the samples are labeled as belonging to two or more classes and you want to learn to predict the class of data without labeling in turn, within this class of algorithms can be made the following classification

- Classification algorithm when data are used to predict a categorical variable, supervised learning is also called classification. This is the case when a label or indicator is assigned. When there are only two labels it is called binary classification and when there are more than two the problems are called multi-class classification.
- Regression algorithm by predicting continuous values, the problems become regression problems.
- Forecast algorithm this is the process of making predictions about the future based on past and present data. It is most commonly used to analyze trends.

Within Supervised Learning we find the following models:

Vector Support Machines

They are a modern and effective AI technique, applied mainly to the processing of large amounts of information that has had a formidable development in recent years. One of the fundamental concepts in this technique is the Support Vector (VS) algorithm, which is a non-linear generalization of the Generalized Semblance algorithm, developed in Russia in the 1960s. The development of VS brings with it the emergence of Vector Support Machines. These are learning systems that use a space of hypotheses of linear functions in a space of larger features, trained by an algorithm from the optimization theory [10].

Minimization of Empirical Risk and [[Vapnik-Chervonenkis Dimension]] are fundamental in Vector Support Machines. Put more simply, the algorithm focuses on the general problem of learning to discriminate between positive and negative members of a given n-dimensional vector class. MSVs belong to the family of linear classifiers. Using a mathematical function called kernel, the original data is resized to look for a linear separability of the data. A characteristic of the MSV is that it maps the input vectors to determine the linearity or non-linearity of the cases which will be integrated into the Lagrange Multipliers to minimize the Empirical Risk and the Vapnik-Chervonenkis Dimension. In general, the Vectorial Support Machines allow to find an optimal hyper plane that separates the classes [10].

The K-Nearest Neighbors Method

It is one of the most known classification algorithms that exists, it belongs to the class of statistical methods of pattern recognition. This algorithm classifies a set of examples from others of which we already know their class and we will call them training set or train, this training set contains positive and negative data. What it does is calculate the distance of each example to classify with all the examples of the train set and classifies the example according to the class to which those closest examples belong, the major sign of this point will determine the classification of the sample. The variable k is used to determine how many of its closest examples of the train set have to be taken into account to classify it [14].

According to [9] the performance of the k-NN model is influenced by three main factors:

- The measure of distance used to locate the nearest neighbors.
- The decision rule used to derive a classification of the k-neighbors.
- The number of neighbors used to classify the new sample.

The Naïve Bayes Method

It is based on Bayes' theorem and the premise of independence of attributes given a class. It is also one of the most widely used supervised learning methods because it can be adapted for the analysis of emotions [15].

The Bayesian classifier is based on Bayes' conditional probability rule, which is used for the classification task. The Bayesian classifier assumes that the predictors are statistically independent, which makes it an effective classification tool that is easy to interpret. It is best used when faced with the problem of the "dimensionality curse", that is, when the number of predictions is very high [9] when Bayesian classifiers have been applied in large databases, they have demonstrated high accuracy and speed.

Neural Networks

Neural networks are sophisticated non-linear modelling techniques that are capable of modelling complex functions. They can be applied to prediction, classification or control problems in a wide range of fields such as finance, cognitive psychology/neuroscience, medicine, engineering and physics [9]. These neural networks are able to detect and learn patterns and characteristics of complex data.

Unsupervised Learning

In this case the machine is presented with unlabeled data. It is asked to discover the intrinsic patterns underlying the data, such as a grouping structure, a low-dimensional variety, or a scattered tree and graph. [13] indicate that in unsupervised learning the training data consist of a set of input vectors without any corresponding labeled value.

1.4.7 Technological Tools

Weka

In the Weka manual by (García Morate) describes Weka, a bird endemic to New Zealand. This endangered Gallinacea is famous for its curiosity and aggressiveness. Brown in appearance and similar in size to a hen, the wekas feed mainly on insects and fruit (Fig. 2).

Fig. 2 New Zealand endemic bird. *Source* (García Morate). Retrieved from https://knowledgesociety.usal.es/sites/default/files/MANUAL%20WEKA.pdf

This bird gives its name to an extensive collection of Knowledge Machine algorithms developed by the University of Waikato (New Zealand) implemented in Java [16]; useful to be applied on data through the interfaces it offers or to embed them within any application. In addition, Weka contains the necessary tools to perform data transformations, classification tasks, regression, clustering, association and visualization. It is a complete tool and its graphic interface is quite intuitive so it is easy to adapt to the user, its interface is composed of 4 elements, the explorer which allows to easily load a set of data and apply a learning algorithm giving several results to be interpreted, there is the Knowledge Flow which allows to drag elements which allows to understand how the data cycle is, the experimenter which allows to analyse different algorithms with different parameters helping the decision making, finally there is the workbench which integrates the 3 previous ones, being developed in java it allows easy connection with databases both structured and unstructured [17]. Weka has a file format with an arff extension which has a simple structure: It consists of an @RELATION label which defines the name of the dataset, with @ATTRIBUTE where the attributes contained in each row of the dataset are defined, these attributes can be of any type (numeric, strings, characters, among others) and the @DATA label where it tells weka where the data is located in order to work with it, It is important to note that the attributes are separated by commas so it is recommended to put the String type attributes between single or double quotes, also allows you to work with Excel files and a large number of other file types [17].

Tableau

It is a data analysis software with an excellent visualization and presentation layer, considered by many as one of the best programs for the visual presentation of data and with a very high classification in the ease of use, so it follows very closely Microsoft Excel. Especially used for Business Intelligence it simplifies data to present it in a compressible and intuitive format.

Tableau Features

Tableau allows you to drag and drop large amounts of data onto a digital "canvas" and make graphs instantly. The idea behind how Tableau works is that it's easier to manipulate what's going on in the interface so you can see what you're doing right and, by extension, what you're doing wrong. Tableau has been defined as a more advanced form of the pivot table (arithmetic's).

Tableau Tools

- Desktop Tableau. The desktop version of this software.
- Tableau Server. You can share the information generated in the desktop version.
- Tableau Online. The server cloud version. You do not need to install the program to work with the software.

1.5 Legal Framework

The general regulations presented in the study are generally collected through the following concept map:

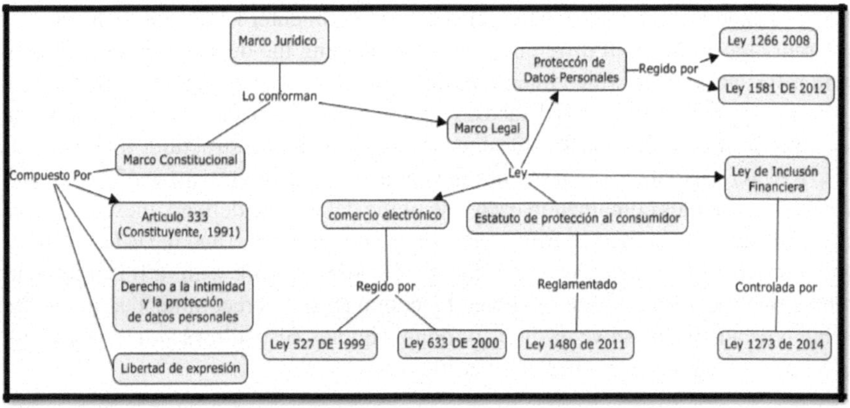

Conceptual map of the legal framework.

1.5.1 Constitutional Framework

It describes the rights that people have to carry out economic activities in e-commerce, information and communication technologies.

Article 333

Economic activity and private initiative are free, within the limits of the common good. No one may demand prior permission or requirements for their exercise without the authorization of the law. Rules establishing limits for electronic commerce:

In principle, no rules are required in addition to those that already exist to regulate existing economic activities.

- Civil and commercial law (contractual, corporate, etc.),
- Misleading advertising,
- Habeas data.

Exceptionally, there are some special rules applicable to e-commerce. Problem of regulating e-commerce and, in general, information and communication technologies:

- Constant progress
- The process of legislation and regulation is static
- Advances in technology and computing are dynamic.

Right to Privacy and Protection of Personal Data

- Right to privacy and good name
- Right to habeas data (protection of personal data)
- Inviolability of correspondence and other forms of private communication
- They can only be intercepted or recorded by court order, in the cases and with the formalities established by law.

Freedom of Expression

- Freedom to express thoughts and opinions.
- Right to inform and receive truthful and impartial information
- Right to establish mass media.
- Right to rectification under fair conditions.
- There will be no censorship [18].

1.5.2 Special Rules

Each of the national authorities with consumer protection functions has special regulations, which are the fundamental basis for the development of their powers and activities in this area. (Consumer., n.d.).

Law on Electronic Commerce

Law 527 of 1999—E-Commerce Law

It establishes the "principle of functional equivalence", Equivalence between:

- Electronic signature and autograph signatures
- Data messages and written documents
- It establishes rules for the certification of digital signatures and for the creation of Certification Entities.

Law 633 of 2000

All web pages and Internet sites of Colombian origin that operate on the Internet and whose economic activity is of a commercial, financial or service provision nature must be registered with the Commercial Registry and provide the National Tax and Customs Directorate (DIAN) with information on economic transactions under the terms required by this entity.

Consumer Protection Statute

Where the laws that favor and protect consumers in Colombia are related.

Law 1480 of 2011

Special obligations for suppliers of goods and services that are offered using electronic means:

- Special information obligations (identification of the supplier, characteristics of the goods, available means of payment, text of the contract, etc.)
- Duties to retain information
- Form for filing requests, complaints and claims
- Link to Consumer Protection Authority

  Reversal of payment:

- When the consumer is the object of fraud, or corresponds to an unsolicited operation, or the product acquired is not received, or the product delivered does not correspond to what was requested or is defective, the consumer will have the right to a reversal of payment.
- Decree 587 of 2016

  Protection of children and adolescents in electronic commerce.

- The supplier shall take all possible measures to verify the age of the consumer.
- Advertising aimed at minors: Decree 975 of 2015

Contact portals (persons who make available to the public an electronic platform where people offer products for marketing and consumers can contact them by the same mechanism).-) It must require from all offerors' information that allows their identification, which can be consulted by anyone who has purchased a product in order to file a complaint or claim and must be provided to the competent authority upon request [19].

Financial Inclusion Law

Law 1735 of 2014

- Creates a Society Specialized in Electronic Deposits and Payments (SEDE)
- Under capital.
- More flexible financial institutions.
- SEDPEs will offer low-cost wire transfers using new technologies.
- It will allow payments and transfers through cell phones, internet and debit cards.
- With accounts that are exempt from the tax on financial movements (four per thousand).

Personal Data Protection

The laws that constitute the general framework for the protection of personal data in Colombia are described.

Law 1266 of 2000

Special regime for financial services - databases for calculating the credit risk of individuals.

- Decree 1727 of 2009.
- Decree 2952 of 2010.

Law 1581 of 2012

General regime for the protection of personal data (Table 2).

- Decree 1377 of 2013.
- Decree 886 of 2014.

Research
Name of the job:

Table 2 Analysis of related research

Trabajos relacion ados	Orientación del estudio		Técnicas de Predicción de ventas con machine learning										
			Fuentes de datos corpus	Conjunto de datos Dataset									
	Cuantitativa	cualitativa		Fuentes de datos Dataset	TF	TF-IDF	Tokenizar	Data	csv	Sinonimos	Crawier	Genetico	Multiagente
(Adamsme and Salómon Díaz [22])	✓	✓	Tienda	Compras de Tienda por departamentos				✓					
(Pallres Cabrera [23])	✓	✓	Hoteleria	BD Hotel				✓					
(Paz Villaseca, [24])	✓		Demanda. Eléctrica	Distribución Eléctrica					✓				

Técnicas de Predicción de ventas con machine learning						
Aprendizaje supervisado						
Bayes	Algoritmo A	Arboles	Redes	FP growth	algoritmo GSP	Herramientas de minería de Datos
	✓			✓	✓	Rapid Miner
			✓			Desarrollo Propio
			✓			Minitab-matlab-R-Studio

Own elaboration

1.5.3 Development of a Model Based on Machine Learning for the Prediction of Room Demand and Occupancy in the Hotel Sector

Year and place:

2014, Cartagena Colombia.

Author: Fabián Pallares Cabrera.

Objectives: To develop and validate a model based on Machine Learning for the prediction of room demand and occupancy in the hotel sector.

Method: Quantitative—qualitative

Data Collection Technique:

For the application of the techniques, three sets of data were constructed, one set for training, one for validation and the last one for testing. The data are collected from the operations of reservations and daily occupation of a hotel, so that the results of this work are based on real information. The sample is 7 years old, starting from July 1, 2008 to June 30, 2014 for a total of 2191 days of occupancy in analysis. Additionally, other independent variables are included in the modelling, such as the days of the week, holidays, months of the year, and the number of holidays before and after the date of occupancy. Finally, information on reservations made 7, 10, 15, 20, 30, 60, 90 days in advance is included.

Results:

The results found show the superiority of the Ridge regression technique in the occupation forecast, using the dataset construction scheme based on the reservations observed with 90, 60, 30, 20, 15, 10 and 7 days of anticipation, adding information of holidays, seasons, weekdays and months of the year. The ASM error found is 8.2012 on the validation set and 8.656144 on the test set, these results are very approximate so it is a good estimate that there is no overfitting. Other findings of this research show that adding to the time series model additional variables such as the day of the week, season and holidays, there are more accurate forecasts (it should be clarified that the improvement in forecasts was quite modest), also there is a better performance of the Kernel Ridge Regression technique against multi-layer perceptron type Neuronal Networks and Neuronal Networks Radial Base Function in time series modelling. Likewise, the results show the superiority of Radial Base Function Neural Networks over Multilayer Perceptron Neural Networks in the context of our application. Name of the research.

1.5.4 Deep Learning Tools for Predictive Analytics

Year and place: 2017, Concepción—Chile.

Author: Daniel Isaac Paz Villaseca

Objectives: To propose a Neural Network application that works within the scope of Predictive Analytics with Deep Learning, whose purpose will be to establish guidelines or decisions regarding the resulting information, using computer tools creating the algorithm of the Neural Network Perception, taking into account a set of data from a Time Series.

Method: Quantitative

Data Collection Technique:

The study database corresponds to Monthly Series of Electrical Generation and Distribution from December 1997 to December 2011 (National Institute of Statistics 2017).

Results:

An evaluation was made regarding the performance of the neural networks and their capacity to work on a series, obtaining as a result an important technological tool to make predictions. The results are satisfactory when checking a better prognosis using a recurrent neuronal network over a traditional Multilayer Perceptron with Backpropagation since the recurrent network, presents a lower processing time. It is observed that when selecting the training algorithm and the optimal network topology, there is a limited but useful insight into the shortcomings of neural network-based models, but it does not eliminate the fact that when the precise neural network is available, it manages to be an "optimal" tool for prognosis.

Name of the research.

1.5.5 Analysis of Retail Shopping Patterns Using Business Intelligence Year and Place: 2013, Santiago de Chile

Authors: Daniel Adasme Alarcón, Eduardo Salomon Diaz

Method: Quantitative—Qualitative

Data Collection Technique:

The database was extracted from the purchases of a department store in London for the years 2009 and 2010. It is an anonymous database so no personal data of the customer is shown. It has 41821 observations classified in 17 columns.

Results:

As a result of this process, the sets of attributes that are most frequently found in the database can begin to anticipate the purchasing behaviour of consumers in the store. In 88% of the database, the category "Single Order" is present within the attribute Order Quantity, and in 72.8% of the database, the category "Dark" and in 63.9%, the relationship between these two categories, that is to say, that one occurs and the other, Certain attributes have categories that have too much presence within the database, and although this presence provided fundamental primary data to be able to describe certain general characteristics of the store's customers, the presence of these categories hinders the analysis of the relationships of the rest of the attributes, so a new analysis was carried out, analysing separately the categories of attributes with high presence in the database, performing different analyses. The majority of the products purchased are from the lower part of the body, with dark colours predominating in this case.

- If light colours are chosen, customers prefer to buy clothes towards the upper body, such as T-shirts and shirts. However, buying light colours in the shop is rare.
- Among the products dedicated to the upper body, the category Shirts, T-shirts and Casual Tops stands out.
- Shopping is predominant in the spring season.
- Both shoes and accessories are not a determining factor in the products that consumers regularly buy, however, they are products that are regularly found in a generalised clothing store.
- Clearly dark colours are preferred by consumers in the store, so much so that once they have purchased dark products, shoppers in the future may purchase products (including upper body products) in dark shades, especially the colour black.
- Most people buy large size clothes; this may be due to the incipient obesity that is facing England that has quadrupled in the last 25 years reaching 23.1% of the population in 2012.
- There are few loyal shoppers. This is reflected in the low customer rebound rate. Only 12% of customers bought more than once in the period.
- The colour black when trapping the light diminishes the silhouette formed by the body, thus stylising the figure. This corroborates the thesis of loose clothing in dark colors, which in turn is represented in the data.
- There are several studies that show that the mental state of a person is affected by climatic conditions (among other biological, organic and genetic factors). London being one of the rainiest cities in the world with 754 mm of annual rainfall and where the skies are regularly cloudy, consumers are more likely to wear dark-colored clothing as shown in the data. Light colors are also more likely to be purchased in the spring as well as in the summer when temperatures increase and the climate improves.

2 Methodology

A predictive analysis of the sales and fashion trends of the Zara Company will be carried out, based on a database provided on the official website of the company under study, using Machine Learning tools.

2.1 General Objective

Predict e-commerce sales and fashion trends in the ZARA company according to the location areas of each of the stores in the world using correlations between variables through data sets containing products, color, size, quantity of online sales, stock, quality of perception, fashion style, average age and symbolic capital, store country and continent provided on the official website of the company.

2.1.1 Particular Objectives

To carry out analysis through machine learning, to determine the future behavior of products, color, size, quantity of online sales, stock, quality of perception, fashion style, average age, symbolic capital, country of the store, and continent of the Zara company.

To determine the correlation between the variable's products, color, size, quantity of online sales, stock, quality of perception, fashion style, average age, symbolic capital of the Zara company establishing the sales and fashion trend.

Establish which is the most assertive supervised learning machine through the highest correlation levels in the application of prediction of Linear Regression models, K-nearest neighbors, neural networks and Naïve Bayes for the company Zara.

2.2 Participants

Online buyers of the different textile products of the company Zara in 127 countries around the world, generating a database containing 10,001 records of product sales transactions. This data set contains attributes such as products, color, size, quantity of online sales, stock, quality of perception, fashion style, average age, symbolic capital of the company under study. For this analysis, a 95% confidence level and a 5% margin of error are considered.

2.3 Scenario

Data was taken from the Zara website https://www.zaratalent.com/data.html, where the different textile products manufactured by the company under study are registered. The company is located in 127 countries, and there is a history of transaction data for the sales of the different products offered by this company worldwide.

2.4 Information Collection Instruments

Data is collected through a web page offered by Zara to its customers, which is accessible through electronic equipment such as computers, tablets or mobile devices, thus recording the transactions of online purchases generated by consumers.

2.5 Procedure

For the research work, the database was consulted through Zara's official web pages, which details the records of its product sales and other attributes such as products, color, size, quantity of online sales, stock, quality of perception, fashion style, average age, symbolic capital, country of the store, continent, among others, provided on the company's official web page. that contribute to the analysis of the case study.

2.6 Method Design

The type of research design in this case study is non-experimental, according to the definition of [6] where it indicates that it will be a process in which there will be no deliberate manipulation of the variables, therefore only the phenomenon will be observed which in this case are the data that are recorded in the data set to be later analysed through the different correlation models.

2.6.1 Design

Defined since (Supo 2016) as the criterion for the external or internal evaluation of the results obtained in the instrument and the design according to [6] it is then a non-experimental, transectional correlational causal design, which is used is the following (Table 3).

It is necessary to indicate that relationships are established between variables without a sense of chance or to analyze the causal relationships [6], therefore, if a non-causal relationship is sought, the correlational approaches and hypotheses are being taken into account, and will be developed through a single moment.

2.6.2 Timing of the Study

As it is a non-experimental research, it will be developed under a transversal design since data is collected in a single moment, data that was registered in Zara's database

Table 3 Correlational-Causal transectional design [6]

Variable X	Variable Y
X_1 →	Y_1
X_2 →	Y_2
X_3 →	Y_3

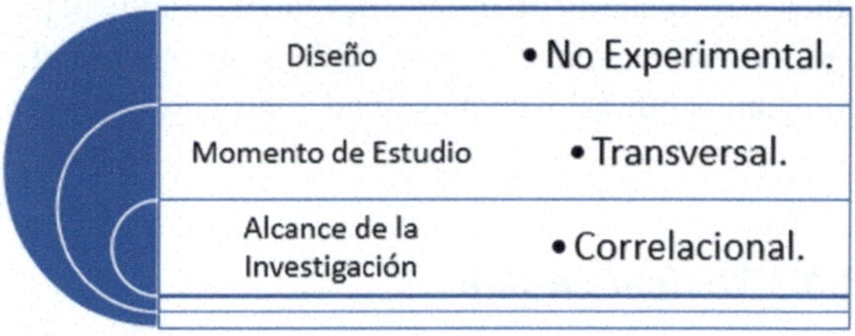

Fig. 3 Experimental design of the research. Own Elaboration

during several years through the online purchases made by the users, having as purpose to describe and analyze those variables to interrelate through a given moment.

2.6.3 Scope of the Study

This study is not experimental but transactional and its scope is to correlate the variables containing products, colour, size, sales, average age, location, price, ideal price, quality perception, quantity of online purchases, symbolic stock capital and fashion style of the company ZARA according to the areas of location of each of the stores using data set provided on the official website of the company (Fig. 3).

When running the Weka platform we must enter the Explorer option (Fig. 4).

The Explorer mode is the most used and most descriptive mode*. It allows you to perform operations on a single data file. The main window is shown in Fig. 5.

2.7 Operationalisation of the Variables

The study variables described in Table 4 have been treated at the ordinal-cardinal measurement level with the existence of several categories and with the maintenance of a logical order [6], where the labels indicate hierarchy and intervals with equidistant values between labels (Coronado Padilla 2007).

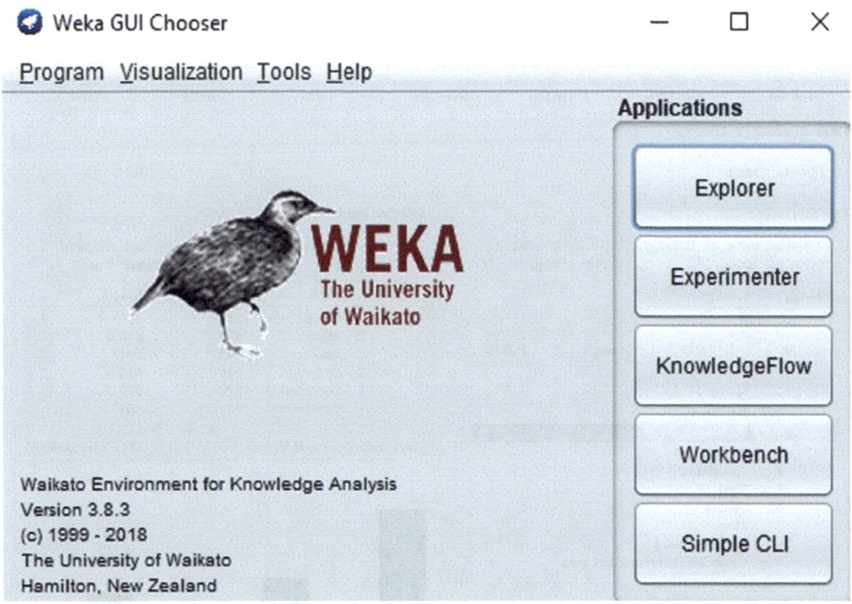

Fig. 4 Initial screen of the Weka application. *Source* [21]

The data was taken from Zara's website, https://www.zaratalent.com/data.html, being a historical data of transactions of the sales of the different products offered by this company worldwide.

Product Preference Determine in the system the type of Product ID, ColorBase and Size_id that the customer buys online? 1, 2, 3, and 4.

Fashion trend What is the buyer's age, seasonal preferences and the product's symbolic capital? 1, 8, 9, and 10.

Quality of price Do you consider that the price of the garment is in line with the quality? 1,5,6, and 7. Own elaboration.

2.8 Data Analysis

In the present investigation data analysis will be used under the use of the program weka being a tool that allows the experimentation of data analysis by means of the application, analysis and evaluation of the most relevant techniques of data analysis, mainly those coming from the automatic learning, on any set of data of the user [20] in which supervised learning will be carried out by carrying out K nearest neighbours, neural networks and Naive bayes, a vectorial support machine. Linear Regression.

They are a modern and effective AI technique, applied mainly to the processing of large amounts of information that has had a formidable development in recent

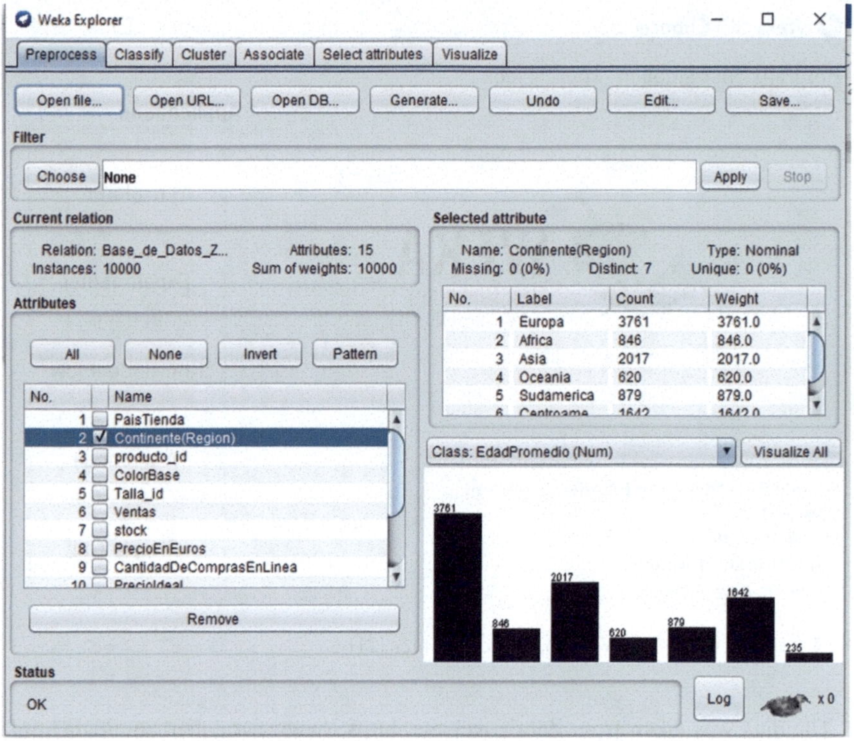

Fig. 5 The explorer mode. Main Window

Table 4 Dependent and independent variables

Independent Dependent Instrument Dimensions Indicators Item
CountryStore continent
Product_ID
ColorBase
size_id
PriceinEuros
QuantityshopOnline
QualityPerception
SymbolicCapital
FashionStyle
AverageAge
Online Experience
Difficult to buy online

years. One of the fundamental concepts in this technique is the Support Vector (VS) algorithm is a non-linear generalization of the Generalized Semblance algorithm, developed in Russia in the 1960s. The development of VS brings with it the emergence of Vector Support Machines. These are learning systems that use a space of hypotheses of linear functions in a space of larger features, trained by an algorithm from the optimization theory [10]. The Minimization of Empirical Risk and the Vapnik-Chervonenkis Dimension are fundamental in Vector Support Machines. Put more simply, the algorithm focuses on the general problem of learning to discriminate between positive and negative members of a given n-dimensional vector class. MSVs belong to the family of linear classifiers. Using a mathematical function called kernel, the original data is resized to look for a linear separability of the data. A characteristic of the MSV is that it maps the input vectors to determine the linearity or non-linearity of the cases which will be integrated into the lagrange multipliers to minimize the Empirical Risk and the Vapnik-Chervonenkis Dimension. In general, Vector Support Machines allow finding an optimal hyper plane separating classes [10]. The k-Nearest Neighbors method (k-NN): It is one of the best known classification algorithms that exists, belongs to the class of statistical methods of pattern recognition. This algorithm classifies a set of examples from others of which we already know its class and which we will call training set or train, this training set contains positive and negative data. What it does is calculate the distance of each example to classify with all the examples of the train set and classifies the example according to the class to which those closest examples belong, the major sign of this point will determine the classification of the sample. The variable k is used to determine how many of its closest examples of the train set have to be taken into account to classify it [14]. According to [9] the performance of the k-NN model is influenced by three main factors:

- The measure of distance used to locate the nearest neighbors.
- The decision rule used to derive a classification of the k-neighbors.
- The number of neighbors used to classify the new sample.

The Naïve Bayes method:

It is based on Bayes' theorem and the premise of independence of attributes given a class. It is also one of the most widely used supervised learning methods because it can be adapted for the analysis of emotions [15].

The Bayesian classifier is based on Bayes' conditional probability rule, which is used for the classification task. The Bayesian classifier assumes that the predictors are statistically independent, which makes it an effective classification tool that is easy to interpret. It is best used when faced with the problem of the "dimensionality curse", that is, when the number of predictions is very high [9] when Bayesian classifiers have been applied in large databases, they have demonstrated high accuracy and speed.

Neural Networks:

Neural networks are sophisticated non-linear modeling techniques that are capable of modeling complex functions. They can be applied to prediction, classification or control problems in a wide range of fields such as finance, cognitive psychology/neuroscience, medicine, engineering and physics [9]. These neural networks are able to detect and learn patterns and characteristics of complex data.

Using the tableau program, which will also be considered for data analysis, the geographic location of each of Zara's stores on each of the continents will be carried out by correlating each of the variables between the continental location of each of the references.

The first step to start working with the explorer is to define the origin of the data. Weka supports different sources such as CSV, C4.5 and others. In this case the source file is in WEKA's default format which is ARFF (Fig. 6).

Each of these supervised learning machines allows me to establish the levels of correlation that exist between the variables of the data set contributing to the fulfillment of the proposed objectives within the research. Execution of supervised learning machine K-neighbors closer.

The browser allows to perform tasks of:

```
@RELATION  Base_de_Datos_Zara

@ATTRIBUTE ID                          NUMERIC

@ATTRIBUTE PaisTienda               {1,2,3,4,5,6,7,8,9,10,11,12,13,14,15,16

@ATTRIBUTE Continente(Region)       {Europa,Africa,Asia,Oceania,Sudamerica,

@ATTRIBUTE producto_id                 NUMERIC

@ATTRIBUTE ColorBase                   NUMERIC

@ATTRIBUTE Talla_id                    NUMERIC

@ATTRIBUTE Ventas                      NUMERIC

@ATTRIBUTE stock                       NUMERIC

@ATTRIBUTE PrecioEnEuros               NUMERIC

@ATTRIBUTE CantidadDeComprasEnLinea    NUMERIC

@ATTRIBUTE PrecioIdeal                 NUMERIC

@ATTRIBUTE DificultaddeCompraenLinea   NUMERIC

@ATTRIBUTE CalidadDePercepcion         NUMERIC

@ATTRIBUTE CapitalSimbolico            NUMERIC

@ATTRIBUTE EstiloDeModa                NUMERIC

@ATTRIBUTE EdadPromedio                NUMERIC

@DATA
1,118,Centroamerica,310130,69,7026791,1,93,80.26,647,1.67,4.39,6,6,3.47,45.79
2,65,Asia,310130,92,8015276,2,62,74.13,999,1.25,6.42,6,4,5.42,48.11
3,56,Europa,310130,77,452052101,1,18,99.55,565,1.03,4.82,6,1,2.73,19
4,47,Europa,310130,86,541898038,2,42,22.35,747,1.04,1.56,5,5,3.88,28.98
```

Fig. 6 Structure.arff

1. Pre-processing of data and application of filters.
2. Sorting.
3. Clustering.
4. Search of associations.
5. Attribute selection.
6. Data visualization.

The first step to start working with the explorer is to define the origin of the data. Weka supports different sources such as CSV, C4.5 and others. In this case the source file is in WEKA's default format which is ARFF.

Figure 7 shows the order in which the data are structured in WEKA for analysis. In the first section you have the label @RELATION is the one that allows you to assign the name to the database in this case with the name Base_de_Datos_Zara, in the second section you see the label @ATTRIBUTE in this part is listed the amount of attributes with its data type, the database contains 16 attributes, in the third section this the label @DATA contains the amount of data to do analyzed for this data set is counted with 10. In WEKA, it is essential that this data is separated by commas and that it is found in the order established for each of the attributes. In order to access

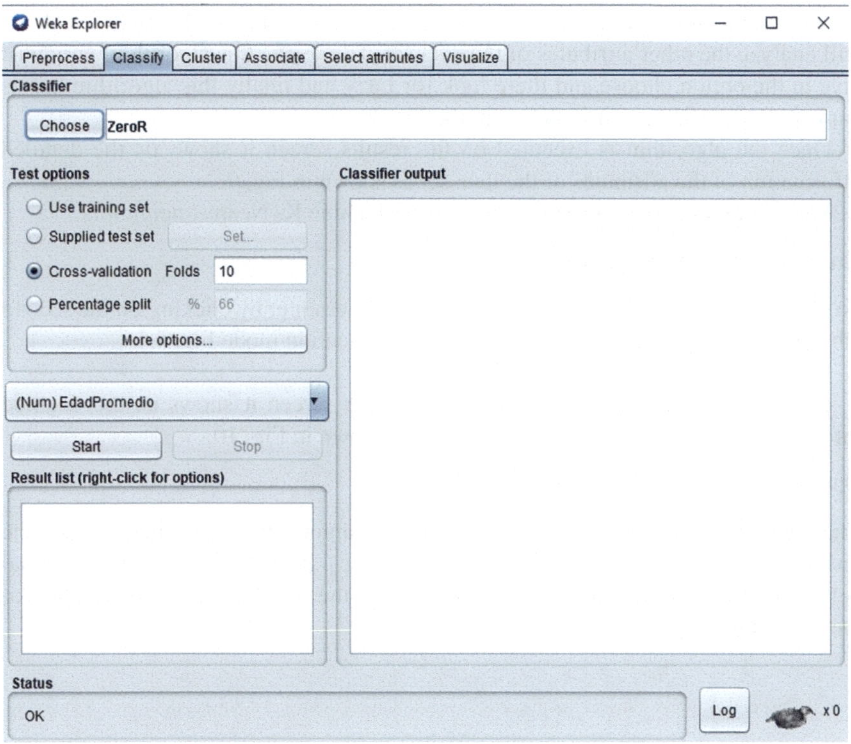

Fig. 7 Structure.arff. Own elaboration

the properties of each classifier we must double-click on the label mentioned above. When you click on it, a new window will appear with the properties along with a brief explanation of it. Once the classifier and its characteristics have been chosen, the next step is to configure the training mode (Test Options). Weka provides 4 test modes.

Use training set: With this option Weka will train the method with all the available data and then apply it again on the data.

Supplied test set: By checking this option you will have the opportunity to select, by pressing the Set button, a data file with which to test the classifier obtained with the classification method used and the initial data.

Cross-validation: Pressing the Cross-validation Weka button will perform a stratified cross-validation of the given number of partitions (Folds). Cross-validation consists of: given a number n, the data is divided into n parts and, for each part, the classifier is built with the remaining n-1 parts and tested with that. Thus, for each of the n partitions. A cross-validation is stratified when each of the parts retains the properties of the original sample (percentage of elements in each class).

Percentage split: A percentage is defined with which the classifier will be built and tested with the remaining part.

We will execute the Supervised Learning Machine K- Nearest neighbors.

The continent (Region) attribute will be our independent variable with which we will analyze the other attributes of the data set. To execute this algorithm, you must give in the option choose and there look for Lazy and finally this algorithm has as name in WEKA IBK as it is shown in the Fig. 8.

Once the algorithm is executed on the results screen it shows us the detailed information of the attributes of the data set as shown in Fig. 9.

We will execute the Supervised Learning Machine K- Nearest neighbors.

Execution of the K-Neighbour Algorithm.

To run the algorithms once the data set is loaded, we enter by clicking on the second tab (top zone) of the explorer we enter the classification mode Fig. 8.

The classification mode within the browser.

Once the algorithm is executed in the results screen it shows us the detailed information of the attributes of the data set as shown in Fig. 10.

Data set information.

The columns are the instances classified by the algorithm and in the rows are the instances or reports predicted for each of the classes. It can be seen that the diagonal is thc correctly sorted values. Next, you can see the results of the classification of the algorithm Fig. 11.

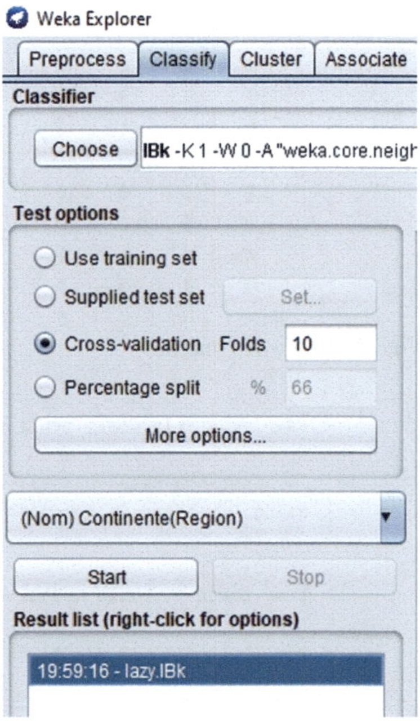

Fig. 8 Execution of the K-Neighbour Algorithm

Fig. 9 Data set information

```
=== Run information ===

Scheme:       weka.classifiers.lazy.IBk -K 1 -W 0
Relation:     Base_de_Datos_Zara
Instances:    10000
Attributes:   16
              ID
              PaisTienda
              Continente(Region)
              producto_id
              ColorBase
              Talla_id
              Ventas
              stock
              PrecioEnEuros
              CantidadDeComprasEnLinea
              PrecioIdeal
              DificultaddeCompraenLinea
              CalidadDePercepcion
              CapitalSimbolico
              EstiloDeModa
              EdadPromedio
Test mode:    10-fold cross-validation
```

```
=== Confusion Matrix ===

    a    b    c    d    e    f    g   <-- classified as
 3754    1    3    0    0    3    0 |    a = Europa
    0  844    1    1    0    0    0 |    b = Africa
    0    0 2017    0    0    0    0 |    c = Asia
    1    0    0  618    0    1    0 |    d = Oceania
    0    0    0    0  879    0    0 |    e = Sudamerica
    2    0    1    1    0 1638    0 |    f = Centroamerica
    0    1    0    0    0    0  234 |    g = Norteamerica
```

Fig. 10 Confusion matrix

```
=== Classifier model (full training set) ===

IB1 instance-based classifier
using 1 nearest neighbour(s) for classification


Time taken to build model: 0 seconds

=== Stratified cross-validation ===
=== Summary ===

Correctly Classified Instances          9984               99.84   %
Incorrectly Classified Instances          16                0.16   %
Kappa statistic                            0.9979
Mean absolute error                        0.0006
Root mean squared error                    0.0214
Relative absolute error                    0.2935 %
Root relative squared error                6.4378 %
Total Number of Instances              10000
```

Fig. 11 Results of the classification

In Fig. 12, you can see a matrix of values, the first 2 values correspond to the test instances correctly and incorrectly classified, there you can see how many instances were classified in that category and the percentage that corresponds, to find this percentage simply divide the number of instances of the class on the total of instances classified in the example the instances correctly classified are: 9984/10000 = 99.84

=== Detailed Accuracy By Class ===

	TP Rate	FP Rate	Precision	Recall	F-Measure	MCC	ROC Area	PRC Area	Class
	0,998	0,000	0,999	0,998	0,999	0,998	0,999	0,998	Europa
	0,998	0,000	0,998	0,998	0,998	0,997	0,999	0,995	Africa
	1,000	0,001	0,998	1,000	0,999	0,998	1,000	0,998	Asia
	0,997	0,000	0,997	0,997	0,997	0,997	0,998	0,994	Oceania
	1,000	0,000	1,000	1,000	1,000	1,000	1,000	1,000	Sudamerica
	0,998	0,000	0,998	0,998	0,998	0,997	0,999	0,996	Centroamerica
	0,996	0,000	1,000	0,996	0,998	0,998	0,998	0,996	Norteamerica
Weighted Avg.	0,998	0,000	0,998	0,998	0,998	0,998	0,999	0,997	

Fig. 12 Detailed precision matrix per class

The same procedure is done for the incorrect ones:

16/10000 = 0.16%

The Kappa value must be found to be 2 values, observed precision and expected precision, and represents the degree of agreement between the classifications and the true classes (concordance) taking into account the confusion matrix.

Observed accuracy Found as follows:

(3754 + 844 + 2017 + 618 + 879 + 1638 + 234)/10000 = 0.9984

Expected accuracy It's like this:

3761*3757/10000 = 1413.0077

The total number of instances in a class is taken and multiplied by the total number of instances classified in a class over the total number of communications.

846*846/10000 = 71.5716

2017*2022/10000 = 40.6422

620*620/10000 = 38.44

879*879/10000 = 77.2641

1642*1642/10000 = 269.6164

235*234/10000 = 5.499

The values obtained are then added up and divided over the total number of instances or short communications. (1413.0077 + 71.5716 + 40.6422 + 38.44 + 77.2641 + 269.6164 + 5.499)/10000 = 0.1916 Then we apply the formula and find the value of Kappa:

By replacing

Kappa = (0.9984−0.1916)/(1−0.1916) = 0.9980

Kappa is a measure of how closely the instances classified by the machine learning classifier matched the data labelled true, it provides a better indicator of how well the classifier performed in all instances following a standard, if kappa is in: >0.75 is excellent, 0.40–0.75 is taken as fair to good, and <0.40 as poor or not correctly classified.

Kappa's result was 0.9979, which means that it has an excellent ranking of true instances with respect to the database. To find the absolute error MAE (Mean Absolute Error), the following formula must be applied, where

is: expected accuracy

en: observed accuracy

It is: the total of the data or the total of a set of predictions or communications.

MAE is used to measure the average magnitude of errors in a set of releases, regardless of their direction or sign. MAE is a linear score, meaning that all individual differences are weighted equally in the average. The next value observed is the root mean squared error (RMSE) which is calculated with the following equation:

$$\mathrm{RMSE} = \sqrt{\frac{1}{N}\sum_{i=1}^{N}(x_i - y_i)^2} = 0.214$$

RMSE is a quadratic scoring rule that is used to measure the average magnitude of the error and in explaining the formula: square root is applied to the average of the square differences between the expected precision and the observed precision all this over the sample.

MAE and RMSE can be used at the same time to diagnose the variation in error in a set of predictions. The RMSE will always be greater than or equal to the MAE; the greater the difference between them, the greater the variance (dispersion) in the individual errors in the sample. If RMSE = MAE, then all errors are of the same magnitude. The relative absolute error RAE (Relative Absolute Error) is found by the following formula:

$$\mathrm{RAE} = \frac{\frac{1}{N}\sum_{i=1}^{N} |X - \theta_i|}{\frac{1}{N}\sum_{i=1}^{N} |\alpha - \theta_i|} = 0.2935$$

This error indicates the quality of the measurement, the lower the value means that the measurement is better, alpha represents the average of the data.

The root of the relative quadratic error RRSE, is calculated with the following formula

$$\mathrm{RRSE} = \sqrt{\frac{\frac{1}{N}\sum_{i=1}^{N} (X - \theta_i)^2}{\frac{1}{N}\sum_{i=1}^{N} (\alpha - \theta_i)^2}} = 6.4378$$

RRSE is the square root of the mean squared error and it is a precision measure, its result never has a negative sign and a value similar to 0, it informs a perfect adjustment of the data. Detailed precision matrix per class.

The accuracy value is shown in Fig. 13 and is calculated with the following formula:

This value is given per class and indicates the percentage of how many predicted values in a class are actually of that class.

The recall value or also called sensitivity is calculated with the following formula:

$$Precision = \frac{TP}{TP + FP}$$

=== Detailed Accuracy By Class ===

	TP Rate	FP Rate	Precision	Recall	F-Measure	MCC	ROC Area	PRC Area	Class
	0,998	0,000	0,999	0,998	0,999	0,998	0,999	0,998	Europa
	0,998	0,000	0,998	0,998	0,998	0,997	0,999	0,995	Africa
	1,000	0,001	0,998	1,000	0,999	0,998	1,000	0,998	Asia
	0,997	0,000	0,997	0,997	0,997	0,997	0,998	0,994	Oceania
	1,000	0,000	1,000	1,000	1,000	1,000	1,000	1,000	Sudamerica
	0,998	0,000	0,998	0,998	0,998	0,997	0,999	0,996	Centroamerica
	0,996	0,000	1,000	0,996	0,998	0,998	0,998	0,996	Norteamerica
Weighted Avg.	0,998	0,000	0,998	0,998	0,998	0,998	0,999	0,997	

Fig. 13 Detailed precision matrix per class

Like the precision, this value is referenced in each class and indicates how many values of the class that should be selected were selected.

The F-Measure is calculated in the following way:

$$Recall = \frac{TP}{TP + FN}$$

This measure represents a balance between accuracy and recall and represents how well the algorithm predicts.

The MCC value is calculated:

$$F - Measure = 2 * \frac{Precision + Recall}{Precision + Recall}$$

This value indicates the quality of the classification of each class.

The ROC curve gives an idea to the user of how is the general performance of the algorithm, to make this graph is used the TP versus FP, the value that indicates in the table between closer to 1 means that the general performance of the algorithm is good.

The PRC area is focused on each class, this value tells how is the behavior of each class for this uses precision versus recall, as well as the ROC curve the closer to 1 the better the behavior.

2.9 Ethical Considerations

Certification In Human Research Work Ethics

In accordance with the principles established in Resolution 008430 of October 4, 1993) and due to the fact that this investigation was considered a risk-free investigation in accordance with Article 10 of the present resolution, and in compliance with the aspects mentioned in Article 6 of the same resolution, this study will be developed in accordance with the following criteria In accordance with international and national regulations, Resolution 008430/93 and the provisions of Chapter 1, Articles 5 and 6 of the resolution 008430 of October 4, 1993. Research that is developed on a population of ages between 18 and 49 years old, belonging to the countries of the world where the company Zara has physical and online stores at the service of the world community, where the levels of correlation between variables have to be measured through a supervised learning machine such as K-Neighbors Nearest, Bayes Native Method, neural networks, and vectorial support machines, which obtains information through a Likert questionnaire and the data is processed by the WEKA program establishing the relationship between continents, store country, products, base color, size, stock, price in euros, quantity of online purchases, ideal price, difficulty of online purchase, quality of perception, symbolic capital, fashion style, average age The population under study will not be at risk from the research, since the information provided by Zara from its databases for the research process must be confidential and will not have repercussions on the online purchasing process, since the information will only be handled by the researcher, who will in turn provide information that will only be of her own handling and relevant for research purposes. By providing Zara with the information through databases that it supports online for the management of various users in a free and spontaneous manner, it has its own informed consent for the processing of the data by the research subject for the relevant purposes, and likewise, the authorization of the buyers for the processing of personal data according to law 1581 of 2012 of the colombian data protection law which regulates the right to personal data protection and aims to: "(…. to develop the constitutional right of all persons to know, update and rectify the information that has been collected on them in databases or files, and the other constitutional rights, freedoms and guarantees referred to in Article 15 of the Political Constitution; as well as the right to information enshrined in Article 20 thereof" of the colombian data protection law. The line of research to which the investigation is ascribed is Advanced Statistics, Mathematics and Data Science, of the Universidad Cuauhtémoc de Aguascalientes Mexico, which has advisors with extensive research experience, as they are: Alejandra Espinoza Guzmán Mtro. Luis Ramón Gutiérrez Medina, they are in charge of orienting and guiding the researcher within each one of the processes with the same entails as far as multivariate analysis, data mining, machine learning, artificial intelligence, neural networks, statistical-analytical, statistical models, computational algorithms, algebraic reasoning, predictive analysis, multivariate analysis, spatial interaction analysis, data warehouse, business intelligence, numerical methods and optimization.

3 Research Results

Results of the research

The company under study is located in several countries as shown in Fig. 14.

When the supervised learning machines are executed, the results are divided into 3 sections: the first is the summary section, the second is the detailed precision for each class and the third is the confusion matrix.

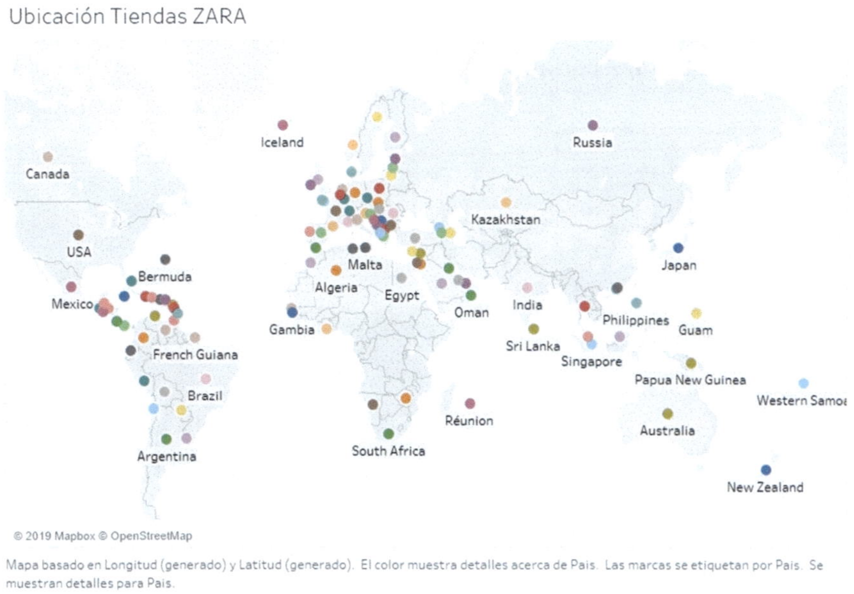

Fig. 14 Map of Zara stores

3.1 Result Supervised Learning Machine K-Nearest Neighbors

In the Fig. 15 can be seen the result obtained from the algorithm K-neighbors closest in which was obtained a 99.84% this percentage corresponds to 9,984 instances correctly classified of 10,000 instances, with respect to the value kappa which is 0.9979 showing that the machine had an excellent agreement with the predicted values, with respect to the measure weighted PRECISION reports that in 99.The weighted measure of RECALL informs that the TP (True Positive) of each one of the classes were classified in a 99.9% of correct way, with the weighted measure of MCC informs that the quality of classification of the classes was 99.9% and the weighted measure of PCR informs that the classifiers behaved in a 99.9% of correct way, the F-Measure of the algorithm is of 99.9%. 4.2 Result Learning Machine Vector Support Machine.

In Fig. 16 you can see the result obtained from the vector support algorithm in which 100% was obtained this percentage corresponds to the total of all instances correctly classified, with respect to the kappa value which is 1 showing that the machine had 100% agreement with the predicted values, with respect to the weighted measure of PRECISION reports that 100% of the values of the class if they correspond to it, the weighted measure of RECALL reports that the TP (True Positives) of each of the classes were classified 100% correctly, with the weighted measure of MCC reports that the quality of classification of the classes was 100% and the weighted

```
=== Classifier model (full training set) ===

IB1 instance-based classifier
using 1 nearest neighbour(s) for classification


Time taken to build model: 0 seconds

=== Stratified cross-validation ===
=== Summary ===

Correctly Classified Instances        9984          99.84   %
Incorrectly Classified Instances        16           0.16   %
Kappa statistic                          0.9979
Mean absolute error                      0.0006
Root mean squared error                  0.0214
Relative absolute error                  0.2935 %
Root relative squared error              6.4378 %
Total Number of Instances            10000
=== Detailed Accuracy By Class ===

               TP Rate  FP Rate  Precision  Recall  F-Measure  MCC    ROC Area  PRC Area  Class
               0,999    0,000    1,000      0,999   0,999      0,999  0,999     0,999     Europa
               0,999    0,000    0,999      0,999   0,999      0,999  0,999     0,998     Africa
               1,000    0,000    1,000      1,000   1,000      1,000  1,000     1,000     Asia
               1,000    0,000    0,997      1,000   0,998      0,998  1,000     0,997     Oceania
               1,000    0,000    0,999      1,000   0,999      0,999  1,000     0,999     Sudamerica
               0,999    0,000    0,999      0,999   0,999      0,999  0,999     0,998     Centroamerica
               1,000    0,000    1,000      1,000   1,000      1,000  1,000     1,000     Norteamerica
Weighted Avg.  0,999    0,000    0,999      0,999   0,999      0,999  1,000     0,999

=== Confusion Matrix ===

    a    b    c    d    e    f    g   <-- classified as
 3754    1    3    0    0    3    0 |    a = Europa
    0  844    1    1    0    0    0 |    b = Africa
    0    0 2017    0    0    0    0 |    c = Asia
    1    0    0  618    0    1    0 |    d = Oceania
    0    0    0    0  879    0    0 |    e = Sudamerica
    2    0    1    1    0 1638    0 |    f = Centroamerica
    0    1    0    0    0    0  234 |    g = Norteamerica
```

Fig. 15 K-Neighbourhood watch results

measure of PCR reports that the classifiers behaved 100% correctly, the F-Measure of the algorithm is 100%.

3.2 *Naive Bayes Learning Machine Result*

In the Fig. 17 can be seen the result obtained from the Bayes Nail in which was obtained a 99.41% this percentage corresponds to 9,924 instances correctly classified of 10,000 instances, with respect to the value kappa which is 0.9924 showing that the machine had an excellent agreement with the predicted values, with respect to the measure weighted PRECISION reports that in 99. The weighted measure of RECALL informs that the TP (True Positive) of each one of the classes were classified in a 99.4% of correct way, with the weighted measure of MCC informs that the quality of classification of the classes was 99.3% and the weighted measure of PCR informs

```
=== Stratified cross-validation ===
=== Summary ===

Correctly Classified Instances       10000              100      %
Incorrectly Classified Instances         0                0      %
Kappa statistic                          1
Mean absolute error                      0.2041
Root mean squared error                  0.3012
Relative absolute error                 92.5611 %
Root relative squared error             90.7103 %
Total Number of Instances            10000

=== Detailed Accuracy By Class ===

                 TP Rate  FP Rate  Precision  Recall   F-Measure  MCC      ROC Area  PRC Area  Class
                 1,000    0,000    1,000      1,000    1,000      1,000    1,000     1,000     Europa
                 1,000    0,000    1,000      1,000    1,000      1,000    1,000     1,000     Africa
                 1,000    0,000    1,000      1,000    1,000      1,000    1,000     1,000     Asia
                 1,000    0,000    1,000      1,000    1,000      1,000    1,000     1,000     Oceania
                 1,000    0,000    1,000      1,000    1,000      1,000    1,000     1,000     Sudamerica
                 1,000    0,000    1,000      1,000    1,000      1,000    1,000     1,000     Centroamerica
                 1,000    0,000    1,000      1,000    1,000      1,000    1,000     1,000     Norteamerica
Weighted Avg.    1,000    0,000    1,000      1,000    1,000      1,000    1,000     1,000

=== Confusion Matrix ===

    a    b    c    d    e    f    g   <-- classified as
 3761    0    0    0    0    0    0 |    a = Europa
    0  846    0    0    0    0    0 |    b = Africa
    0    0 2017    0    0    0    0 |    c = Asia
    0    0    0  620    0    0    0 |    d = Oceania
    0    0    0    0  879    0    0 |    e = Sudamerica
    0    0    0    0    0 1642    0 |    f = Centroamerica
    0    0    0    0    0    0  235 |    g = Norteamerica
```

Fig. 16 Results of the vector support machine

that the classifiers behaved in a 99.3% of correct way, the F-Measure of the algorithm is of 99.4%.

3.3 Result Learning Machine Neural Network

In Fig. 18 you can see the result obtained from the neural network algorithm in which it was obtained a 100% this percentage corresponds to the total of all instances correctly classified, with respect to the kappa value which is 1 showing that the machine had 100% agreement with the predicted values, with respect to the weighted measure of PRECISION reports that 100% values of the class if they correspond to it, the weighted measure of RECALL reports that the TP (True Positives) of each of the classes were classified 100% correctly, with the weighted measure of MCC reports that the quality of classification of the classes was 100% and the weighted

```
Time taken to build model: 0.07 seconds

=== Stratified cross-validation ===
=== Summary ===

Correctly Classified Instances        9941              99.41   %
Incorrectly Classified Instances        59               0.59   %
Kappa statistic                          0.9924
Mean absolute error                      0.024
Root mean squared error                  0.0504
Relative absolute error                 10.8912 %
Root relative squared error             15.1912 %
Total Number of Instances            10000

=== Detailed Accuracy By Class ===

                 TP Rate  FP Rate  Precision  Recall   F-Measure  MCC      ROC Area  PRC Area  Class
                 0,994    0,000    1,000      0,994    0,997      0,996    0,997     0,998     Europa
                 1,000    0,002    0,979      1,000    0,989      0,989    1,000     0,990     Africa
                 0,998    0,004    0,985      0,998    0,991      0,989    0,998     0,981     Asia
                 0,990    0,000    1,000      0,990    0,995      0,995    0,996     0,995     Oceania
                 0,998    0,000    0,999      0,998    0,998      0,998    0,998     0,998     Sudamerica
                 0,987    0,000    1,000      0,987    0,993      0,992    0,994     0,994     Centroamerica
                 0,983    0,001    0,963      0,983    0,973      0,972    0,987     0,970     Norteamerica
Weighted Avg.    0,994    0,001    0,994      0,994    0,994      0,993    0,997     0,993

=== Confusion Matrix ===

    a    b    c    d    e    f    g   <-- classified as
 3740    5   14    0    1    0    1 |    a = Europa
    0  846    0    0    0    0    0 |    b = Africa
    0    1 2013    0    0    0    3 |    c = Asia
    0    2    3  614    0    0    1 |    d = Oceania
    0    0    2    0  877    0    0 |    e = Sudamerica
    0    8   10    0    0 1620    4 |    f = Centroamerica
    0    2    2    0    0    0  231 |    g = Norteamerica
```

Fig. 17 Results Naive Bayes

measure of PCR reports that the classifiers behaved 100% correctly, the F-Measure of the algorithm is 100%.

In Table 5, we can validate that the algorithms with the best classification are the vectorial support machines (SVM) and the neural network. These two machines have a classification level of 100%, with a Kappa value of 1, and we can report that these two algorithms present the highest agreement in the classification of the classes for this case study: 10,000. To make the respective predictions of online purchases of customers and fashion style in the case study will take the method of vector support machine, once trained the model must load the test data set in this case the data set is named Base_de_Datos_Zara_Test, which contains 16 attributes and 31 records taken at random to test the method. To perform the method test, random data was taken from the original database set, the instances to be predicted should be assigned the character (?) so that the method understands that they are the data to which the continent should be assigned according to the characteristics of the other records as well (Fig. 19):

```
Time taken to build model: 692.65 seconds

=== Stratified cross-validation ===
=== Summary ===

Correctly Classified Instances       10000              100      %
Incorrectly Classified Instances         0                0      %
Kappa statistic                          1
Mean absolute error                      0.0007
Root mean squared error                  0.0011
Relative absolute error                  0.3107 %
Root relative squared error              0.3316 %
Total Number of Instances            10000


=== Detailed Accuracy By Class ===

                 TP Rate  FP Rate  Precision  Recall   F-Measure  MCC      ROC Area  PRC Area  Class
                 1,000    0,000    1,000      1,000    1,000      1,000    1,000     1,000     Europa
                 1,000    0,000    1,000      1,000    1,000      1,000    1,000     1,000     Africa
                 1,000    0,000    1,000      1,000    1,000      1,000    1,000     1,000     Asia
                 1,000    0,000    1,000      1,000    1,000      1,000    1,000     1,000     Oceania
                 1,000    0,000    1,000      1,000    1,000      1,000    1,000     1,000     Sudamerica
                 1,000    0,000    1,000      1,000    1,000      1,000    1,000     1,000     Centroamerica
                 1,000    0,000    1,000      1,000    1,000      1,000    1,000     1,000     Norteamerica
Weighted Avg.    1,000    0,000    1,000      1,000    1,000      1,000    1,000     1,000

=== Confusion Matrix ===

    a    b    c    d    e    f    g   <-- classified as
 3761    0    0    0    0    0    0 |    a = Europa
    0  846    0    0    0    0    0 |    b = Africa
    0    0 2017    0    0    0    0 |    c = Asia
    0    0    0  620    0    0    0 |    d = Oceania
    0    0    0    0  879    0    0 |    e = Sudamerica
    0    0    0    0    0 1642    0 |    f = Centroamerica
    0    0    0    0    0    0  235 |    g = Norteamerica
```

Fig. 18 Neural Network

Table 5 Results of supervised learning machines

KNN 99.84
SVM 100%
Naive Bayes 99.41%
100% Neuronal Network

In the following illustrations you can see the predictions of the evaluated and trained method (Figs. 20 and 21).

As can be seen in Fig. 22, instance number 6 specifies that the record of the test data set corresponds to the continent of Central America. This can be verified by opening the original database with which the model was trained and Central America should appear in its record:

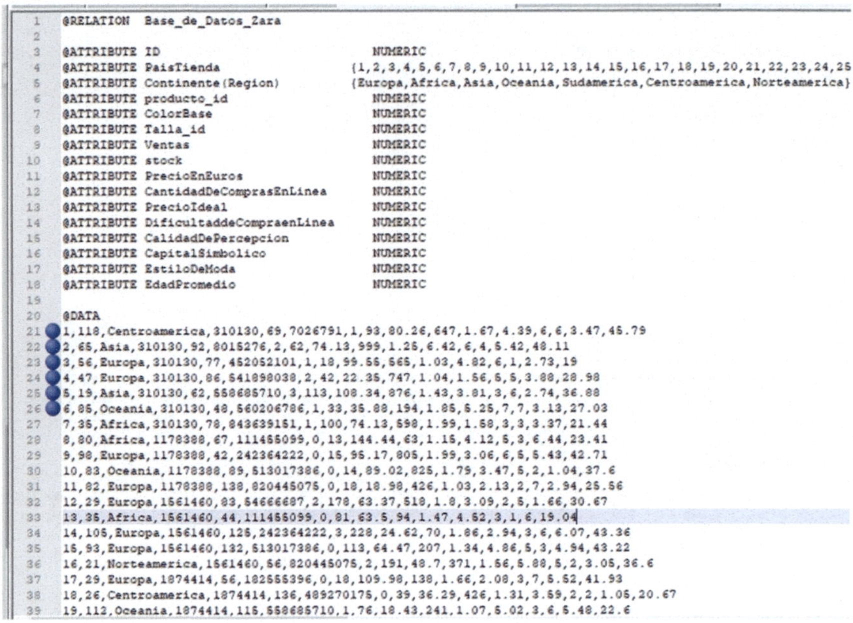

Fig. 19 Data structure for testing the method

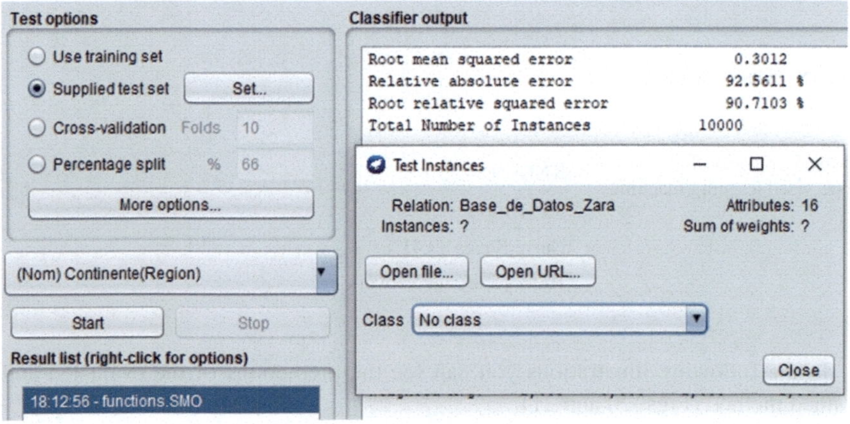

Fig. 20 Configuration and loading of the test data set

The trained method performs the prediction process correctly, where it can be inferred that a buyer with the characteristics as they appear in the records of the Fig. 23, with the attributes products, color, size, quality of perception, fashion style, average age, symbolic capital, difficulty of buying online belong to some region and thus the company Zara has the knowledge to launch its new collections of

```
=== Predictions on test set ===

 inst#     actual  predicted error prediction
     1        1:?       3:Asia       0.286
     2        1:?       3:Asia       0.286
     3        1:? 6:Centroamerica         0.286
     4        1:? 7:Norteamerica        0.286
     5        1:?       3:Asia       0.286
     6        1:? 6:Centroamerica         0.286
     7        1:? 5:Sudamerica        0.286
     8        1:?     1:Europa       0.286
     9        1:?       3:Asia       0.286
    10        1:?     1:Europa       0.286
    11        1:?       3:Asia       0.286
    12        1:?     1:Europa       0.286
    13        1:?    4:Oceania       0.286
    14        1:?       3:Asia       0.286
    15        1:? 5:Sudamerica        0.286
    16        1:? 5:Sudamerica        0.286
    17        1:? 6:Centroamerica         0.286
    18        1:? 6:Centroamerica         0.286
    19        1:? 5:Sudamerica        0.286
    20        1:?     1:Europa       0.286
    21        1:? 6:Centroamerica         0.286
    22        1:?     1:Europa       0.286
```

Fig. 21 Results of the method's predictions

```
@DATA
65,92,?,3817767,34,558685710,2,126,139.17,318,1.74,2.73,1,2,5.68,41.65
66,62,?,3817767,64,642148717,0,86,76.54,620,1.01,5.29,7,4,2.18,41.2
561,48,?,19599326,61,111455099,2,33,91.24,997,1.59,1.16,2,4,5.42,41.73
562,21,?,19599326,131,452052101,0,29,43.03,13,1.11,4.54,5,1,2.93,20.89
563,72,?,19599326,136,541898038,0,47,14.66,910,1.13,3.1,7,2,6.71,32.8
5554,26,?,202760561,42,422108434,1,40,74.27,120,1.59,5.39,4,4,4.78,35.61
```

Fig. 22 Method prediction test

seasons according to the location of their customers and purchases recorded by their customers.

As we can see in the illustration the colour blue belongs to Europe they are the ones who have more amount of online purchases, these purchases are more reflected and marked from the years 37 this means that the more advanced in age buy more products through the Internet, probably this is because they already have an economic stability, It can also be observed that there is no age limit for this type of product purchases, in the age range in this case study which is 18–48 years approximately all actors make purchases, in the smaller purchases in younger ages the Asians are

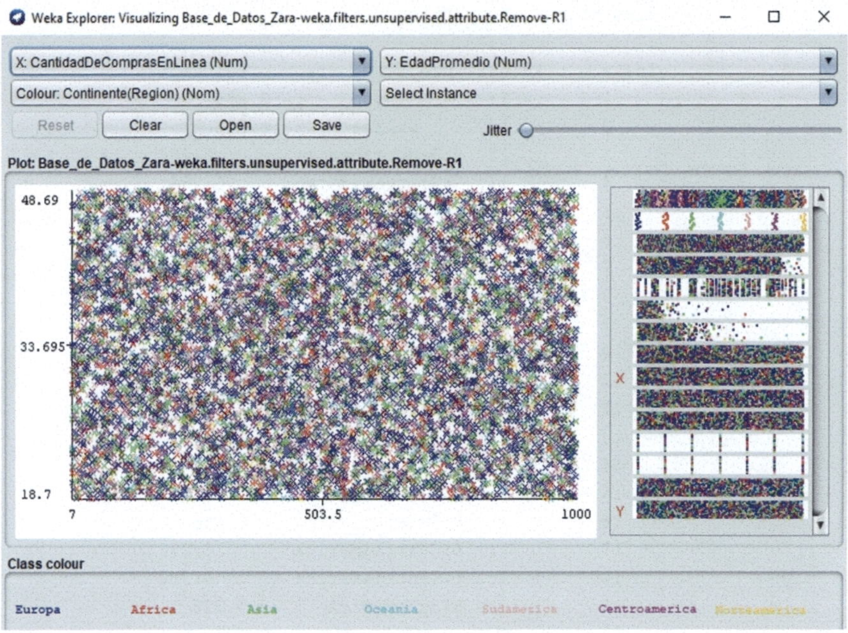

Fig. 23 Quantity of online purchases. X: Number of online purchases, Y: Average Age, Class: Continent (Region)

the ones who monopolize the market more, the Africans as you can see are the ones who buy less online (Figs. 24 and 25).

Certain colours are represented more times according to the age and the place where people live. Colors from 1 to 30 are light colors as yellow and white are concentrated with the Asians, Europeans are more concentrated towards the dark colors (Figs. 26 and 27).

4 Experimentation

In order to organize the most efficient arrangement of garments in a dressing table, we develop an atmosphere capable of storing the data of each one of those representing these garments, in order to distribute in an optimal way each of the garments. evaluated. One of the most interesting features observed in this experiment was the diversity of the cultural patterns established by each garment with respect to its symbolic capital. The structured scenarios associated with the agents cannot be reproduced in general, since they represent only a few times given in the space and time of the different garments. These represent a unique and innovative form of adaptive behavior that solves a computational problem that does not attempt to group the garment only with a factor associated with its external appearance (attributes of

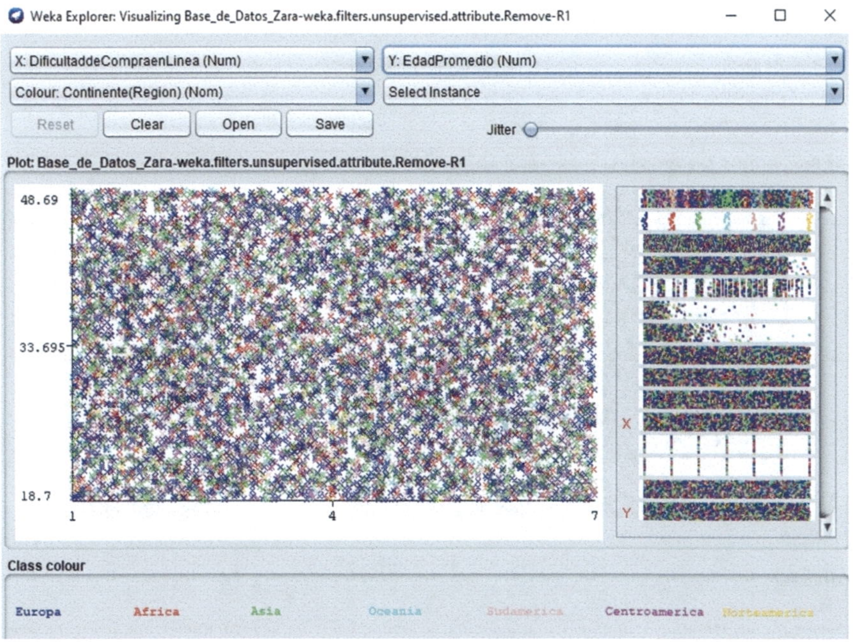

Fig. 24 Difficulty of online shopping. X: Difficulty in buying online, Y: Average Age, Class: Continent (Region)

each attire), trying to solve a computational problem that involves a complex change between the existing representations. The configurations generated can be related to the knowledge of the behavior of a potential client with respect to an optimization problem (to culturally select similar clothes, without being of the same type [19]). The main experiment was to detail each of the garments of a collection, with 500 agents and a condition of 50 times, which allowed us to generate the best selection of each type of garment and its possible location in a commercial dresser. It was obtained after comparing the different cultural and social similarities of each dress, and evaluating each of them with the multiple coincidences model. The research allowed to classify each of the garments belonging to each type, with different clothing for garments with cultural identity only, which allows to identify changes in time with respect to other garments. Figure 3 shows the final sideboard (exterior-interior) constructed according to up to eight different attributes, respectively (see Fig. 3). The design of the experiment consists of an orthogonal matrix test, with the interactions between the variables: emotional control, fighting ability, intelligence, agility, strength, resistance, social leadership and speed. These variables are studied in a range of color (1 to 64). The orthogonal arrangement is LN (2 ** 8), in other words, 8 factors in N executions, N is defined by the combination of the possible values of the 8 variables and the possible color range (See Fig. 28). According to the results obtained for the lodging time in the transport, it must be done to accommodate the download time in

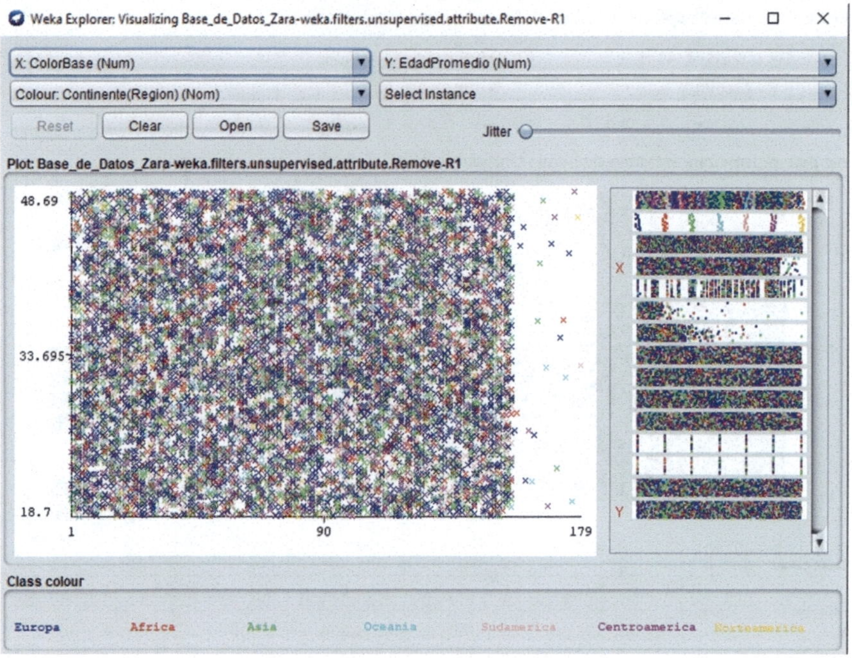

Fig. 25 Base colour. X: ColorBase, Y: Average Age, Class: Continent (Region)

the following sign. Based on the results obtained in the experiment, the average similarity between the garments was found to be 4.0625; This means that you must have to improve in the management of the use of the combination of colors and standard according to the climate. Other factors that would affect the use of accessories such as: bracelets, necklaces, pins and charms, an implementation to solve this problem is the use of fashion visualization. As you can see to maintain this type of materials it is necessary to use direct and indirect merchandise. The indirect merchandise secures the floor, gives resistance to the lateral and longitudinal movement of the clothes. While the direct ties anchored to the mannequins in the chest of drawers. To carry out this process, it is necessary to have the correct dimensions and the distance of the street and the height of these, cables or cadences, taking into account the material of these mannequins.

5 Conclusions and Future Research

Through the use of the ICA the understanding is improved to obtain the change of the "best paradigm", because the communities of agents are properly classified based on an approach to the relation maintained by their attributes, this allowed to understand

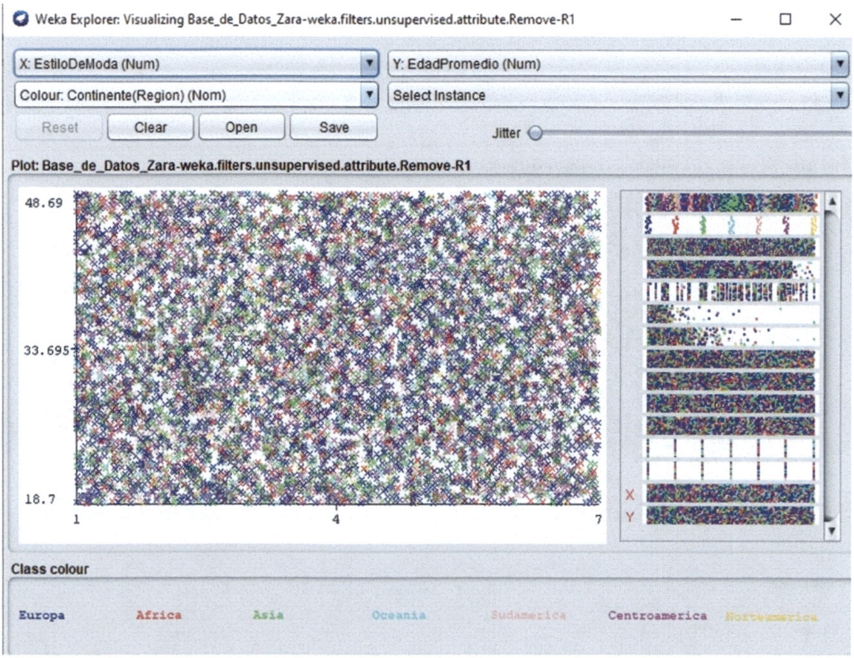

Fig. 26 Fashion Style. X: Fashion Style, Y: Average Age, Class: Continent (Region)

that the concept of "fashion based on symbolic capital" exists based on the determination of the acceptance function by the rest of the clothes to the proposed place for the rest of them. ICA offers a powerful alternative to the problems of optimization and redistribution of the clustering technique. For this reason, this technique offers a fairly comprehensible picture of the cultural phenomenon represented [1]. This technique allows to include the possibility of generating experimental knowledge created by the community of agents for a new application domain. The analysis of the level and degree of cognitive knowledge of each community is an aspect that we wish to evaluate for future work. The answer may lie between the similitude that exists in the communication between two different cultures and how they are perceived. On the other hand, understanding the true similarities that different societies have based on the characteristics that make them cluster contributors and also allows them to maintain their own identity, shows that the small variations go beyond the phenotypic characteristics and are mainly associated to tastes and similar characteristics developed over time.

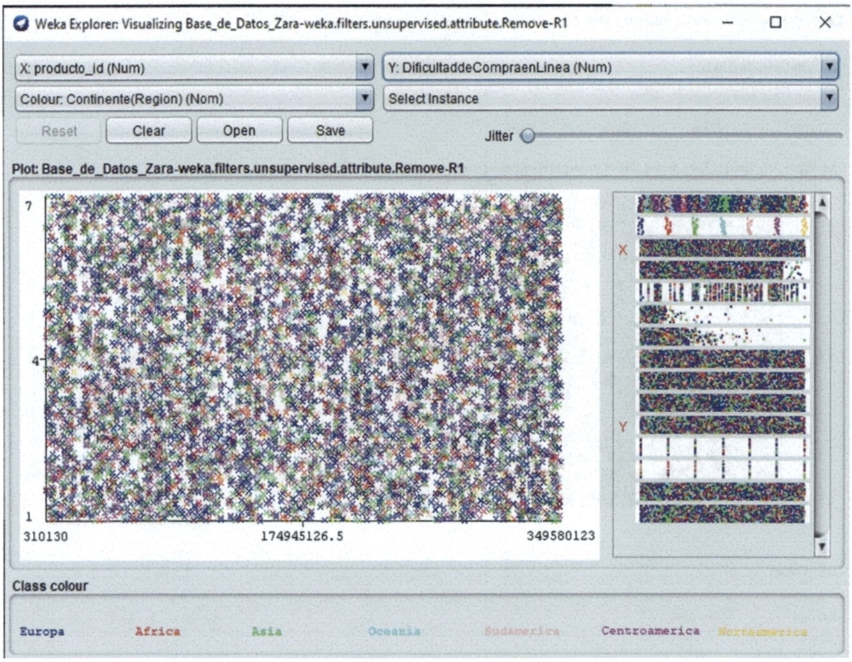

Fig. 27 Product. X: Product ID, Y: Average Age, Class: Continent (Region)

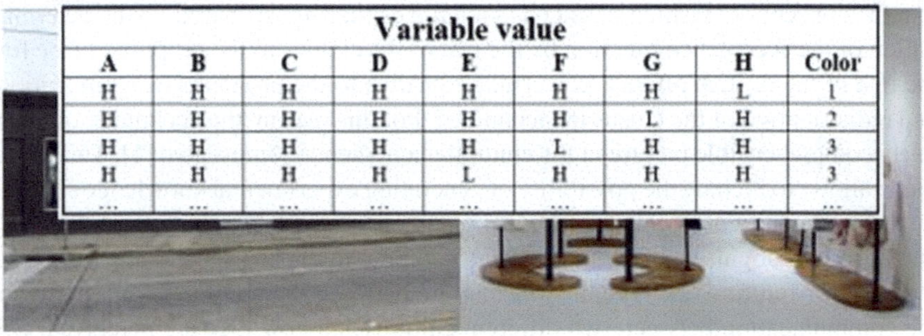

Variable value								
A	B	C	D	E	F	G	H	Color
H	H	H	H	H	H	H	L	1
H	H	H	H	H	H	L	H	2
H	H	H	H	H	L	H	H	3
H	H	H	H	L	H	H	H	3
...	...	...	...	...	...	...	...	...

Fig. 28 Final sideboard (interior-exterior) built according to up to eight different attributes based on the competitive imperialist algorithm (Multiagent System)

References

1. Blasco Sánchez J (2015) Zara y su análisis estratégico. Universitas Miguel Hernández, España
2. Casado Diaz A, Sellers Rubio R (2006) Dirección de Marketing teoria y práctica. Alicante
3. Echevarria Sanfeliu M (2015) Aálisis de la estrategia empresarial de Zara. Universidad de Barcelona, Barcelona España
4. Talent Z (11 de Octubre de 2019) www.zaratalent.com. Obtenido de zara: https://www.zaratalent.com/data.html

5. Alvarez LA (2000). Vistiendo a tres continentes: la ventaja competitiva del gripo Inditex - zara 193-1999. Historia Industrial, 157-179
6. Hernández Sampieri R, Fernández Collado C, Baptista Lucio P (2010) Metodología de la Investigacón. M c graw Hill, México
7. Guersenzvaig A, Adell F (2013) «Big Data » y los nuevos métodos de visualización de. Universitat Oberta de Catalunya, España
8. Talburt J, Zhou Y (2015). Entity information life cycle for big data: master data. Morgan Kaufmann
9. Espino Timón C (2017) Análisis predictivo: técnicas y modelos utilizados y aplicaciones del mismo -herramientas Open Source que permiten su uso. Universitat Oberta de Catalunya, Catanlunya, España
10. Resendiz Trejo J (09 de 2006) www.ctrl.cinvestav.mx. Obtenido de https://www.ctrl.cinvestav.mx/~yuw/pdf/MaTesJAR.pdf
11. Lanzt B (2015) Machine learning with R Second Edition. Packt Publishing Ltd
12. Hui L (12 de 04 de 2017) blogs.sas.com. Obtenido de https://blogs.sas.com/content/subconsciousmusings/2017/04/12/machine-learning-algorithm-use/
13. Gago Utrera R (2017) Uso de algoritmos de aprendizaje automático aplicados a bases de datos genéticos. Univertitad Oberta Catalunya, España
14. Berástegui Arbeloa G (2018) Implementación del algoritmo de las K vecinos más Cercanos (k-NN) y estimación del mejor valor local de K para su cálculo. Universidad Pública de Navarra, Pamplona
15. Songbo T, Xueqi C, Yuefen W, Hongbo X (2009) Adapting Naive Bayes to domain adaptation for sentiment analysis. Springer-Verlag, China
16. Eibe F, Ian HW (2002) Data Mining: practical machine learning tools with Java implementations. Morgan Kaufmann, San Francisco
17. Eibe F, Hall MA, Witten IH, Pal CJ (2016) Data mining: practical machine learning tools and techniques. Morgan Kaufmann
18. Constituyente A (1991) Constitución Politica de Colombia
19. Colombia C. d. (1999) Ley de comercio electrónico
20. García Jiménez M, Álvarez Sierra A (05 de 01 de 2016) http://www.it.uc3m.es/. Obtenido de http://www.it.uc3m.es/jvillena/irc/practicas/06-07/28.pdf
21. Of Waikato T, New Zelanda H (2018) Obtenido de https://www.cs.waikato.ac.nz/
22. Adasme AD, Salomón Díaz E (2013) Análisis de Patrones de Compra de Tiendas Retail utilizando Business Intelligence. Facultad de economia y negocios Universidad de Chile, Santiago de Chile
23. Pallres Cabrera F (2014) Desarrollo de un Modelo Basado en Machine Learning para la predicción de la demanda de habitaciones y ocupacion en el sector hotelero. Universidad Tecnológica de Bolivar, Catagena
24. Paz Villaseca DI (2017) Herramientas deep learning para predictive analytics. Universidad de Concepción Chile, Chile
25. arimetrics. (s.f.). arimetrics.com. Obtenido de https://www.arimetrics.com/glosario-digital/tableau

Analysis of Mental Fatigue Under Delivery Pressure and Considering Creativity and Precision to Organize and Distribute a Diorama to Represent Social Issues Based on Cultural Algorithms

Alberto Ochoa-Zezzatti, José Mejia, Jose Diaz, Patricia Sánchez-Solís, Vicente García, Gilberto Rivera, and Rogelio Florencia-Juárez

Abstract Mental fatigue is characterized by excessive work under pressure, which occurs when creativity, precision and organization of the material produced are required. Considering a case study, the behavior of a group of people in a scenario similar to the one that occurs in Industry 4.0 considering people of generation Z was analyzed by means of a metaheuristic. This case of study attempts to explain this innovative representation and location of societies. The creation of a diorama and, therefore, the correct arrangement of issues in it represents a manufacturing model that implies precision, creativity, and optimization of space, all considering the restriction of time and the representation of the symbolic capital of the issues in a physical space. This research aims to reveal and blur how we adapt to quality standards during work involve several tasks under pressure. In addition, this research addresses a problem in the literature on social modeling associated with the selection of adequate locations in decoration to make the correct distribution of a diorama of imaginary groups of societies to specify the relationships between them, associated with their main attributes to represent symbolic capital of each one. The case study presented deals with the diversity of cultural patterns described in Memory Alpha. We use several main attributes and a range of diverse adequate locations. The purpose of this research is to apply the computational solution of cultural algorithms to solve the problem and later represent the diorama distribution decoration. Memory Alpha is organized by 1,287 societies -Memory Alpha repository-, which shows that the pairing of social themes allows a correct selection of adequate locations.

Keywords Analysis of mental fatigue · Cultural algorithms · Social modeling · Multi-combinatorial problem · Pattern recognition

---

A. Ochoa-Zezzatti (✉) · J. Mejia · J. Diaz · P. Sánchez-Solís · V. García · G. Rivera · R. Florencia-Juárez
Universidad Autónoma de Ciudad Juárez, 32310 Juárez, Chihuahua, México
e-mail: alberto.ochoa@uacj.mx

© The Author(s), under exclusive license to Springer Nature Switzerland AG 2021

A. Ochoa-Zezzatti et al. (eds.), *Technological and Industrial Applications Associated with Intelligent Logistics*, Lecture Notes in Intelligent Transportation and Infrastructure,
https://doi.org/10.1007/978-3-030-68655-0_20

1 Introduction

In this research, we focused our attention on a practical problem from literature related to societies modeling, which is the selection of visual representation including adequate locations and design of features associated with their symbolic capital, in this case, using a diorama which allows include the position that keeps relationships between societies these to focus on design decoration distribution diorama. Our proposed solution uses a hybridization of two techniques: adequate location graph (for representation) and cultural algorithms (to define the distribution and social relationship among them), subsequently represent the adequate location graph in a diorama. As it is described in [1, 2], cultural algorithm agents can only select from a limited repository of adequate locations and forms. For this research, we use the societies described in [3], which is a repository of information about artificial societies [4] that are divided into four quadrants. The selection of each attribute and its visual representation involves the skill evaluation of several individuals from diverse quadrants. The social networking representation requires the development of similarity measures that allow establishing locations in dioramas to carry out a comprehensive analysis of behavioral changes in societies [5]. In addition, we establish an index to measure the workload and mental fatigue associated with the development of the diorama and the times established for it.

2 Cultural Algorithms

Cultural Algorithms (CAs) were developed by Reynolds [6, 7] as a complement to the metaphor used by evolutionary algorithms, which had focused on the concepts of genetics and natural evolution. CAs are based on the theories of anthropologists, sociologists, and archaeologists, who have tried to model the evolution as a process of cultural evolution [8]. The belief space characterizes the CAs of other evolutionary algorithms; it is used to store the knowledge acquired from previous generations. The information in this space must be accessible to any individual who may use it to change their behavior and their respective proposed solution. To join the belief space and the population is necessary to establish a communication protocol, which dictates rules of the type of information to be exchanged between the spaces. This protocol defines the acceptance and influence functions. The acceptance function is responsible for accepting the information or experience that the individuals have obtained in the current generation and bring it to belief space. On the other hand, the influence function is responsible for "influence" like variation operator of individuals (such as crossover and mutation in the case of genetic algorithms). This means that this function has a type of pressure on individuals resulting from the application of variation operators to approach the desirable behavior and away from the undesirable, according to the information stored in the belief space.

Figure 1 shows the interaction between the belief space and population space. The population space has been worked in the same way that evolutionary algorithms, i.e., the population consists of a set of individuals where each has an independent feature set for others; it is possible to determine their suitability (fitness). Over time, such individuals could be replaced by some of their descendants, obtained from a set of operators applied to the population. In the literature about CAs, this space is defined based on an evolutionary algorithm that has operators of mutation and selection. Adding information from the problem domain knowledge to the evolutionary stage, this algorithm must worry about the interaction of both spaces. The interaction between the two spaces makes the CA more complex than a traditional evolutionary algorithm. Algorithm 1 shows a CAs.

Algorithm 1. Cultural Algorithm

```
Begin
 t=0;
 Initialize POP(t); // Initialization of population
 Initialize BLF(t); // Initialization of believing space
 Evaluate POP(t);
 While (Do not condition of term t=t+1)
  Vote (BLF (t), Accept (POP(t)))) ;
  Adjust (BLF (t));
  Evolve (POP(t), Influence (BLF(t)));
  t = t +1;
  Select POP(t) from POP(t-1);
 End While
End
```

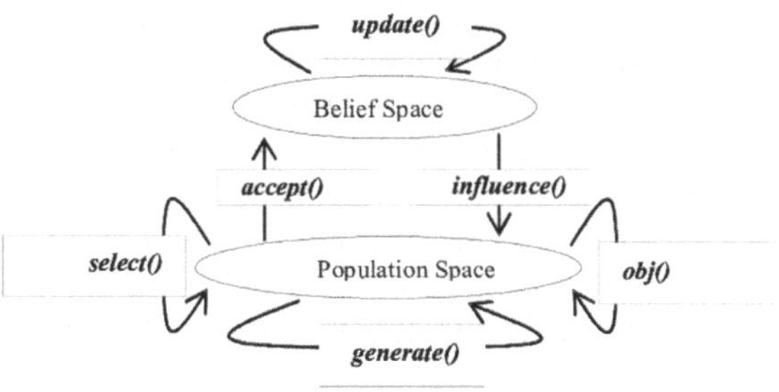

Fig. 1 Components and functions in a cultural algorithm

3 Selection of Relationships in a Workforce to Determine Mental Fatigue

A social network is a theoretical construct useful to study social relationships. As it is mentioned in [9], the diorama and the graphs are used to represent social networks. The representation of a social network can consist of one or more graphs where these graphs conceptualize the network. Social modeling is a combinatorial problem that can be solved with CA, determining a specific solution according to the different perspectives of the population space (see Eq. 1). The main reason for using this innovative algorithm is the potential to describe different application domains with uncertainty solutions in random situations such as classification of languages, architecture, transportation, race competitions, logistics, mental fatigue, and many different aspects and algorithmic problems such as:

- Cooperative Interactive Based on Dynamic Knowledge and Alliance to find the solution in a Diorama.
- Multi-population Cooperative belief space.
- Cooperative interactive information to adopt knowledge migration.

$$\beta = \sum_{i}^{n} abs\left[corr\left(W_i Attribute_i^A,\ W_i Attribute_i^B\right)\right] \tag{1}$$

where:

β	Index associated with creativity and mental fatigue.
W_i	Vector of weights that represent the importance of attribute i.
A, B	They represent the societies whose similarity we want to compare.
$Attribute_i$	Vector of the attributes for each society.
$corr$	Correlation de Pearson.
abs	Absolute values from relative data.

A graph is an abstract representation of a set of objects where some pairs of the objects are connected by links. For this work, each object in the graph represents one society, and the link represents the relationship between societies. In the CA to measure the similarity between societies from Memory Alpha, the Pearson correlation function is used. This measure allows us to relate each society to the others; also, to organize the societies in clusters (64 adequate locations). We have a weight for each attribute, which represents the importance of one attribute to compare among two societies. It means that the sum of the weighted attribute correlation between two societies gives the measure of similarity, as it is shown in Fig. 2. The greater the correlation increase, the stronger the relationship between societies. If the correlation value is less than a threshold of 0.1, we assume an empty relationship between the societies (zero value); the vector of weights and attributes are described. The correct location selection is used (1) only in the societies that have the same adequate location.

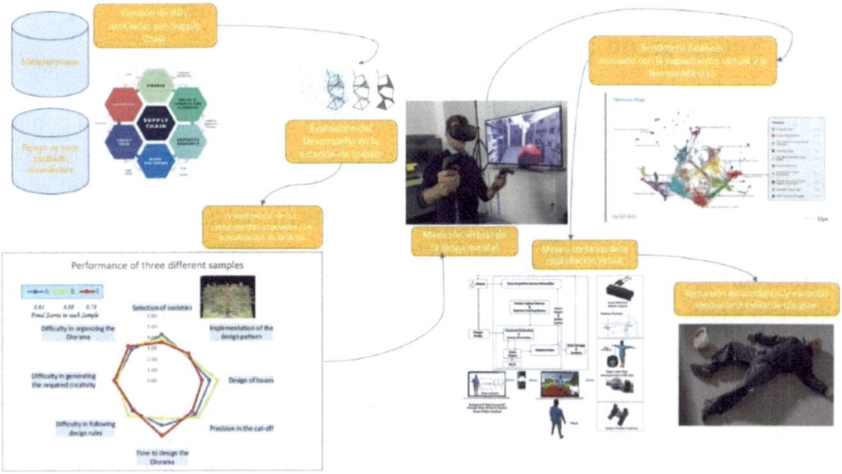

Fig. 2 Conceptual diagram detailed diverse aspects to determine mental fatigue in a workforce produce issues in a Diorama and its social representation

Therefore, each agent in the CA has the links between societies and their adequate locations, where Eq. (1) is used as a fitness function.

4 Adequate Location Selection Based on Visual Representation

From the point of view of agents, this combinatorial optimization problem is very complex, since the best location of an individual representing a society with respect to the other representatives is unknown. In the algorithm proposed for cultural change, individuals in the belief space (beliefscape) through their best paradigm (BestParadigm), are set to zero to represent the fact that culture increases the number of expectations associated with the location of a society with respect for others, encouraging behavior associated with the best paradigm (BestParadigm). For this, we selected 1244 societies described in [3] and characterized their social behavior based on eight attributes: emotional control, ability to fight, intelligence, agility, force, resistance, social leadership, and speed. These characteristics make it possible to describe both society and the individual. When the comparison between societies is analyzed, the main factors are associated with the most representative adequate locations like a country with the adequate locations of its national flag in a scenario.

5 Multiple Matching to Determinate and Evaluate the Selection on a Location in a Diorama

The multiple matching is a series of seven evaluations according to different combinations of adequate locations and wardrobe over 50 runs in different scenarios. The bidding process for the final selection of one wardrobe to each issue is done when different features are analyzed. In the interest of proportional development, individuals from all quadrants should be represented in the diorama based on the interest generated in each one. In the evaluation phase, preference will be given to the topics with the most sociocultural similarities; 47 topics will be selected to compete in these comparisons. Each issue agrees and will contract to participate in exactly seven of these evaluations. Issues must rank their preference of tournaments once the final list of multiple matching is evaluated, and the algorithm evaluates these. The hybrid algorithm reserves the right to assign societies to evaluate their wardrobe according to organizational needs and societies for each comparison and assign the societies list before the cycle begins. Each evaluation will have until 120 societies playing over a schedule of seventeen runs. In view of an objective of harmonization by part of a hybrid algorithm, it will be programmed to meet the comparative time of different similarities using a round of multiple matching analyses and based on the gender assigned to an issue, as it is shown in Fig. 3. The societies that qualify for selection in a diorama will be chosen on the following prioritized basis. For the first cycle of similarity, all societies in Memory Alpha (i.e., Aenar, Ha'Dara, or Ventu Societies) and similarity matches will be invited to participate in different comparators. Given the organization of each society, these matches in each round in the algorithm. All these societies will be asked to commit their participation in the evaluation of each series. If any of these societies decline to participate in the series, the algorithm may nominate one society to replace it, and this society must be ranked among the best societies on Memory Alpha. Based on an average calculation to two decimal places, in the rating list of the comparator series before the start of the cycle, twenty qualified will be selected (excluding the seven societies that will compare the matches). If societies have the same average rating, the number of similarities played (rating period) will

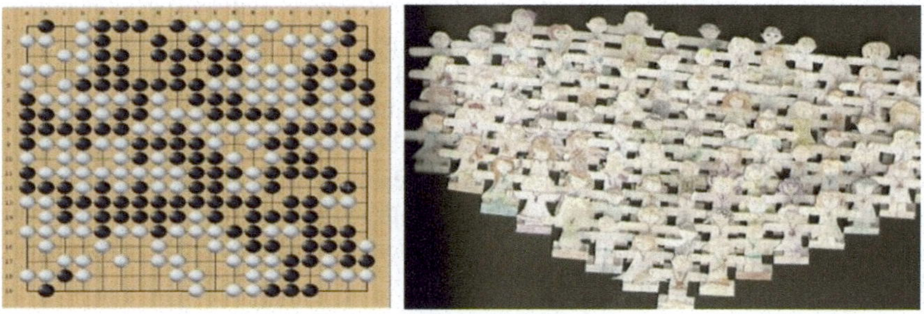

Fig. 3 Comparative model for assigning locations in a diorama, the black dots represent lass issues and the white dots represent lad issues and the adequate clustering

be used to determine ranking. To ensure active participation in the future, a minimum of 25 games is being recommended in the four rating lists, including and prior to the rating list. If any society does not accept to play in the multiple matching series, the selection process using the average rating plus the number of games played in the rating period will be adopted until the number of qualifiers required for the multiple matching series is completed. Using this technique of comparative and classification is possible to organize un adequate ranking based on all possible possibilities to participate in the final issues accommodation.

6 Experimentation

In order to be able similar, the most efficient arrangement of individuals in a social network is developed an atmosphere able to store the data of each one of the representing individuals of each society based on CA, with the purpose of distributing by an optimal form each one of the evaluated societies [2, 10]. The main experiment consisted of detailing each one of the 1244 communities in the CA, with 500 agents and 200 different believes in the belief space, one stop condition of 75 epochs; this allowed to generate the best selection of each quadrant and their possible location in a diorama, which was obtained after comparing the different cultural and social similarities from each community, and to evaluate with Multiple Matching Model each one of them as in [11]. The vector of weight used in the fitness function is $W_i =$ [0.6, 0.7, 0.8, 0.5, 0.6, 0.4, 0.9,0.5, 0.4, 0.9, 0.6, 0.7], which represents the importance of diverse attributes: emotional control, ability to fight, intelligence, agility, force, resistance, social leadership, and speed -in another's-, in the same order, as the design of experiments proposes in [2]. The CA will select the adequate location and features of each society based on the similarity of attributes and determine a kind of society. The time related to building a diorama is shown in Fig. 4. We use multivariable analysis to determine creativity and mental fatigue in our three samples.

Each attribute is represented by a discrete value in a range from 0 to 5, where 0 means absence and 5 the highest value for the attribute. The design of the experiment consists of an orthogonal array test with interactions between the variables (the eight attributes mentioned in a range from 0 to 5). These variables are studied within an adequate location range (1 to 64 diverse positions in the diorama). An orthogonal array is L-N (2**8); in other words, 8 factors in N executions. The value of N is defined by the combination of the 8 possible values of the variables, also values in the adequate location range. Table 1 lists some possible scenarios result from combining the values of the attributes and the specific adequate location to represent this issue (society). The results permit us to analyze the effect of the variables on the adequate location selection in all the possible combinations of values, as was proposed in [2].

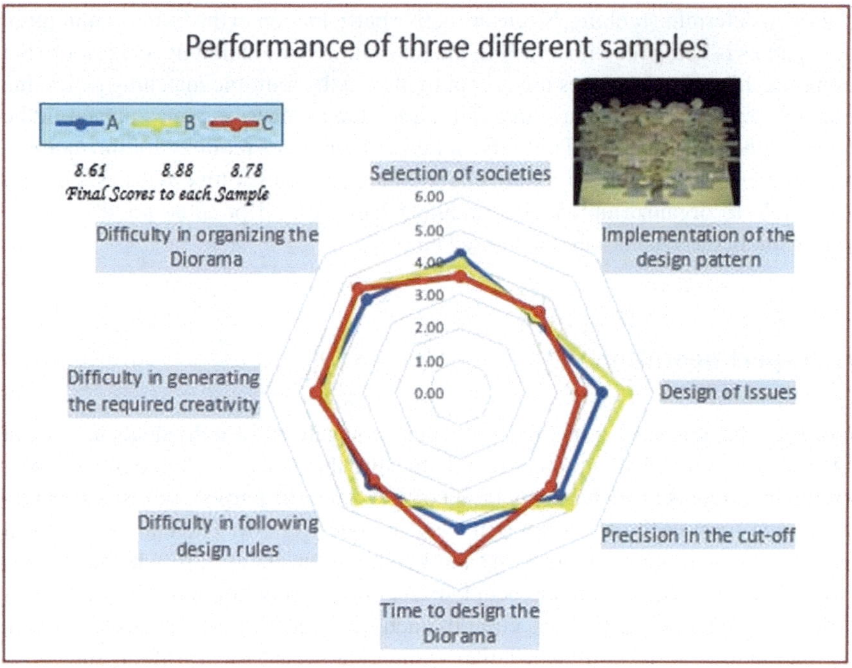

Fig. 4 Multivariable analysis of three samples analyzed to determine mental fatigue

With this orthogonal array test, we try to reorganize the different attributes and specify the best possibilities for adequate correct solutions (skills) in each society. The different attributes were used to identify the real possibilities of improving society in each and potential environment (see Fig. 5), and to specify the correlations with other societies.

7 Conclusions and Future Research

An interesting characteristic observed in this experiment was the diversity of the cultural patterns established by each community, which represents a unique and innovative form of adaptive behavior that solves a computational problem that does not attempt to group societies only with a factor associated with their external appearance. (The attributes of each society). The generated configurations can be metaphorically related to the knowledge of the behavior of the community with respect to an optimization problem (to select culturally until 47 similar societies, without being of the same quadrant of [2, 3]). The developed tool classified each of the societies belonging to each quadrant, with different costumes for societies that included linguistic and

Table 1 Orthogonal array test

Emotional control	Ability to fight	Intelligence	Agility	Force	Resistance	Social leadership	Speed	Negotiation level	Natural resources	Technology implementation	Adequate location
0	1	2	2	3	3	4	5	2	3	3	1
0	1	2	2	3	4	5	5	2	3	4	1
1	1	3	2	4	4	2	1	3	3	5	2
1	1	3	2	5	3	2	1	4	4	5	2
1	0	0	3	4	2	3	5	5	4	5	3

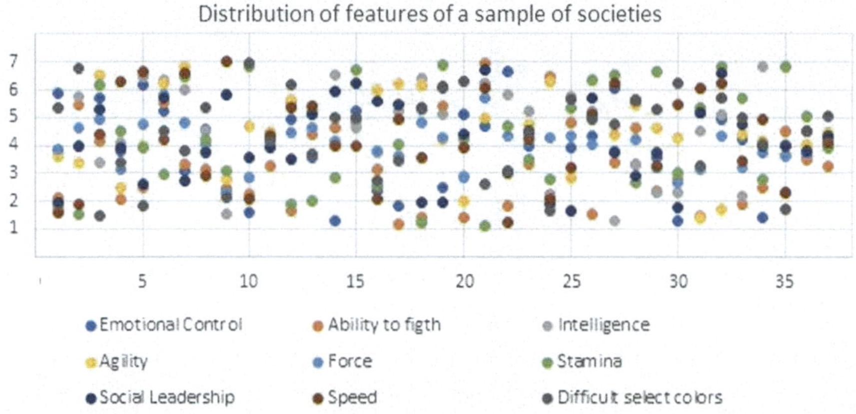

Fig. 5 Values after applying orthogonal array to evaluate the dataset associated with some features of 37 societies to organize in the Diorama

cultural identity. This permit identifies changes over time with respect to other societies (see Fig. 6). A decisive and extremely important aspect in order to correctly evaluate the final product developed in this case study was the proposal of an index to obtain the correlation between the aesthetics of the diorama and the time dedicated to it, considering the different restrictions needed to develop it.

Using CAs, we improved the understanding substantially to obtain the change of "best paradigm", because we appropriately classified the agent communities basing to us on an approach to the relationships that keeps their attributes. This allowed us to understand that the concept of "cooperative collaborative information" exists with

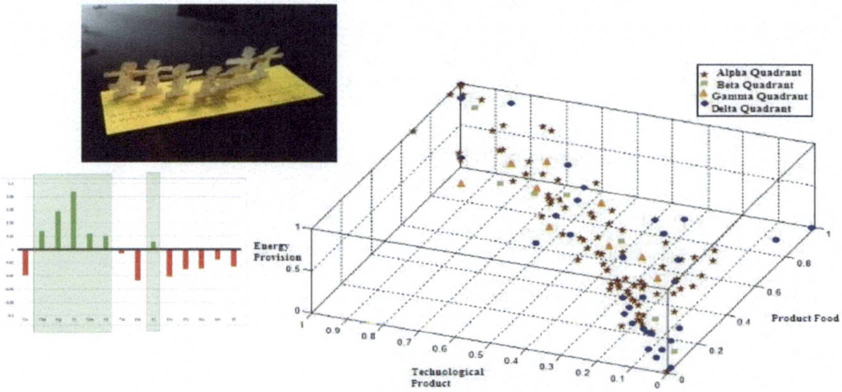

Fig. 6 Visual representation of societies analyzed according to the orthogonal array test and multivariable analysis of specific variables associated with mental fatigue

base in the determination of the function of acceptance on the part of the rest from the communities to the proposed location for the rest of others [2, 12].

Future Research

The analysis of the level and degree of cognitive knowledge of each community is an aspect that is desired to evaluate for future research. On the other hand, in order to understand the true similarities that different societies have based on the characteristics that make them cluster contributors and, in turn, allow them to keep their own identity, in [6] it shows that the small variations go beyond the characteristics phenotypic and are mainly associated with tastes and similar characteristics developed over time. The CAs can be used in the Evolutionary Robotic field, when the problem needs social interaction and decision, for example, in the training phase described in [7], and to organize a group of robots for collaborative tasks; also, could be used to distribute workgroups, social groups or social networking [9]. One future research using CAs is pattern recognition in a social database, for example fashion styling to determine criminal behaviors [13], as it is possible to see in Fig. 7. In addition, the CAs can help in the construction of the Neuromarketing and Cyberbullying model by determining and adjusting correctly the social behavior of a group of people over time and their particular way of changing their way of thinking.

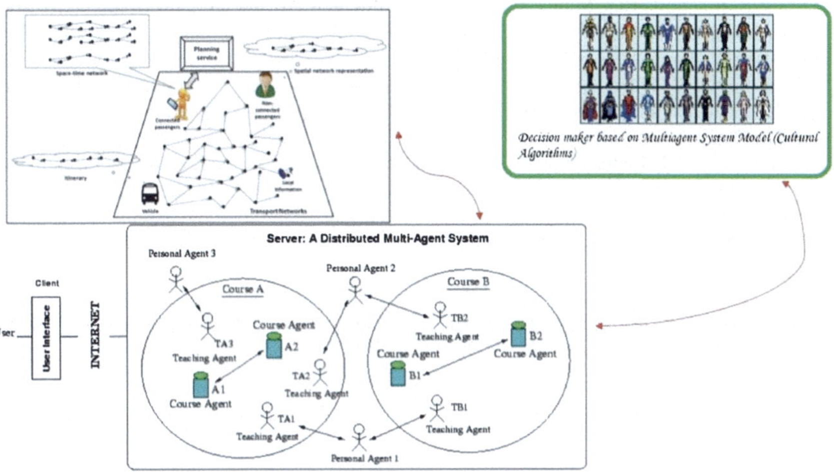

Fig. 7 Future enhancements to cultural algorithms include modeling a collaborative soul sense to make decisions considering future side effects

References

1. Reynolds RG (2018) Culture on the edge of chaos—cultural algorithms and the foundations of social intelligence. Springer. https://doi.org/10.1007/978-3-319-74171-0
2. Cruz-Álvarez V.R., Montes-Gonzalez F., Ochoa A., Palacios-Leyva R.E. (2012) Distribution and Selection of Adequate locations on a Diorama to Represent Social Issues Using Cultural Algorithms and Graph Adequate locationing. In: Omatu, S., De Paz Santana, J., González, S., Molina, J., Bernardos, A., Rodríguez, J. (eds.) Distributed Computing and Artificial Intelligence. Advances in Intelligent and Soft Computing, vol 151. Springer, Berlin, Heidelberg. https://doi.org/10.1007/978-3-642-28765-7_8
3. Zheng Y, Zhong B, Yang F (2018) When algorithms meet journalism: the user perception to automated news in a cross-cultural context. Comput Hum Behav 86:266–275. https://doi.org/10.1016/j.chb.2018.04.046
4. Guo YN, Zhang P, Cheng J, Wang C, Gong D (2018) Interval multi-objective quantum-inspired cultural algorithms. Neural Comput Appl 30(3):709–722. https://doi.org/10.1007/s00521-016-2572-5
5. Hué, T. et al.: The Emergence of Social Network Hierarchy Using Cultural Algorithms, VLDB'06, Seoul, Korea (2006)
6. Reynolds, R.G., Sverdlik, W.: Problem solving using cultural algorithms. In: International Conference on Evolutionary Computation. IEEE World Congress on Computational Intelligence, vol. 2, pp. 645–650. Orlando (1994). https://doi.org/10.1109/ICEC.1994.349983
7. Selvarajah K, Kobti Z, Kargar M (2020) Cultural algorithms for cluster hires in social networks. Proc. Comput. Sci. 170:514–521. https://doi.org/10.1016/j.procs.2020.03.117
8. Ponce J, Karahoca A (2009) Data Mining and Knowledge Discovery in Real Life Applications. IN-Tech, Vienna
9. Kobti Z, Reynolds RG, Kohler TA (2006) The emergence of social network hierarchy using cultural algorithms. Int J Artif Intell Tools 15(6):963–978. https://doi.org/10.1142/S0218213006003065
10. Song A, Niu C, Ding X, Xu X, Song Z (2019) Mental fatigue prediction model based on multimodal fusion. IEEE Access 7:177056–177062. https://doi.org/10.1109/ACCESS.2019.2941043
11. Herlambang MB, Taatgen NA, Cnossen F (2019) The role of motivation as a factor in mental fatigue. Hum Factors 61(7):1171–1185. https://doi.org/10.1177/0018720819828569
12. Ochoa A, Hernández A, Jons S, Montes F, Ponce J, Janacek L (2009) Dyoram's representation using a mosaic image. Int. J. Virtual Reality 8(3):1–4. https://doi.org/10.20870/IJVR.2009.8.3.2734
13. Li, F., Chen, C.-H., Xu, G., Khoo, L.P., Liu, Y.: Proactive mental fatigue detection of traffic control operators using bagged trees and gaze-bin analysis. Adv. Eng. Inform. **42**, 100987 (2019). https://doi.org/10.1016/j.aei.2019.100987

Medicine Inventory Control System Through Fuzzy Logic and Genetic Algorithms: Applied to a Biopharmaceutical

Christian A. Mejía Ramírez, Martín Montes Rivera, Rodolfo R. Medina Ramírez, Rosa M. Ramírez Prado, Carlos M. Gaitán Mercado, and Alberto Ochoa-Zezzatti

Abstract The control of the pharmacy inventory of a public hospital is one of the tasks that the managers of the area face since the inexistence or shortage of medicine incurs the aggravation of a patient. Recently, in a medical services institution, located in the state of Aguascalientes, Mexico, there is an area of opportunity in the inventory control of the drug laronidase. In this institution, a certain number of units of this biopharmaceutical are used weekly, destined to treat an enzymatic deficiency. The area of opportunity for the control of the drug inventory is related to the fact that it is of high cost, its life span is six months after its elaboration in the presentation used in the hospital and it is supplied for life, since it is a congenital disorder. Due to the imprecision of the time needed for the procedures in the purchase management of the biopharmaceutical, the inventory of the drug is affected causing that sometimes there is no immediate availability in the institution. Currently, optimization techniques applied to inventory control are a practical way to make decisions regarding reorder schedule and the number of units to buy (batch), so that the least cost is incurred in maintenance and acquisition. However, fuzzy control techniques are an alternative to such procedures when high-precision information is not available, or when only the

C. A. Mejía Ramírez (✉) · M. Montes Rivera · R. R. Medina Ramírez · R. M. Ramírez Prado
Universidad Politécnica de Aguascalientes, Aguascalientes, Mexico
e-mail: christian.mejia@upa.edu.mx

M. Montes Rivera
e-mail: martin.montes@upa.edu.mx

R. R. Medina Ramírez
e-mail: rodolfo.medina@upa.edu.mx

R. M. Ramírez Prado
e-mail: rosa.ramirez@upa.edu.mx

C. M. Gaitán Mercado
Hospital Hidalgo, Aguascalientes, Mexico
e-mail: unimedint@hotmail.com

A. Ochoa-Zezzatti
Universidad Autónoma de Ciudad Juárez, Ciudad Juárez, Mexico
e-mail: alberto.ochoa@uacj.mx

© The Author(s), under exclusive license to Springer Nature Switzerland AG 2021
A. Ochoa-Zezzatti et al. (eds.), *Technological and Industrial Applications Associated with Intelligent Logistics*, Lecture Notes in Intelligent Transportation and Infrastructure,
https://doi.org/10.1007/978-3-030-68655-0_21

experience of the expert in the area is available. On the other hand, the optimization by means of genetic algorithms (GA) is an heuristic search alternative for finding the vertices in the fuzzy inference system; Therefore, in this paper is proposed, the use of an hybrid system (fuzzy control and genetic algorithms) for programming the date order and the size of a lot for the medicine laronidase given its stock error and its derivative.

1 Introduction

1.1 Description of Problem

Currently, in a medical services institution, located in the state of Aguascalientes, Mexico, three patients with lysosomal disease are being treated. They consume 13 units of biopharmaceutical laronidase weekly; this drug is costly and it is administered to patients with this condition for life.

Once that laronidase has been prepared, the medicine has a six-month shelf life. Due to the variability in the time required for purchasing, the permits, and the release of studies required by both the manufacturer and the distributor, laronidase inventory can be affected in such a way that sometimes there is no immediate availability at the institution.

The head of the pharmacy department at this institution is responsible for requesting the purchase of the drug from the corresponding department; however, sometimes there is not enough budget to purchase the requested units, making a partial purchase with fewer units than the required ones, and in that sense affecting the purchase plan.

Considering the previous problem and the efficiency of the diffuse inference systems tuned to Genetic Algorithms, described in the related works, in this work it is designed an inventory control system for the laronidase drug, taking as reference the registration of the 100 weeks biopharmaceutical unit level.

This case study looks for an inventory control system that proposes the moment to request the acquisition and the number of units to order. The proposed fuzzy system considers the restrictions in expiration, the purchasing time, as well as the following recommendations given by the head of the department of pharmacy:

- There is a storage space for 500 units, which is not fully used now.
- The reorder point occurs when the inventory level has enough units to satisfy the demand of two months.
- The planned order lot size is equal to the number of units to satisfy three months of demand and is ideally requested when the inventory level reaches the reorder point; however, when a sufficient budget is not available, the corresponding department makes a purchase for fewer units.

- There is no exact date when the budget for the acquisition of the biopharmaceutical arrives; however, when units are required, a purchase is made for one month or less, and when the budget arrives, it is purchased as much as possible, considering the space restrictions and the expiration date of the biopharmaceutical.
- The average time for the drug to arrive at the hospital, once the purchase request has been made, is four to five weeks; however, sometimes it has taken up to two months, or it has arrived in one day.
- The reserve (safety stock) is equal to the drug amount needed to cover one month of demand.
- There is only one supplier with an exclusive charter in the whole country.

1.2 Related Jobs

Recent research on drug inventory control systems in public hospitals is based on the use of deterministic and stochastic optimization techniques, which mostly seek to schedule the resupply date and minimize both acquisition and maintenance costs, based on the demand and the duration of the period, as seen in the articles by Uthayakumar and Priyan of 2013 and Priyan and Uthayakumar of 2014, where they apply Lagrange multipliers for the solution of a non-linear programming model in the decision making of the inventory control of three drugs [1, 2].

Another fundamental tool used in hospital pharmacies is the waiting line theory, which is applied in the optimization of the waiting time of a patient, as well as the improvement of the care by the staff in charge. This tool is useful for service managers, as it can increase the satisfaction of all relevant groups, patients, employees, and managers. In the 2013 article by Lakshmi and Sivakumar, several of these applications are mentioned and how they have been helpful for the benefit of patients who go to a public health institution's pharmacy [3].

Particle Swarm Optimization (PSO) techniques have been helpful in supply chain issues. In the 2014 article by Mousavi, Bahreininejad, Musa, and Yusof, a two-tier supply chain network was investigated for a seasonal inventory control problem. The main objective of the problem was to find quantity of products to purchase by buyers, as well as to determine the optimal location of suppliers, so that the total cost of the supply chain including transportation, retention and purchase be minimized. To solve their proposed model, a Modified Particle Swarm Optimization (MPSO) algorithm was used and validated by a Genetic Algorithm (GA) [4].

In the article by Aengchuan and Phruksaphanrat of 2015, a Fuzzy Inference System (FIS) with Artificial Neural Networks (ANNs) called Adaptive Neuro-Diffusion Inference System (ANFIS) is used for the inventory control of materials in the manufacture of three wood products in a Thai furniture company. In the ANFIS system, the functions of trapezoid, Gaussian, and bell memberships are used. The model with Gaussian membership function achieves total cost savings compared with the stochastic economic model [5].

A review of publications on inventory management of surgical supplies and sterile instruments is presented in the 2018 article by Ahmadi, Masel, Metcalf, and Schuller. In this article, it is mentioned that the use of deterministic methodologies and in some cases stochastic, could conduct impractical models, since they do not consider the operational and interruption risks, therefore, but if the models developed consider this elements, they will reflect the real problem and with the results of better level of service in practice [6].

As mentioned above, the use of deterministic and stochastic methodologies in problems related to inventory management in hospitals prevents the models from being practical, since they require highly accurate information. Therefore, one proposal is to work with heuristic methodologies, but instead of linking them directly to the problem, they must be used for tuning a fuzzy system since they together provide outstanding results in practice as described in [6].

1.3 Theoretical Framework

An inventory is defined as assets or materials held in reserve by an organization or institution for future use. In the sense of public health institutions, it refers to items such as medicines, surgical supplies, and substances for sterilization, among others [7].

One of the reasons why a health care institution maintains an inventory is because of the difficulties in predicting medication use needs and demand. Therefore, the inventory serves as a reserve against fluctuating and uncertain use and keeps items available in case they are required by an organization or institution [7].

The demand is a factor that influences the size of an inventory, which is defined as the number of attempts to purchase an item or service, depending on the situation faced by the organization or institution [7].

A lot is defined as the set of counted, weighed, or measured units that form the ordered quantity of a purchase order. In an inventory control system, the size of the lot is expressed in the number of kilograms, meters, liters, units, or its monetary value [7].

Purchasing time is defined as the number of days, weeks, or months it takes for a purchase order to arrive in the warehouse after it has been requested from the vendor or buyer [7].

The reserve or safety stock is defined as the quantity of materials or products that are kept in stock as a safety precaution, or for cases in which the quantities calculated for consumption during the delivery period become exhausted, either due to a delay in delivery or an increase in demand [7].

A reorder point is defined as the inventory level that indicates to the purchasing department, the need to order by quantity to recover the maximum stock level, depending on the available storage space [7].

The term classical set (crisp) is not defined mathematically but is intuitively understood as a collection or group of objects with one or more characteristics in common. Sets are commonly denoted by capital letters of the alphabet and their elements are written between braces, { }, separated by commas, "," [8].

The set of all objects considered in a situation is called a universal set or universe and is denoted by U. We say that a set A is a subset of B if and only if every element of A is an element of B [8].

Let U be a universe. A fuzzy set $\mathcal{A}$ is defined as a set of ordered pairs as follow [9]:

$$\mathcal{A} = \{(x, \mu_{\mathcal{A}}(x)) : x \in U, \mu_{\mathcal{A}}(x) \in [0, 1]\} \tag{1}$$

where $\mu_{\mathcal{A}}$ is called the membership function (MF), and it specifies the degree to which any $x \in U$ belongs to the set $\mathcal{A}$, that is, the definition (1) associates each x element in U with a single real number $\mu_{\mathcal{A}}(x)$ in the interval [0, 1].

In comparison with fuzzy sets, a crisp set is a set in which all its elements have a level of belonging equal to 1; that is, the elements belong entirely to the set. The elements of such sets are known as total belonging [8].

An example of a belonging function is the generalized bell function, which is specified by three parameters $\{a, b, c\}$, and defined in the Eq. (2).

$$\text{bell}(x; a, b, c) = \frac{1}{1 + \left|\frac{x-c}{a}\right|^{2b}} \tag{2}$$

where parameter b is usually positive. (If b is negative, the shape of this MF becomes an inverted bell) [10].

Given the parameters a, b and c, with $a > 0$, of the bell function, the degree of belonging equal to $\frac{1}{2}$ is obtained when $x = c - a$ and $x = c + a$ and the slope of the line tangent to the curve at those values is equal to $m = \frac{b}{2a}$ and $m = -\frac{b}{2a}$, respectively. The derivative of this function in the set $(-\infty, c) \cup (c, \infty)$, with $a > 0$, is given by the function (3) [10].

$$\text{bell}'(x; a, b, c) = \begin{cases} \dfrac{\frac{2b}{a}\left(-\frac{x-c}{a}\right)^{2b-1}}{\left[1+\left(-\frac{x-c}{a}\right)^{2b}\right]^2} \text{if} x < c \\ \dfrac{-\frac{2b}{a}\left(\frac{x-c}{a}\right)^{2b-1}}{\left[1+\left(\frac{x-c}{a}\right)^{2b}\right]^2} \text{if} x > c \end{cases} \tag{3}$$

Now, if $x = c - a$ and $x = c + a$, with $a > 0$, then $\text{bell}'(x; a, b, c) = \frac{b}{2a}$ and $\text{bell}'(x; a, b, c) = -\frac{b}{2a}$, respectively, which is just the slope of the line tangent to the curve when the value of belonged is 0.5 [10].

Let $\mathcal{A}$ and $\mathcal{B}$ two fuzzy sets with membership function $\mu_{\mathcal{A}}(x)$ and $\mu_{\mathcal{B}}(x)$, respectively. The intersection operation of $\mathcal{A}$ and $\mathcal{B}$ is defined as the set of $x \in U$ with a belonging function given by Eq. (4) and is denoted by $\mathcal{A} \cap \mathcal{B}$ [8].

$$\mu_{\mathcal{A}\cap\mathcal{B}}(x) = \min\{\mu_{\mathcal{A}}(x), \mu_{\mathcal{B}}(x)\} \tag{4}$$

The union operation of $\mathcal{A}$ and $\mathcal{B}$ is defined as the set of $x \in U$ with a belonging function given by Eq. (5) and is denoted by $\mathcal{A} \cup \mathcal{B}$ [8].

$$\mu_{\mathcal{A}\cup\mathcal{B}}(x) = \max\{\mu_{\mathcal{A}}(x), \mu_{\mathcal{B}}(x)\} \tag{5}$$

Fuzzy logic models use fuzzy sets to handle and describe vague and complex phenomena and use logical operations to reach a conclusion. Fuzzy logic applied to control problems is related to a field of knowledge called Fuzzy Logic Control (FLC). These are control problems in an environment of uncertainty and imprecision. It is very effective when high precision is not required, and the control object has variables available for measurement or estimation [11].

The inference system of Mamdani is one of the methods used in this control technique. It was proposed by Ebrahim Mamdani in 1975. The process is carried out in four steps [11]:

1. Fuzzification of the input variables.
2. Evaluation of rules.
3. Aggregation of rule outputs.
4. Defuzzification.

The Fuzzification process establishes a relationship between entry points (real numbers) to the system and their corresponding fuzzy sets. It is a mathematical procedure in which each element that enters is assigned a value of belonging according to each of the sets considered [11].

Fuzzy controllers use rules for its operation, these rules combine one or more fuzzy input sets called antecedents or premises and associate a fuzzy output set called consequent or consequence. These rules are called fuzzy rules, they are statements of the type If…, then. The fuzzy sets of the antecedent are associated by fuzzy logical operations $\wedge$ (conjunction, and) and $\vee$ (disjunction, or) [11].

Mamdani's (1975) fuzzy rule is a conjunction rule expressed by the minimum operation given in Eq. (6): If r_k is the conclusion or consequent and p_i and q_j are the premises or background, then:

$$\left(p_i \wedge q_j\right) \to r_k = \min\left\{\mu_{\mathcal{A}_i}(x), \mu_{\mathcal{B}_j}(y), \mu_{\mathcal{C}_k}(z)\right\}, r_k = r_{ij} \tag{6}$$

with $i = 1, 2, \cdots, n$, $j = 1, 2, \cdots, m$, $k = 1, 2, \cdots, l$ y $(x, y, z) \in U_1 \times U_2 \times U_3$ [9].

When you evaluate the rules, you get as many fuzzy sets as there are rules. Aggregation is the methodology used to decide the control action as the result of the evaluation of the rule set. One criterion previously used is to group the inferred sets by means of the (maximum) binding operation [10].

The next step in a fuzzy control system is defuzzification; this is a mathematical process used to convert the fuzzy set obtained in the aggregation stage into a real number, which will be the action to perform given inputs of the real problem [10].

A defuzzification procedure previously used with Mamdani's methodology is the centroid or center of gravity (z_{CG}), given by Eq. (7), with which is found the location of the center of gravity of the diffuse output set given in the aggregation stage [11].

$$z_{CG} = \frac{\sum_{k=1}^{l} z_k \mu(z_k)}{\sum_{k=1}^{l} \mu(z_k)} \tag{7}$$

Fuzzy inference systems described with n inputs and a single output, are known as Multiple Input Single Output (MISO) systems, while those with multiple outputs (greater than 1) are known as Multiple Input Multiple Output (MIMO) systems. For the latter systems, the previous format of rules can be generalized or broken down into k MISO-type systems [11].

A Genetic Algorithm (GA) simulates some aspects of the evolutionary theory of species proposed by drawing and the law of heredity proposed by Gregory Mendel. The first genetic algorithm was developed by John Holland in 1975 at the University of Michigan; the pseudocode for his work is listed below [12].

1. Define an objective function (Aptitude).
2. Randomly generate an initial population.
3. Evaluate the elements of the population.
4. Codify the elements of the population.
5. Select two elements of the population (parents) to reproduce.
6. Crossbreeding operation.
7. Mutation of the offspring.
8. Replace elements of the next-generation population with the best elements obtained in steps 6 and 7.
9. Check if the stop condition is met, otherwise return to step 4.

The selection of parents by the tournament method consists of randomly choosing a certain number of individuals from the population, the tournament size (with or without replacement), then the two best individuals from that sample are selected to be crossed. In the article by Zhong, Hu and Zhang in 2005, it is mentioned that the tournament selection is more efficient compared to the roulette proportional selection method, as it has better convergence rate. In this article, a size of 6 individuals is proposed for this method [13].

GAs, as described above, employ crossover points, in which the two individuals selected for the crossing are recombined by means of the selection of cut-off points, to later exchange their coded elements and in this way generate one or more individuals (offspring) that will replace individuals of the previous generation. In De Jong's work of 1975, the behavior of multiple crossing points is investigated, concluding that the crossing in two points represents an improvement in comparison with more than two points; however, having multiple crossing points, allows an exploitation of the search space [14].

The mutation is made by minor changes in the mutated chromosomes. There are different types of mutation, but common algorithms change randomly selected alleles (single gene mutation) or multiple alleles (multiple gene mutation). In the 1986 article of Grefenstette, it is mentioned that a mutation rate above 5% is detrimental to optimal GA performance [15].

If the chromosomes are binary matrices, then the mutation only requires alternating the mutated bits. The amount of total mutations, t_m, is given by the Eq. (8).

$$t_m = \left(N + 2 * n_g\right) * n_a * p_m \tag{8}$$

where N is the population size, n_g is the number of generations, n_a is the number of alleles and p_m is the percentage mutation [12].

2 Methodology

Inventory control is performed by means of a MIMO system with the fuzzy inference system of Mamdani, which estimates the time to make a purchase request and the biopharmaceutical lot size each week, based on the inventory level and the change in the inventory level with respect to the reorder point (reference).

Given the information by the department head of the hospital pharmacy, at the moment an average of 52 units of laronidase are required monthly, therefore, the reserve or safety stock is 52 units, and since the estimated average time of acquisition of the biopharmaceutical is four to five weeks, then the reorder point is 104 units, which is what is necessary to satisfy the current demand of approximately two months.

To make the controller, the following input variables are considered: error (e) and change in error ($\grave{e}$); the error variable measures the difference between the level of inventory n at the end of week t and a reference r, which is the reorder point; that is, the level of inventory that indicates the moment to make a purchase. The error derivative measures the change in error at the end of week t. If this is positive, it indicates that the error at the end of week t increased compared to the error at the end of week $t - 1$; if it is negative, it indicates that the error at the end of week t

decreased compared to the error at the end of week $t - 1$; and if it is zero, it indicates that the error remains the same at the end of both weeks.

The value of the reference in the error variable is $r = 104$, and this indicates the current point of reorder; however, this value may increase or decrease, depending on the weekly demand for units or the increase or decrease of required units of laronidase.

The output variables considered are when to order (c) and how much to order (o). The variable when to order estimates the time (in days) to reach the point of reordering at the end of each week. The variable how much order estimates the number of units of medication required at the end of each week when the point of reorder is reached, or when it is below the reference to complete the maximum inventory level.

For this control system, the plant is the inventory, and the input variables are the number of units that arrive at the hospital pharmacy each week and have already been reviewed by the department head. This variable is called input (I) and is equal to zero when no units arrive. The second input variable is the number of units consumed each week (u), and the third input variable is the inventory level at the end of the previous week ($n(t-1)$). Finally, plant output is the level at the end of each week ($n(t)$).

The difference equation that describes the inventory level at the end of each week is:

$$n(t) = n(t-1) + I(t) - u(t)$$

Since now 13 units are used weekly for three patients, then $u(t) = 13$, then the above equation is written:

$$n(t) = n(t-1) + I(t) - 13$$

Since the inventory level at the end of each week is always non-negative, then $n(t) = 0$, when $n(t-1) + I(t) - u(t) < 0$ and $n(t) = n(t-1) + I(t) - u(t)$ otherwise.

Figure 1 shows the plant to be controlled, which in this case is the biopharmaceutical inventory, with its respective inputs and outputs.

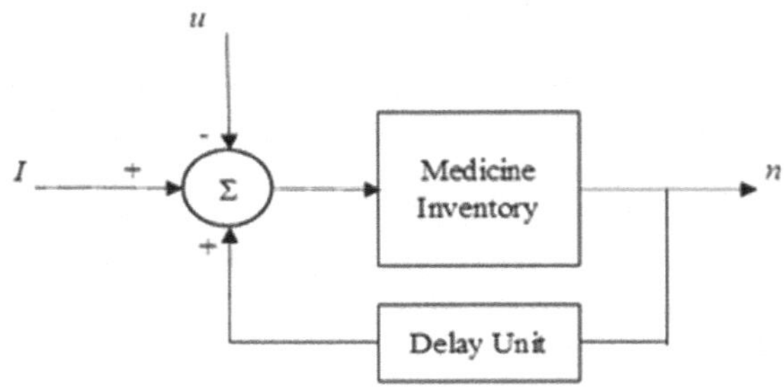

Fig. 1 Diagram of the plant, which in this case is the inventory of the biopharmaceutical

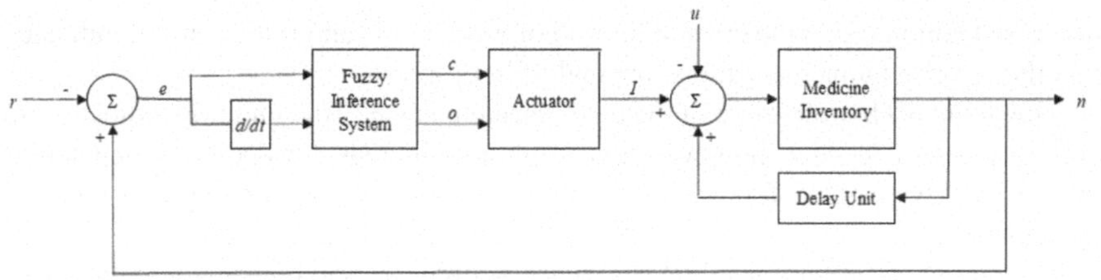

Fig. 2 Diagram of the Fuzzy Inference System in charge of inventory control

Figure 2 shows the fuzzy inference system, in charge of carrying out the control, integrated with the plant to be controlled, which in this case is the inventory of medicine. The system works as follows: At the end of week t, the error is calculated ($e = n - r$) and the change in the error ($\dot{e}$), then the controller will give an estimate of when to make a purchase request (c) and how much will be ordered at that time (o), this process is done every week and when an output c equal to 0 days or less than 7 days is obtained, it is time to request the number of units of drug that the output o.

In the diagram shown in Fig. 2, the actuator is made up of the head of the pharmacy department and the staff in charge of the purchasing department, since they are responsible for requesting and managing the purchase of the drug units indicated by the controller at the time the output indicates when to order, and thus restore the level of inventory.

In the case of this control system, the aim is for the inventory level to always be above the reserve or safety stock, but without reaching excessive quantities, since the life span of the drug is six months, so the value of the reference r indicates only the point of reorder.

Once the purchase is requested, the system will continue to estimate when (c) and how much (o) to order at the end of each week, indicating the output c values in the interval $[0, 7)$ and increasing the value of the variable o, to recover the units used while the requested batch arrives.

When having a I income in the plant at some point, the level of the inventory will be reestablished for estimating when and how much to order at the end of the following weeks. If during any week the requested units have not arrived, or the variable when to order does not indicate a moment to make a purchase, then the actuator gives as output an income, I, equal to zero units.

Since there is storage space available for 500 units of the drug, the universes $U_{\mathcal{E}} = \{x \in \mathbb{Z} : -200 \leq x \leq 300\}$ and $U_{\mathcal{D}} = \{x \in \mathbb{Z} : -500 \leq x \leq 500\}$, respectively, were considered for the error and the error derivative. Due to the expiration of the drug (6 months after its elaboration) and the imprecision of the time it takes to manage the purchase (from immediately to 60 days), the universe was considered for the purchase request $U_{\mathcal{C}} = \{x \in \mathbb{Z} : 0 \leq x \leq 105\}$. For the output "how much order" is considered the universe $U_{\mathcal{O}} = \{x \in \mathbb{Z} : 0 \leq x \leq 500\}$.

To find the control rules and the parameters of the membership functions, a GA was used. For this purpose, 60 data were randomly selected for training, 20 for validation, and 20 for testing from a 100-week database.

Since the universe and parameter search sets are different at each of inputs and outputs, they must be normalized; therefore, the search space is restricted to$\mathcal{C} = \{(x_1, x_2, \cdots, x_n) \in \mathbb{R}^n | x_i \in \mathbb{R}, 0 \leq x_i \leq 1\}$.

The bell function is specified by three parameters: a, b, and c, and it is not necessary for these to comply with a specific order, therefore, the belonging function used for each of the fuzzy sets corresponding to both input and output variables is this function.

Every three consecutive elements of the vector where the optimal one is held, represent the parameters of each function belonging to the diffuse input and output sets used in the control system.

The search intervals for parameters a and c of the bell belonging function for the sets of the error input are $[0, 500]$ and $[-200, 300]$, respectively; and for the derivative of the error are $[0, 1000]$ and $[-500, 500]$, respectively. The functions that map the $[0, 500]$, $[-200, 300]$, $[0, 1000]$ $[-500, 500]$ and intervals to the $[0, 1]$ interval are $f_1(x) = \frac{1}{500}x$, $f_2(x) = \frac{1}{500}x + \frac{2}{5}$, $f_3(x) = \frac{1}{1000}x$ and $f_4(x) = \frac{1}{1000}x + \frac{1}{2}$, respectively.

The search interval for the a and c parameters of the bell membership function for the output sets when to sort is $[0, 150]$ for both, and for the output sets how much to sort is $[0, 500]$ for both. The functions that map the $[0, 150]$ and $[0, 500]$ ranges to the $[0, 1]$ range are $f_5(x) = \frac{1}{105}x$ and $f_6(x) = \frac{1}{500}x$, respectively.

It will then be shown that applying the above mappings to the intervals does not change the membership value for the normalized value. If $a_2 = \alpha a_1$, $c_2 = \alpha c_1 + \beta$ and $y = \alpha x + \beta$, then $\text{bell}(x; a_1, b, c_1) = \text{bell}(y; a_2, b, c_2)$.

$$\begin{aligned}
\text{bell}(x; a_1, b, c_1) &= \frac{1}{1 + \left|\frac{x - c_1}{a_1}\right|^{2b}} = \frac{1}{1 + \left|\frac{\alpha(x - c_1)}{\alpha a_1}\right|^{2b}} = \frac{1}{1 + \left|\frac{\alpha x - \alpha c_1}{\alpha a_1}\right|^{2b}} \\
&= \frac{1}{1 + \left|\frac{\alpha x + \beta - \alpha c_1 - \beta}{\alpha a_1}\right|^{2b}} \\
&= \frac{1}{1 + \left|\frac{(\alpha x + \beta) - (\alpha c_1 + \beta)}{\alpha a_1}\right|^{2b}} \\
&= \frac{1}{1 + \left|\frac{y - c_2}{a_2}\right|^{2b}} = \text{bell}(y; a_2, b, c_2)
\end{aligned}$$

On the other hand, it is known that the slope of the line tangent to the curve of the bell function at the point when the belonging value is equal to 0.5 is given by $-\frac{b}{2a}$ or $\frac{b}{2a}$, if θ is the angle of the line tangent at that point, then $\tan\theta = -\frac{b}{2a}$ or $\tan\theta = \frac{b}{2a}$

The function $\tan\theta$ is increasing in the interval $\left(-\frac{\pi}{2}, \frac{\pi}{2}\right)$, then if $-\Phi \leq \theta \leq \Phi$, with $0 \leq \Phi < \frac{\pi}{2}$, it is obtained that:

$$\begin{aligned} &\tan(-\Phi) \leq \tan(\theta) \leq \tan(\Phi) \\ &-\tan(\Phi) \leq \tan(\theta) \leq \tan(\Phi) \\ &0 \leq |\tan(\theta)| \leq \tan(\Phi) \end{aligned} \tag{9}$$

Now as $\tan\theta = -\frac{b}{2a}$ or $\tan\theta = \frac{b}{2a}$, then $|\tan\theta| = \left|\frac{b}{2a}\right|$, but if $a > 0$, you have to $|\tan\theta| = \frac{|b|}{2a}$, then substituting in the inequality (9) you get that $0 \leq \frac{|b|}{2a} \leq \tan\Phi$, therefore

$$\begin{aligned} &0 \leq \frac{|b|}{2a} \leq \tan(\Phi) \\ &0 \leq |b| \leq 2a\tan(\Phi) \\ &-2a\tan(\Phi) \leq b \leq 2a\tan(\Phi) \end{aligned}$$

Therefore, parameter b of the bell function can be found in the interval $[-2a\tan\Phi, 2a\tan\Phi]$, where Φ is the maximum value of the inclination angle of the line tangent to the curve of the graph of the bell function in the point corresponding to the value of belonging equal to 0.5, and also $0 \leq \Phi < \frac{\pi}{2}$.

The search interval for parameter a is [0, 1], then $0 \leq a \leq 1$, then the search interval for parameter b is

$$-2\tan(\Phi) \leq b \leq 2\tan(\Phi)$$

Since the search interval of each parameter of the bell membership function is [0, 1], it is considered a mapping from the interval [0, 1] to the interval $[-2\tan\Phi, 2\tan\Phi]$ using the matching rule $b(\lambda) = (4\tan\Phi)\lambda - 2\tan\Phi$. Then, considering this mapping, the bell membership function will be modified as follows:

$$\mathrm{bell}(x; a, \lambda, c) = \frac{1}{1 + \left|\frac{x-c}{a}\right|^{2[(4\tan\Phi)\lambda - 2\tan\Phi]}}$$

Since $0 \leq \Phi < \frac{\pi}{2} = 1.578\ldots$, then a limit used for the search interval of b is $\Phi = 1.5$, which is a value close to $\frac{\pi}{2}$. Once the above modifications are made for the fuzzy inference system parameter search region, dimensionless outputs with values in the [0, 1] range will be generated. The suitability for GA is given in the Eq. (10), which is the mean absolute error (MAE) [16].

$$z = \frac{1}{60}\sum_{i=1}^{60}\left(0.5|c_i - \hat{c}_i| + 0.5|o_i - \hat{o}_i|\right) \tag{10}$$

where c_i and o_i, with $i = 1, 2, \cdots, 60$, are the normalized historical values of the order scheduling and lot size, respectively; and $\hat{c}_i$ and $\hat{o}_i$, with $i = 1, 2, \cdots, 60$, are the estimated normalized output values of the order scheduling and lot size.

The dimension of the search space depends on the number of sets used for the system inputs, that is, if C is the number of fuzzy sets used for each of the two system inputs, and each membership function has three parameters, then the dimension of the search space is given by the Eq. (11).

$$D = 2 * (3 * C) + 3 * C^2 * 2 \tag{11}$$

Algorithm 1 shows the series of steps to evaluate a data vector in the modified bell membership function, it returns another vector with its corresponding membership values from the parameters a, λ and c taken by arguments.

Algorithm 1 Modified bell membership function.

Function: bell
Arguments:
x: normalized data vector
a, λ, c: modified bell function parameters
Return:
y: vector with membership level for each element of x
Data:
n: size of vector x
Φ: upper bound of the angle (in radians)
for $i = 1$ **to** $i = n$

$$y_i \leftarrow \frac{1}{1+|\frac{x_i-c}{a}|^{2[(4\tan\Phi)\lambda-2\tan\Phi]}};$$

end for

The fitness function used for training is shown in Algorithm 2, it has as arguments the normalized error vectors, derived from the error, when to order and how much to order, and the parameter vector for the belonging functions.

Then the Fitness function returns the average of the mean absolute errors of each normalized output. The fixed data required for this function are the normalized universes for the output fuzzy sets, when and how much order, the number of elements in these universes, and the number of fuzzy sets used for the inputs.

Algorithm 2 Fitness used for fuzzy inference system training.

Function: Fitness
Arguments:
e: normalized error vector
d: vector of derivatives of the normalized error
c: normalized when ordering vector
o: normalized how much order vector
z: vector of parameters for membership functions
Return:
f: average of the mean absolute errors of each output
Data:
U_c y U_o: Normalized universes for when to order and how much to order, respectively.
C: number of fuzzy sets
H: vector size of $\boldsymbol{e}$ or $\boldsymbol{d}$
M y N: size of universes U_C and U_o, respectively
To assign:
$s \leftarrow 1$; // Initialize counter 1
$t \leftarrow 1$; // Initialize counter 2
for $i = 1$ **to** $i = C$
    $A \leftarrow \text{bell}(e, z_s, z_{s+1}, z_{s+2})$;
    $B \leftarrow \text{bell}(d, z_{3C+s}, z_{3C+s+1}, z_{3C+s+2})$;
    $E_i \leftarrow A$;
    $D_i \leftarrow B$;
    $s \leftarrow s + 3$;
end for
To assign:
$t \leftarrow 1$; // Reset counter 1
$s \leftarrow 6 * C + 1$; // Initialize counter 2 in $6 * C + 1$
for $m = 1$ **to** $m = H$
  for $j = 1$ **to** $j = C$
      for $k = 1$ **to** $k = C$
        $R_{t,:} \leftarrow \min(\min(E_j, D_k), \text{bell}(U_c, z_s, z_{s+1}, z_{s+2}))$;
      //Choose the minimum between each element of the vector $\text{bell}(U_c, z_s, z_{s+1}, z_{s+2})$ and the number $\min(E_j, D_k)$//
        $S_{t,:} \leftarrow$
$\min(\min(E_j, D_k), \text{bell}(U_o, z_{3C^2+s}, z_{3C^2+s+1}, z_{3C^2+s+2}))$;
      //Choose the minimum between each elements of the vector $\text{bell}(U_o, z_{3C^2+s}, z_{3C^2+s+1}, z_{3C^2+s+2})$ and the number $\min(E_j, D_k)$//

$t \leftarrow t + 1;$
$s \leftarrow s + 3;$
end for
end for
Choose maximum element F_p, $p \in \{1, 2, \cdots, M\}$, in each column of R;
Choose maximum element G_q, $q \in \{1, 2, \cdots, N\}$, in each column of S;
$\hat{c}_m \leftarrow \frac{U_c * F^t}{\sum_1^M F_p};$
$\hat{o}_m \leftarrow \frac{U_o * G^t}{\sum_1^N G_q}$; ($F^t$ and G^t are the transpositions of F y G, respectively)
To assign:
$t \leftarrow 1;$
$s \leftarrow 6C + 1;$
end for
$f \leftarrow 0.5 * \frac{1}{H} \|c - \hat{c}\|_1 + 0.5 * \frac{1}{H} \|o - \hat{o}\|_1 ;$

3 Results

The parameters used for the GA were:
Population size: 700.
Number of bits: 8
Percentage of mutation: 1%
Parent Selection: Tournament.
Tournament size: 6
Crossover points: 5
Generations: 5000.
Seed: 8 (MATLAB).

In addition to the MAE, other metrics used in training, validation, and testing are the coefficient of determination (R^2) and the mean relative error (MRE).

The R^2 given in the Eq. (12), is defined as a proportion of the total variance of the variable explained by the regression; that is, it reflects the goodness of fit of a model to the variable it seeks to explain, and its main purpose is to predict future outcomes [16].

$$R^2 = 1 - \frac{\sum_{i=1}^{n} \left(y_i - \hat{y}_i\right)^2}{\sum_{i=1}^{n} \left(y_i - \bar{y}\right)^2} \tag{12}$$

In the Eq. (12), y_i is the i-th expected real value, $\widehat{y}_i$ is the i-th result estimated by the model and $\overline{y}$ is the mean of the data sample. The value of R^2 is always in the interval $(-\infty, 1]$, when this value is negative, it means that the model is worse than predicting the mean, a value close to 1 indicates that the model is able to explain or predict a high percentage of the variability of the data used in training, validation or testing, as the case may be [16].

The MRE is used to determine the precision of the estimated value, it indicates the proportion of the error with respect to the value of the data series, it is given in percentage and defined by Eq. (13), where y_i is the i-th expected real value and $\widehat{y}_i$ is the i-th result estimated by the model [16].

$$MRE = \frac{100\%}{n}\sum_{i=1}^{n}\left|\frac{y_i - \widehat{y}_i}{y_i}\right| \quad (13)$$

Once the initial parameters of the GA are fixed, we proceed to vary the number of sets, C, with the following values $C = 2, 3, 4, 5, 6, 7, 8$, and 9, and execute the training algorithm for each of them.

By performing the runs for each of the C values, the training and validation results shown in Tables 1 and 2, respectively, were obtained. Figure 3 shows the graph of the mean absolute error against the proportional value of speed for both training and validation, when the number of sets for each of the system's inputs is varied, and there we see that the best result is obtained when $C = 8$.

Once the fuzzy inference system in charge of performing the medicine inventory control system has been formulated, the test data is used to measure the efficiency of the proposed model.

When the normalized test data are entered into the model, an average absolute error of 0.0081 is obtained, a relative error of 2.10% and the R^2 of 0.9838 and 0.9661 for the variables when and how much to order, respectively. Therefore, the model predicts 98.38% and 96.61% of the variability of the test data when and how much to order, respectively.

Table 1 Results of training through the GA

C	Dimension	MAE	MRE (%)	R^2 for output when ordering	R^2 for output how much order
2	36	0.012818	2.75	0.982	0.7979
3	72	0.016005	2.86	0.9658	0.9218
4	120	0.026059	5.11	0.9687	0.1917
5	180	0.010193	2.2	0.9808	0.9238
6	252	0.028016	4.38	0.8638	0.9587
7	336	0.019231	5.32	0.9537	0.4891
8	432	0.0064445	1.58	0.9886	0.9736
9	540	0.013495	2.01	0.966	0.9881

Table 2 Results of the validation through the GA

C	Dimension	MAE	MRE (%)	R^2 for output when ordering	R^2 for output how much order
2	36	0.01363	2.38	0.9809	0.8211
3	72	0.019402	2.76	0.946	0.8996
4	120	0.027053	4.59	0.9511	0.1515
5	180	0.018007	4.42	0.817	0.8948
6	252	0.033269	4.6	0.7875	0.9571
7	336	0.025874	6.7	0.8893	0.3981
8	432	0.0084515	1.6	0.9874	0.9617
9	540	0.014786	1.41	0.9464	0.9941

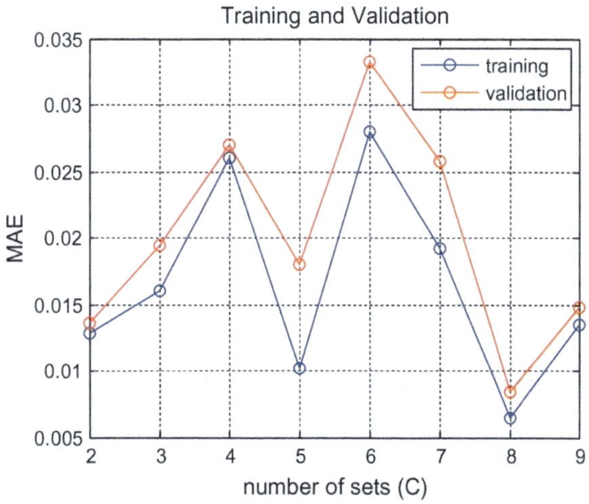

Fig. 3 Graph of the absolute error against the number of C sets for training and validation

4 Conclusion

In this research, a MIMO inference system is proposed to control the inventory of the laronidase drug used in patients with lysosomal disease, who resort to a hospital located in the state of Aguascalientes, Mexico. This system determines the moment to request the biopharmaceutical and the units to be ordered from the input variables: error and error derivative.

The input variable error measures the level of the inventory with respect to a reference, which in this case is the point of reordering. The variable derived from the error quantifies the rate of change in the inventory level each week.

The proposed fuzzy inference system that controls the level of drug units (plant), works as follows: At the end of each week, the values of the input variables to the system are calculated: error and derived from error, later the system gives with two outputs an estimated date for ordering and how much to order.

The fuzzy inference system is tuned by means of an GA, with this procedure the parameters of the bell functions used in the inputs and outputs of the system are found, as well as the rules (laws) of control that govern it.

When executing the algorithm with the established variations, the best model is obtained when the number of fuzzy sets for each one of the inputs is $C = 8$, since this one gives the smaller validation error. In this way, the fuzzy inference system tuned to an GA algorithm is elaborated and a control strategy (laws) is proposed for the biopharmaceutical inventory system.

Using the data set is tested with the selected model and the determination coefficient (R^2) as an efficiency measure, 98.38% is achieved to predict the moment to order and 96.61% to predict the quantity to order. The contributions that are achieved with this research are:

- A methodology to control the inventory of medicines.
- A methodology for tuning a fuzzy MIMO inference system through a GA algorithm, by performing the normalization of training, validation, and test data, and using the differentiability properties of the bell membership function.
- A system capable of identifying the moment to perform an acquisition of the laronidase drug used in patients with lysosomal disease who recur to the hospital where this study was performed, as well as the units to be ordered.

References

1. Uthayakumar R, Priyan S (2013) Pharmaceutical supply chain and inventory management strategies: Optimization for a pharmaceutical company and a hospital. Oper Res Heal Care 2(3):52–64. https://doi.org/10.1016/j.orhc.2013.08.001
2. Priyan S, Uthayakumar R (2014) Optimal inventory management strategies for pharmaceutical company and hospital supply chain in a fuzzy-stochastic environment. Oper Res Heal Care 3(4):177–190. https://doi.org/10.1016/j.orhc.2014.08.001
3. Lakshmi C, Sivakumar AI (2013) Application of queueing theory in health care: A literature review. Operat Res Health Care 2(1–2):25–39. https://doi.org/10.1016/j.orhc.2013.03.002
4. Mousavi SM, Bahreininejad A, Musa SN, Yusof F (2017) A modified particle swarm optimization for solving the integrated location and inventory control problems in a two-echelon supply chain network. J Intell Manuf 28(1):191–206. https://doi.org/10.1007/s10845-014-0970-z
5. Aengchuan P, Phruksaphanrat B (2018) Comparison of fuzzy inference system (FIS), FIS with artificial neural networks (FIS + ANN) and FIS with adaptive neuro-fuzzy inference system (FIS + ANFIS) for inventory control. J Intell Manuf. 29(4):905–923. https://doi.org/10.1007/s10845-015-1146-1
6. Ahmadi E, Masel DT, Metcalf AY, Schuller K (2019) Inventory management of surgical supplies and sterile instruments in hospitals: a literature review. Heal Syst 8(2):134–151. https://doi.org/10.1080/20476965.2018.1496875
7. García Cantú A (2013) Enfoques prácticos para planeación y control de inventarios, 4th ed. Trillas
8. Nguyen H, Walker C, Walker E (2019) A first course in fuzzy logic, 4th ed. CRC Press
9. Bojadziev G (2007) Fuzzy logic for business, finance, and management
10. Jang J-SR, Sun C-T, Mizutani E (2017) Neuro-Fuzzy and Soft Computing: A Computational Approach to Learning and Machine Intelligence. Prentice Hall

11. Jantzen J (2013) Foundations of Fuzzy Control: A Practical Approach: Second Edition. Wiley
12. Weise T, Global Optimization Algorithms – Theory and Application, 2nd Ed
13. Jinghui Z, Xiaomin H, Min G, Jun Z (2005) Comparison of performance between different selection strategies on simple genetic algorithms. In: Proceedings - international conference on computational intelligence for modelling, control and automation, CIMCA 2005 and international conference on intelligent agents, web technologies and internet, 2005, vol 2, pp 1115–1120, https://doi.org/10.1109/cimca.2005.1631619
14. DeJong KA (1975) Analysis of the behavior of a class of genetic adaptive systems
15. Grefenstette JJ (1986) Optimization of Control Parameters for Genetic Algorithms. IEEE Trans Syst Man Cybern 16(1):122–128. https://doi.org/10.1109/TSMC.1986.289288
16. James G, Witten D, Hastie T, Tibshirani R (2017) An Introduction to Statistical Learning with Applications in R. Springer

Technical Analysis of Shipments in an Automotive Company to Forecast Sales Volumes

Fernando Anaya-Villalvazo, Alberto Ochoa-Zezzatti, Oliverio Cruz-Mejía, and Jose Diaz

Abstract This research presents an analysis of the shipments forecast of automotive sensors in three regions where such products are sold and shipped of 15,000 unique products and more than 100 million of units shipped last year. The Sensors company is one of the world's leading suppliers of sensing solutions for automotive brands with operations and business centers in 11 countries so it's very important to have a forecasting analysis based on time series as historical data is basement and estimating for 18 + months decisions. Models used in this paper are Holt-Winters, Cross Correlation and Simulation. Basically, as output of the Holt-Winters model, results show a constantly increasing forecast for the coming years when done a seasonable additive algorithm. With this model output, a prediction is performed having the forecasting till 2028 showing a stable increased of the quantity of auto sensors that are manufactured delivering to all the regions where the automotive sensors are shipped to.

Keywords Forecasting · Shipments · Holt-winters method · Predictive forecasting · Cross correlation

F. Anaya-Villalvazo
Universidad Cuauhtémoc Aguascalientes, Aguascalientes, Mexico

A. Ochoa-Zezzatti (✉) · J. Diaz
Juárez City University, Ciudad Juárez, Mexico
e-mail: alberto.ochoa@uacj.mx

O. Cruz-Mejía
Universidad Autónoma del Estado de México, Toluca, Mexico

© The Author(s), under exclusive license to Springer Nature Switzerland AG 2021

A. Ochoa-Zezzatti et al. (eds.), *Technological and Industrial Applications Associated with Intelligent Logistics*, Lecture Notes in Intelligent Transportation and Infrastructure,
https://doi.org/10.1007/978-3-030-68655-0_22

1 Introduction

The Sensors company being a world leader and early innovator in mission-critical sensors designed to make the world cleaner, safer and more efficient. Mission-critical means products that are essential and difficult to do. You'll find 15,000 unique products in many automotive systems—anywhere from automotive braking systems to heavy off-road vehicles oil pressure sensors which includes, among others, the primary sensor technologies in use today are classified in three major areas of applications—powertrains, chassis and body [1]. 100 million of units are shipped per year with high revenue generated during 2018 fiscal year and now with electric vehicles rise in popularity and demand, they require sensors to monitor and optimize everything from battery systems to thermal management systems. This Sensors company is developing sensors that will enable light passenger cars, off-road vehicles and material handling equipment to operate autonomously as well as developing smart, connected sensors that enable actionable insights for commercial vehicle operators [2]. For this reason, making a forecasting analysis of sales and shipments is crucial to meet the company goals as electrification is having popularity plus big benefit to ecosystems. Then predicting the future necessity and the demand of such products, is fundamental for the company's growth as estimating the additional required capacity in the company is vital.

2 Auto Market Sensors

For the manufacturing of Electric Vehicle Technology, most company are developing flexible devices to support the requirement for more efficient electric vehicles. For this end, the processes for product assembly and calibration are machined in high volumes [2], Fig. 1. Among the requirements for these sensors are precision, efficient performance, low cost, high capacity, lightweight and compact size.

For automotive sensors, cost is a big driver in future designs. Improved signal processing will enable more information obtained from existing sensing elements and system-level software techniques will be used along with virtual sensing methods [3]. Examples of such sensors are Thermal and pressure Sensors, direct Tire Pressure sensors, battery gas sensors, and magnetic sensors, see Fig. 2 [2].

In next graph, the monthly shipment quantity is shown for sales months during 2018 and 2019 where the purpose is to forecast the shipments for following years (Fig. 3) [4].

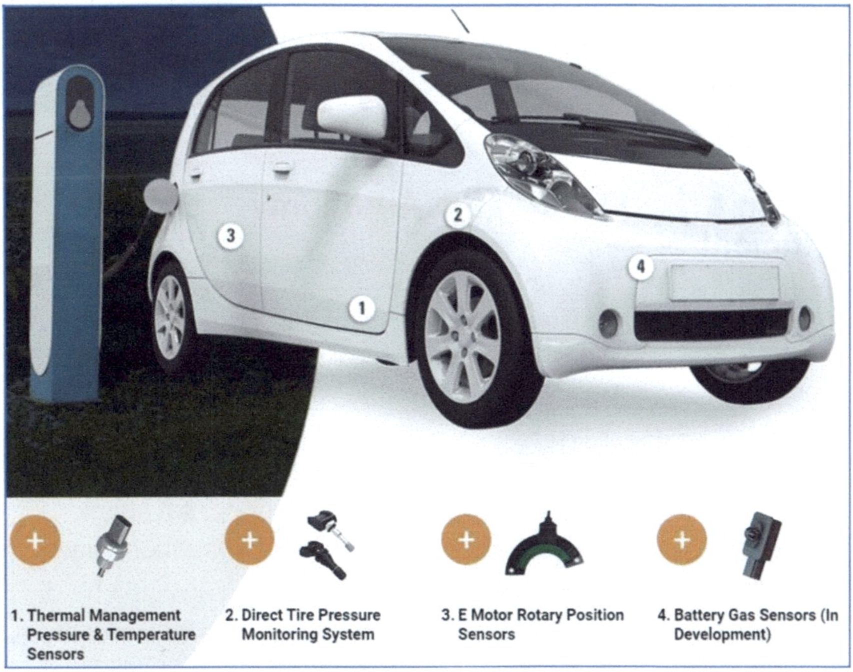

Fig. 1 Sample of some sensors in an electric vehicle [2]

3 Methodology

Prediction or forecasting models are useful when time series are required to analyze along with other models [5]. Forecasts are important for decision-making in businesses and other organizations, like in this particular case. The predictive validity of forecasting method is assessed by comparing the accuracy of forecasts from the method with the accuracy of forecasts from currently used methods, or from simple benchmark methods such as the naive no-trend model [6]. In this paper, a time series based on Shipment Quantity of sensors built, sold, and shipped globally in the automotive market segment, is the baseline for the analysis, starting on January 2019 in a monthly basis. The regions where the items are shipped to, are Americas, Europe and Asia and for different vehicles makes. The data set consists of 12,062 rows and 10 columns. The time series is as follow where is grouped by 48 End Customer, like Audi, BMW, Volkswagen, etc. Table 1 (R generated, 2020) shows the cumulative shipments per end customer followed by the time series plot.

This data set also contains the Shipment Amount columns so doing a multivariable analysis becoming each column or variable, a temporal series where we apply the *ts*() function, the data frame cmes the *mts* class [7] and plotting it, we can see a regular trend for each column.

Fig. 2 Four buckets of the Automotive sensors manufactured and shipped [2]

Fig. 3 Monthly shipments during 2018 and 2017 [4]

Table 1 Cumulative shipment quantity per end customer (48) (R generated, 2020)

	Jan	Feb	Mar	Apr	May	Jun	Jul	Aug	Sep	Oct
2019	4,009,587	11,408,068	11,410,179	12,345,993	12,435,151	13,490,171	13,509,036	24,147,259	24,274,895	28,117,002
2020	28,498,614	28,533,393	28,533,884	31,983,527	46,836,165	469,667 59	47,444,004	71,581,407	72,193,891	72,819,274
2021	96,530,064	96,532,800	96,587,758	96,622,678	96,689,596	97,429,545	97,453,431	97,539,228	100,064,967	100,912,716
2022	133,473,378	134,468,575	135,240,650	135,331,033	137,868,799	138,749,419	139,228,636	140,509,179	140,934,629	145,179,441
	Nov	Dec								
2019	28,229,754	28,497,609								
2020	72,819,850	78,455,935								
2021	101,270,769	113,096,167								
2022	166,415,884	169,340,542								

Now the projection of the time series has gone to 3024 when exploring the full data set. There are some periods with high peaks but then comes to regular forecast, See Figs. 4 and 5 (R generated, 2020).

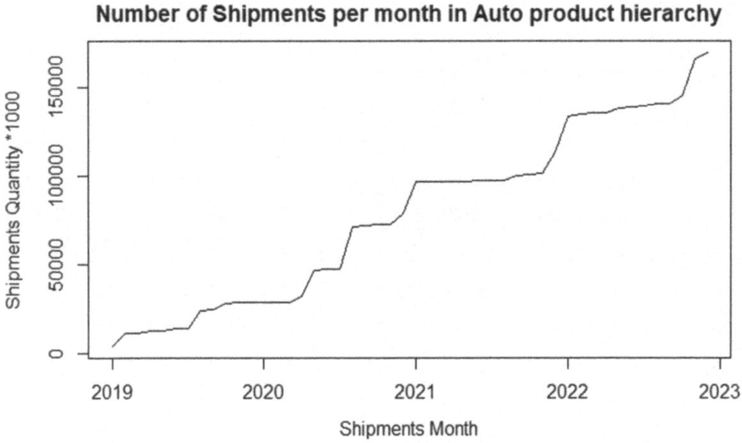

Fig. 4 Time series for shipments quantity (R generated, 2020)

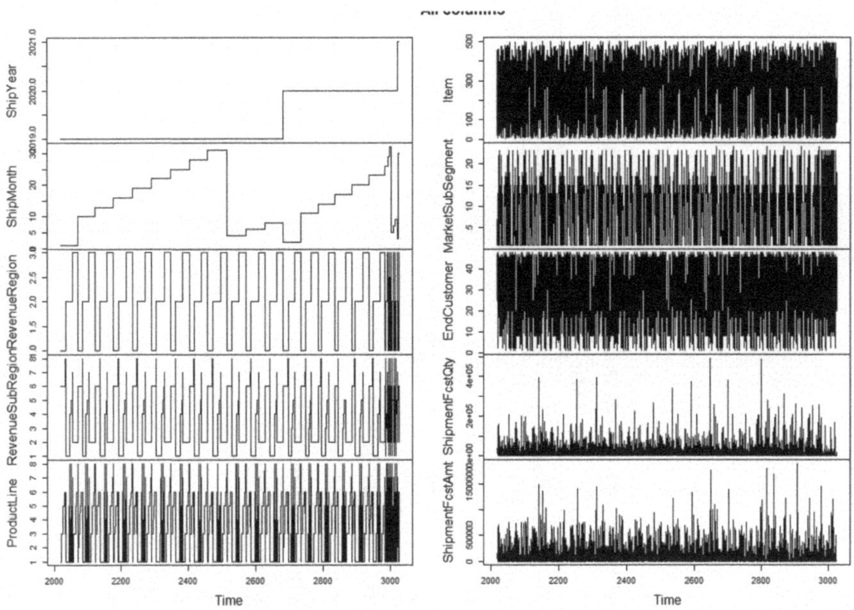

Fig. 5 Multivariable análisis (R generated, 2020)

4 Forecasting Results

For categorical forecasts accuracy is typically measured as a variation of percent correct. For quantitative forecasts, accuracy is assessed by differences between ex ante forecasts and data on what actually transpired [6]. To continue the analysis, applying autocorrelation functions on the time series for just the shipment quantity and other for all the variables, Fig. 6 shows the result (R generated, 2020).

Table 2 (R generated, 2020) shows the autocorrelation parameters in the time series for shipment quantity (Fig. 7) (R generated, 2020).

Cross correlation among the two time-series is shown in Fig. 8 (R generated, 2020) and Table 3 (R generated, 2020) include the correlation details.

Decomposing the time series, we get additional inputs, this is shown in Fig. 9. Figure 10 (R generated, 2020) shows the seasonal graph (as a seasonal plot allows the underlying seasonal pattern to be seen more clearly and is especially useful in identifying years in which pattern changes [8]) from where it can be seen that depending on the month, the shipments can vary based on the end customer and the

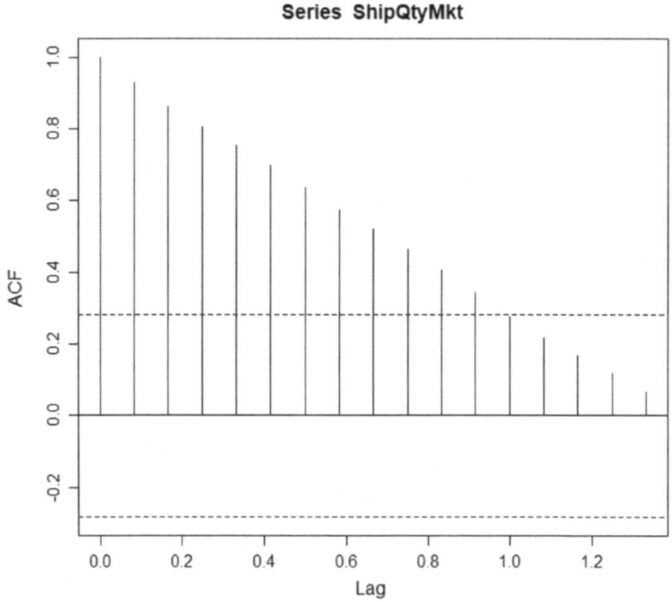

Fig. 6 Autocorrelation for Shipment Quantity (R generated, 2020)

Table 2 Autocorrelation for shipment quantity (R generated, 2020)

Autocorrelations of series 'ShipQtyMkt', by lag													
0.0000	0.0833	0.1667	0.2500	0.3333	0.4167	0.5000	0.5833	0.6667	0.7500	0.8333	0.9167	1.0000	1.0833
1.000	0.929	0.861	0.807	0.754	0.697	0.637	0.576	0.521	0.463	0.405	0.342	0.276	0.218
1.1667	1.2500	1.3333											
0.168	0.117	0.066											

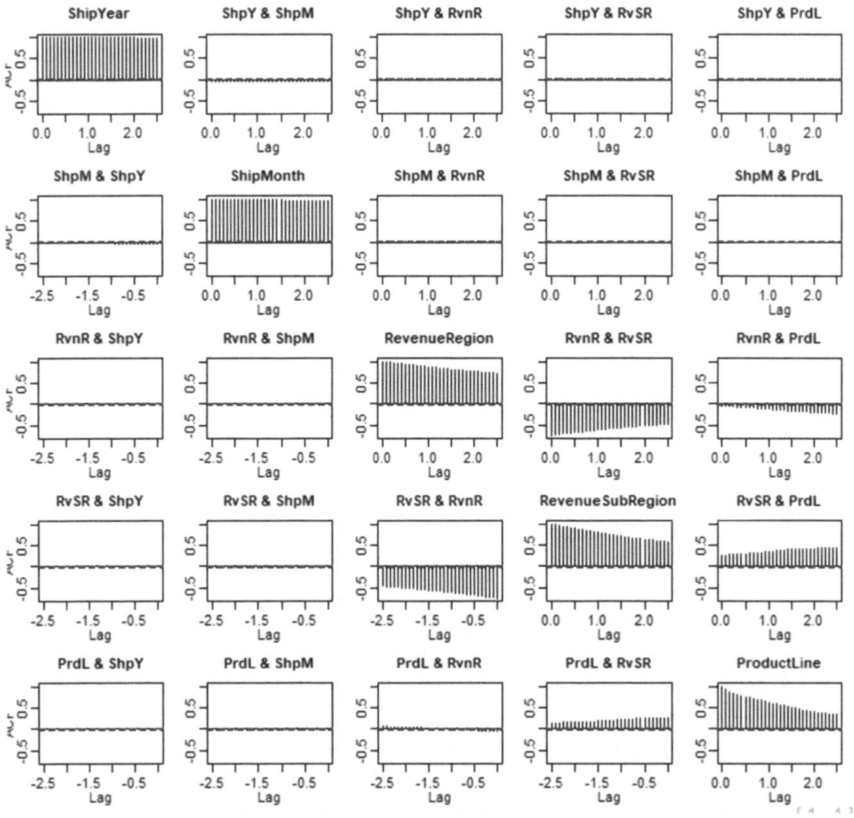

Fig. 7 Autocorrelation for all variables in the data set (R generated, 2020)

region but in general, the trend keeps showing good forecast, like the observation of the time series.

Now, using the Holt-Winters model, which simplifies both obtaining maximum likelihood estimates of all unknowns, smoothing parameters and initial conditions and the computation of point forecasts and reliable predictive intervals [9], we can observe that the additive model is the one with most adjustments as parameters are smaller so adjusting better to the observation of the time series.

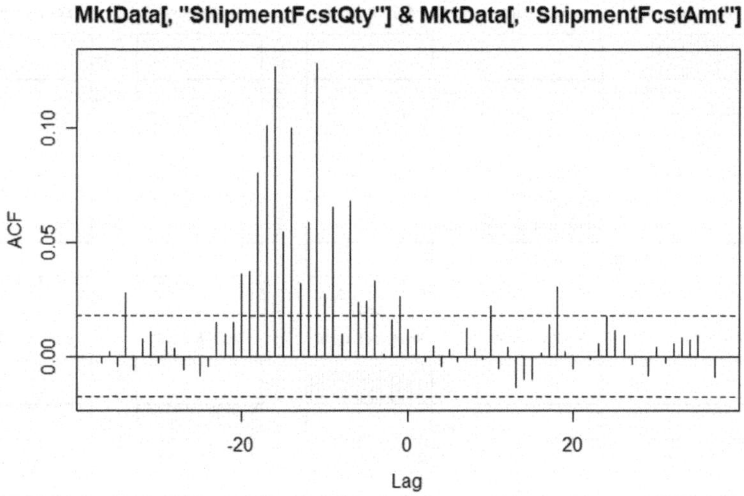

Fig. 8 Cross correlation plot (R generated, 2020)

The predictions from the model Holt—Winters Additive for the period from 2020 to 2023 with a confident interval of 95%, indicates that the growth trend continues for the following years with not many gaps but still, need to be cautious with the decisions to build enormous quantity of sensors as many factors are involved in the global market. Figure 11 (R generated, 2020) shows prediction for 2029.

Finally, we can use a data simulation or synthetic series for this case to identify the autocorrelation of the simulated shipments time series is only significant at lag 0 and with practically no variation, Fig. 12 (R generated, 2020).

Table 3 Cross correlation details (R generated, 2020)

Autocorrelations of series 'X', by lag													
−37	−36	−35	−34	−33	−32	−31	−30	−29	−28	−27	−26	−25	−24
−0.003	0.002	−0.004	0.028	−0.006	0. 008	0.011	−0.003	0.007	0.004	−0.006	0.000	−0.008	−0.004
−23	−22	−21	−20	−19	−18	−17	−16	−15	−14	−13	−12	−11	−10
0.015	0.010	0.015	0.036	0.037	0. 080	0.101	0.127	0.055	0.100	0.032	0.059	0.129	0.027
−9	−8	−7	−6	−5	−4	−3	−2	−1	0	1	2	3	4
0.065	0.010	0.068	0.024	0.024	0. 033	0.001	0.016	0.026	0.012	0.009	−0.002	0.005	−0.005
5	6	7	8	9	10	11	12	13	14	15	16	17	18
0.003	−0.002	0.012	0.003	−0.001	0. 022	−0.005	0.004	−0.014	−0.010	−0.010	0.001	0.014	0.031
19	20	21	22	23	24	25	26	27	28	29	30	31	32
0.002	−0.005	0.000	−0.001	0.005	0.017	0.011	0.009	−0.003	0.000	−0.008	0.004	−0.003	0.006
33	34	35	36	37									
0.008	0.007	0.009	0.000	−0.009									

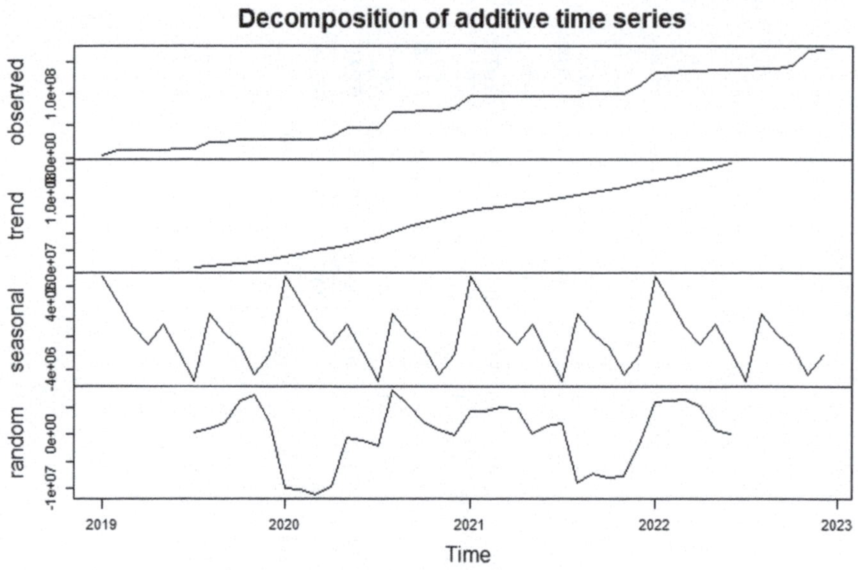

Fig. 9 Decomposition of the time series (R generated, 2020)

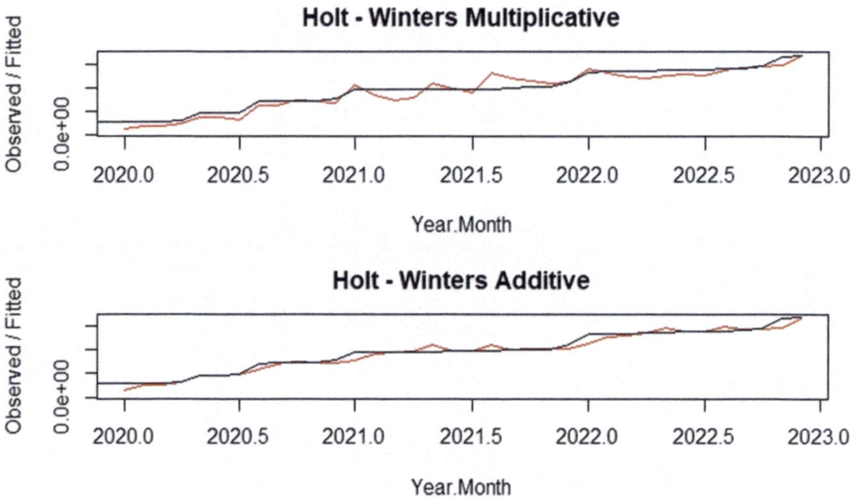

Fig. 10 Holt-Winters comparative analysis (R generated, 2020)

5 Conclusions

The predictions done by the Holt-Winters model show that the shipment quantity in the decade will continue growing based on the tendency from last years. For this reason, the Sensors company needs to consider investing in the Automotive

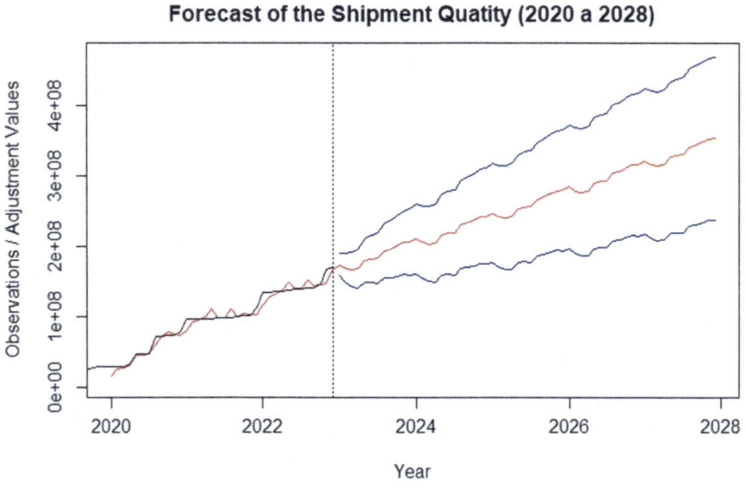

Fig. 11 Prediction of shipment quantity (R generated, 2020)

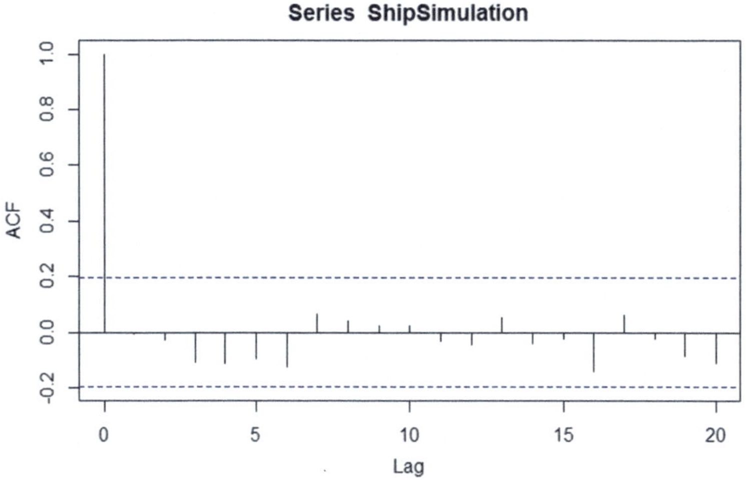

Fig. 12 Simulated autocorrelation (R generated, 2020)

market segment specially in the electrified vehicles sector. The pace of development of self-powered sensing devices is brisk, and it would have a substantial impact on automotive sensing systems [10]. But also need to be aware that this was based on a 95% confident interval, so business decisions would make the difference if not handle carefully. Current economical global factors are a consideration that need to pay attention with, so the forecasting prediction need continue under analysis and even additional models would be a good decision to run.

References

1. Fleming W (2001) Overview of automotive sensors. IEEE Sens J 1. https://doi.org/10.1109/7361.983469
2. Sensata Technologies web portal. https://sensata.com/about
3. Sparks D (2013) MEMS for automotive and aerospace applications. Woodhead publishing series in electronic and optical materials, pp 78–105. https://doi.org/10.1533/9780857096487.1.78
4. Oracle Business Intelligence. Marketing, Sales & Forecasting dashboard.
5. Cowpertwait PS, Metcalfe AV (2009) Introductory Time Series with R. Springer, New York. https://doi.org/10.1007/978-0-387-88698-5
6. Armstrong JS, Green KC (2018) Forecasting methods and principles: evidence-based checklists. J Glob Scholars Mark Sci 28(2):103–159
7. University of Texas at El Paso, El Paso, TX, USA, ISSN 1870–4069 Santana A, Nieves C, Objetos en R: Series temporales
8. Hyndman RJ, Athanasopoulos G (2018) Forecasting: principles and practice, 2nd edition. OTexts, Melbourne, Australia. https://otexts.com/fpp2/
9. Bermudez J, Vercher E, Segura J (2007) Holt-winters forecasting: an alternative formulation applied to UK air passenger data. J Appl Stat. https://doi.org/10.1080/02664760701592125
10. Askari H, Khajepour A, Behrad M, Lin Z (2019) Embedded self-powered sensing systems for smart vehicles and intelligent transportation. Nano Energy 66. https://doi.org/10.1016/j.nanoen.2019.104103

Distributed Programming Applied for the Optimization of Hydraulic Networks Through a Web Application

Beatriz Martínez-Bahena, Juana Enriquez-Urbano, and Jesús del Carmen Peralta-Abarca

Abstract In this work, it was proposed the design and implementation of a web application using distributed computing with the goal to optimize the design of hydraulic networks. The application required for its operation only of the meta-heuristics parameters, algorithm parameters to evaluate the optimization model, the instance to be solved and the execution parameters at cluster level. The application shows graphically that is fulfilled with the constraint satisfaction model and optimization model realized in a distributed way. Running the web interface allows users to optimize the costs of different elements (valves, pumps, pipes, etc.), which are added to the hydraulic network and analyze the results graphically, easily and friendly without need to have extensive experience in programming and management of clusters or grids.

Keywords Parallel computing · Distributed programming · Grid · MPI · Hydraulic network design

1 Introduction

Water is an essential element for life and for the human development, reasons for which why the ancient Civilizations were located along the rivers. With the advance of technologies the man was able to transport and store water, as well as extract from the subsoil and with this they managed settlements in areas away from rivers, consequently, having an efficient and effective hydraulic distribution network (HDN) has

B. Martínez-Bahena (✉) · J. Enriquez-Urbano · J. C. Peralta-Abarca
Universidad Autónoma del Estado de Morelos (UAEM), Av. Universidad 1001, Cuernavaca, México
e-mail: bmartinezb@uaem.mx

J. Enriquez-Urbano
e-mail: juana.enriquez@uaem.mx

J. C. Peralta-Abarca
e-mail: carmen.peralta@uaem.mx

© The Author(s), under exclusive license to Springer Nature Switzerland AG 2021
A. Ochoa-Zezzatti et al. (eds.), *Technological and Industrial Applications Associated with Intelligent Logistics*, Lecture Notes in Intelligent Transportation and Infrastructure,
https://doi.org/10.1007/978-3-030-68655-0_23

become a constant challenge for man. An HDN is a set of pipes (which are joined in various points called nodes or joints), accessories (valves and pumps ...) and structures, which leads the water from the storage tanks or reservoirs to the different intakes, in sufficient quantity at adequate pressures. There are different factors, such as: geographical conditions, various configurations among all the components of the distribution network that determines the complexity of the network. Being one of the main challenges to find an HDN that comply with restricted restrictions of pressure and flow at reasonable cost [1]. To solve the problem different methodologies have been developed to give optimal solutions. Khatavkar and Mays [2] proposed a methodology to monitor the real-time operation of water distribution systems (WDS) under critical conditions, lack of electrical power and/or water and other man made conditions. The objective of his work was to optimize WDS operations in critical conditions minimizing the difference between the requested demands and the demands satisfied, while meeting the requirements of system pressure. To determine optimal operation (pump and valve operations) of the distribution system optimization (genetic algorithm) with a simulator (EPANET). The algorithm presented is a methodology to determine optimal operations for water distribution systems under normal and emergency conditions. One disadvantage is that their model solves their stay in a time of more than 40 h and does not optimize the design; in their future work they already consider the development of an interface to work your model. Edwin et al. [3] developed a computer program called Magmoredes, based on the use of a microalgorithm multiobjective genetic, proposed by Coello and Toscano, they adapted it for the application of water distribution networks, that met the economic requirements, and under the limits pressure and speed according to Peruvian regulations. The program was developed in the Java programming language and gave several solutions, the network designer chose the lowest cost but with low reliability and its flow rates did not satisfy the restrictions. Saldarriaga, Pulgarín, Cuero and Duque [4] designed a software (NETWORKS) in the Pascal language and object oriented programming architecture for the optimized design of water distribution networks, they applied 5 methodologies: Optimal Power Use Surface (OPUS), Rapid Design which is a combinatorial optimization algorithm, Genetic Algorithms, Search for Harmony and Optimal Surface of Hydraulic Gradient (SOGH) and with EPANET, it has a graphical interface that allows visualizing the characteristics of the network obtained. Sarbu and Ostafe [5] developed a linear optimization model and graph theory together with a computational iterative procedure of optimal discharges through pipes on the basis of linear programming for the design of distribution networks of water, their model compares the linear optimization model with some others, such as the classical model of economic speeds, the Moshnin model and a non-linear model. The optimization model they propose offers the possibility of optimal design of water supply networks; it is an algorithm that is executed on a computer. De Corte and Sörensen [6] worked with an iterated local search algorithm (ILS), it generates an initial solution and improves it until it meets a stop criterion. Each candidate solution generated by the IL algorithm is sent to the hydraulic solver; EPANET 2.0 (hydraulic simulator) to check hydraulic feasibility. Two different configurations are compared: one that minimizes the cost of the solution and one that minimizes computational time. Abdelbaki et al. [7] worked with Software Mapinfo Geographic information systems (GIS) GIS (8.0), EPANET 2.0 and they obtained satisfactory results. Dongre and Gupta [8] formulated a fuzzy optimization model based on genetic algorithms for the design that would give a lower cost solution of a water distribution, the optimization model with fuzzy restrictions becomes a deterministic model. The methodology provides a solution that meets pressure requirements in extreme conditions. According to the results found in the literature, algorithms have been implemented, some combining hydraulic simulators (EPANET) to solve these types of problems, however, so far no web application has been found that applies these elements. In the literature there are also different commercial software such as WaterCad, WaterGems among others; which allow the analysis and design of a water network, however when network optimization is required, it is necessary to implement heuristics or algorithms that allow a more detailed analysis, evaluating the different configurations in which each of the elements that make up the network can be accommodated, and ensuring that it complies with the appropriate pressures and flows. A correct configuration of these or of each of the parameters, correspond to feasible solutions. To reach these feasible solutions, a high computational and temporal cost is required due to the complexity of the problem, which according to the theory of computational complexity is classified as an NP- hard problem [9] because the computational effort grows exponentially as the size of the instance grow; an alternative solution is the implementation of mechanisms or methods that solve it in a reasonable computational time, as is the case of metaheuristics algorithms with distributed programming. That is why in this work the design and implementation of a web application using distributed computing is presented, which allows us to obtain an optimized and feasible solution for the design of the hydraulics network and due to the easy configuration of the web application, users with basic knowledge of distributed programming can use it, since they only need the algorithm parameters, the data instance to be solve and the parameters of the execution at the cluster level.

2 Heuristics and Metaheuristics

The word "heuristics" received from the Greek heuriskein, which means "find" or "discover", a heuristic adapts and designs to solve a specific problem [10]. Maroto, Alcáraz and y Ruíz [11] what defined as the techniques or procedures to solve problems, specified in intuition, knowledge or experience to find correct solutions or improve an existing solution; The term metaheuristic is obtained from put before the term meta that means "beyond" or "at a superior level" [12]. The metaheuristic term was introduced by Glover in the year of 1986, which defines it as an algorithm of general purpose that can be applied to solve almost any optimization problem, because they combine traditional heuristics with efficient strategies of search space exploration [13–15].

Figure 1 shows the different techniques that exist to solve different problems, such as optimization and design of an HDN. In the literature it can be observed

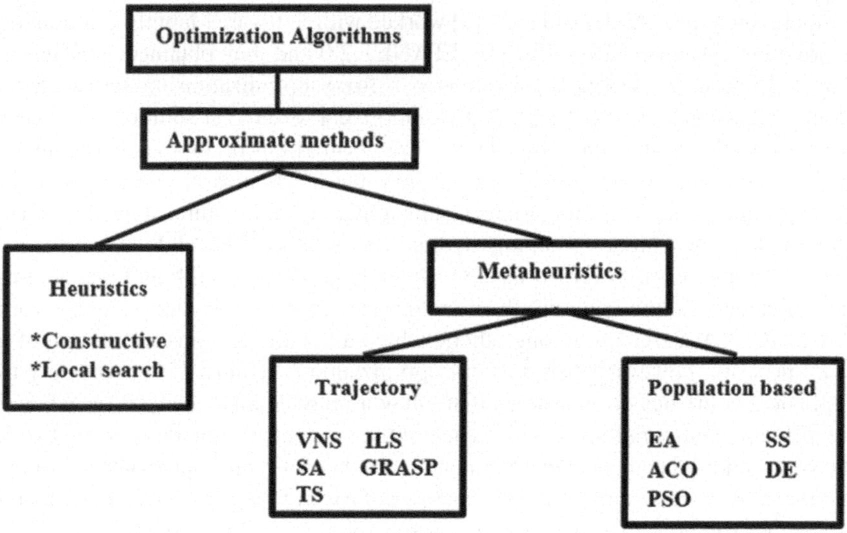

Fig. 1 Optimization techniques classification

that numerous optimization techniques have been applied to the design of water distribution networks such as: Simulated Annealing (SA) [16], Tabu Sarch (TS) [17], Particle Swarm Optimization [18], and Genetic algorithms [19]. In this work we used a Genetic algorithm working with distributed programming, the use of these techniques allows the search for different solutions beyond the local optimum, broadening the search and therefore giving rise to the possibility of obtaining better solutions [1].

3 Distributed Programming (DP)

In recent years the problems that arise in the economic life of today's society are not simple problems to solve due to their complexity or the time required to solve them, these problems vary from the design of routes, calculation of shipments, up to the allocation of resources to processes, among others. Computationally the existence of these complex problems requires the manufacture of powerful computers that allow solving them in an adequate time. In sequential computers to increase your productivity, techniques such as segmentation or vector computing are used; technological advances are also made, such as in the integration of circuits, which allows them to accelerate their computing speed [20].

Parallel Computing (PC), which is a very cheap and also faster way, is then used to solve these problems. Parallelism is the execution of several activities at the same time and that have an interrelation. Parallel computation is the execution of more than one calculation at the same time using more than one processor [21]. The technique for programming or computation in parallel, allows fragmenting a problem into subproblems to be solved, its objective is to divide the calculations associated with the problem and the data on which the algorithm operates in an equitable way, and thus reduce to a minimum the total computation time by distributing the workload among multiple processors. In the literature, it is considered as parallel computing all those techniques that use multiple computing resources to do processing at the same time.

Within the parallel computation there is a classification: Flynn's 1966. And in that classification, there is an option called MIMD (Multiple Instruction Multiple Data), and the MIMD option is subclassified into 2 options: 1) shared memory architectures and 2) architectures of distributed memory [11]. If you use openMP, thread or CUDA you are programming on a shared memory architecture. If you use MPI or PVM you are programming on a distributed memory architecture [20]. The acronym MPI stands for 'Message Passing Interface' and is an implementation of a standard that defines the syntax and semantics of message passing designed and used in programs that use parallelism in multi-processor architectures [22, 23]. Distributed Programming (DP) is an application that has different components that run in separate environments through a network, it is also a type of computing where a computer network communicates and coordinates work through the passage of messages to achieve a common goal, it uses synchronization algorithms and shared resources [24, 25]. Another definition indicates that a distributed memory architecture is made up of a set of processors, each with its own local memory and address space. It is also known as a cluster and is defined as a set of computers (called nodes), which are interconnected through a high-speed network that work together to solve a problem, through special configurations that enable parallel programming. Through the use of message passing libraries. When an algorithm distributed in a cluster, its instructions and data are distributed among the processors in order to obtain a good performance [20].

For this project, we worked with the master–slave model, which works as follows: a process (master process) imposes the direction of program control on one or more processes (slaves) that process the information, the only communication they have with the teacher is the delivery of the results of the assigned task. Figure 2 represents this model.

It can be said that there are only two communications between the master node and the slaves; the first when the master node delivers the information and the second when the slaves communicate with the master to deliver the results. The MPI library was used to carry out the communication between the processes.

4 Problem of the Water Distribution Network

The problem is defined as a network that contains hydraulic elements: pipes, pumps, valves, supply sources (reservoirs or tanks), and depending on all these elements make an adequate distribution of water. The pipes allow water to be drawn from the supply sources to the points of consumption. Supply sources are external sources

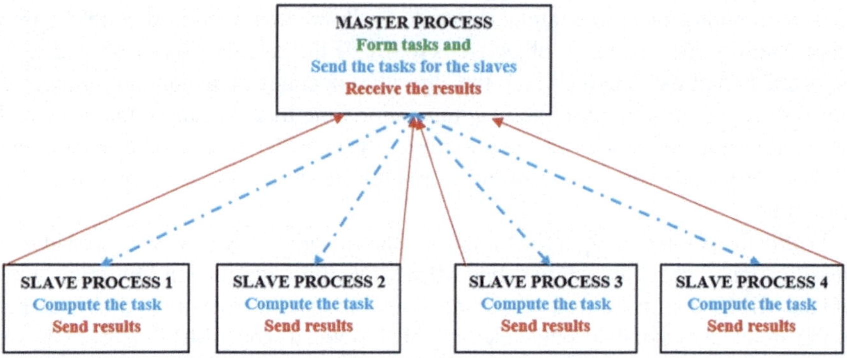

Fig. 2 Master–Slave Model

(wells, rivers, streams) for a system. Pumps raise the water from the surface to storage tanks or directly to the distribution system. Valves enable adequate control of the pressure in a system [26, 27]. Currently, the HDN in some regions does not make a satisfactory distribution, due to some factors such as the increase in demand based on new settlements of the population and geographic areas, therefore, the needs of the users are not accomplished. It is important to mention that for water distribution system to operate correctly it must consider a set of conditions. The main condition is to have the minimum pressures necessary to guarantee consumers an adequate service, avoiding water leaks due to pipe breaks. The problem to be solved is to improve water distribution, supplying all users through a redesign of the existing HDN. One of the options is to add new storage tanks, also considering pressure reducing valves (PRVs), pipes or pumps.

4.1 Symbolic Representation of the Problem

A hydraulic distribution network can be represented by a graph. A graph is denoted as $G = (N, E)$, where N is a set of nodes and E is a set of edges [28]. Figure 3 shows the graph of the hydraulic network (obtained from Epanet), where the nodes represent supply sources (reservoirs, tanks) or consumption points (homes, shops, industries, etc.). The nodes are represented with a different symbol. The edges represent the connection elements such as pipes, valves, pumps, among others. Each edge has an associated cost that can represent the cost of the pipe diameter, the length of a pipe, the cost of the valve, the cost of the pump, etc.

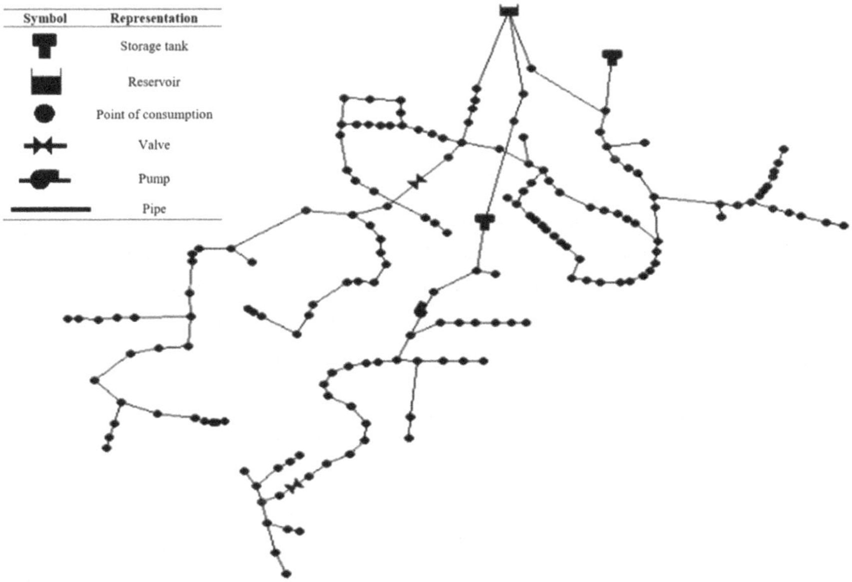

Fig. 3 Graphical representation of a hydraulic network

4.2 *Optimization Model*

The described optimization model is proposed to determine the optimal design of any network, with values of a known instance, which distributes water from the main supply points (reservoirs or pumping stations, of which it is known, respectively, the water level or the manometric height). The following is a simplified description of the model in a water distribution system [29].

$$\min C(P, T, V) = \sum_{\tau=1}^{nT} \sum_{i=1}^{nd} \sum_{j=1}^{p} C_i L_{ij} X_{ij\tau} + \sum_{T=1}^{nT} C_T Y_T + \sum_{V=1}^{nV} \sum_{j=1}^{nP} C_V K_{VJ} \quad (1)$$

Subject to

$$\sum_{i=1}^{nd} \sum_{j=1}^{p} X_{ijT} = 1 \quad \forall T \quad (2)$$

$$d_{min} \leq d_{ij} \leq d_{max} \quad (3)$$

$$X_{ijT} \in \{0, 1\} \quad (4)$$

$$Y_T \in \{0, 1\} \tag{5}$$

$$K_{VJ} \in \{0, 1\} \tag{6}$$

The objective function to minimize (1) represents the cost of the distribution system, according to the changes made in the network; where j is the index of the pipes; i is the index of the commercial diameters; *nd* is the number of commercial diameters, p is the number of pipes for the new tanks, nV is the number of valves, nT is the number of tanks and nP is the total number of pipes in the network. C_i, is the unit cost of the new pipes which depends on the diameter i. L_{ij}, is the length of each pipe j with diameter i. The cost, C_T, of the tanks, and valves, C_V.

According to the cost function, the problem to be solved consists in determining the number of pipes, tanks and valves that make it a minimum, complying with certain restrictions. Constraint (2) indicates that for each pipe segment, only one diameter is used. Constraint (3) ensures that the chosen diameter is included on the list of available commercial diameters. Constraint (4) indicates whether a diameter is being used or not. For example, if X_{ijT}= 1, the diameter is being used. Otherwise, X_{ijT}= 0. It should be noted that the network may have repeated diameters. Constraint (5) indicates that a tank is added if Y_T= 1. Otherwise, Y_T= 0. The tanks can have different characteristics (diameter, initial water level, maximum water level and minimum water level). Constraint (6) indicates whether a valve V is inserted into a pipe, J. If a valve is added into the pipe, then K_{VJ}= 1. Otherwise, K_{VJ}= 0.

4.3 Constraints Satisfaction Model

The constraint satisfaction model consists of a set of hydraulic constraints that must be met to obtain the proper functioning of the system and satisfy the needs of the users [29].

$$\sum_{i} \mathrm{Q_{in}} - \sum_{i} \mathrm{Q_{out}} = \mathrm{Q_e} \tag{7}$$

$$\sum_{c} \mathrm{h_f} = \sum_{c} \mathrm{E_p} \tag{8}$$

$$\mathrm{H_{min}} \leq \mathrm{H_i} \leq \mathrm{H_{max}} \tag{9}$$

Equation (7) represents the physical law of mass conservation. The constraint indicates that the flow that enters a node, Q_{in}, minus the flow leaves a node, Q_{out}, is equivalent to the demand Q_e. Equation (8) represents the law of energy conservation. It indicates that the sum of the frictional energy losses h_f in any circuit c must be equal to the power energy, E_p, supplied by a pump. Equation (9) refers to the minimum and maximum pressure requirements that satisfy water users. The constraint requires

Table 1 Nomenclature of the models

Nomenclature	
nd	Number of commercial diameters
nT	Number of tanks
nV	Number of valves
nP	Total number of pipes in the network
p	Number of pipes for the new tank
C_i	Unit cost of the new pipes
L_{ij}	Length of each pipe j with diameter i
C_T	Cost of tanks
C_V	Cost of valves
d_{min}	Minimum diameter
d_{max}	Maximum diameter
d_{ij}	Diameter of pipe
Q_{in}	Flow that enters a node
Q_{out}	Flow leaves a node
Q_e	Demand flow
h_f	Frictional energy loss
E_p	Power energy supplied by a pump
H_i	Pressure at a node
H_{min}	Minimum pressure
H_{max}	Maximum pressure

that the pressure at a node H_i be greater than the required minimum pressure H_{min} and lower than the maximum pressure H_{max}. The required minimum pressure is 10 mca (meters for water column) and the maximum pressure is 60 mca (Table 1).

5 Solution Methodology

5.1 Genetic Algorithm

A GA is a search procedure for the optimum of a function (maximum or minimum) based on natural selection and genetics, which allows the survival of the best adapted individual, through the use of genetic operators (selection, crossing and mutation). Individuals (solutions) are the ones with the greatest probability of reproducing, with the appropriate parameters (which are the values of the decision variables) that make them more suitable, passing to the following generations, while the less suitable may become extinct. Through the application of GA it is possible to determine the

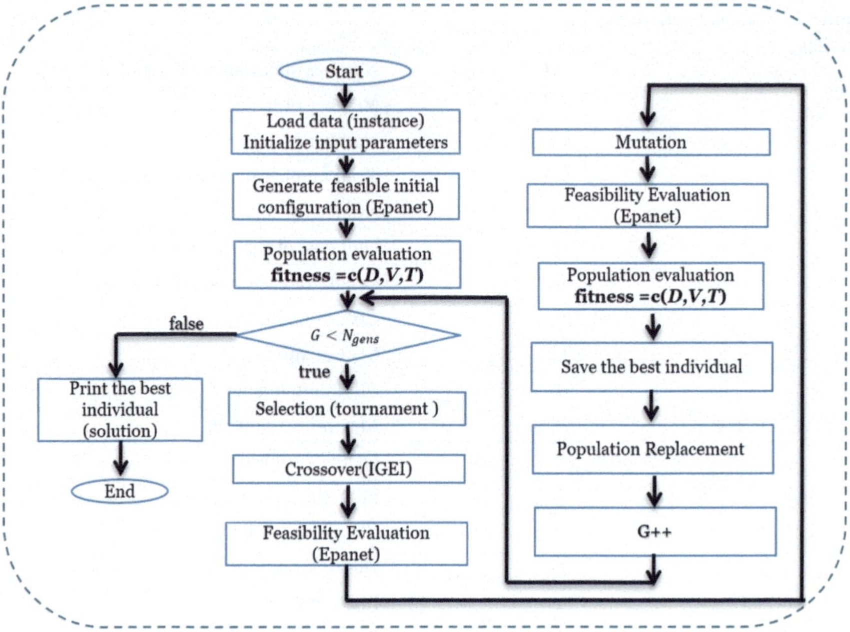

Fig. 4 Flow chart of the genetic algorithm

minimum cost network (best adapted individual) after a series of iterations (generations), thanks to transition rules (genetic operators) in charge of choosing the best solutions (lowest costs), according to the value of the objective function (cost) of each one of them, starting from a set of initial designs established at random [30].

Figure 4 shows the flow diagram of the genetic algorithm used for the web application. The algorithm starts with entering the HDN (instance) data to generate an initial configuration. The values that this initial configuration takes depend on the sensitivity analysis that is carried out for the input parameters that are: probability of crossing (P_c) and probability mutation (P_m), percentage of selection (P_s) and percentage mutation (P_{mut}), population size (T_p) and number of generations (N_{gens}). The initial configuration provides a population which is evaluated to verify its feasibility (comply with the constraint satisfaction model), using Epanet functions. Subsequently, the fitness value of each individual is evaluated. Individuals evolve for generations through three operators such as: selection, crossing and mutation. The selection method used is by tournament, where you select the most suitable individuals. Subsequently, the selected parents are recombined to form descendant individuals who inherit the characteristics of their parents using the IGEI crossover operator, due to a probability factor, which determines whether the crossover takes place or not. After crossing, the feasibility of the individuals is evaluated using Epanet. Later, the mutation is made to a percentage of individuals in the total population. After the mutation, the feasibility of the individuals is again evaluated. Feasible individuals are added to the new population until the population size is complete. The fitness value of the new population is evaluated. The new population is replaced, and the generation number is increased, and the procedure is repeated until the stop criterion is reached (number of generations defined by the user), then the best solution is printed (individual).

The development of this type of algorithms can result difficult if you do not have the required experience, due to the fact that clusters or grids are used, which implies intensive computing programming, in addition to having knowledge of the commands in the Linux environment.

5.2 *Web Application*

For the development of the Web application the following requirements were used:

- Development environment: Linux, Java, Tomcat and MySQL.
- Portability: Windows, Linux.
- Security: Servlets and JDBC.
- Interface: Java Server Pages (JSP) and HTML.
- JSch library.

5.2.1 Design

The use case diagrams describe the procedure and communication of a system with users and/or other systems. Figure 5 shows the use case diagram of the Web application, which describes the activities that the user will be able to perform.

5.2.2 Implementation

User validation, this allows access to the Web page only to users who are previously registered, Fig. 6 shows the ingress window. Parameter configuration, the user enters the necessary parameters (Probability of crossing and mutation, percentage of selection and mutation, population size and number of generations) for the correct functioning of the algorithm genetic. Enter the data of the instance to be resolved (hydraulic network). Choose the parameters for executing the algorithm in a distributed way at the cluster level. These parameters correspond to the number of nodes, the number of processes per node and the number of tests to be performed for a certain instance. Figure 7 shows the parameter configuration page.

Algorithm execution. Once the data has been chosen, it is executed. In this procedure, the web application generates a file in.txt format (Fig. 8a), which contains the chosen parameters by the user. The first six values correspond to the parameters of the genetic algorithm. The last three values indicate the parameters of the algorithm execution in a distributed way. For example, two nodes, two processes per node, and

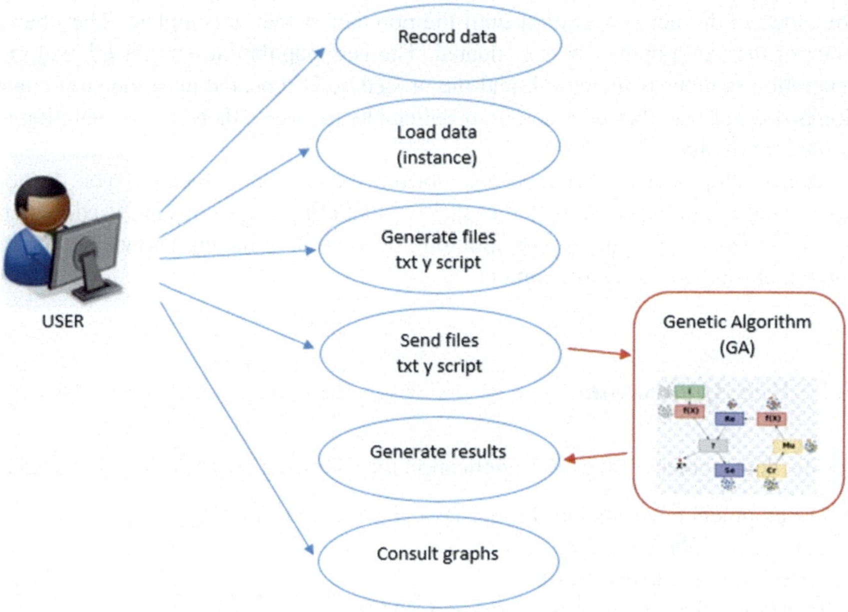

Fig. 5 User case diagram of the Web applications

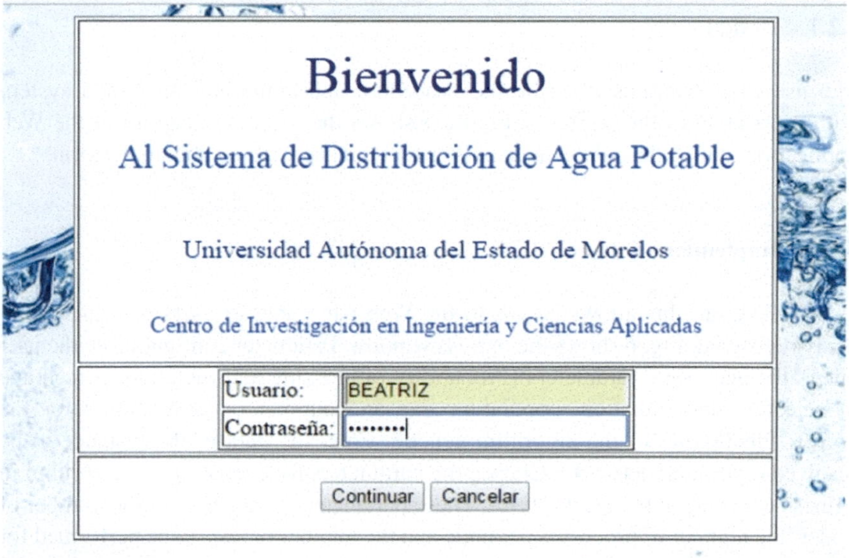

Fig. 6 User login page

Fig. 7 Parameter setting page

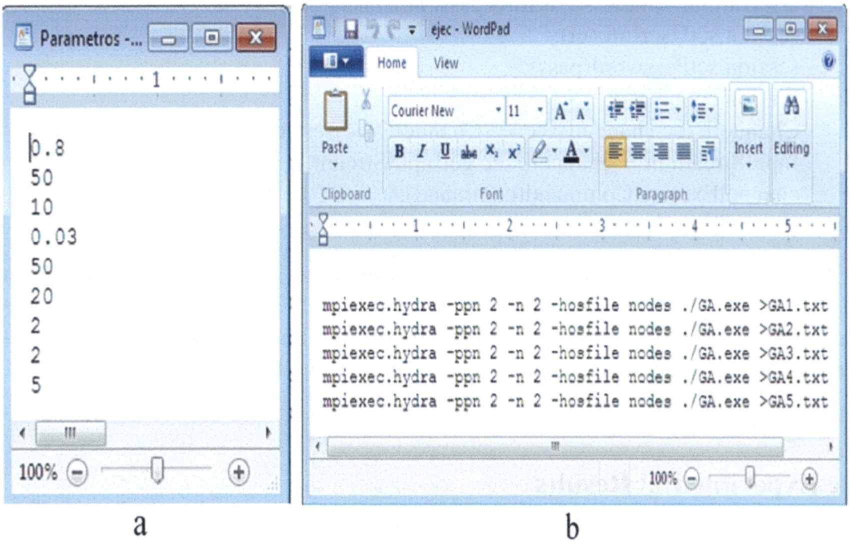

Fig. 8 Output files: (**a**) parameters and (**b**) script with the execution command

five tests. It also generates a script file in.sh format (Fig. 8b), which contains the command to execute the genetic algorithm in a distributed manner, and according to the number of tests is the number of times the following command is repeated.

Command syntax:

- mpiexec.hydra –ppn 2 –n 2 –hostfile nodes./GA.exe < GA1.txt.

Arguments:

- mpiexec.hydra: library to activate processes
- ppn: number of processes per node
- n: number of nodes
- hostfile nodes: name of the files that contains the node list
- ./GA.exe: executable file name
- > GA1.txt: name of the results file, a different file for each test number.

Execution of the script file. For the execution of the script, knowledge of the commands of the Linux environment is required, or to connect to the cluster remotely and securely using an SSH client (putty, winSCP, FileZilla …) from Windows. This application facilitates the procedure through the JSch library, written in java. This library, in addition to connecting to an SSH server, also allows file transfer.

Syntax:

```
1:  JSch jsch = new JSch();
2:  Session session = jsch.getSession(user, host, port);
3:  UserInfo ui = new SUserInfo(pass, null);
4:  session.setUserInfo(ui);
5:  session.setPassword(pass);
6:  session.connect();
7:  ChannelExec channelExec = (ChannelExec)session.openChannel("exec");
8:  InputStream in = channelExec.getInputStream();
9:  channelExec.setCommand(command);
10: channelExec.connect();
```

Line 1 creates an object from the JSch library. Line 2 establishes a user session to connect to the server. The data required are: user, host and port. Line 5 sets the password of the user account to make the connection (line 6). Line 7 opens the Shell terminal. Line 9 executes the command (script file, Fig. 8b).

6 Experimental Results

The experimental tests were carried out in the TEXCAL Cluster, located in the State of Morelos, Mexico.

- Cluster Features:
- 2 Intel Xeon Six Core 3.06 GHz processors, 12 MB cache
- Distributed memory: 120 GB RAM
- Nodes: 5
- Processing cores: 60
- 14 TB HD

- Operating system: Linux
- Compiler: gcc, mpicc.

The experimental tests were performed with an instance of 350 nodes, 364 pipes, 6 storage tanks and a reservoir. The minimum required pressure is 10 mca and maximum required pressure is 60 mca. Pipe diameters from 22.7 to 101.6 mm, with a standard demand of 0.003 L/S. The behavior analysis of the genetic algorithm is carried out with a total of 30 tests, and the following parameters were used: number of generations: 100, population size: 50 individuals, percentage of mutation: 10%, selection percentage: 50%, crossing probability: 0.8 and mutation probability: 0.03. Once the execution of the algorithm has been completed and the data has been stored in the database, the user will be able to analyze the results produced by the genetic algorithm through the option "Graphs", where pressure graphs are shown, of generations and statistics. Figure 9a allows selecting the test number so that the graph corresponding to the pressures obtained for the mentioned instance is presented. The behavior of the pressures is shown in Fig. 9b, it is observed that the nodes comply with the minimum and maximum pressure required; therefore the solution is feasible, since it satisfies the conditions (constraints) of the constraint satisfaction model.

Figure 10a allows selecting the test number to show the corresponding graph based on the number of generations of the genetic algorithm. The convergence behavior of the algorithm is shown in Fig. 10b, where it is observed that as the number of generations increases, the cost of the solution improves until it achieves a stable value.

Figure 11a shows the different types of statistical graphs that the user can choose, for example graphs average cost, execution time and convergence of the algorithm. Figure 11b shows the behavior of the genetic algorithm based on the number of generations and the average cost of a total of 30 executions.

Table 2 presents the best solution found with respect to the objective function. It also shows the worst solution, the average of the30 executions and their standard deviation. The standard deviation is small; therefore the mean is a reliable value of the solution that the algorithm produces. The median is within $\pm\ \sigma$ which confirms the central tendency of the data. The best cost of the hydraulic distribution network (HDN) was 244,500 (currency units). In this best cost scenario, three storage tanks, three pipes, and eight pressure reducing valves would have to be added.

Finding the optimum location of the valves, the proper configuration of the storage tanks and the suitable diameters for the pipes allow the HDN to improve. Figure 12 shows the pipes where a valve is added in order control pressure throughout the system. Also where three tanks and three pipes were added in different sections of the network. The simulation was obtained by Epanet. According to the results, the configuration is adequate to obtain the lowest possible cost while achieving a feasible solution. Figure 13 shows the pressure obtained at each node. All nodes reached the minimum pressure required defined as 10 mca and the maximum pressure, defined as 60 mca. This indicates an acceptable network performance since the pressures are adequate.

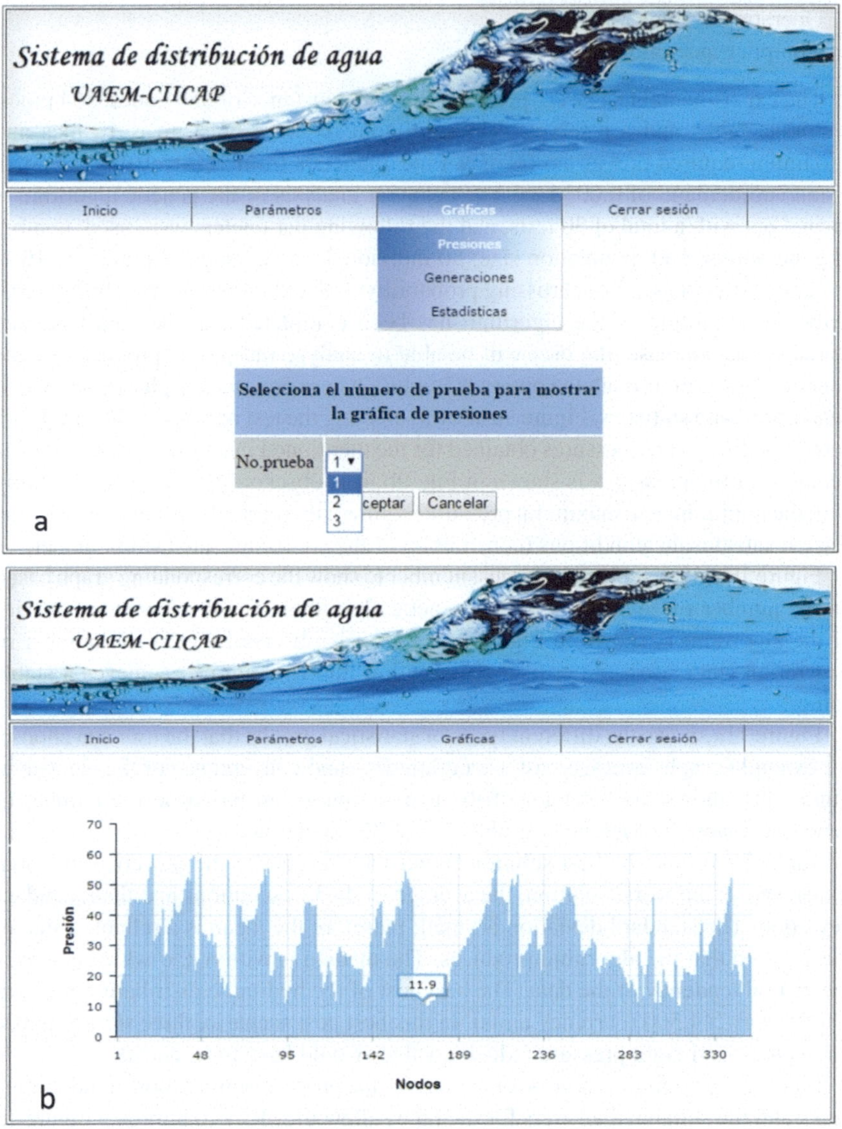

Fig. 9 Pressure graphs: (**a**) choose test number and (**b**) pressure graph

Tables 3–5 show the characteristics of the elements added to the HDN. These are the characteristics suitable for the best solution found by the algorithm with a cost of 244,500 (currency units). The optimal solution has eight pressure reducing valves (PRVs) (Table 3) with different parameters. For example, valve 1 connects node 21 to node 22, with a diameter of 38.10 mm and valve setting of 24.67 m. Valve 7 connects node 220 to node 221, with a diameter of 76.20 mm and valve setting of 11.92 m.

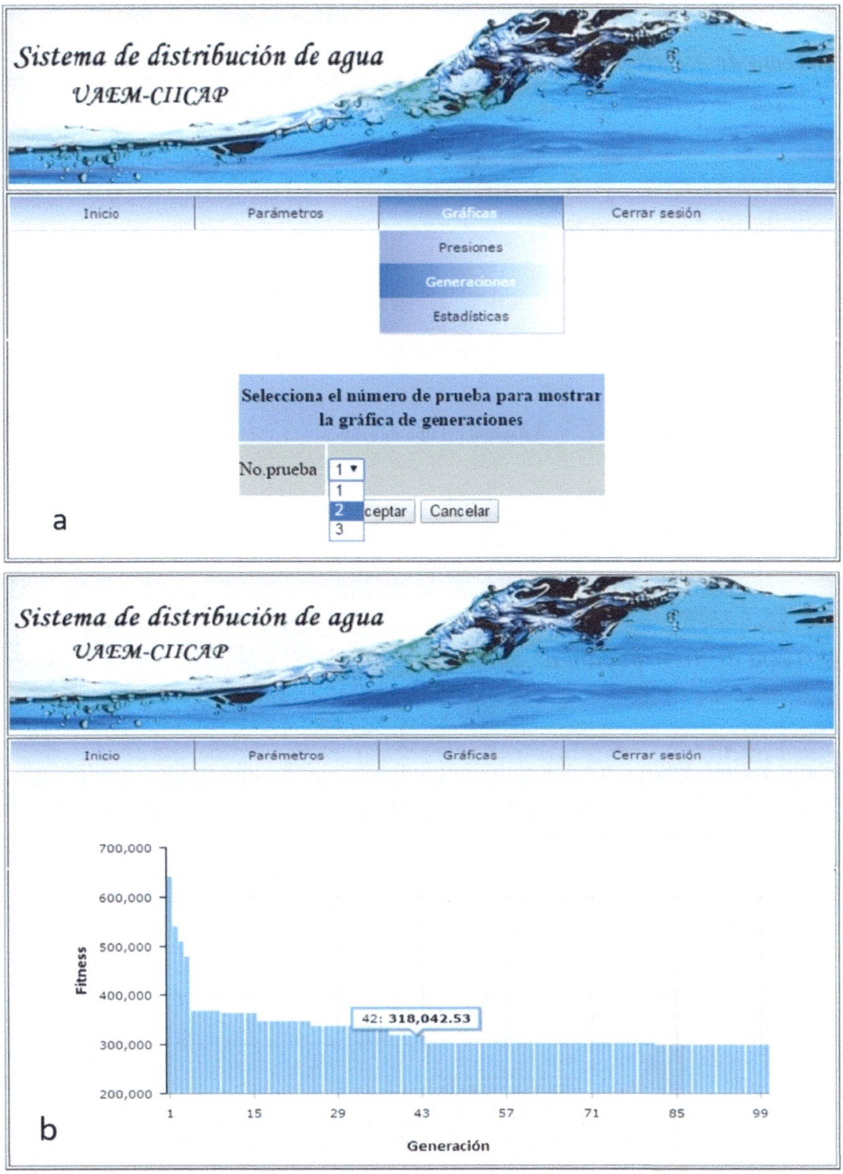

Fig. 10 Generation graphs: (**a**) choose test number, (**b**) generation graph corresponding to a certain test

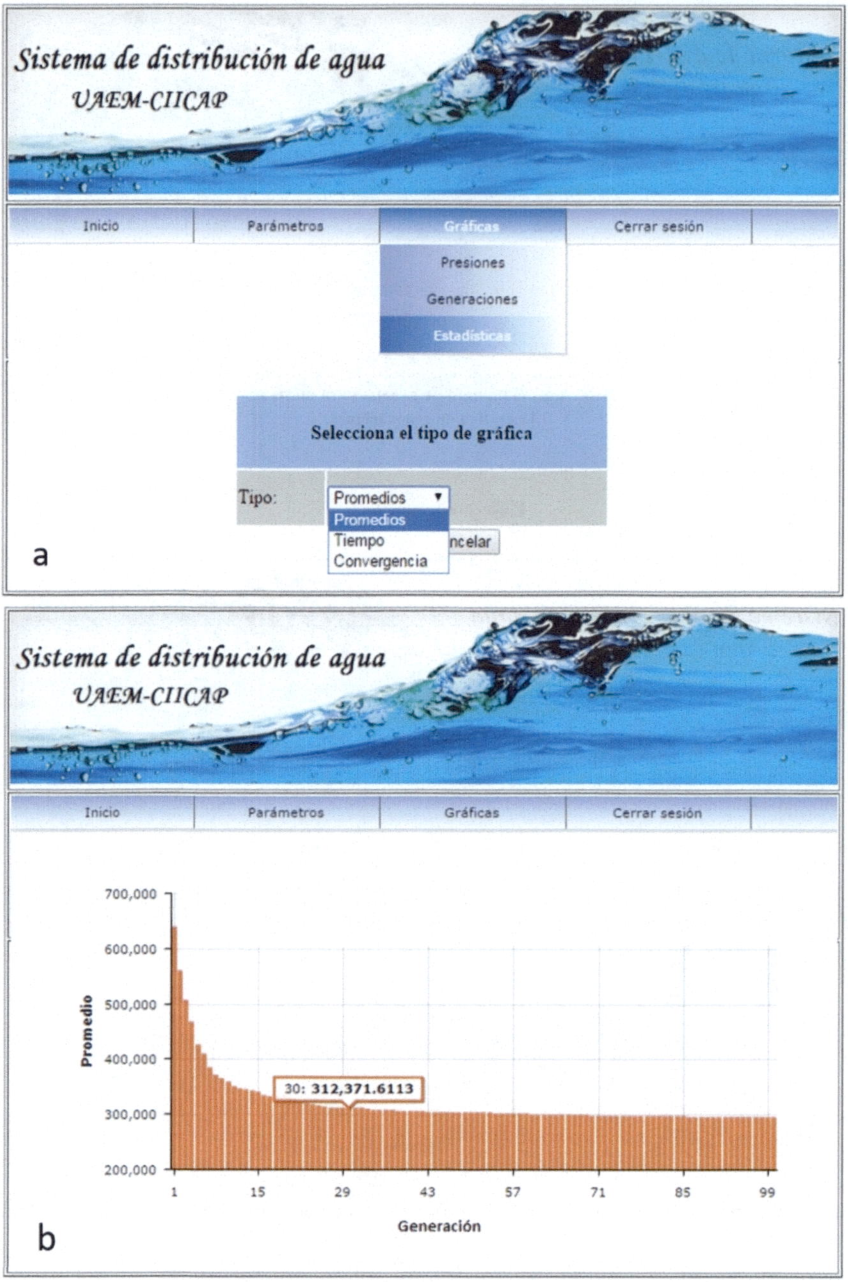

Fig. 11 Statistical graphs: (**a**) choose the type of graph, (**b**) graph of averages

Table 2 Results obtained by the genetic algorithm for the HDN

Parameter	Cost (Currency Units)
Worst	301,324
Mean	253,631
Best	244,500
Mean ± Standard deviation	253,631 ± 16,780
Median	244,800

Fig. 12 Hydraulic simulation of the best solution

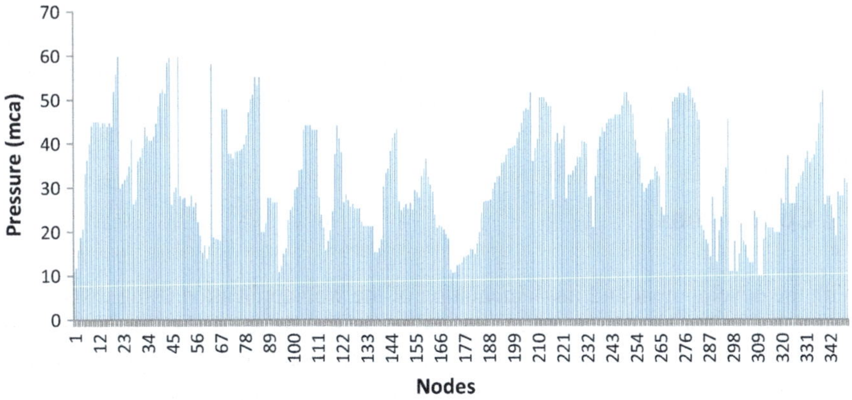

Fig. 13 Pressure in each node of the best solution

Table 3 Characteristics of new PRVs added to the HDN

ID_Pipe	Start node	End node	Diameter (mm)	Valve setting
1	21	22	38.10	24.67
2	44	45	38.10	27.85
3	48	49	50.80	37.90
4	63	64	19.05	27.74
5	164	165	50.80	34.17
6	207	208	50.80	36.02
7	220	221	76.20	11.92
8	339	340	31.70	22.19

Table 4 Characteristics of new tanks added to the HDN

ID_Tank	Elevation(m)	Initial level (m)	Minimum level (m)	Maximum level (m)	Diameter (m)
1006	2385	2.50	1.00	5.00	5.86
1007	2428	3.00	1.00	5.80	6.50
1008	2324	3.00	1.00	6.00	6.76

Table 5 Characteristics of new pipes added to the HDN

ID_Pipe	Start node	End node	Length (m)	Diameter (mm)
365	1006	118	200.00	101.00
366	1007	283	240.00	76.00
367	1008	170	250.00	31.00

Three tanks with different characteristics have been added (Table 4). Tank 1006 has a diameter of 5.86 m, a maximum level of 5 m and an initial level of 2.50 m. In the proposed solution, three new pipes were added (Table 5), which connect each of the tanks with the rest of the network. For example, pipe 365 is connected to tank 1006 at node 118. The pipe length is 200 m with a diameter of 101 mm.

7 Conclusions

With the development and implementation of the web application using distributed computing and according to the results shown, it is observed that optimal solutions to the problem of hydraulic network design were obtained. Besides that its use is very simple since it only requires having defined the parameters of the algorithm, the data of the instance and the execution parameters.

It is also observed that the genetic algorithm proposed as an alternative solution is an ideal method of searching for optimal solutions for engineering design problems, in this work used for the design of water distribution networks. Its efficiency has been demonstrated based on the results previously presented, as it can be seen graphically that the constraint satisfaction model and the optimization model performed in a distributed way are fulfilled and a *.INP file is also generated which is it enters a graphical application (EPANET) to view pipes, tanks, pumps and valves of the optimized network as shown in Fig. 3.

Some of the advantages of this work is that it allows to reduce the time and computational resources for the design of an HDN and the flexibility to execute the application *n* number of times with different input parameters of the algorithm, according to the needs of the instance to work in addition, which is a portable application, as long as you have an internet connection. Unlike works found in the literature, this work allows finding a solution through a graphical interface, in addition to applying an optimization algorithm through distributed programming.

Future work. Develop a module for the graphic visualization of the network where can be seen the pipes, pumps, valves, tanks, etc. Also test and compare the performance of other metaheuristics for these types of problems.

Acknowledgements We thank Dr. Marco Antonio Cruz Chávez for the facilities granted for the use of the cluster for the development of the work.

References

1. Pereyra G, Pandolfí D, y Villagra A. (2016) Diseño y Optimización de redes de distribución de agua utilizando algoritmos genéticos. ICT-UNPA-158–2017, N° 0287/17- R-UNPA
2. Khatavkar P, Mays LW (2019) Optimization-simulation model for real-time pump and valve operation of water distribution systems under critical conditions, Urban Water J
3. Edwin PV, Angely VC, Franz CP, Jesus MM, Eduardo CV, Luis AR (2017) Diseño Óptimo de Redes de Distribución de Agua Usando Un Soft-ware Basado En Microalgoritmos Genéticos Multiobjetivos. Ribagua 4(1):6–23. https://doi.org/10.1080/23863781.2017.1317087
4. Saldarriaga J, Pulgarín L, Cuero P, Duque N (2017) Software para la enseñanza de hidráulica de tuberías (Pipeline Hydraulics Academic Software). In: Ibero-American Seminar on Water and Drainage Networks (SEREA 2017)
5. Sarbu I, Ostafe G (2015) Optimal design of urban water supply pipe networks. Urban Water J https://doi.org/10.1080/1573062X.2014.994007
6. De Corte A, Sörensen K (2016) An iterated local search algorithm for water distribution network design optimization. Networks 67(3):187–198
7. Abdelbaki C, Benchaib MM, Benziada S, Mahmoudi H, Goosen M (2017) Management of a water distribution network by coupling GIS and hydraulic modeling: a case study of Chetouane in Algeria. Appl Water Sci 7(3):1561–1567
8. Dongre SR, Gupta R (2017) Optimal design of water distribution network under hydraulic uncertainties. ASCE-ASME J Risk Uncertainty Eng Syst Part A: Civ Eng 3(3):G4017001
9. Gupta I, Bassin JK, Gupta A, Khanna P (2000) Optimization of water distribution system. Environ Softw 8:101–113
10. Talbi EG (2009) Metaheuristics: from design to implementation. https://doi.org/10.1002/978 0470496916

11. Maroto C, Alcáraz J. y Ruíz R (2002) Investigación operativa: modelos y técnicas de optimización. Valencia, España: Universidad Politécnica de Valencia
12. Melián B, Moreno Pérez JA, Moreno Vega JM (2003) Metaheurísticas: una visión global. Inteligencia Artificial, Revista Iberoamericana de Inteligencia Artificial. N° 19, pp 7–28, ISSN:1137–3601
13. Glover F (1986) Future paths for integer programming and links to artificial intelligence. Comput Oper Res 5:533–549
14. Glover F (1990) Tabu search part I. ORSA. J Comput 2:4–32
15. Blum C, Roli A (2003) Metaheuristics in combinatorial optimization: Overview and conceptual comparison. ACM Comput Surv 35(3):268–308
16. Kirkpatrick S, Gelatt CD, Vecchi MP (1983) Optimization by Simulated Annealing. Science. 220:671–680. https://doi.org/10.1126/science.220.4598.671
17. Glover F (1995) Tabu Search Fundamentals and Uses. University of Colorado, Boulder, Colorado, USA, p 85
18. Kennedy J. Y Eberhart R. 1995. Particle swarm optimization. Proceedings of IEEE International Conference on Neural Networks, Perth, Australia:1942–1948. https://doi.org/10.1109/ICNN.1995.4889668
19. Holland JH (1962) Outline for a logical theory of adaptive systems. J ACM 3:297314. https://doi.org/10.1145/321127.321128
20. Thomas Rauber, Gudula Rünger. Parallel programming: for multicore and cluster systems. Springer-Verlag
21. García Regis OR, Cruz Martínez E (2003) Paralelización. UNAM, México, DF
22. Pacheco P (2011) An introduction to parallel programming. Elsevier
23. Blaise Barney LL (15 de 12 de 2014). Message Passing Interface (MPI). Recuperado el 28 de 04 de 2015, de https://computing.llnl.gov/tutorials/mpi/
24. Difference Between Parallel and Distributed Computing (2018) Retrieved (date), from https://differencebetween.com/difference-between-parallel-andvs-distributed-computing/
25. Kshemkalyani AD, Singhal M (2011) Distributed computing: principles, algorithms, and systems. Cambridge University Press
26. Rossman LA (2000) EPANET 2 User's Manual: national risk management research laboratory office of research and development. Cincinnati, OH, USA, US Environmental Protection
27. National Water Comission (2007) Manual of drinking water, sewerage and sanitation, 2007th edn. Mexico City, Mexico, National Water Comission
28. Korte B, Vygen J (2008) Combinatorial optimization theory and algorithms. In: Series Algorithms and Combinatorics, vol. 21, 4th edn. Springer: Berlin/Heidelberg, Germany
29. Martínez-Bahena B, Cruz-Chávez MA, Ávila-Melgar EY, Cruz-Rosales MH, Rivera-Lopez R (2018) Using a genetic algorithm with a Mathematical programming solver to optimize a real water distribution system. Water 10:1318. https://doi.org/10.3390/w10101318
30. Montesinos MP, García-Guzmán A, Ayuso JL (1997) Optimización de Redes de Distribución de Agua utilizando un Algortimo Genético. Ingeniería del agua, vol. 4 N° 1. Pág. 71–77

Diverse Kind of Logistics in Amalgamed Application Domains

What is the Best Location of a Smart Airport in Juarez, Mexico?

Aida-Yarira Reyes-Escalante, Alberto Ochoa-Zezzatti, Diego-Adiel Sandoval-Chávez, and Karla-Stephania Venegas-Ortiz

Abstract Location of airports is a key factor for the sustainable and competitive development of the smart cities. Airport location projects influence many aspects, such as: growth of the city, manoeuvres of airlines, transport of passengers, and soil preservation. They also bring a variety of social, environmental and economic benefits. The location of an airport that comply with land use and airspace rules and regulations is of utmost importance. This paper presents the problem of: What is the best location of an airport in Ciudad Juarez? The designed a multi-criteria analysis approach to locate a new smart airport in a border area of Northern Mexico. The study bases on the model of indicators for the competitiveness of the city and inter-national regulations for locating airports in border areas. The study also bases on the medium smart airport model and considers four location alternatives: two to the Northwest (1) and (2), one to the Southeast (3) and the fourth in the South (4). We used TOPSIS multi-criteria model to obtain the best alternative to allow for the most convenient location. The analysis resulted in alternative 2, located to the northwest, being the best, followed by location 1 and location 4.

Keywords Airport location · Border regions · Multi-criteria analysis · Intelligent airports · Smart cities

A.-Y. Reyes-Escalante (✉) · A. Ochoa-Zezzatti · K.-S. Venegas-Ortiz
Universidad Autónoma de Ciudad Juárez, Ciudad Juárez, Mexico
e-mail: aida.reyes@uacj.mx; yarizue@gmail.com

D.-A. Sandoval-Chávez
Tecnológico Nacional de México Campus Ciudad Juárez, Ciudad Juárez, Mexico

© The Author(s), under exclusive license to Springer Nature Switzerland AG 2021

A. Ochoa-Zezzatti et al. (eds.), *Technological and Industrial Applications Associated with Intelligent Logistics*, Lecture Notes in Intelligent Transportation and Infrastructure,
https://doi.org/10.1007/978-3-030-68655-0_24

1 Introduction

The optimal location of an airport is of the most important for smart cities because it greatly defines the competitive edge and long-term development. Such a decision involves a variety of actors ranging from government agencies, airlines, transport agencies, food services, reservation agencies, and tourism companies to local stakeholders.

2 Cities and Development

According to Dimuro [10] the cities went throughout three stages in their historic evolution. The first stage characterized for a high dependence on nature; during the second, cities experience the arrival of technology and urban sprawl; finally, during the third stage cities have come back to appreciate nature, particularly in public spaces.

Due to their entropic nature, urban cities consume a great deal of materials and energy just to maintain the order within them. As urban sprawl increases, the city needs additional amount of materials, fuel and public services. As a result of the urban metabolism, a great amount of degrades energy and waste are deposited in the environment [27, 30].

Two milestones were decisive for sustainability to become a premise for being a key competitive advantage for countries in a global market. One is the United Nations Conference on the Human Environment, also referred as to Stockholm Summit, which was held in 1972 to emphasize the need to address environmental deterioration. The other is a dossier that Club of Rome generated in which they warned about the eventual impact of the microelectronic industry. Both initiatives trigged the interest in sustainability and promoted public policy to address it at a high level. Since then, sustainability has been a critical issue which is considered a key factor in the decision-making process both in public as well as in private organizations [15, 23].

Very soon, the growing interest in sustainability brought a transgenerational commitment Left [19] mentioned that factors like entropic degradation, nature biochemical cycles, economic crisis and technological innovation are now issues that organization need to address to cope with sustainability demands. In addition, Baker [1] indicate that sustainability models must consider development seen as a whole, including regulations, governance, technology and policy. In order to be able to meet the difficulties of this new scenarios, it is imperative to stimulate innovation, creativity and experimentation [12].

3 Smart Cities: Competitive and Sustainable

Technological development as it concerns to public spaces reflects in the quality of these entities. Public space infrastructure and policy decant down to performance indicators for both urban and rural areas.

Lever [20] states that competitiveness is paramount for cities to promote investment and economic growth. Competitiveness among cities takes place for many reasons

(a) To catch mobile investment for creating industrial jobs and commercial developments
(b) Economic growth measured as gross value added or Gross Economic Product (GEP), which includes growth of existing entities and the arrival of new ones.
(c) Populations level, based on income, human capital, demand and political power.
(d) Public funds at national or regional level.
(e) Distinctive events and infrastructure.

Cites with better competitiveness indexes achieve better market share. Initially, this vision concentrated at national level base natural resources, human skills, capita and productivity, but nowadays nations no longer compete, but cities [11].

A highly-wired, tele informational-based city is identified with the quality of intelligent [14]. In this kind of city government policy emphasizes sustainability when instrumenting long-term planning programs within which technology suppliers, government offices and academia conflux. This approach proves to be one of competitive shape that reinforces itself monitoring pertinent indicators, such as infrastructure systems, intelligent framework, urban space y renewable energies [16].

Hollands [16] Fig. 1 depicts the balance among energy, transportation and technologies of information (TICs). This portray gives place to the idea of the holistic

Fig. 1 Smart cities elements [14]

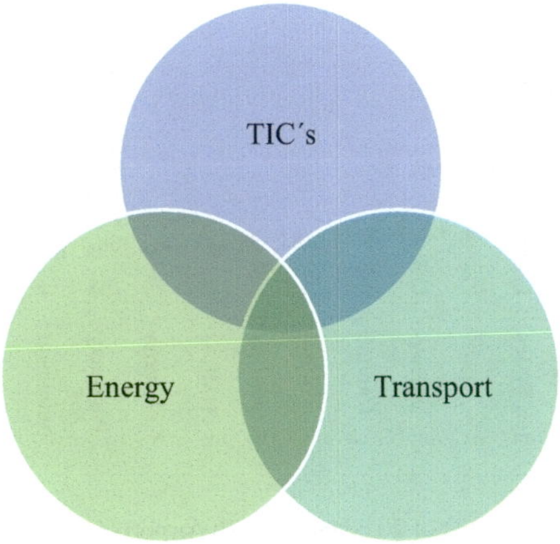

approach in the intelligent city when consider it as a multifunctional, complex system in which governance dictates the decision-making process [15].

According to Valderrama [31] there are six areas in which intelligent cities must focus: citizen participation, governance, mobility (all kinds, water, land and air) as a public service, attention to environmental issues, competitiveness and quality of life.

Transportation is crucial to attain sustainability in urban cities. Its importance reinforces the idea that vehicles automation and collective systems reduces vehicular park, thus improving sustainability. It is also important to increase commuter efficiency by establishing routes at an optimal velocity.

4 Intelligent Airports: Competitive and Sustainable

Design and location of new airports face increased regulations as customer's demand better services at lower cost. This calls for new, technology-based intelligent airports not only to meet customers' demands and expectations but also to position branding and reputation. An intelligent airport, optimally located reduces waste, promotes investment and is not energy-intensive,

The Central Intelligence Agency (CIA) reports there are 41,788 airports in 236 countries in the world. USA ranks first with 13,513; Brazil, México and Canada follow with 4,093, 1,714 and 1,467, respectively. Figure 2 shows the number of airports or airfields recognizable from the air. The runway(s) may be paved or unpaved and may include closed or abandoned installations [4].

The International Airport Council (AIC) reports that air traffic increased 6.6% in 2017 with respect to 2016. At the same time, international traffic increased 8.4%.

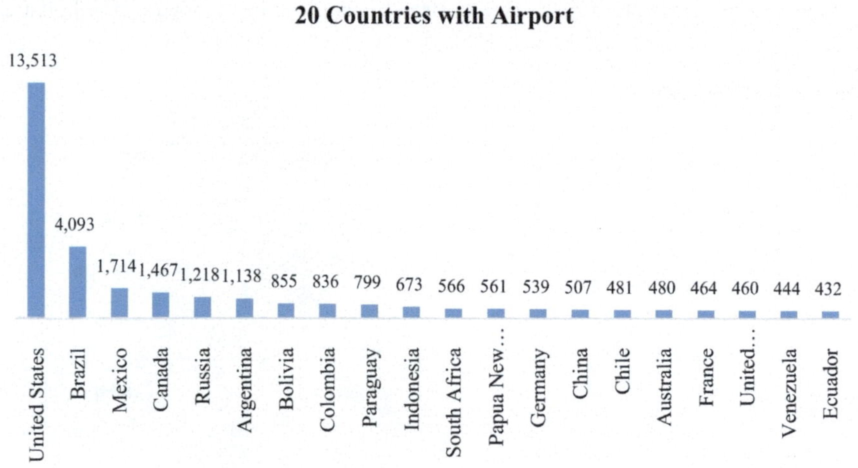

Fig. 2 List of 20 countries with Airports [4]

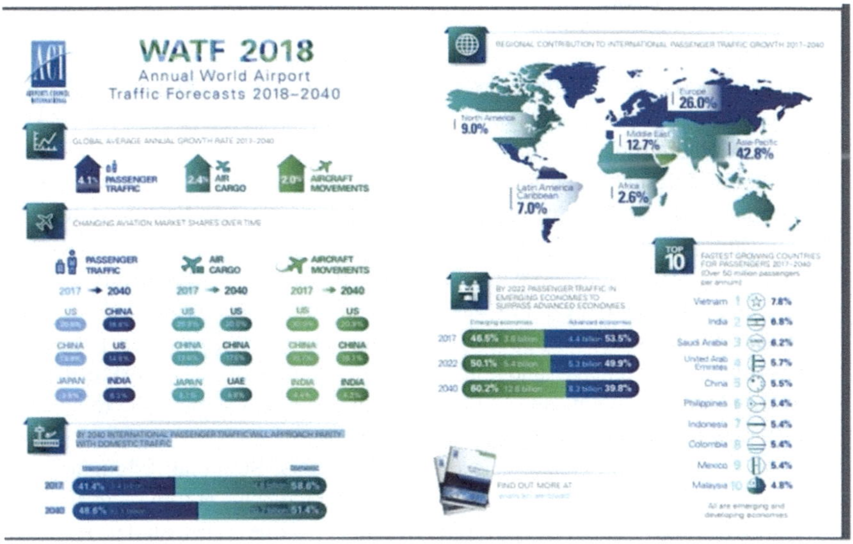

Fig. 3 Annual World Airport Traffic Forecast 2019, [4]

Due to operations consolidation, increased occupation and higher-capacity aircrafts, the number of flights had less dynamism with 2.4% of growth [4].

According to the Annual World Airport Traffic Forecast (WATF) 2018–2040, passenger transportation will increase from 2.4 to 4.2% for cargo flights and 2% for aircraft movements, this growth will be world-wide, see Fig. 3. Mexico expects a growth of 5.4% for both domestic and international flights, Fig. 3 [4].

The European Airport Trade Association (ACI Europe) releases its traffic report for July 2019, during which average passenger traffic in geographical Europe expanded by +2.2% compared with the same month in 2018. Standing at less than half the growth of the preceding month (June: +4.7%), this is the weakest monthly performance so far this year. Meanwhile, Freight traffic declined for the 9th month in a row at −2.3%. Growth in aircraft movement was subdued at +1.2%—compared to +3.8% at the start of the year [5].

Due to the importance of airports for competitive edge, the following research avenues have been identified:

(a) Airport privatization. Voguel [32] points out about the importance of private airports, including the impact of the investment process. After studying the airports in Saudi Arabia, Chaouk, Pagliari, Miyoshi [6] recommended a policy for future airport privatizations to increase market share and reduce losses.

(b) Analysis of airport impacts: Bringmann et al. [2] planned the study related about the cooperative arrangements between European airports by focusing on their financial intertwinement. The findings indicate a high degree of stability in investment relations in the airport industry over time. Further-more, it is suggested that the formation of inter-organisational linkages is facilitated by spatial proximity.

(c) Also, Gasco et al. [13] remark the noise is negative environmental effect that must be addressed when planning an airport. Martin-Domingo and Martín [21] analyses the airport mobile internet an innovation, also, studies the adoption of mobile Internet by airports. Using a new theoretical model, the study tests whether early adopters of mobile Internet for airports can be considered real innovators.

(d) Airport classifications: Mayer [22], indicted the airport classification based on cargo characteristics, 114 airports are grouped according to their cargo business characteristics. Applying a hierarchical cluster analysis, the paper uses absolute (cargo tonnage) and relative measures (share of cargo workload units, of freighter movements and of international cargo) to establish the reliance of different airport types and groupings on air cargo.

(e) Borders and requirements. Gasco et al. [13] also, Sulmona, Edgington and Denike [29] study the role of Advanced Border Controls at Canadian airports, with government between 1985 and 2010 and how this contributed to a balance between trade and national security imperatives.

(f) Tourism: Debbage [9] mentioned the international tourist arrivals accounts for approximately 25% of worldwide in The United States–European Union market, that landing slot policy and the manner in which airport capacity is allocated among airlines across the north Atlantic is likely to underpin the future geographic structure of the tourism industry.

5 Location Airport Analyses

The importance of the location of airports is derived from the development of air transport and the use of airspace in order to operate aircraft without the physical existence of the airport and land surface facilities. The air transport system is integrated by aircraft and airports, together they allow to offer the service to users who are mostly passengers, organizations, delivery services and movement of products. Airport facilities today seek to turn these into smart airports so as not to lag behind and ensure their market share by being competitive and novel.

The passenger terminal is one of the main elements of the cost of an airport's infrastructure, in addition to the current technological implementations, which offer it through service. There are a lot of airports in the world that have been built as architectural monuments sizes vary: large, medium-small, to small runways. Others have also been remodelled to stay in the market and passengers have become accustomed to a comfort ostentation of design and services, and there are investments in small airports such as ending them to vent the weight of airports large, and also, to have more mobility and access to the population in large cities or megacities.

Within international regulations there is the International Civil Aviation Convention signed in Chicago in December 1944, each participating State undertook to work together to achieve uniformity of regulations relating to features of airports and landing areas. This allows airports in their proposals for new designs to have the minimum established in the Convention and the application of all changes within the building.

Airport location-related research is scarce. Cheng-Lung Wu and Andy Lee [33] conducted a research related to the impact of airline alliance terminal co-location on airport operations and terminal development, they address questions about the notion of co-locating alliance carriers while benefits on the part of airlines are made, the tangible benefits to airport operators are less clear. Analyses cases of London Heathrow, Paris Charles de Gaulle and Tokyo Narita Airport, and applies their operational practices to a medium-sized airport in Asia Pacific to evaluate the universal applicability of alliance member co-location.

To the extent of our knowledge, a multicriteria approach has not been adopted for the location of an airport. Studies on the subject matter are perceived fragmented and incomplete. The research question is ¿what is the best location for new airport in Juarez, Mexico? Thus, the objective of this work is to determine such a location.

6 Juárez, Mexico Airport

Built in 1968 in an area of 4,736,247.74 m^2 (Fig. 4) Juarez, Mexico Abraham Gonzalez International Airport (CJS) has undergone several modifications and improves. CJS encompasses a series of modern facilities, including a new traffic control tower, a renovated track, hangars, lobby and a new telecommunications baseline. The carriers that operate in CJS are Vivaaerobus, Aeromexico, Volaris, Interjet, TAR, cargo airlines and private flights.

CJS experiences air saturation which causes frequent delays that lead to increased costs. These problems give place to the idea that a new airport is an urgent requirement for Juarez to continue being a competitive Smart city.

As the urban sprawl took place in Juarez, the airport saw itself surrounded by main avenues, habitational developments and manufacturing firms. These factors made impossible the airport to expand (see Figs. 5 and 6).

Restricted areas for the location of hovers are established according to the Secretary of Communications and Transport in 2016, are presented in Fig. 7 and the restrictions are documented in AIP of Mexico, see Fig. 8. The most restricted area is stablished base on the geographical land located on the South Oriented started on the km 20, this problem affected the visibility and is not approved to airports.

7 Method

The study relies on an exploratory analysis carried out in 2019 in Juarez, Mexico. We considered three different sectors: East, West and South. Also, this study involves a

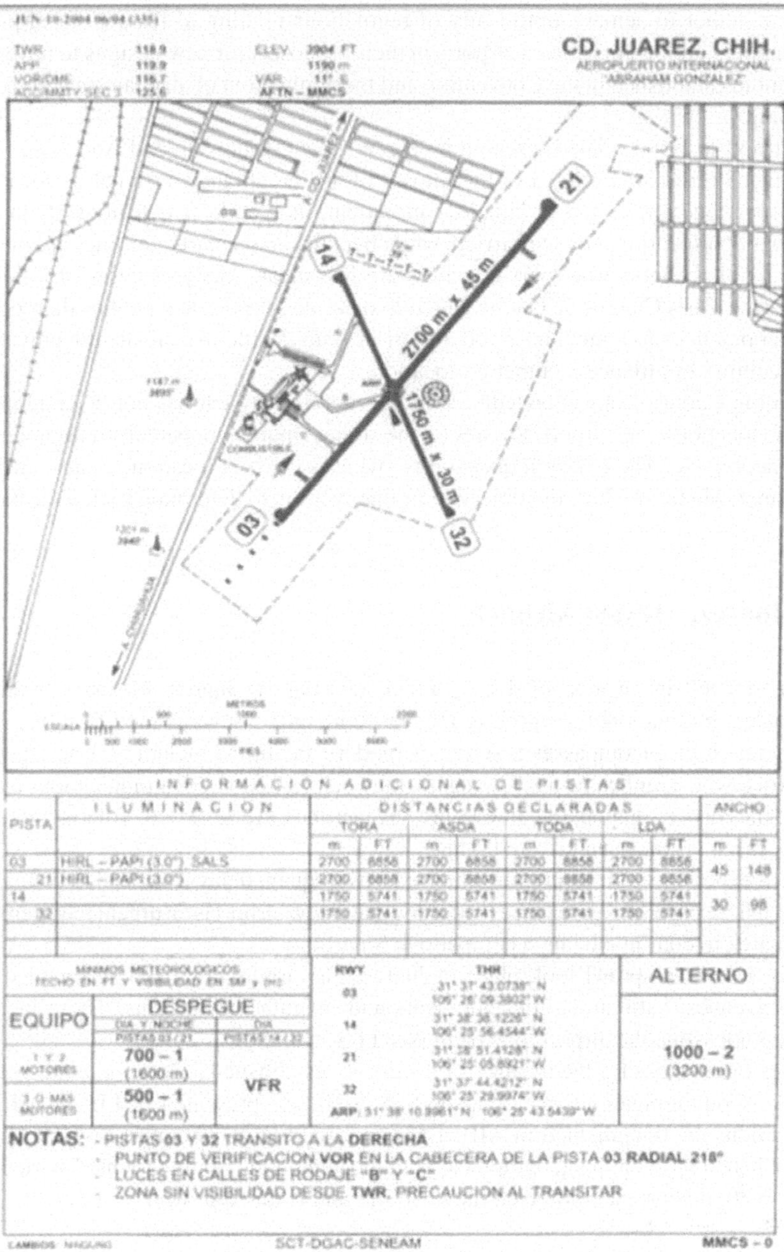

Fig. 4 International Abraham González Airport [3]

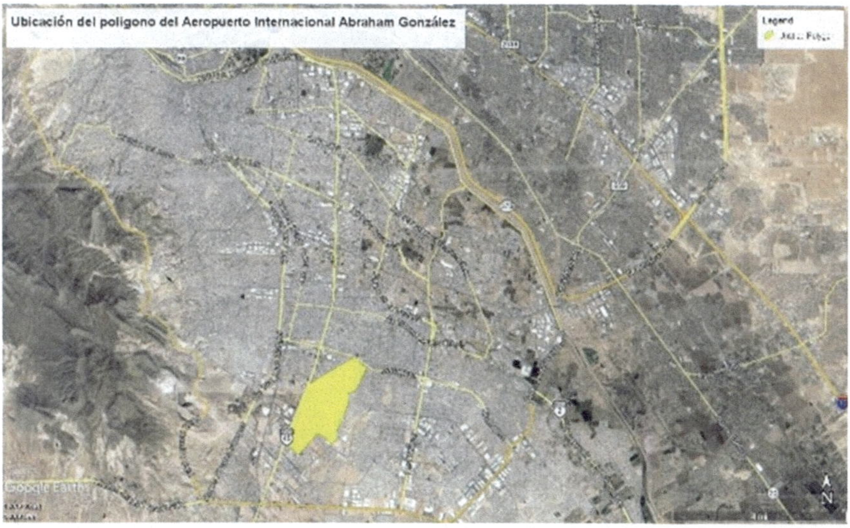

Fig. 5 Geography location International Abraham González Airport

Fig. 6 Geography polygon

quantitative design, adopting a multi-criteria decision methodology to determine a basis for planning the location of an International Smart AirPort.

In order to determine the category of the smart airport appropriated for Juarez, this work looked at different regions and their related airports. Four airports were considered: Changi (Singapur), Incheon (South Korea), Hamad (Doha) and Carrasco (Uruguay). After taking into account a set of factors the analysis concluded that the

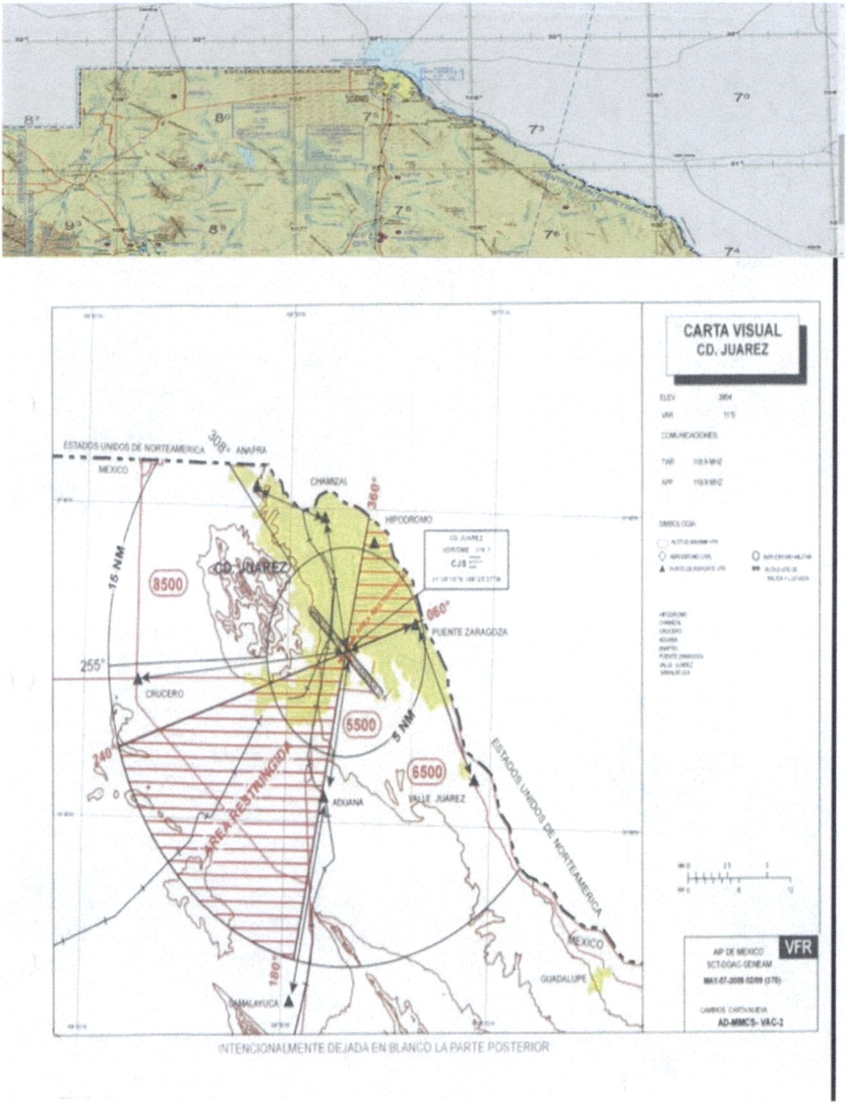

Fig. 7 Restricted Air Conditional Juarez Municipal [28]

International Airport of Hamad (DOH) met the characteristics to be considered as the baseline for the study, (see Table 1).

The polygon dimension of DOH will be the reference to identify spaces beyond the restricted areas to locate a new airport in Juarez (see Figs. 9 and 10).

Multiple-criteria decision-making (MCDM) also referent as multiple-criteria decision analysis (MCDA) is a sub-discipline of operations research that explicitly evaluates multiple conflicting criteria in decision making. The MCDA is the

AIP DE MEXICO AD-MMCS VAC-0

REGLAS Y PROCEDIMIENTOS PARA VUELOS VFR CD JUAREZ.

1. **Restricciones**

a) Queda prohibido el vuelo VFR arriba de la(s) altitud(es) máxima(s) establecida(s) para cada sector, dentro de un radio de 15 millas náuticas del Aeropuerto MMCS

b) No se permite el vuelo VFR dentro de las áreas restringidas N entre la radial 360° y radial 060° hasta la línea fronteriza, y S entre la radial 180° y radial 240° hasta 15 MN a menos que se encuentre con autorización del Control de Aeródromo (Torre de Control)

2. **Zona de tránsito de aeródromo (ATZ).**

a) Esta zona esta reservada para las aeronaves que vayan a despegar o aterrizar en el aeropuerto y sólo podrá ser penetrada con autorización de la Torre de Control, dicha zona comprenden un radio de 5 MN con centro en CJS dentro del territorio nacional

b) Se establecen las rutas visuales de salida y llegada VFR descritos en la Carta Visual para efectos de sobrevolar en el aeropuerto de El Paso, Tex (KELP) y del aeropuerto de Santa Teresa, NM (K5T6) hacia el interior del territorio nacional y viceversa, así mismo las rutas VFR que deberán utilizarse para integrarse a los circuitos de transito para aterrizar en el aeropuerto de CD. Juárez Chih (MMCS)

c) Las aeronaves que utilicen el corredor VFR deberán hacerlo a una altitud no mayor de 6500 FT, dicho corredor se extiende perpendicular a la pista 03-21, con referencia en el VOR CJS en los radiales 120 y 300 con 3 MN de longitud

3. **Procedimientos de vuelo VFR.**

Las aeronaves VFR planearán su vuelo dentro del area Terminal de CD Juárez, tanto para salir o **llegar al** aeropuerto, utilizando las rutas VFR mostradas en la carta y respetando las altitudes especificadas para cada sector

3.1 Llegadas

a) Las aeronaves en vuelo VFR deberán notificar su posición en la frecuencia de la Torre de Control para que el controlador le asigne la ruta para sobrevolar hacia ELP o en su defecto aterrizar en CJS

3.2 Salidas

a) El controlador instruirá a las aeronaves en vuelo VFR la ruta a seguir después de su despegue

b) Las aeronaves en vuelo VFR mantendrán radiocomunicación con la Torre de Control CD Juárez hasta recibir autorización de esta para abandonar la frecuencia

4. **Rutas VFR de llegada / salida**

SALIDAS

IDENTIFICADOR	RUTA	DESTINO
CHAMIZAL	JUAREZ-CHAMIZAL-ANAPRA	SANTA TERESA, N.M
PUENTE ZARAGOZA	JUAREZ-PUENTE ZARAGOZA-VALLE DE JUAREZ	DURANGO
CRUCERO	JUAREZ-CRUCERO	HERMOSILLO SON
SAMALAYUCA	JUAREZ-ADUANA-SAMALAYUCA	CHIHUAHUA

LLEGADAS

CHAMIZAL	ANAPRA-CHAMIZAL JUAREZ	SANTA TERESA N.M
PUENTE ZARAGOZA	VALLE DE JUAREZ-PUENTE ZARAGOZA -JUAREZ	DURANGO
CRUCERO	CRUCERO –JUAREZ	HERMOSILLO
SAMALAYUCA	SAMALAYUCA-ADUANA-JUAREZ	CHIHUAHUA

SCT-DGAC-SENEAM MAY-07-2009 02/09 (370)

Fig. 8 Restriction location government request for the Internacional Abraham Gonzalez Airport [28]

process of ranking discrete candidate alternatives and finding the best compromise solution based on the decision maker's subjective assessments of multiple evaluative criteria [17]. Ting Yu Chen [7] indicated the MCDA problems becoming increasingly complicated, exact assessments of the choices based on evaluative criteria may be difficult to measure or quantify along the MCDA cycle, see Fig. 11

Table 1 Evaluation smart airports

Aspects evaluations	Smart Airports			
	Changi (Singapur)	Incheon (Corea del Sur)	Hamad (Doha)	Carrasco (Uruguay)
Scanning systems (delivery at home)	x		x	
Luggage processing by radiofrequency	x	x	x	x
Auto service boots	x	x	x	x
Smart PathTM de SITA	x			x
Customer service robots	x			
Advanced systems for parking information	x	x	x	x
Interterminal transportation			x	
Bird dispersion acoustic system	x	x	x	
Wall retention for out-of-control landings or takes off	x	x	x	x
Instrument-based landing system	x	x	x	x
Portatile system for dangerous cargo	x	x	x	x
Ionized molecule-based device for detection of explosives and narcotics	x	x	x	x
T-ray technology for inspecting clothes	x	x	x	x
Retractile tunnels for boarding and onboarding	x			
Noise-mitigation walls combined with exhaust gases deflectors	x	x	x	x
Aircraft retrieval systems				
Security points	x	x	x	x

Source Own elaboration

The Technique for Order of Preference by Similarity to Ideal Solution (TOPSIS) is a MCDA tool. The TOPSIS Model is considered low criteria but is complete ranking with close score, according with Hwang and Yoon the Analysis Multicriterial Model have different level to evaluated and utility. It was primarily established by Hwang and Hwang and Yoon [17] for ranking based on resemblance to perfect solution, with advancements done by Yoon in 1987, and Hwang, Lai and Liu in 1993 (2016). TOPSIS is a prevalent method suitable for taking a multiple criteria decision for rank ordering by comparison. It is a technique for rank ordering based on closeness to

Fig. 9 Polygon International Hamad Airport

Fig. 10 International Hamad Airport

perfect outcomes. The ultimate option is the one that is nearest to the perfect positive outcome and extreme from the negative perfect outcome [34].

This study uses the TOPSIS method. A positive ideal solution maximizes the benefit criteria or attributes and minimizes the cost criteria or attributes, whereas a negative ideal solution maximizes the cost criteria or attributes and minimizes the benefit criteria or attributes. The TOPSIS method expresses in a succession of six steps as follows:

Step 1: Calculate the normalized decision matrix.

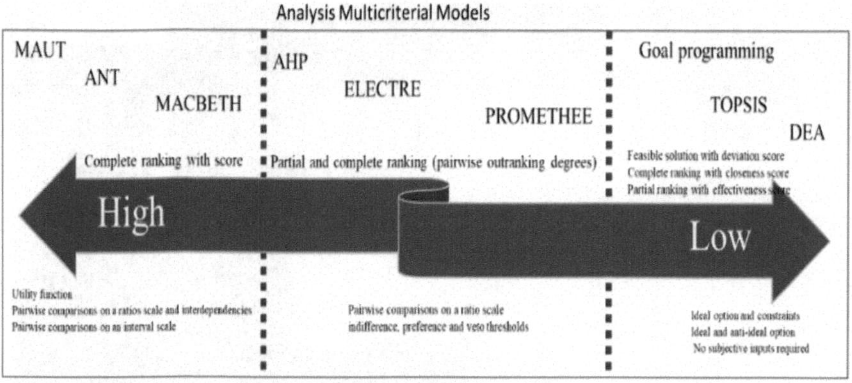

Fig. 11 Multiple-criteria decision-making (MCDM) [7]

The normalized value *rij* is calculated as follows:

$$r_{ij} = x_{ij} \sqrt{\sum_{i=1}^{m} x_{ij}^2} \, i = 1, 2, \ldots, m \, and \, j = 1, 2, \ldots, n. \quad (1)$$

Step 2: Calculate the weighted normalized decision matrix. The weighted normalized value *vij* is calculated as follows:

$$Vij = rijXWj \quad i = 1, 2, \ldots, m \, and \, j = 1, 2, \ldots, n. \quad (2)$$

where W_j is the weight of the jth criterion or attribute and $\sum_{j=1}^{n} Wj = 1$

Step 3: Determine the ideal (A^*) and negative ideal (A^-) solutions.

$$A^* = \left\{ \left(\max_i v_{ij} | j \in C_b \right), \left(\min_i v_{ij} | j \in C_c \right) \right\} = \left\{ v_j^* | j = 1, 2, \ldots, m \right\} \quad (3)$$

$$A^- = \left\{ \left(\min_i v_{ij} | j \in C_b \right), \left(\max_j v_{ij} | j \in C_c \right) \right\} = \left\{ v_j^- | j = 1, 2, \ldots, m \right\} \quad (4)$$

Step 4: Calculate the separation measures using the m-dimensional Euclidean distance.

The separation measures of each alternative from the positive ideal solution and the negative ideal solution, respectively, are as follows:

$$S_i^* = \sqrt{\sum_{j=1}^{m} \left(v_{ij} - v_j^* \right)^2}, j = 1, 2, \ldots, m \quad (5)$$

$$S_i^- = \sqrt{\sum_{j=1}^{m} \left(v_{ij} - v_j^-\right)^2}, j = 1, 2, \ldots, m \tag{6}$$

Step 5: Calculate the relative closeness to the ideal solution. The relative closeness of the alternative A_i with respect to (A^*) is defined as follows:

$$RC_i^* = \frac{S_i^-}{S_i^* + S_i^-}, i = 1, 2, \ldots, m \tag{7}$$

Step 6: Rank the preference order.

The studies carried out using the two methods of MCDA using AHP and TOPSIS in the analysis of green spaces is increase. The review of the literature allows visualizing how the methods of MCDA are used comparatively to determine the best decisions through AHP and TOPSIS, finding that there is no research where they are used for the decision of Green spaces and urban parks location [8, 17, 18, 24–26].

8 Criterial Evaluation Areas

Furthermore, we included Beta-values, or compliance (or suitability or fitness) value judgments, with the following arbitrary Lickert scale:

The Beta-values, or compliance (or suitability or fitness) value judgments, based on the following arbitrary licker scale:

(a) Worst (no compliance)
(b) Very (low compliance
(c) Undesirable compliance
(d) Slightly undesirable compliance
(e) Neutral compliance
(f) Slightly desirable compliance
(g) Desirable compliance
(h) Very desirable compliance
(i) Best possible compliance.

Federal office, such as *Federación Mexicana de Pilotos y Propietarios de Aeronaves* (FEMPPA), *Organización de Aviación Civil Internacional* (OACI) and *Secretaría de Comunicaciones y Transporte* (SCT), dictate the normative airport restrictions for both Mexico as well as for the US-Mexico border region (see Table 2).

Table 2 Evaluations items

ID	Items	Level
FEMPPA1	The VFR flight above the maximum latitude(s) established for each sector within a 15 nautical mile radius of the MMCS airport is prohibited	9
FEMPPA2	VFR flight is not permitted within the N-restricted areas between radial 360° and radial 060° to the border line, and S between radial 180° and radial 240° up to 15 MN unless you are authorized by the Airfield Control (Control Tower)	9
OACI1	Uses of noise-sensitive terrain should be restricted; most buildings are not allowed	9
OACI2	When planning land use, at least two zones should be established with regard to aircraft noise in the vicinity of airports: A (high) and B (moderate)	9
OACI3	Parallel tracks must be built at least 1534 m between center lines	8
SCT1	The airport must have technical and infrastructure characteristics adapted to the needs of its users	9
SCT2	The airport must be complemented by a transport infrastructure that communicates it with the territory where it is located so that an efficient multimodal transport structure is achieved	9
SCT3	The airport must be self-sufficient, self-regulate its growth and not represent an economic burden on society	7
SCT4	Temperature, pressure, air humidity, wind rate, in the field should be taken into account	9
SCT5	The location must have a record of the quantity and distribution of rainfall and evaporation of the place	9
SCT6	Physical characteristics of the soil its use, its type of permeability and surface storage capacity	9
SCT7	You should not interrupt locally in the hydrological cycle	9
SCT8	Presence of other airports and availability of airspaces	9
SCT9	Surrounding obstructions (the surrounding area of the airport must be restricted and protected from any future building)	9
SCT10	Nearby access roads	9
SCT11	Terrain size (land availability for possible expansion)	7
SCT12	The identification and measurement of pollutants generated by the airport at this location are required for the location	7

8.1 *Proposed Location Zones*

Four locations were identified for the location of the new airport in Juarez. All of the are out of the actual urban spectrum and comply with the normative requirements and with the minimum area of 27 km^2 (see Figs. 12 and 13).

Fig. 12 Location areas

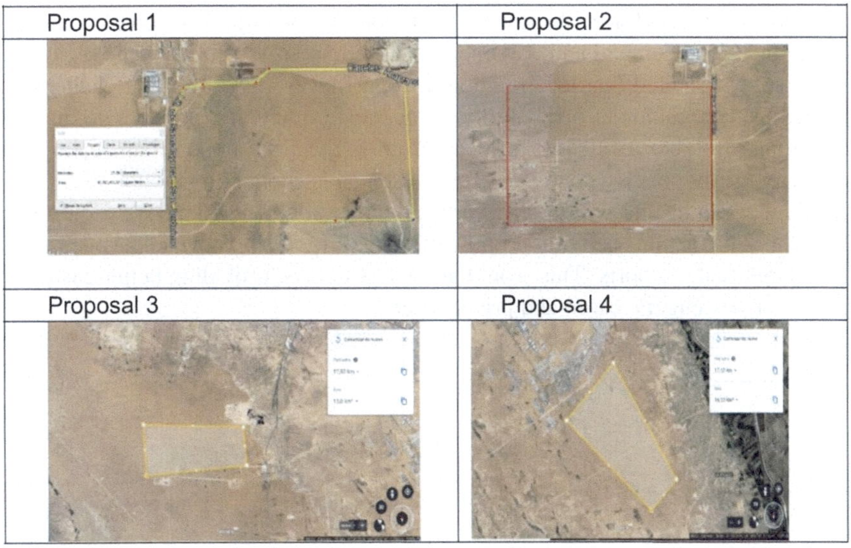

Fig. 13 Four proposition areas

9 Results

The results presented according to the TOPSIS method. The criteria used to determine the values are as follows: (1) Worst (no compliance), (2) Very (low compliance, (3) Undesirable compliance, (4) Slightly undesirable compliance, (5) Neutral compliance, (6) Slightly desirable compliance, (7) Desirable compliance, (8) Very desirable compliance, (9) Best possible compliance.

The four locations proposal were evaluated under the criteria previously defined. The initial results are shown in Table 3.

TOPSIS Evaluation according with four steps, Tables 4, 5, 6, and 7 show the results.

The algorithm proved that the best location for the new airport is proposal 2, see Fig. 14.

This polygon encompasses 30 km, the state highway towards Santa Teresa border cross would be the main Access to this location. There are no building developments in the surroundings. The normative requirements can be easily met. Moreover, the location has the potential to increase initial airport capacity.

10 Conclusions

TOPSIS allowed to identify the best location for a new airport in Juarez, Mexico. The proposed location meets all federal and international regulations. These regulations served as the criteria for the locations evaluation. A new airport in Juarez will cope with an increased demand and will enhance the competitive profile of the region.

The study revealed that locating a new airport faces a great deal of challenges, ranging from regulations, availability of adequate empty land to environmental restrictions. Be able to identify a location in the nearby of the urban realm in a smart city brings positive conditions for a sustainable economic growth.

This work is in line with the recommendations of Cheng-Lung Wu, Andy Lee [25], who state that it is better to propose mid-size airports to alleviate the operations of the large scale airports. This would allow less delays, providing better customer service for passengers, cargo and private users.

Table 3 Proposal evaluations

	FEMPP	FEMPP	OACI1	OACI2	OACI3	SCT1	SCT2	SCT3	SCT4	SCT5	SCT6	SCT7	SCT8	SCT9	SCT10	SCT11	SCT12
	9	1	5	8	8	7	9	7	7	7	8	4	9	3	7	2	9
	9	9	9	8	9	7	9	7	7	9	8	9	9	9	7	9	9
	9	9	9	8	9	7	1	3	6	9	8	9	3	1	3	9	9
	9	9	7	8	8	5	1	3	5	6	8	8	9	9	1	3	6

Source Own

Table 4 Step 1 Calculate Normalized Matrix

WEIGHTAGE	0.062	0.062	0.062	0.062	0.048	0.062	0.062	0.048	0.062	0.062	0.062	0.062	0.062	0.062	0.062	0.048	0.048
ZONE	FEMPPA1	FEMPPA2	OACI1	OACI2	OACI3	SCT1	SCT2	SCT3	SCT4	SCT5	SCT6	SCT7	SCT8	SCT9	SCT10	SCT11	SCT12
PROPOSAL1	9	1	5	8	8	7	9	7	7	7	8	4	9	3	7	2	9
PROPOSAL2	9	9	9	8	9	7	9	7	7	9	8	9	9	9	7	9	9
PROPOSAL3	9	9	9	8	9	7	1	3	6	9	8	9	3	1	3	9	9
PROPOSAL4	9	9	7	8	8	5	1	3	5	6	8	8	9	9	1	3	6

Table 5 Step 2 Calculate the weighted normalized decision matrix

ZONE	FEMPPA1	FEMPPA2	OACI1	OACI2	OACI3	SCT1	SCT2	SCT3	SCT4	SCT5	SCT6	SCT7	SCT8	SCT9	SCT10	SCT11	SCT12
PROPOSAL1	0.027	0.003	0.015	0.024	0.024	0.021	0.027	0.021	0.021	0.021	0.024	0.012	0.027	0.009	0.021	0.006	0.027
Proposal 2	0.027	0.027	0.027	0.024	0.027	0.021	0.027	0.021	0.021	0.027	0.024	0.027	0.027	0.027	0.021	0.027	0.027
Proposal 3	0.027	0.027	0.027	0.024	0.027	0.021	0.003	0.009	0.018	0.027	0.024	0.027	0.009	0.003	0.009	0.027	0.027
Proposal 4	0.027	0.027	0.021	0.024	0.024	0.015	0.003	0.009	0.015	0.018	0.024	0.024	0.027	0.027	0.003	0.009	0.018

Table 6 Step 3 Calculate the ideal best and ideal worst value

ZONE	FEMPPA1	FEMPPA2	OACI1	OACI2	OACI3	SCT1	SCT2	SCT3	SCT4	SCT5	SCT6	SCT7	SCT8	SCT9	SCT10	SCT11	SCT12
Proposal1	0.001	0.000	0.001	0.001	0.001	0.001	0.001	0.001	0.001	0.001	0.001	0.000	0.001	0.000	0.001	0.000	0.001
Proposal2	0.001	0.001	0.001	0.001	0.001	0.001	0.001	0.001	0.001	0.001	0.005	0.001	0.001	0.001	0.001	0.001	0.001
Proposal3	0.001	0.001	0.001	0.001	0.001	0.001	0.0002	0.0004	0.001	0.001	0.001	0.001	0.0006	0.0002	0.0006	0.001	0.001
Proposal4	0.001	0.001	0.001	0.001	0.001	0.001	0.0002	0.0004	0.001	0.001	0.001	0.001	0.001	0.001	0.0002	0.0004	0.0009
Step 4 Calculate the separation measures using the m-dimensional Euclidean distance																	
V+	0.001	0.001	0.001	0.001	0.001	0.001	0.001	0.001	0.001	0.001	0.001	0.001	0.001	0.001	0.001	0.001	0.001
V-	0.001	0.000	0.001	0.001	0.001	0.001	0.0002	0.0004	0.001	0.001	0.001	0.0008	0.0006	0.0002	0.0002	0.0003	0.0009

Table 7 Final results. Step 6 Calculate the relative closeness to the ideal solution, Step 7 Calculate the Rank

ZONE	Si+	Si-	Pi	Rank
Proposal 1	0.002531	0.002450	0.49189	3
Proposal 2	0	0.003665	1	1
Proposal 3	0.002641	0.002408	0.476862	4
Proposal 4	0.002	0.002599	0.518525	2

Fig. 14 Proposal 2, Santa Teresa area

References

1. Baker S (2015) Sustainable development. London; New York
2. Bringmann K, De Langhe K, Kupfer F, Sys C, Van de Voorde E, Vanelslander T (2018) Cooperation between airports: a focus on the financial intertwinement of European airport operators. J Air Transp Manag 69:59–71. https://doi.org/10.1016/j.jairtraman.2018.02.004
3. Camara de Diputados del H. Consejo de la Unión (2018) Ley de Aeropuertos 30
4. Central Intelligence Agency (2019a) The World Factbook [WWW Document]. https://www.cia.gov/library/publications/the-world-factbook/rankor-der/2053rank.html. Accessed 09 May 20
5. Central Intelligence Agency (2019b) The world factbook—annual world airport traffic forecasts 2018–2040 [WWW Document]. URL https://www.cia.gov/library/publications/the-world-factbook/rankor-der/2053rank.html. Accessed 09 May 2020
6. Chaouk M, Pagliari R, Miyoshi C (2019) A critical review of airport privatisation in the Kingdom of Saudi Arabia: case study of Medina Airport. Case Stud Transp Policy 7
7. Chen T-Y (2018) An interval-valued pythagorean fuzzy compromise approach with correlation-based closeness indices for multiple-criteria decision analysis of bridge construction methods. Complexity 2018:1–29. https://doi.org/10.1155/2018/6463039
8. Dammak F, Baccour L, Alimi AM (2015) A comparative analysis for multi-attribute decision making methods: TOPSIS, AHP, VIKOR using intuitionistic fuzzy sets, In: 2015 IEEE International Conference on Fuzzy Systems (FUZZ-IEEE). Presented at the 2015 IEEE International

Conference on Fuzzy Systems (FUZZ-IEEE), pp 1–5. https://doi.org/10.1109/FUZZ-IEEE.2015.7338059
9. Debbage KG (2002) Airport runway slots: limits to growth. Ann Tour Res 29:933–951. https://doi.org/10.1016/S0160-7383(02)00004-X
10. Dimuro G (2009) os ecosistemas como laboratorios. La busqueda de modos de vivir para una operatividad de la sostenibilidad [WWW Document]. https://www.eumed.net/libros-gratis/2009b/542/indice.htm. Accessed 09 Apr 2020
11. Dirks WF (2009) A vision of smarter cities: how cities can lead the way into a prosperous and sustainable future 20
12. Gallopín GC (2003) Sostenibilidad y desarrollo sostenible: un enfoque sistémico. CEPAL
13. Gasco L, Asensio C, de Arcas G (2017) Communicating airport noise emission data to the general public. Sci Total Environ 586:836–848. https://doi.org/10.1016/j.scitotenv.2017.02.063
14. González SFG (2017) Smart Cities, La evolución de las ciudades 18. Google Earth [WWW Document], n.d. URL https://www.google.com/intl/es/earth/. Accessed 09 May 2020
15. Gunter F, Schaff A (1982) Microelectronics and society: for better or for worse: a report to the Club of Rome. Pergamon Press
16. Hollands RG (2008) Will the real smart city please stand up? City 12:303–320. https://doi.org/10.1080/13604810802479126
17. Hwang C-L, Yoon K (1981) Multiple attribute decision making: methods and applications a state-of-the-art survey, lecture notes in economics and mathematical systems. Springer-Verlag, Berlin Heidelberg. https://doi.org/10.1007/978-3-642-48318-9
18. Instituto Municipal de Investigación y Planeación I (2010) Atlas de riesgo de Ciudad Juárez [WWW Document]. URL https://www.imip.org.mx/imip/node/35. Accessed 09 May 2020
19. Leff E (2004) Racionalidad Ambiental. la Reapropiación. Siglo XXI Ediciones Sony Electronics [distributor, México; Los Ángeles]
20. Lever WF (1999) Competitive Cities in Europe. Urban Stud 36:1029–1044. https://doi.org/10.1080/0042098993349
21. Martin-Domingo L, Martín JC (2016) Airport mobile internet an innovation. J Air Transp Manag 55:102–112. https://doi.org/10.1016/j.jairtra-man.2016.05.002
22. Mayer R (2016) Airport classification based on cargo characteristics. J Transp Geogr 54:53–65. https://doi.org/10.1016/j.jtrangeo.2016.05.011
23. Meadows D, Meadows D, Randers J, Behrens, W (1972) The limits to growth: a report for the club of rome's project on the predicament of mankind [WWW Document]. URL https://www.dartmouth.edu/library/digital/publishing/meadows/ltg/. Accessed 09 Apr 2020
24. Pazand K, Hezarkhani A (2015) Porphyry Cu potential area selection using the combine AHP - TOPSIS methods: a case study in Siahrud area (NW, Iran). Earth Sci Inform 8:207–220. https://doi.org/10.1007/s12145-014-0153-7
25. Saaty TL (2014) Analytic heirarchy process, In: Wiley StatsRef: Statistics Reference Online. American Cancer Society. https://doi.org/10.1002/9781118445112.stat05310
26. Saaty TL (2008) Decision making with the analytic hierarchy process. Int J Serv Sci 1:83. https://doi.org/10.1504/IJSSCI.2008.017590
27. Sandoval-Chávez D, Cordova A, Cervantes E, Cervera L (2018) Aproximación conceptual a un modelo de evaluación de parques urbanos con criterios de sustentabilidad [WWW Document]. ResearchGate. URL https://www.researchgate.net/publication/327755369_aproximacion_conceptual_a_un_modelo_de_evaluacion_de_parques_urbanos_con_criterios_de_sustentabilidad. Accessed 09 Apr 2020
28. Secretaría de Transporte F (2016) Carta Navegación Ciudad Juárez [WWW Document]. URL https://femppa.mx/wp-content/uploads/2017/06/CdJuarez-1M-oct-2016.jpg. Accessed 09 May 2020
29. Sulmona LG, Edgington DW, Denike K (2014) The role of Advanced Border Controls at Canadian airports. J Transp Geogr 39:11–20. https://doi.org/10.1016/j.jtrangeo.2014.06.006
30. Tojo JF, Sánchez JR (2011) Orden, desorden y entropía en la construcción de la ciudad. Urban 0, 8–15
31. Valderrama N (2017) Ciudades Inteligentes Conceptos Básicos 12

32. Vogel H-A (2019) Foundations of airport economics and finance. 1st edn
33. Wu C-L, Lee A (2014) The impact of airline alliance terminal co-location on airport operations and terminal development. J Air Transp Manag 36, 69–77. https://doi.org/10.1016/j.jairtraman.2013.12.006
34. Yang G, Wang Q, Liu J (2016) TOPSIS and AHP model in the application research in the evaluation of coal. In: Huang D-S, Bevilacqua V, Pre-maratne P (eds) Intelligent computing theories and application. Lecture Notes in Computer Science. Springer International Publishing, Cham, pp 145–152. https://doi.org/10.1007/978-3-319-42291-6_14

Colombian Coffee Price Forecast via LSTM Neural Networks

Yoe A. Herrera-Jaramillo, Johana C. Ortega-Giraldo, Alejandro Acevedo-Amorocho, and Duwamg Prada-Marin

Abstract This work deals with the contributions Machine Learning techniques can bring into the coffee growing conglomerate, committees and other points in the production and marketing chain involved in the dynamics of this commodity. It is well known that the different variables that interact with prices both nationally and internationally have a direct, dramatic affect on the sector under study. In this work, we summarize an extensive review of the coffee price dynamics and the forecast techniques used in this eld. In addition, the internal coffee price in Colombia has been modeled using a long short-term memory (LSTM) recurrent neural network that was chosen as the one of better performance out of three original models. The archetype that evidenced a pertinent superiority of fitness within the parameters specified for this type of model is composed of a linear self-regressive component, plus a multi-layer perceptron-type artificial neural network with twenty (40) LSTM cells neurons in the hidden layer. This epitome captures the chaotic coffee price dynamics. The normalized residuals of the model are uncorrelated and homoscedastic and follow a normal distribution. The results indicate that the current price depends on the prices that occurred in the last four (4) years. This tool can be used to help the coffee growing community to better design alternatives to overcome difficulties with the price of the grain, and this makes it a Logistics solution for them.

Y. A. Herrera-Jaramillo (✉)
Tecnológico de Antioquia - Institución Universitaria, Medellín 050034, Colombia
e-mail: yoe.herrera@tdea.edu.co

J. C. Ortega-Giraldo
Universidad Cuauhtemoch, Aguascalientes, Mexico

A. Acevedo-Amorocho · D. Prada-Marin
Universidad Pontificia Bolivariana, Bucaramanga, Colombia
e-mail: alejandro.acevedoa@upb.edu.co

D. Prada-Marin
e-mail: duwamg.prada@upb.edu.co

© The Author(s), under exclusive license to Springer Nature Switzerland AG 2021
A. Ochoa-Zezzatti et al. (eds.), *Technological and Industrial Applications Associated with Intelligent Logistics*, Lecture Notes in Intelligent Transportation and Infrastructure,
https://doi.org/10.1007/978-3-030-68655-0_25

1 Introduction

Every single science, engineering or social discipline, requires of forecasting, with no exception, as this component is vital for the senior management of the organizations when making decisions. Such decisions should be made in contexts of seasonality, extreme dynamics of price behaviors in the markets, climate changes direct effects on commodities, how market professionals bahave in terms of speculation, among others. It is also necessary to indicate that families who make their living on their growing coffee are directed affected due to such conduct, and that can be improved by the application of instruments minted by Machine Learning, which allow to estimate to some extent the prices ranges at which such asset can fluctuate in the international markets, thereby being able to establish mitigation strategies against these impacts in the national territory.

For the development of this work, the variables known as the Alpha Factors were holistically reviewed, which, in the experts opinion, have the most correlation and influence on the behavior of the coffee price dynamics. Among these variables prevail the following: the internal price (that is, in Colombia) of a coffee 125 kg sac, the amount of coffee cultivated area in the national territory, technification level 1, 2 and 3, the coffee price futures found on the ICE platform, the representative rate of the TRM market, BRENT and WTI prices, and the US Prime Benchmark Rate.

Researchers such as [1] stated in their work that the development of forecasts for the price of financial assets in stock markets is one of the most demanding tasks in terms of research, precisely because of the dynamics and complexities of the financial markets. Nowadays, modern finances analyze the movements of market prices from a spectrum of rational behavior, appearing in this scenario the concept of market efficiency, term coined in the financial literature by [2]. Market efficiency refers to markets delivering all information of interest in the prices, in other words, values have memory and therefore their trends and prices can indeed be predicted.

Along these lines, [3], reviewed thoroughly the existing proposals on the axioms of efficient markets, finding categories among which the weak form, the semi-strong form and the strong form are postulated. The genesis of the these proposals is born from the imagination of a market where agents act rationally, an action that makes asset prices adjust in a versatile way to the perspectives of relevant information. Ramiah et al. [4] stated that this type of current is established as a new alternative of economic and financial thinking under a behavioral type by the stock market agents.

The aforementioned postulates made above have been successfully nested in the thoughts and diligence of researchers, postulates which are oriented to the observation of market phenomena and movements at the international level. This is why tools and techniques around the prediction of these assets and the potential prices they can reach have been born. Predictions in confidence intervals can help producers and federations, as is the case in the coffee sector, to have diagnostic and risk management tools that help improve quality of life of the large number of families who make their living on the coffee production.

The tools that have been produced as a technical support for the development of forecasts are categorized into two large groups. In the first group we have the classical tools such as the Box & Jenkins proposals (ARIMA, ARMA, AR, MA), exponential smoothing, econometric models born in the eighties which are called heteroscedastical self-regressive such as the ARCH, GARCH and self-regressive family. The second group consists of the models of Intelligent Computing which can be subdivided in three classes: Class 1: Heuristic and Meta-heuristic models such as GA, SA, SO, SCATTER, TS, ANT, among others [5]; Class 2: Artificial Neural Networks and Deep Learning Networks and Class 3: Hybrid models, that is, combinations of the above.

Given the previous argument, we looked for models of the neural network type, obtaining a model of long short-term memory features. Therefore, this work harmonizes a discrete nonlinear dynamic structural system by means of a forecasting algorithm to predict the coffee price in the national geography. This scheme has allowed us to move towards the identification of variables that can be studied in subsequent proposals taking into account the climate changes, phenomenon much studied but not correlated with the prices of the commodity under study.

Lastly, it should be noted that the body of the document is constituted as follows: Sect. 2 outlines the theoretical background pointed out by means of a structured review of equations in the technological surveillance processes to specify empirical contributions regarding forecast analysis. In Sect. 3, the mathematical and computational background needed for the implementation of the forecasting models are provided. The results of the implementation of the proposed models are discussed in Sect. 4 as well as the predictions of the coffee price. The final Sect. 6, contains the concluding remarks.

2 Literature Review

For this work to have a solid foundation and robustness, and at the same time be virtuous in the scientific community and real sector, it became necessary for it to be built on the theoretical contributions and developed postulates about the subject under study. This led us to an extensive look for instruments that help define what the price trend is in international markets. All of this search helped establish what the scopes, limitations, tools and reasoning presented in previous works.

2.1 Coffee Price Dynamics

In order to access the premises above stated, a holistic review of the existing literature was developed. The reviewing of different databases and virtual resources aligned with the needs of this research. We do this by means of equations and Boolean operators queries in the Scopus and Web of Science databases. The query equations were structured as follows TITLE-ABS-KEY ("coffee prices") AND ("price formation")) and TS = (("coffee prices") AND ("price formation")). The query results were 147 and 71 articles respectively. For systemic query on Google Scholar platform we used the same basic terms and we got a total of 60,700 results. After a further filtering the pile of papers were reduced through successive search refinement processes to 219 articles, which in the opinion of the researchers had a potential contribution to the investigative work. At the end, 53 papers were analyzed more closely, those which were substantially more correlated with the purpose of the work, and in addition, conferral theoretical reflections, conceptualizations and structured postulates.

From the execution of the previous work it was evident that four are the points on which authors agree on. Here we show some of the most revelant findings related to the scope of this paper. The first aspect is that coffee is a commodity which provides the economic means for sustainability of millions of people in the entire globe [6].

A second rendezvous point is that climate is an important agent in coffee price dynamics [7]. In the same vein the [8] indicated that climate fluctuations are directly responsible for fluctuations in coffee grain performance worldwide. The latter implies that such a change in climate will alter latitudes and at the same time the coffee production and consumption in the future. And all of this will affect the vast majority of producers and other links of the chain involved without a doubt, and also to the very final consumer, event which undoubtedly represents a challenge of great proportions for the coffee industry.

A third meeting point is the fact that price instability in international markets impinges directly the coffee value in domestic markets. These consequences are seldom created by the speculation generated in market dynamics, speculation caused due to the lack of international agreements, climate variability, lack of labor, among others [6]. On the other hand, studies on volatility analysis suggest that, while structured stimuli to intervene prices and delimit supply are vigorous, these efforts fall short. The latter in part because of the consumer countries do not take part actively, and this eventually hampers the market dynamics in the long term and this hinders more higher and more stable prices [9]. Nonetheless, [10] found that the effects of medium- and long-term prices can be complex and sensitive to particular policy scenarios, hence fine tuned economic analysis should be carried out in order to explore the coffee markets.

The fourth point on which most researchers accord is the suggesting of the need to develop tools for observance and quantification of the commodity performance, which has a potential for timely assistance in mitigating impacts. The quest for the address this issue has brought benefits to the coffee-growing families and other links of the coffee chain [11].

Given the above considerations, it can be inferred that the work and contributions of academics and the research community have had a noticeable impact in the social and business sectors. More specifically, they have contributed with their providing of tools to help stakeholders who are affected by the dynamics and dramatic changes in international market, and whose decisions generate collateral effects to the communities that live and develop around the coffee production.

2.2 *Forecast of Coffee Prices*

Regarding the forecast of coffee prices we performed an analogous approach as the one for the coffee price dynamics. This time, the equations and Boolean operators used in the queries for searching on Scopus and Web of Science databases were: TITLE-ABS-KEY ("forecast") AND ("coffee price")) and TS = (("forecast") AND ("coffee price")). The obtained results were 1,305 articles in total. Further filtering led us to 43 papers or book chapters that resemble the focus of our work.

From the above mentioned review we can highlight the fact that the coffee prices are chaotic, hence when predicting its future values care must be taken. It is worth highlighting the work [12], where a hybrid model consisting of a combination of an autorgressive model and a multi-layer multi-perceptron model is usde for modeling the Colombian coffee price in the New York Stock Exchange. On the other hand, in [13] the authors forecast the prices of the Ethiopian coffee by using a Kalman Filtering Algorithm. In [11] the authors proposed a modified multiscroll Chua system to forecast the coffee price based on the resemblence in the oscillations of the chaotic multiscroll Chua system and the dynamics of the Colombian coffee price within the established observation window.

3 Methodology

In this section we provide the background for our model. We start by defining time series, which we use to model the internal coffee price through the last 17 years. Later we summarize the concepts and results needed from neural networks and their training.

3.1 *Time Series*

A time series is a collection of random variables indexed according to the order they are obtained in time [14], time measured in a uniform scale time series can be written in the form $s = \{t, s_t\}_t = \{s_t\}_t$, where s_t is the value of the variable at time t. The applications of time series abound in many fields of knowledge, almost everything can be modeled by a times series. The process of forecasting deals with the estimating of the "future" terms of a time series. There are many techniques as mentioned earlier on this work, but given the chaotic behavior of our dataset (coffee daily internal prices), we decided to implement a deep learning model, more specifically, a long short-term memory neural network.

3.2 Neural Networks

Supervised Artificial Neural Networks is an alternative to statistical modeling for making predictions from a labeled dataset $\{(x_i, y_i)\}_i$, where x_i's are the explanatory variable vectors and the y_i's are the response variable vectors. In the case of a time series $\{t, s_t\}_t$, the time t is the explanatory variable and s_t is the response variable, respectively.

A perceptron or (artificial) neuron is an information processing unit consistent of a vector of weights $\boldsymbol{w} = (w_1, w_2, \ldots, w_n)$, an activation function ϕ and a bias b. Given an inpuct vector $\boldsymbol{x} = (x_1, x_2, \ldots, x_n)$ the output of the neuron is $\hat{y} = F(b + \boldsymbol{w} \cdot \boldsymbol{x})$, where $\boldsymbol{w} \cdot \boldsymbol{x} = \sum_{i=1}^{n} w_i x_i$ (see Fig. 1b). Some of the most commonly used actividation functions are the rectified linear unit function, relu(x) = max(0, x) and tanh(x). A multiperceptron or artificial neural network (NN) is a collection of reunions arranged in layers. The first layer is the input layer and the last one is the output layer, the rest of layers are the hidden ones. If the NN has more than two (2) hidden layers we call it a deep neural network. Figure 1a displays a DLNN whose input and output layers have 2 units each and with two 3-unit hidden layers (colored gray).

A Feed-Forward Neural Network (FFNN) is a NN where the activations ow in the same direction, that is, the inputs of the neurons on a hidden layer L^k are the outputs of the units of the previous layer L^{k-1}. Given an input x, the outcome of the process of the passing through the layers of NN is known as the prediction and this process is called Feedfordward propagation. In mathematical terms it can be expressed by Eqs. (1) and (2).

$$\boldsymbol{h} = \sigma(W_{xh}\boldsymbol{x} + b_h), \tag{1}$$

$$\hat{y} = \phi(W_{hh}\boldsymbol{h} + b_y), \tag{2}$$

Fig. 1 A deep learning network and a neuron

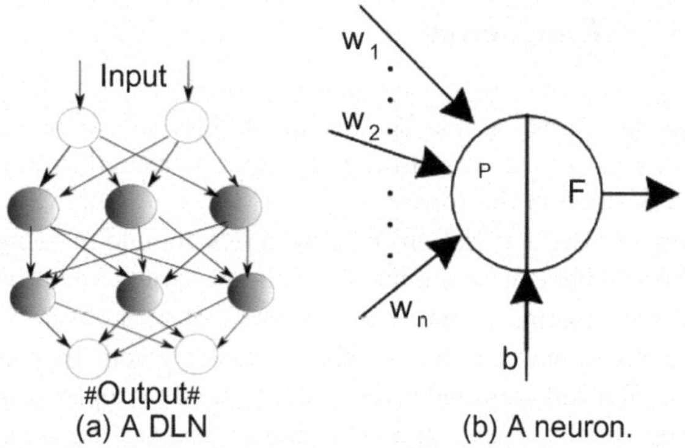

where W_{xh} is the weight vector of the connection from $\boldsymbol{x}$ to the hidden layers and W_{hh} is the weight matrix among units of the hidden layers, σ and ϕ are hidden and output layer activation functions, and b_h and b_y the corresponding bias matrices, respectively.

A cost or loss function is a function that measures how the NN performs in terms of prediction. Some of the most commonly used loss functions are the Mean Squared Error (MSE), the Maximum Absolute Error (MAE) and the Root of Mean Squared Error (RMSE).

The training or learning of the NN is the process by which the weights and biases of all the units of the NN are determined in a way that the loss function is the smallest possible. The most popular algorithm used for training NNs is the Backpropagation Algorithm, proposed by [15], which uses the chain rule to calculate the derivative of the loss function with respect to each parameter in the network, and weights are then adjusted by gradient descent. Backpropagation refers to the process of propagating the errors measured at the output layer backwards through all previous layers in order to update the weights of the NN [16]. One last component of the training of a NN involves an optimization problem, for this one of the most popular optimizers is the Adaptive Momentum Estimation Algorithm (ADAM) proposed by [17].

3.3 *Recurrent Neural Networks*

A Recurrent Neural Network (RNN) is a NN which allows feedback cycles inside the network. What differentiates a RNN from a FFNN is that the latter maps one input vector to an output vector, and the first maps from the entire history of the inputs into each output. This feature makes RNN the best model structure for sequences. [18] proved that an RNN, with enough hidden units, is an universal approximator that can learn to map any two arbitrary-length sequences to a certain accuracy.

Figure 2 illustrates the RNN and the unfolded version. The forward pass formula for RNNs is like the one in FNN; the only difference is the hidden layer activation. To consider previous hidden units in each time step, Eq. (2) is changed to

$$\boldsymbol{h}^{(0)} = \sigma(W_{xh}\boldsymbol{x}^{(0)} + b_h) \tag{3}$$

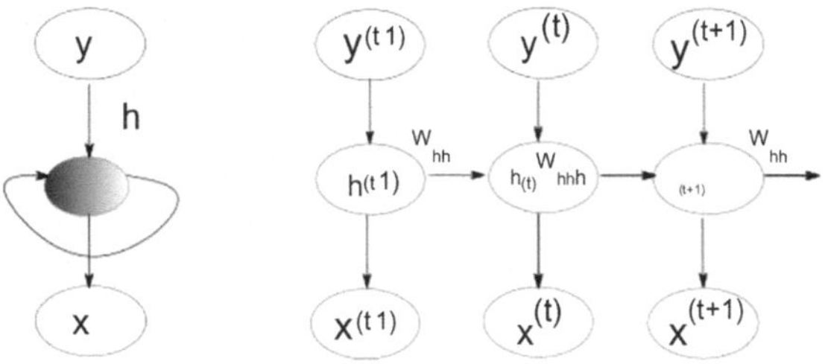

Fig. 2 A RNN and its unfolding

$$h^{(t)} = \sigma(W_{hx}x^{(t)} + W_{hh}h^{(t-1)} + b_h) \tag{4}$$

$$\hat{y}^{(t)} = \phi(W_{yh}h^{(t)} + b_y),$$

where $x^{(t)}$ is the input in time t and the vectors b_h and b_y are biases which allow each node to learn an o set.

3.3.1 LSTM Neural Networks

One drawback of the RNNs is the "vanishing gradient" problem which refers to the exponential loss or gain of information in a RNN during the back-propagation process. The cause of this problem occurs when the gap between the points where the relevant information is located and the point where this information is used becomes too large. In this case, the updating of the weights will have almost no impact on the output, and the NN does not actually learn. For solving the problem of the vanishing gradient [19] proposed the implementing of an architecture of a RNN equipped with gated cells. The function of the gates was to decide which information will be remembered or which one will be forgotten. The three gates are nonlinear summation units that collect activations from inside and outside the block, and control the activation of the cell via multiplications (small black circles). The input and output gates multiply the input and output of the cell while the forget gate multiplies the cell's previous state. No activation function is applied within the cell. The gate activation function 'f' is usually the logistic sigmoid, so that the gate activations are between 0 (gate closed) and 1 (gate open). The cell input and output activation functions ('*i*' and '*o*'). The weighted 'peephole' connections from the cell to the gates are shown with dashed lines. All other connections within the block are unweighted (or equivalently, have a xed weight of 1.0). The only outputs from the block to the rest of the network emanate from the output gate multiplication.

The gated cell was named long short-term memory cell and a RNN equipped with this type of cell learns over a long-time interval thanks to the gate switch. With this type the temporal correlations between previous information and current circumstances can be established. A LSTM cell is illustrated in Fig. 3.

The math formulation of LSTM is as below [20].

$$f(t) = \varphi(W_f(h^{(t-1)}, x^{(t)} + b_f)$$

$$i(t) = \varphi(W_i(h^{(t-1)}, x^{(t)}) + b_i)$$

$$o(t) = \varphi(W_o(h^{(t-1)}, x^{(t)}) + b_o)$$

$$\tilde{C} = \tilde{\varphi} W_c(h^{(t-1)}, x^{(t)}) + b_c$$

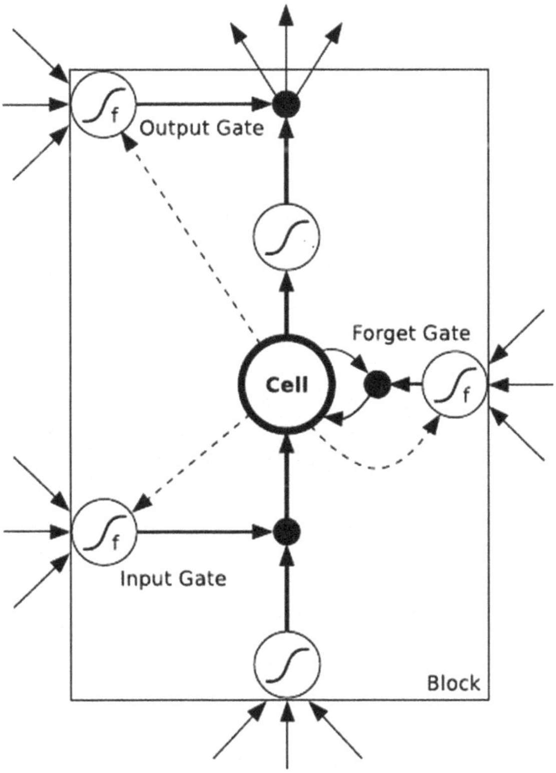

Fig. 3 Structure of a LSTM cell ([21])

$$C(t) = f(t)C(t) + i(t)\tilde{C}(t)$$

where i, o and f correspond to the input, output and forget gates of the cell and φ is the activation function and $\tilde{\varphi}$ the one of the cell state.

4 Implementation

As preliminary component it is necessary to enumerate the assumptions needed to be able to develop the work in terms of conditions and parametrization required for obtaining and proposed prediction model. It should be made very clear that time series have components such as trend, cyclicity, seasonality and irregularity which consists unpredictable or random fluctuations. In Fig. 4, the historical behavior of the prices and its returns can be seen.

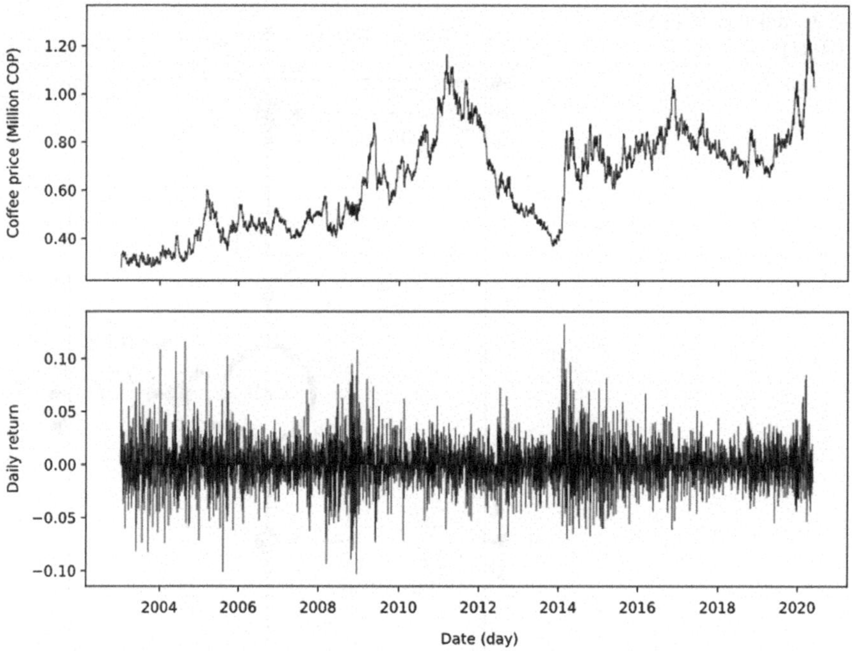

Fig. 4 National coffee price and its daily return

4.1 Data Pre-Processing

The data for the time series used in this study corresponds to the internal price (in Colombia) of a 125 kg sac of Colombian coffee measured in a daily basis. The data were extracted from the database of the National Federation of Coffee Growers (https://federaciondecafeteros.org), downloaded for a time window between January 4, 2010 and May 22, 2020 in Colombian Pesos (COP). All the procedures executed in this work were done in Python with the help among other of the modules of Pandas, Numpy, Skilearn, keras and Tensor ow. Figure 4 shows internal coffee price as well as its returns in the period of time chosen. A first visual inspection allows to identify the most representative price increases and lows over time which are associated with different historical milestones.

Since LSTM NNs are sensible to the range value scale of the data, it is necessary to normalize our data. Hence, our first step is to normalize the time series of coffee prices. In this case, we normalize it to be in the range of [0, 1]. The normalization is done via the formula (6):

$$z_t = z_{min} + \frac{x_t - x_{min}}{z_{max} - x_{min}}(z_{max} - z_{min}), \tag{6}$$

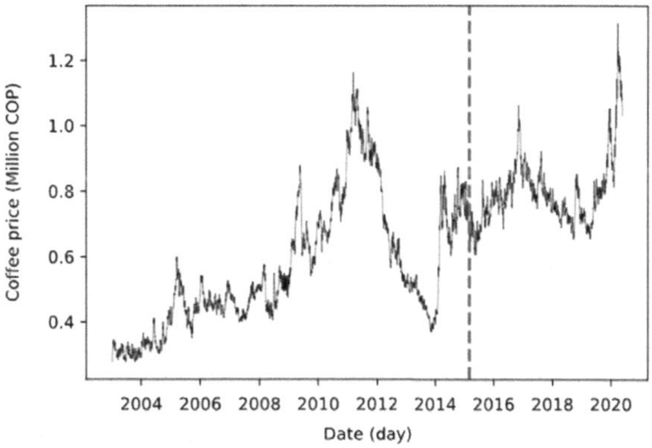

Fig. 5 Train and test datasets

where x_t is the time series term, $x_{\min}$ is the minimum value of the series and $x_{\max}$ its maximum value, z_i is the normalized term and $z_{\min}$ is the minimum value that will exist in the series when normalized, and $z_{\max}$ the maximum. Equation (6) produce a time series whose values are in the range [0,1], which will allows to work with any type of activation function.

In order to reverse the process we use Eq. (7).

$$x_i = x_{min} + \frac{z_i - z_{min}}{z_{max} - z_{min}}(x_{max} - x_{min}) \tag{7}$$

The next step in the process consists of the partition of the data in two subsets one for training and the other for testing. This is done to have a criterion of comparison between the experimental models. For this work, given that the amount of data is 6347 we decided to have the first 4442, that is the 70% of the total data, for the training and the remaining 1905 records for testing. Figure 5 shows the specified partition.

4.2 *Model Archichectures*

As mentioned earlier, the chosen model was the one with better performance out of three (3) final different LSTM models. In regards to the structure of neural networks, the number of layer units and the type of units are to be determined. In particular, these three models and their parameters calibration were determined through an exhaustive exploration and experimentation.

After experimenting with all kinds of options and combinations of parameters such as optimizers, learning rates, batch size, number of epochs, callbacks, layers such as Dense, Dropout, LSTM and Bidirectional, loss, metrics and callbacks. It is noticeable how sensible is a model to small changes in its architecture. For instance, adding more units to one layer may cause values to grow to infinity, or can make

Table 1 Common parameters of the models

Parameter	Value
Lookback	200
Batch size	40
Number of epochs	40
Input shape	(200, 1)
Activation	relu
Learning rate	0.001
Loss	MSE
Optimizer	Adam
Metrics	MAE

Table 2 Trainable parameters of the models

	Model 1	Model 2	Model 3
Number of hidden layers	1	1	2
Number of LSTM units in each layer	20	40	20
Total number of hyperparameters	1,781	6,761	5,061

the loss to converge more slowly. The values of these options which yielded the best results are the one given in Table 1.

The last step was the selection the number of units in each layer. The first model was chosen with 20 LSTM units in its hidden layer, the second model with 40 LSTM units in its hidden unit, and lastly, the third model was a stacked vanilla neural network with two hidden layers with 20 LSTM units each. Table 2 gives the specifications of numbers of units in the hidden LSTM layer as well as the total number of trainable parameters of the models.

5 Results

In order to compare the performance of these models we consider the following loss functions:

$$\text{The Mean Squared Error, MSE} = \frac{1}{n}\sum_{i=1}^{n}(y_i - \hat{y}_i)^2 \tag{8}$$

and the Mean Absoulte Error

$$\text{The Mean Absolute Error, MAE} = \frac{1}{n}\sum_{i=1}^{n}\left|y_i - \hat{y}_i\right|. \tag{9}$$

In Formulae (8) and (9), n is the total number of testing examples (i.e. training set size), y_i is the target output and $\hat{y}_i$ is the prediction obtained by the corresponding model.

As can be seen in Fig. 6, the three models captured the behavior of the time series satisfactorily. The predictions of the three (3) models in the test dataset are close to the actual coffee price time series. We can also observe that the model that reached the lowest values of loss functions was Model 1.

Nonetheless, the graphs of the train and test loss (Fig. 6 bottom) of the models revealed that the one with lower error is the Model 3. This implies that the number of units improve the performance of the level. It is also worth noticing that as we tried a greater number of units in the layer, the error did not improve significantly, but the time compiling did. This can also be deduced from Table 3 where the values of the different errors for the train and test datasets of each model.

A further calibration consisted in gradually increasing the number of epochs and the batch size and decreasing the learning rate. More specifically, we use the same parameters, but changing the learning rate to 0.0001 and the number of epochs to

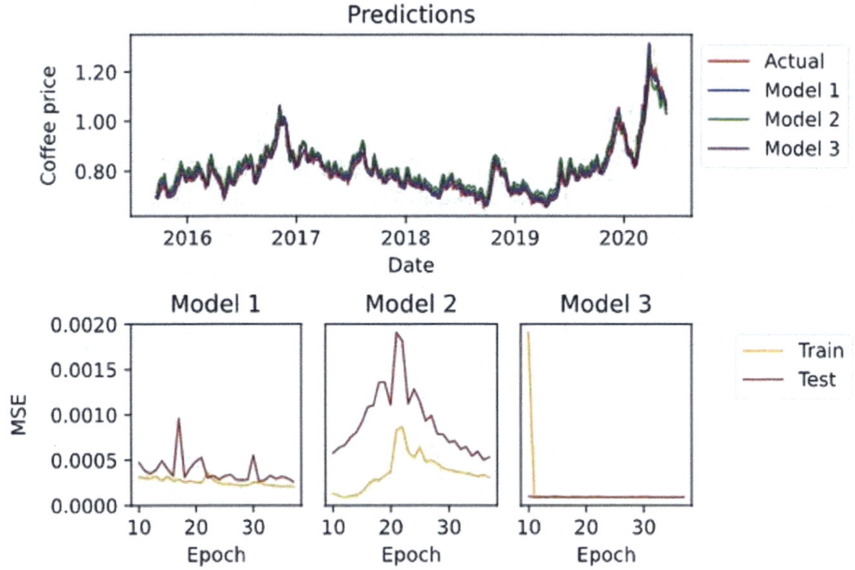

Fig. 6 Predictions, and train and test losses

Table 3 Comparison of the three models

Model	MSE	MAE	MSE validation	MAE validation
Model 1	0.000205	0.010180	0.000253	0.011753
Model 2	0.000314	0.000314	0.000314	0.000449
Model 3	0.000571	0.000571	0.000571	0.000643

Table 4 Final modified models

Model	MSE	MAE	MSE validation	MAE validation
Model 1	0.000125	0.007260	0.000214	0.009894
Model 2	0.000114	0.000114	0.000114	0.000191
Model 3	0.000131	0.000131	0.000131	0.000226

1000, and the batch size to 200. In addition, a change in the size of the test and training sets were used: 65% of the data for the training set. The larger the number of epochs the lower the error, but with not a significantly low value. Thus, we decided to leave it at 1000 epochs. These changes produced the results displayed in Table 4.

Hence, it is clear that the best model is the modified Model 2.

5.1 *Forecasting*

Finally, we get the forecasting of the coffee prices using Model 1, the one with the best performance. To do this we can roll the window to forecast one value, the one corresponding to May 23, 2020. We then iterate this process to get a total de 140 days of forecast, so that we can foresee the coffee price for the season of sorrow (beginning of November 2020).

The forecast obtained with the modified Model 2 is given in Fig. 7 and its description is given in Table 5. We can observer that according to this model coffee price will tend to decrease slowly reaching a minimum of 933,399 and a mean value of 970,145 COP.

Fig. 7 Coffee price forecast

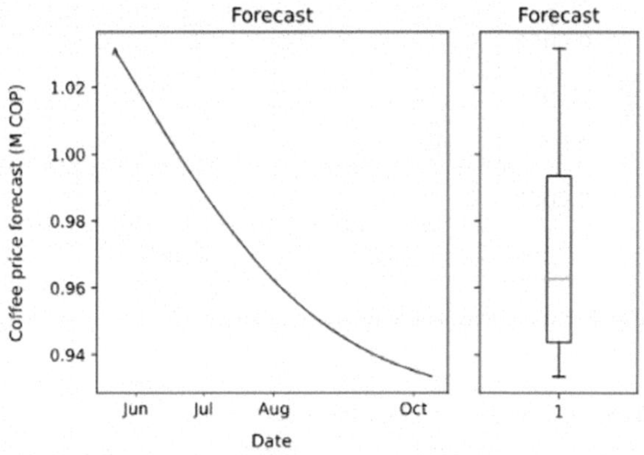

Table 5 Forecast obtained with the chosen model

Forcast description	
Count	140
Mean	970,145
Std	29,626
Min	933,399
25%	943,696
50%	962,933
75%	993,600
Max	1,031,750

6 Conclusions

Since the opening of the markets to neoliberal currents commodity producers and especially coffee growers have had to wage a strong battle in the value chains because it is well known that current practices are not based on Fair Trading. In addition, this is also due to the undeniable and totalizing presence of market logic forces that end up conditioning prices both nationally and internationally.

Furthermore, coffee-growing families have been affected by pests such as rust and berry borer, climate change, regional violence that exists in Colombia, among other variables that affect the margins for these families. The latter goes hand in hand with weaknesses given by little or no research, technological development and innovation, poor farming practices. In addition, the idiosyncrasies of coffee growers also affect the almost 56,000 families in the national geography.

In this work we can build a long-short term memory neural network which can be used to forecast the Colombia internal coffee price. Model chosen from three possible configurations. It turned out that the one with best performance consisted of a hidden layer with 40 LSTM units.

Tools like the one designed in this work can be helpful firstly for under-standing what market dynamics affect the prices of this commodity, and what their behavior might be. Thus, this can help to take courses of proactive adjustments that allow them to largely circumvent the collateral damages that develop in such markets.

The instrument shows us that the forecast price for the first quartile is 970,145 COP per 125 kg sac of coffee, having a peak value of 1,031 Million (COP)), and that when compared to the prices reported by the National Federation of coffee growers (National Federation of Coffee Growers of Colombia, 2020) they are very close to the realities lived in the buying and selling de-posits of the same institution at the national level. On the other hand, the second quartile is 962,933 pesos per load, and being reviewed by the figures presented by the guiding body of the grain for the middle of June 2020 shows us that this price closes at June 15 at a price of 974,000 COP per 125 kg sac. Hence, the model that developed has a great level of accuracy even in times of pandemic, since such quarantine impact was feared a variation accentuated in consumption correlated with a proportional decrease in prices at the domestic level.

The authors consider that with the awareness of the coffee growers and by their understanding of the dynamics by which the coffee price fluctuates, they can take the necessary adjustments to reduce the affectations.

Finally, it is necessary to structure financial derivatives that allow coffee producers to mitigate the impacts by not only price risk, but also exchange rate risk, especially for agribusiness entrepreneurs who are personally engaged in their export because it can lead to decreased margins and expected profitability from the sale of coffee.

References

1. Wei L-Y (2013) A hybrid model based on anfis and adaptive expectation genetic algorithm to forecast taiex. Econ Modelling, 33:893–899. ISSN 0264-9993. https://doi.org/10.1016/j.econmod.2013.06.009. http://www.sciencedirect.com/science/article/pii/S0264999313002253
2. Fama EF (1970) Efficient capital markets: a review of theory and empirical work. J Finance 25(2): 383–417. ISSN 00221082, 15406261. http://www.jstor.org/stable/2325486
3. Sierra Suarez KJ, Duarte Duarte JB, Rueda Ortz VA (2015) Predictibilidad de los retornos en el mercado de colombia e hipotesis de mercado adaptativo. Estudios Gerenciales 31(137):411–418. ISSN 0123-5923. https://doi.org/10.1016/j.estger.2015.05.004. http://www.sciencedirect.com/science/article/pii/S0123592315000340
4. Ramiah V, Xu X, Moosa IA (2015) Neoclassical finance, behavioral finance and noise traders: a review and assessment of the literature. Int Rev Financ Anal 41:89–100. ISSN 1057-5219. doi: https://doi.org/10.1016/j.irfa.2015.05.021. http://www.sciencedirect.com/science/article/pii/S1057521915001039
5. Khashei M, Hajirahimi Z (2017) Performance evaluation of series and parallel strategies for financial time series forecasting. Financ Innov 3(1): 24, Nov. ISSN 2199-4730. https://doi.org/10.1186/s40854-017-0074-9
6. Albertus M (2019) The effect of commodity price shocks on public lands distribution: evidence from colombia. World Dev 113: 294–308. ISSN 0305-750X. https://doi.org/10.1016/j.worlddev.2018.09.012. http://www.sciencedirect.com/science/article/pii/S0305750X1830336X
7. Sephton PS (2019) El nino, la nina, and a cup of joe. Energy Econ 84:104503. ISSN 0140-9883. https://doi.org/10.1016/j.eneco.2019.104503. http://www.sciencedirect.com/science/article/pii/S0140988319302841
8. International Trade Centre (2012) Climate change and the coffee industry. ITC, Geneva. http://www.intracen.org/publicacion/Climate-change-and-the-coffee-industry1-en/
9. Odell JS (1997) Open-economy politics: the political economy of the world coffee trade. In: Bates RH. Princeton university press, Princeton, NJ, 221p. 59:00cloth; 18.95 paper. American Political Science Review, 95(1):250–251, 2001. https://doi.org/10.1017/S000305540175201X
10. Mehta A, Chavas J-P (2008) Responding to the coffee crisis: What can we learn from price dynamics? J Dev Econ 85(1):282–311. ISSN 0304-3878. https://doi.org/10.1016/j.jdeveco.2006.07.006. http://www.sciencedirect.com/science/article/pii/S0304387806001349
11. Rodriguez A, Melgarejo M (2020) Identification of colombian coffee price dynamics. Chaos: An Interdiscipl J Non-linear Sci 30(1):013145. https://doi.org/10.1063/1.5119857
12. Juan Velasquez Henao MAD (2007) Modelado del precio del cafe colombiano en la bolsa de nueva york usando redes neuronales artificiales. Revista Facultad Nacional de Agronom a Medellin 60(2):4129–4144. https://revistas.unal.edu.co/index.php/refame/article/view/24463. ISSN 2248-7026

13. Berhane T, Shibabaw N, Shibabaw A, Adam M, Muhamed AA (2018) Forecasting the Ethiopian coffee price using kalman filtering algorithm. J Resour Ecol 9(3): 302–305. https://doi.org/10.5814/j.issn.1674-764x.2018.03.010
14. Shumway RH, Stoffer DS (2005) Time series analysis and its applications (Springer Texts in Statistics). Springer-Verlag, Berlin, Heidelberg ISBN 0387989501
15. Rumelhart DE, Hinton GE, Williams RJ (1986) Learning representations by back-propagating errors. Nature 323(6088): 533–536. ISSN 1476-4687. https://doi.org/10.1038/323533a0
16. Wytho BJ (1993) Backpropagation neural networks: a tutorial. Chemometr Intell Lab Syst 18(2):115–155. ISSN 0169-7439. https://doi.org/10.1016/0169-7439(93)80052-J. http://www.sciencedirect.com/science/article/pii/016974399380052J
17. Kingma DP, Ba J (2015) Adam: a method for stochastic optimization. In Bengio Y, LeCun Y (eds) 3rd international conference on learning representations, ICLR 2015, San Diego, CA, USA, May 7–9. Conference Track Proceedings. http://arxiv.org/abs/1412.6980
18. Hammer B (1998) On the approximation capability of recurrent neural networks. In: International symposium on neural computation, pp 12–4
19. Hochreiter S, Schmidhuber J (1997) Long short-term memory. Neural Comput 9(8):1735–1780. ISSN 0899-7667. https://doi.org/10.1162/neco.1997.9.8.1735
20. Wang Y (2017) A new concept using lstm neural networks for dynamic system identification. In: 2017 American control conference (ACC), pp 5324–5329
21. Graves A (2012) Supervised sequence labelling with recurrent neural networks. In: Studies in Computational Intelligence. Springer, Berlin. https://doi.org/10.1007/978-3-642-24797-2. https://cds.cern.ch/record/1503877

Some Pragmatic Prevention's Guidelines Regarding SARS-CoV-2 and COVID-19 in Latin-America Inspired by Mixed Machine Learning Techniques and Artificial Mathematical Intelligence. Case Study: Colombia

Danny A. J. Gómez-Ramírez, Yoe A. Herrera-Jaramillo, Johana C. Ortega-Giraldo, and Alex M. Ardila-Garcia

Abstract We use an enhanced methodology combining specific forms of AI techniques, opinion mining and artificial mathematical intelligence (AMI), with public data on the spread of the coronavirus SARS-CoV-2 and the incidence of COVID-19 disease in Colombia during the first three months since the first reported positive case. The results obtained, together with conceptual tools coming from the global taxonomy of fundamental cognitive mechanisms emerging in AMI and with suitable contextual information from Colombian public health and mainstream social media, allowed us to stating specific preventive guidelines for a better restructuring of initial safe and stable life conditions in Colombia, and in an extended manner in similar Latin American Countries. More specifically, we describe three major guidelines: (1) regular creative visualization and effective planning, (2) the continuous use of constructive linguistic frameworks, and (3) frequent and moderate use of kinesthetic routines. They should be understood as effective tools from a cognitive and behavioural perspective, rather than from a biological one. Even more, the first two guidelines should be acknowledged in integral cooperation with the third one

D. A. J. Gómez-Ramírez
Institución Universitaria Pascual Bravo, Institución Universitaria ITM (Instituto Tecnológico Metropolitano) and Visión Real Cognitiva S.A.S., Medellín, Colombia
e-mail: daj.gomezramirez@gmail.com

Y. A. Herrera-Jaramillo (✉)
Tecnológico de Antioquia - Institución Universitaria, Medellín 050034, Antioquia, Colombia
e-mail: yoe.herrera@tdea.edu.co

J. C. Ortega-Giraldo
Universidad Cuauhtémoch, Aguascalientes, Mexico
e-mail: jcog2583@outlook.com

A. M. Ardila-Garcia
Visión Real Cognitiva S.A.S., Medellín, Colombia
e-mail: Alex.Ardila.Garcia@gmail.com

© The Author(s), under exclusive license to Springer Nature Switzerland AG 2021
A. Ochoa-Zezzatti et al. (eds.), *Technological and Industrial Applications Associated with Intelligent Logistics*, Lecture Notes in Intelligent Transportation and Infrastructure,
https://doi.org/10.1007/978-3-030-68655-0_26

regarding the global effect of COVID-19 in human beings as a whole, this includes the mind and the body.

1 Introduction

Each country in the world, without exception, faces an outstanding challenge regarding the generation of biological, medical, and chemically based therapeutic guidelines and solutions for overcoming the effects of COVID-19 disease. But also, each country has the challenge of developing effective policies and behavioural actions for minimizing, preventing, and overcoming all the negative effects of the SARS-CoV-2 virus, COVID-19's causing pathogen. In order to solve the first challenge, a wide plethora of techniques from artificial intelligence (AI), machine learning, computational biology, bio-medicine, and public policy initiatives have shown to be very valuable [1, 2, 24].

For the case of the second challenge, a combination of AI-based techniques such as opinion mining, or more specifically sentiment analysis [8], and the new emerging multidisciplinary paradigm of artificial mathematical intelligence [6] have a huge potential for providing contextually-robust and cognitively-sound pragmatic behavioural guidelines.

It is worth noting that most of the existing specialized literature involving COVID-19 is mainly focused on biological, medical, and chemically based therapeutic guidelines, and significantly less e orts are focused on developing effective policies and behavioural actions [2]. In order to illustrate how AI-based techniques can be applied, we will focus our attention in the Latin American context, in particular, in the Colombian society.

Since the behavioural patterns are structurally conditioned not only by the whole taxonomy of cognitive abilities of the mind (seen as both an individual and a collective), but also by the entire anthropological, cultural, environmental conditions (e.g. geographical), it is necessary to integrate a kind of minimal cognitive-anthropological blueprint of the Colombian population. Such a special kind of blueprint should be structured from a global proto-typical perspective in order to enhance the range of effectiveness that such (attitudinal) guidelines should possess.

For the purposes of the present work, we will emphasize mainly the inter-action component of such a blueprint due to the fact that this is one of the most relevant features for the spreading of SARS-CoV-2. In Sect. 2, we will provide and introduction to the origin and spread of SARS-CoV-2 in the world and its arrival and impact in Latin America and Colombia.

In Sect. 3, we will briefly introduce the methods used in our work, in order to give a more solid background to non-experts. This is very important in our current situation, where the whole (non-specialized) community needs access to more specialized information as self-contained as possible, and even more, when research work contains valuable results of preventive nature.

In Sect. 3.1, we give a description of the main qualitative features of the information analysed (e.g. tweets of main trends in the Colombian context).

Furthermore, in Sect. 4 we present the global landscape of findings, together with the taxonomy of main (behavioural) guidelines inspired partially by the Colombian social environment, but with a potential broader scope due to their ontological and universal nature. Such cognitive directions have a great potential to be applied to other countries with a relatively similar (anthropological) blueprint to Colombia.

Lastly, in Sect. 5 we depict the major conclusions of our work and its potential extensions to the Latin American community, and potentially beyond.

2 Understanding the Context

2.1 COVID-19: Origin

The coronavirus disease of 2019 (COVID-19) is a multisystemic disease in humans that is caused by the beta-coronavirus SARS-CoV-2 (Severe Acute Respiratory Syndrome Coronavirus 2) [21]. SARS-CoV-2 is a zoonotic RNA virus that originated from wild bats [10]. SARS-CoV2 is thought to have been transmitted to humans by pangolins, an intermediate host [27], that are frequently traded for its scales in the black markets of Asia.

The first outbreak of SARS-CoV-2 occurred in the city of Wuhan (China) and was reported on December 12th, 2019 [7]. There were 27 cases confirmed in the first report, but that escalated rapidly as over 84,000 people got infected in China in a period of six months [7] (Table 1). The first confirmed case of SARS-CoV-2 infection outside China occurred in Thailand on January 13, 2020. The World Health Organization (WHO) declared a pandemic on March 13, 2020 due to the rapid propagation of SARS-CoV-2 in every continent. As of June 29, 2020, there were over 10.2 million confirmed SARS-CoV-2 infections and over half a million confirmed deaths in the world.

2.2 COVID-19 in Europe and the United States

Italy and Spain were the first two major outbreaks in Europe and the world outside of China. As of June 29, 2020, Italy and Spain were in the world's top ten in the total number of confirmed cases, deaths, and recovered individuals. In Spain, there were 248,970 confirmed cases, 28,346 deaths, and 150,376 recovered individuals by June 29, 2020 (Table 1). In Italy, there were 240,436 confirmed cases, 34,744 deaths, and 189,196 recovered individuals by the same date (Table 1). The outbreaks in Spain and Italy slowed down significantly in the month of June 2020 after strict lock-down

Table 1 COVID-19 confirmed reports in Colombia, The United Sates, China, and a selection of countries from Latin America and Europe. *Sources* The John Hopkins University: Coronavirus Resource Center. June 29, 2020. Online: https://coronavirus.jhu.edu/map.html. *Instituto Nacional de Salud: Coronavirus en Colombia. June 29, 2020. Online: https://www.ins.gov.co/Noticias/Paginas/Coronavirus.aspx

Country	Confirmed	Recovered	Deaths	Deaths/100,000 people
Colombia*	91,769	38,280	3,106	6.56
Brazil	1,344,143	746,018	57,622	27.43
Mexico	216,852	164,646	26,648	21.11
Peru	279,419	167,998	9,317	29.11
Chile	271,982	232,210	5,509	28.99
United States	2,562,921	685,164	125,928	38.37
Spain	248,970	150,376	28,346	60.39
Italy	240,436	189,196	34,744	57.30
United Kingdom	313,467	1,364	43,659	65.50
Germany	194,910	177,812	8,976	10.81
Austria	17,723	16,420	703	7.94
China	84,757	79,610	4,641	0.33

and quarantine measures were imposed in the months of March, April, and May that resulted in significant death-toll reductions.

In the United Kingdom, COVID-19 has also been devastating. There were 313,467 confirmed cases, 43,659 deaths (65.50 deaths/100,000 people), and just 1,364 official reports of recovered individuals as of June 29, 2020 (Table 1). By this date, The United Kingdom was the most affected region in Eu-rope while Germany and Austria were at the opposing end when comparing death-tolls. In Germany, there were 194,910 confirmed cases, 8,976 deaths (10.81 deaths/100,000 people), and 177,812 recovered individuals (Table 1). In Austria, most of the reported cases (17,723) had recovered (16,420) and the death-toll was 703 (7.94 deaths/100,000 people) (Table 1) by June 29, 2020.

The United States was the country most affected in the world by COVID-19 as of June 29, 2020. The reports up to then included over 2.5 million confirmed cases, over 125,928 confirmed deaths (38.37 deaths/100,000 people), and 685,164 recovered individuals (Table 1).

2.3 *COVID-19 in Latin America: Arrival, Propagation, and Mortality*

SARS-CoV-2 arrived in Latin American in February and March 2020 and it expanded rapidly in most countries [22]. It was brought to most countries by multiple infected individuals traveling from Europe (usually Spain, Germany, and Italy) and from the

United States [12, 18]. The first confirmed case of COVID-19 in Latin America was reported in Brazil on February 21, 2020 [18]. This was two months after the initial outbreak in Wuhan, China. The first confirmed infected Brazilian national did not get infected in China [18]. Instead, this individual got infected with SARS-CoV-2 in Italy, which at the time had the largest outbreak outside China. As of June 29, 2020, Brazil was the second most affected country in the world by COVID-19 with over 1.3 million confirmed cases, 57,622 deaths (27.43 deaths/100,000 people), and 746,018 recovered individuals (Table 1).

Peru, Mexico, and Chile were the most affected countries in Latin America after Brazil as of June 29, 2020. Peru was second in Latin America in the total number of confirmed cases (279,419), of which 9,317 died (29.11 deaths/100,000 people) and 167,998 recovered (Table 1). Mexico had the greatest number of confirmed deaths (26,648; 21.11 deaths/100,000 people) in Latin America after Brazil by June 29, 2020 (Table 1). The number of confirmed and recovered cases in Mexico until June 29 2020 was 216,852 and 164,646 respectively. In Chile, there were 271,982 confirmed cases that included 5,509 deaths (28.99 deaths/100,000 people) and 232,210 recovered individuals (Table 1). It is important to note that Brazil, Mexico, and Chile di er greatly in population size, but shared a similar approach to confronting COVID-19. These countries did not impose strict nation-wide lock-downs in the initial stages of their COVID-19 outbreaks.

The propagation and daily mortality rates of the COVID-19 outbreaks accelerated in June 2020 in the majority of Latin American countries. This occurred after easing strict curfew preventive measures. These measures were imposed for one to two months (April and May) by the vast majority of countries in the region after confirming their first SARS-CoV-2 cases.

2.4 COVID-19 in Colombia

The first confirmed case of COVID-19 in Colombia was reported on March 6, 2020 in Bogota, the capital and largest city of Colombia [12]. The affected individual was a Colombian 19-year-old female that had returned to Colombia from Milan, Italy. By March 20, 2020, there were 145 confirmed COVID-19 cases in Colombia and zero confirmed deaths (Fig. 1). The first confirmed COVID-19 death in Colombia was reported on March 21st, 2020. This individual was a 58-year-old male who was a taxi driver in the touristic city of Cartagena. It is suspected that the taxi driver got infected after driving two Italian tourists on March 4. The taxi driver presented the first COVID-19 symptoms on March 6, the same day that the first confirmed COVID-19 positive was confirmed in Bogota, Colombia.

The number of confirmed COVID-19 cases and deaths has increased exponentially in Colombia since March 6 (Fig. 1). There were 1,579 confirmed cases and 46 deaths by April 6 (Fig. 1). These more than tripled by April 25, reaching 5,144 confirmed cases and 233 deaths (Fig. 1). In May 2020, the spread of SARS-CoV-2 continued to increase rapidly. There were 8613 confirmed COVID-19 cases and

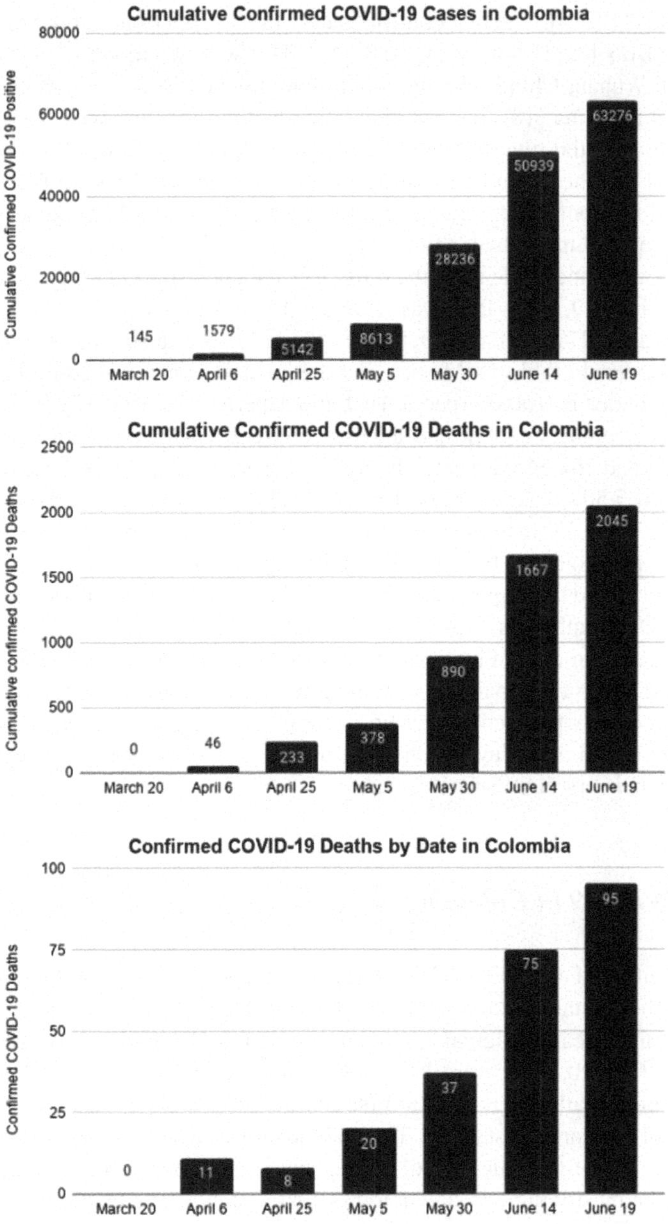

Fig. 1 COVID-19 Colombia from March 6th to June 19th 2020. *Source* Instituto Nacional de Salud: Coronavirus en Colombia. June 19, 2020. Online: https://www.ins.gov.co/Noticias/Paginas/Coronavirus.aspx

378 confirmed deaths on May 5. The figures nearly quadrupled by May 30 reaching 28,236 confirmed cases and 890 confirmed deaths (Fig. 1). These figures doubled by June 14 with 50,939 confirmed cases and 1,667 confirmed COVID-19 deaths (Fig. 1). By June 19, 2020, there were 63,276 confirmed COVID-19 cases and a cumulative total of 2,045 deaths. As of June 29, 2020, there were 91,769 confirmed cases, 3,106 deaths (6.56 deaths/100,000 people), and 38,280 recovered individuals (Table 1).

The Colombian data reported here has been widely available to the Colombian public since the start of the COVID-19 outbreak. The Colombian president reported daily on COVID-19 in an one-hour long live television show that was broadcast live during prime-time by nearly all national television channels, radio stations, and official online platforms. In these reports, the president reinforced guidelines to prevent the spread of the SARS-CoV-2 virus that included mandatory use of face-masks in public, a two-meter distance from other individuals, and frequent hand-washing with soap and water. Official reports were open-access and published online daily by the National Institute of Health (Instituto Nacional de Salud). Official COVID-19 statistics and COVID-19 prevention guidelines were also reported daily by nearly all media on television, radio news, and newspapers.

3 Fundamental Methods

3.1 Opinion Mining

In order to understand the impact of COVID-19 preventive quarantine measures on Colombia's population, we carried out a task of sentiment analysis on one of the most popular platforms of massive use, Twitter. We chose seven important dates to analyze Colombians' reactions by measuring the polarity of their tweets (positive, negative, or neutral) [20]. Sentiment analysis is a branch of opinion mining which is a popular area of research in which the major goal is to find important information on texts or documents, more precisely, as stated by [16], opinion mining deals with the computational treatment of opinion, sentiment, and subjectivity in text (in alphabetical order).

3.1.1 Data Collection

The tweets selected for this work were retrieved from the Twitter platform via a twitter API in Python. The selection criteria was that the tweets appear in one of the queries of the top ten trends on the selected dates at 8:00 p.m. EST according to the website: https://getdaytrends.com/es/colombia/.

Table 2 summarizes the selected dates, the number of tweets collected, and the event that led to selecting each date for the present work.

Table 2 Criteria for the recollection of tweets

Dates	Tweets	Criteria
March 20–21	583	Colombian government announces mandatory COVID-19 preventive lockdown starting on March 24 through April 13
April 6–7	350	First quarantine extension
April 20–21	496	Second quarantine extension
May 5–6	276	Third quarantine extension
May 30–31	272	Decentralization of COVID-19 policy, partial regional policy independence
June 14–15	304	Number of daily cases surpasses 100
June 19–20	687	First Tax-Free day
Total tweets	2968	

3.1.2 Data Pre-processing

A tweet is a concise text containing opinions expressed in many different ways by different users. In November 2017, tweet length was extended to a maximum of 280 characters. A tweet can also contain abbreviations and symbols as emojis. For these reasons, the raw data obtained in these cases are highly susceptible of inconsistency and redundancy, and hence, pre-processing is crucial. This is and should be the first step on any twitter analysis.

In this work, pre-processing is performed in Python with the frameworks of Natural Language Package ToolKit (NLPTK) and sklearn. We proceeded in the following order:

- Remove all URLs, hashtags (#topic) and targets (@username).
- Translate the tweets gathered with googletrans module from Spanish to English.
- Remove punctuation marks, symbols, numbers.
- Tokenization, that is, separate tweets in its constituting words.
- Common stop words are removed, e.g., what, a, the, of,…
- Convert all text to lower case to make the dataset uniform.
- Delete repeated characters, e.g. sweeeet was replaced sweet.

The set of all unrepeated preprocessed words constitute the so-called "Corpus" or vocabulary. This corpus is the base for the sentiment analysis and its the real purpose of the data pre-processing process. It is important to notice that we are working with seven (7) datasets of tweets, so pre-processing and corpus were obtained for each individual of dataset.

3.1.3 Feature Extraction

The following step in the sentiment analysis is the extraction of features from the tweets with respect to the corpus corresponding to each dataset.

Table 3 Vectorization of tweets

Tweet *Ti*	Vectorized tweet *vect*(*Ti*)
"authorities wait declare mandatory quarantine colombia"	[1 1 1 1 1 1 0 0 0 0 0 0 ... 0]
"control measures italy colombia trapped italy outside colombia"	[0 0 0 0 0 2 1 1 2 1 1 0 ... 0]
⋮	⋮

Let $C = \{w_i\}_{i=1}^N$ be the corpus of tweets of a given date, that is, C contains all the words in the tweets from the given date with no repetition. The vectorized tweet $vect(T_i)$ assigned to tweet T_i is defined by $vect(T_i) = [v_k]_{i=1}^N$, where v_k is the number of times the word w_k appears in tweet T_i, we do this by following [15]. The function we used to carry out this process of vectorization in Python is *Countervector* from the Natural Language ToolKit (NLTK) module. As an illustration of this process, consider T_1 the first tweet of March 20 and 21, which reads "authorities wait declare mandatory quarantine Colombia", then the first six (6) entries of $vect(T_1)$ are 1's and the rest are zeroes. The second pre-processed tweet is T_2: "control measures Italy Colombia trapped Italy outside Colombia", then $vect(T_2)$ has its first five (5) entries zero and its sixth entry is 2 because it contains the word Colombia twice and Colombia was the six word in the first tweet, the entries 7, 8, 10, 11 are 1 because they correspond to the words control, measures, trapped and outside, and the entry 9 is 2 because Italy appears twice. These examples are illustrated in Table 3.

In this way, we get a matrix known as the term document matrix (TDM). In TDM, rows correspond to the words from the corpus and the columns correspond to the tweets. Hence, the size of a TDM corresponds to the pair $(N; m)$ where N is the number of words after pre-processing and m is the number of tweets given in Table 2. For each of the datasets considered, the number N obtained was 2914 for March 20–21, 2145 for April 6–7, 2929 for April 20–21, 1968 for May 5–6, 1951 for May 30–31, 2073 for June 14–15, and 2986 for June 19–20.

3.1.4 Sentiment Analysis

Our goal was to determine the polarity and sensitivity of each tweet in our datasets. For computing polarity and sensitivity of a tweet we make use of the TextBlob module in Python. TextBlob is a python library and offers a simple lexicon-based API to access its methods and perform basic Natural Language Programming (NLP) tasks, including sentiment analysis. Polarity of a word is a real number between –1 and 1, where a negative value corresponds to a negative word, 0 to a neutral word, and a positive value to a positive word. The polarity of a sentence or document is then calculated by the average of the polarity of its constituting words. Subjectivity on the other hand, is a value from 0 to 1, where 0 is assigned to the word if it is a fact. The higher the subjectivity the more opinionated the word is. The subjectivity of a sentence or document is then defined as the average of the subjectivity of its words [26].

We take advantage of the TDMs to extract the words with more frequency on each of the tweet datasets. This was performed by adding the column values, which contains the number of times the corresponding word appears in each tweet. By extracting the most common words, we can deduce the most common topics. Nevertheless, we preferred to visualize the most common topics by using the wordcloud module in Python to represent these topics for each of dates chosen.

3.1.5 Analysis of Results

Here we show the graphical results and conclusions of our research in regards to the sentiments perceived in tweets obtained from Colombia's inhabitants during the COVID-19 pandemic. First, we analyze in detail the first and the fourth datasets because of the results obtained. Second, for the most interested reader we analyze in a general fashion the rest of the datasets shown in Appendix.

From the tweet dataset from March 20 and 21, we can highlight from the word-cloud (Fig. 2) that the most common topics were related to COVID-19 and the preventive quarantine decree ordered by the Colombian government to slow down the spread of COVID-19. This dataset also shows the concern among Colombians about the growing number of confirmed cases in the country.

In Fig. 3 (left side), we can observe that there were more positive tweets (blue dots) than negative ones (red dots). This is also displayed in detailed in Table 4, which summarizes the number of positive, neutral, and negative tweets by date. On March 20 and 21, the more prevalent type of tweets were neutral. We can also note in Fig. 3 (right side) that positive and negative tweet posting orders were equally distributed with respect to the posting order.

As it can be appreciated in Table 4, neutral tweets prevailed in all dates selected, except on May 5 and May 6. On these dates, there were 112 negative tweets of the

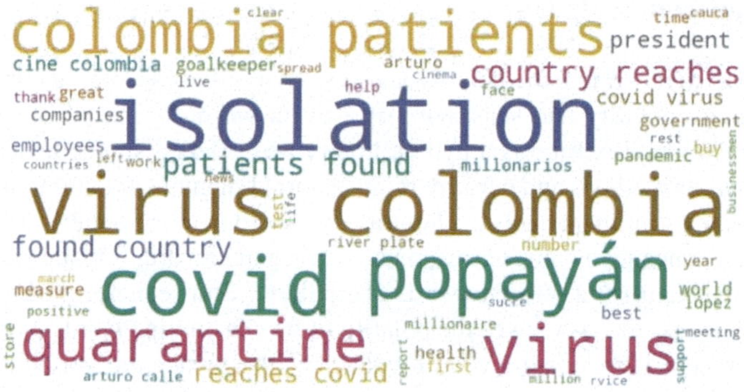

Fig. 2 Wordcloud of March 20 and 21

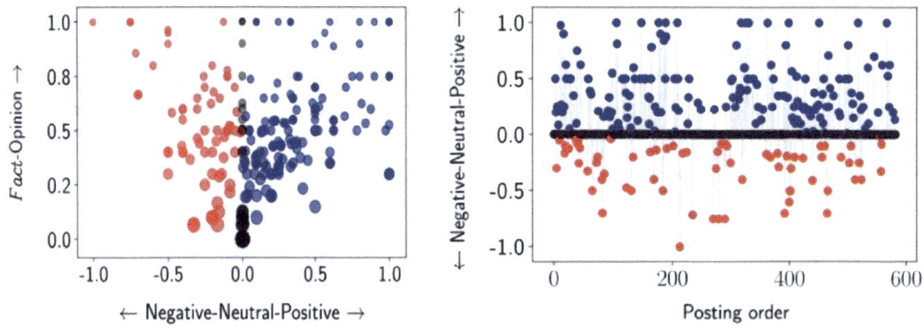

Fig. 3 Polarity and subjectivity of tweets on March 20 and 21

Table 4 Summary of sentiment analysis

Dates	Positive	Negative	Neutral	Total
March 20–21	169	68	346	583
April 6–7	167	56	127	350
April 20–21	136	62	298	496
May 5–6	76	112	88	276
May 30–31	98	46	128	272
June 14–15	115	61	128	304
June 19–20	187	98	402	687
Total	948	503	1517	2968

276 analyzed. That is, 40% of the tweets in this dataset were from people writing on negative feelings. It is worth noticing that a third extension on the mandatory isolation was announced on May 5 and that the main topics in twitter were miserable, Karl Marx, COVID, and grimes (Fig. 4).

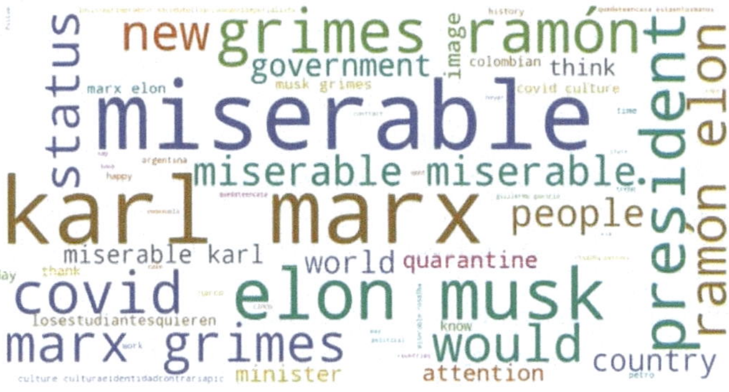

Fig. 4 Wordcloud of tweets on May 5 and 6

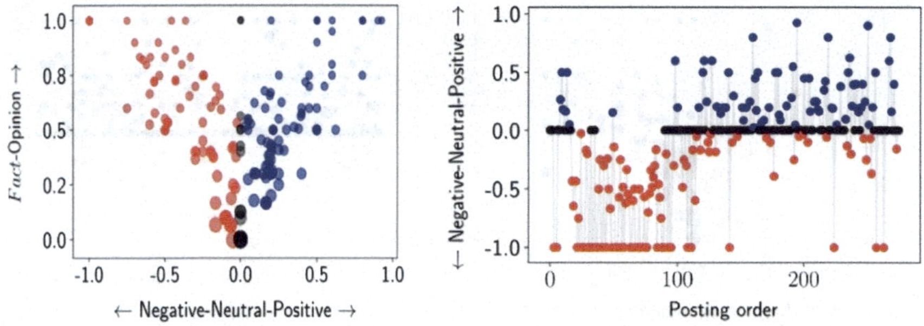

Fig. 5 Polarity and subjectivity of tweets on May 5 and 6

In addition, we notice that approximately Tweets 20–80 were negative as shown in Fig. 5 (right) suggesting that one of the trends was very negative and strong because of their polarities is –1. An example of such tweets was tweet T_{23}, which read: "miserable playing health Colombia, squandering billions pesos armored cars, advertising wash image esmad midst pandemic." The subjectivity of this tweet was 1, meaning that this tweet was absolutely an opinion. This tweet was referring to the Colombian government buying security cars for considerable amounts of money.

Based on the data from the other dates examined, we can state that twitter users in Colombia during the first three months of the COVID-19 outbreak in Colombia were focused on topics other than the pandemic and quarantine. Colombians were centered on the pandemic and quarantine only on the first days (March 20 and 21) we analyzed. As time went by, it was noticeable that Colombians started to express neutral sentiments about the pandemic and that their focus switched to different topics such as corruption, the economy, and entertainment, especially soccer. This occurred in spite of the alarming rate of spread of the COVID-19 pandemic in Colombian and Latin America.

By April 6 and 7, we can observe that even though the total number of confirmed COVID-19 cases was 1,579 and that 46 deaths were reported in total, people were not yet fully aware of the future effect of COVID-19 in their lives. This is shown in the worldcloud (7) where it is evident that Colombians were concerned about topics other than the pandemic. On April 6 and 7, Colombian twitter users were most concerned about a political scandal involving the Colombian ambassador in Uruguay. They were also concerned about losing access to college education and jobs because of the pandemic. In terms of polarity, we can state that polarity was evenly distributed in posting order (7 bottom right). Forty eight percent of the tweets were neutral and the level of subjectivity indicated that the vast majority of tweets were opinions.

For April 20 and 21, the wordcloud (8) showed that the economy was the dominant twitter topic, especially oil prices and the dollar. Polarity in these tweets shows that 60% of tweets were neutral, 13% negative, and 27% positive. The majority of tweets were neutral and positive even though the wordcloud showed that tweets appeared to express Colombians were very concerned about the economy.

We observed that people were not talking much about the coronavirus on May 30 and 31 anymore. On these dates, the decentralization of COVID-19 pandemic's measures was decreed, each state acquired some autonomy to impose or liberate restrictions in their territories. The most trending topics on these dates were the SPACEX shuttle, tourism, and future challenges (Fig. 9). There were only 17% of negative tweets, but 36% were positive and 47% were neutral.

The main topics on Twitter in Colombia on June 14 and 15 were soccer, the Colombian soccer player James Rodriguez and his soccer club (Real Madrid). COVID-19 did not show up in the worldcloud even though there were already an increased number of confirmed COVID-19 cases (1579) and deaths (46). We can also see that by June 14 and 15 there were 20% of negative tweets only and that the majority were neutral. This is surprising because there were 75 COVID-19 deaths reported on June 14 alone and people appeared not to be talking about this issue.

The last dates (June 19 and 20) selected in the present study were the dates the Colombian government had declared a tax exemption in all purchases to reactivate the Colombian economy. The COVID-19 death-toll on June 19 was 95 deaths and a total of 2,045 deaths since the start of the COVID-19 outbreak in Colombia. These numbers did not appear to be of concern for most Twitter users in Colombia. Perhaps, most people were tired of being under lockdown and were more interested in other topics such as the tax-free day. This day became an opportunity and perhaps good reason for most people to leave their homes and go shopping. These resulted in large messy gatherings at shopping malls for which there were not sufficient preventive measures to avoid crowds. Quarantine and social distancing appeared to be something distant in the past. People were very excited and anxious about shopping and this was reflected on Twitter. However, in spite of this, the majority of tweets were neutral (51%), 26% were positive, and 13% were negative (11).

3.1.6 Machine Learning Methods

Supervised machine learning techniques are used for classifying problems. The classification algorithm learns from a training dataset with known categories for each training input and then predicts the category for an unseen input. The performance of the method is measured with a testing set, that is, a dataset of input with known categories but not seen by the machine in the training process. In this work we compare three (3) different sentiment classifiers of the supervised type with the dataset of tweets collected.

For evaluating the performance of a classifier, the confusion matrix $C = [c_{ij}]$ is used. The entry c_{ij} is the number of testing items true class i that were classified as being in class j. The two metrics we used for measuring the performance of the classifier is obtained by taking the weighted average of the accuracies and precisions for each class or category as given by Formulae (1) and (2). *Accuracy*:

$$Acurracy_i = \frac{TP_i + TN_i}{TP_i + TN_i + FP_i + FN_i} \tag{1}$$

$$Precision : \frac{TP_i}{TF_i + FP_i} \quad (2)$$

Here TP_i is the number of testing inputs correctly classified in the ith class; FP_i is the number of inputs which are wrongly classified in the category i; FN_i is the number of inputs from category i that are misclassified and TN_i is the sum of the remaining entries of the diagonal of the confusion matrix.

The are two options for the weighted average: the Macro-level which consists of given the same weight for every class, and the Micro-level evaluation metrics, which gives each item equally, and making classes with more observations heavier than other with lesser items.

Support vector machine (SVM) [14] This method consists of assigning a vector representation of the labeled training dataset as points in space, mapped such that the members of the separate categories are divided by a gap as wide as possible. The linear SVM separates these categories by linear subspaces. The testing dataset vectors are then mapped into the same space and predicted to belong to one of the categories based on which side of the gap they fall in.

In mathematical terms, a support vector machine provides a set of hyperplane in an in nite dimensional space. The regions determined by these hyperplanes become the categories of the data. It is clear that the hyperplane which gives the best separation is the one that is farthest from all the nearest training examples (the support vectors) of each category. Further, the larger the margin, the lower the generalization error of the classifier will be.

Naive Bayes (NB) [14] This method generates a dataset with the frequency tables of the labeled data. With this dataset, a table with the corresponding probabilities of occurrences is created. The algorithm utilizes the Naive Bayes Eq. (3) for the probabilities.

$$P(c|d) = \frac{P(c)P(d|c)}{P(d)} \quad (3)$$

The category or class of an input corresponds to the highest probability obtained. Hence, the NB Algorithm assigns an input d to the class c, that maximizes $P(c|d)$.

***k*-Nearest neighborhood (K-NN)** [9] It is a non-parametric algorithm, usually very effective in most of the cases. It is also one of the most popular for text categorization, data mining, and many more. The k-NN algorithm works by analogy, that is, compares the unknown data point with the training data points to which it is close as measured by Euclidean distance. Through the combination of a number of local distance functions based on individual attributes, a global distance function can be calculated. The distance is precisely the weighted of the point to the k nearest neighbors.

Results. For the implementation of these algorithms we use several configurations. The results of the ones with the best performance are presented in Table 5. We split the collection of all the tweets from the seven datasets into a training part (85%) and

Table 5 Performance evaluation of the machine learning algorithms

Classifier	Accuracy (%)	Precision micro (%)	Precision macro (%)
NB	72.76	72.77	68.15
k-NN	59.15	59.15	54.97
Linear SVM	**82.13**	**82.13**	**78.85**

a test part (the remaining 15%). For the KNN method we use the linear algorithm $k = 3$ neighbors. As it can be appreciated the Linear SVM algorithm outperformed the others soundly.

3.2 *Artificial Mathematical Intelligence*

The research program of artificial mathematical intelligence (AMI) is a multidisciplinary meta-project with the aim of setting all the necessary theoretical and computational foundations for constructing (general versions of) a Universal Mathematical Artificial Agent (UMAA), which essentially is able to serve as co-creative formal assistant for any kind of researcher who used mathematically-based frameworks in his/her research. Specifically, an UMAA will be able to help in the purely conceptual part of the frameworks, e.g. solving a (solvable) mathematical conjecture (i.e. finding a formal proof or giving a counterexample), giving conceptual hints in order to solve faster the conjecture. In other words, the AMI program envisions to create the most wide a powerful form of artificial intelligence using, combining and fusioning techniques and concepts of (computational) cognitive sciences, computational creativity, computational logic, pure and applied (meta)mathematics, theoretical physics, classic AI, computational linguistics and philosophy of mind, among others [6].

The main three pillars of AMI are the following: Firstly, the new cognitive foundations of mathematics program aiming to found new and renewed multi-disciplinary foundations for (working) mathematics and for the way in which the mind generates mathematics from a pragmatic perspective. Secondly, the identification and meta-formalization of a global taxonomy of all the essential cognitive (metamathematical) abilities that the mind uses for generating (abstract) mathematical structures, conjectures and theories. Lastly, the implementation of all the former notions and structures into computational languages and software being able to simulate initial (and local) versions of a UMAA in a pragmatic and human style way.

Due to the wide methodological spectrum of the second pillar of AMI, there are fundamental aspects of this meta-cognitive taxonomy that can be used not only for the original goals of AMI, but that can be applied for understanding and forecasting qualitatively behavioural patters of individual and collectives. In particular, in the context of this chapter we will use some of the seminal results of this taxonomy together with specific contextual information coming from data mining techniques for designing pragmatic guidelines for preventing (and overcoming) and potential infection with SARS-CoV-2 (and COVID-19) in the current Latin American context.

4 Pragmatic Preventive Guidelines

Based on the former qualitative and quantitative results obtained by understanding the (Latin American) context, by doing a relevant sentiment analysis via data mining applied upon a relevant collection of tweets, by using some of the most relevant results of the artificial mathematical intelligence meta-project concerning cognitive fundamental mechanisms, and finally by integrating addition behavioural results, we will present here some pragmatic performance guidelines for increasing the quality of life during (permanent) confinement and gradual return to a public normality. Our view of health is very holistic, meaning that a human being is integrated by a visible (e.g. physical) component and by an invisible (e.g. cognitive) one, and only in a suitable balance of sound conditions between both dimensions a genuine and compact health is truly obtained. Local governments and public health organizations should be cautious about advising people on how to enlighten and integrate both components of health and incorporate them into their policies.

4.1 Regular Creative Visualization and Effective Planning

It is a well-known fact in neuropsychology that when the human brain (some-times referred as our subconscious mind) is exposed either to sensory information (e.g. images) or to purely phenomenological information (e.g. mental self-created images) cannot differentiate between reality and fantasy [5]. Moreover, creative visualization has a wide number of advantages into the cognitive health of human beings. One of the most important is that suitable techniques of visualization possess the power, the speeding up, the fulfillment of personal goals, and of giving a more authentic sensation of self-realization [5, Part III–IV]. Furthermore, in an environment where leaving home turns out to be so challenging and one should be long periods of time inside a building, creative visualization (e.g. involving desired empowering natural environments) would have the positive cognitive effect of giving a genuine sensation of being there (at least from a phenomenological perspective). Such a sensation has, in fact, a positive influence into the body since the neuro-logical networking patterns responsible for processing the sensory-perceptual information coming from having visited such empowering natural environment in a life manner, and for having visited them phenomenologically are quite similar; and so, they affect the rest of the body in a resembled manner.

Now, after using creative visualization for generating more clear and spe-ci c visions and goals in life, the following natural steps is to be able to create realistic (written) short-, medium- and long-term plans for the achieving of these objectives. Effective and specialized planning is not only highly practical for companies and any kind of society, but also for individuals. Even more, in a context of a pandemic, to construct a subjective feeling and a perception of potential achievement of goals turns out to be very useful for the emotional states of the mind, since they replace an excessive focus of attention on the disadvantages of the pandemic and shift the cognitive emphasis on the advantages of the environment (for the fulfillment of the personal visions, for example).

4.2 Using Constructive Linguistic Frameworks

The particular language we use in our daily life, the words we choose to hear and to speak play a crucial role in the way we think. In particular, they play a seminal component in the way we create new ideas and in the way we combine them e ciently. In fact, a huge amount of the things that we do in our daily lives are previously configured in a syntactic manner and pronounced in a phonetic way in our mouths [4]. In particular, one of the most fundamental cognitive-linguistic abilities that our mind uses for constructing a coherent subjective model of reality is metaphorical thinking [13]. Roughly speaking, a metaphor is saying that a concept X is (like) another concept Y. For example, "Life is (like) a Game" is a metaphor. Moreover, metaphorical reasoning is one of the most fundamental cognitive meta-mathematical abilities for generating formal mathematical knowledge, and in an extended manner for doing scientific research [6, Part II].

So, in the context of the COVID-19 pandemic it is fundamental to be able to build constructive, useful and inspiring metaphors for life, development, and personal success based on inspiring works instead of destructive ones. In our case, the Sect. 3.1 shows the morpho-syntactic evolution of hundreds of Colombian tweets for three months. From the evolution of the graphics one sees that the main trends in Colombia use gradually less and less words directly related with the COVID-19 and they started to evolve to other kind of topics like sports and space launches, politics, among others. This evolution into more constructive existential topics and notions is desirable, since an excessive exposition to destructive notions related with COVID-19 and with the increasing number of deaths could lead to premature depressive patterns and to an emotional atmosphere of global panic, which would deteriorate the confinement conditions and the inner harmony of the families. However, one can also appreciate in all the graphics shown in Sect. 3.1 a tendency of amplifying issues of secondary importance related to the pandemic situation and not completely positive tendency of the feelings described in the majority of the tweets. Therefore, here we aim to decisively invite the usage of an explicit positive language aiming to renew the immediate social, mental and financial (subjective) conditions of each person. This would be an initial plus for being able to materialize the desired goals and visions that each person wishes to have in order to overcome all the challenges that the current outbreak has started. In particular, the formulation of empowering and inspiring metaphors of personal (and communal) use possess a huge force for boosting new and better habits for restructuring a more solid, stable and altruistic way of living (as individual and as a collective (e.g. family)). In the Fig. 6, one can see an ideal wordcloud containing the essential aspects described in the last two guidelines.

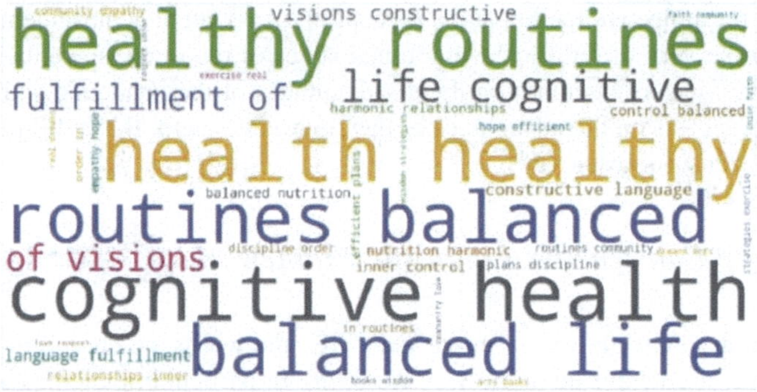

Fig. 6 Desired wordcloud

Note that a similar linguistic behavioural approach can be applied virtually for any kind of cognitive (metamathematical) mechanism like analogical thinking, conceptual blending, conceptual substratum, exempli cation, particularization, generalization, among others [6, Part II].

4.3 Frequent and Moderate Kinesthetic Routines

The beneficial and therapeutic effects of regular exercise routines for a healthy life are very well-known in medicine and psychology (see, for instance [11, 19, 23]). Now, due to the fact that the SARS-CoV-2 is a virus that attacks and debilitates the respiratory system, it seems fundamental, from a preventive perspective, to integrate strengthening routines for the breathing system; so, one of the most natural and cheapest of them is without any doubt regular sporting activities. Even more, in a broader context beyond Latin America, two of the most common co-morbidity factors for people dying of COVID-19 are arterial hypertension (and diabetes) [17, 25], which can also be prevented by doing exercise on a regular basis [3]. Thus, our third prevention guideline is to do intelligent (indoor) exercise routines like continuous moderate-intensity training (CMT) routines with a duration or at least 30 min or more, 3–4 times a week. Here is important to create the routine first with small and pleasant sessions orchestrated with motivating and loved music pieces. Due to the spatial restrictions caused by the pandemic, it is highly recommended to have access to indoor running machines, stationary bicycles, stair-climbers or to following encouraging online exercise routines like aerobics, Zumba, salsa (or any kind of dance style), martial arts, among many others.

The most challenging part requires to create the routine at the beginning, and the more you have repeated the sessions (moderately), the easier it is, the deeper it influences your unconscious mind and the greater of the benefit.

5 General Conclusions

Due to the wide spectrum of anthropological dimensions that the human being possesses, a genuine state of holistic health can be achieved taking care not only of the pure biological and physiological dimension of people (understood as individuals and as communities), but also taking really precise care of the cognitive dimension. Thus, based in a suitable combination of techniques coming from data mining, artificial mathematical intelligence and implicitly from cognitive sciences we over simple and, at the same time, powerful behaviour techniques or guidelines for maintaining the mental health while preserving (longer) public policies of confinement due to the spread of COVID-19 in Colombia, and in other Latin American countries with similar contextual conditions. In fact, based on the universality of laws governing the functioning of the mind, it is worth to study the wide potential of applying the same kind of preventive guidelines to virtually any kind of society.[1]

Specifically, we describe guiding principles relates with suitable visualization and planning techniques,[2] the use of inspirational linguistic schemes and the generation of habits related with empowering kinesthetic practices.

In this chapter, we focus only on the former three main preventive guidelines. However, due to the wide spectrum of complexities of the human mind, a lot of additional guidelines could be analyzed in more detail in subsequent works.

On the other hand, the former cognitive standards are completely suitable to be integrated by local governments into their prevention policies along with the corresponding bio-security measures. So, both dimensions, the biological and the cognitive, can work together for the improvement of human health globally.

This is a small effort towards the development of more sophisticated techniques of a new kind of academic discipline that can be called "behavioural medicine".

Acknowledgements Danny A. J. Gómez-Ramírez would like to thank the Institución Universitaria Pascual Bravo and to the Institución Universitaria ITM (Instituto Tecnológico Metropolitano) and to Visión Real Cognitiva S.A.S. for all the support. He also wishes to thank William Restrepo, Elizabeth Yepes and Yury Baena for all their kindness and support. Yoe and Johana want to thank their three daughters Salomé, Sofía and Sara for the love and happiness they bring to their lives. Yoe Herrera also thanks Universidad Autonóma de Bucaramanga for their support.

Appendix

See Figs. 7, 8, 9, 10 and 11.

[1] Assuming, of course, a suitable contextualization of each of them to the corresponding local policies and cultural conditions.

[2] These kind of techniques possess a wide and strong support from groups of people going beyond the academic field, like, for example, particular communities of entrepreneurs or athletes.

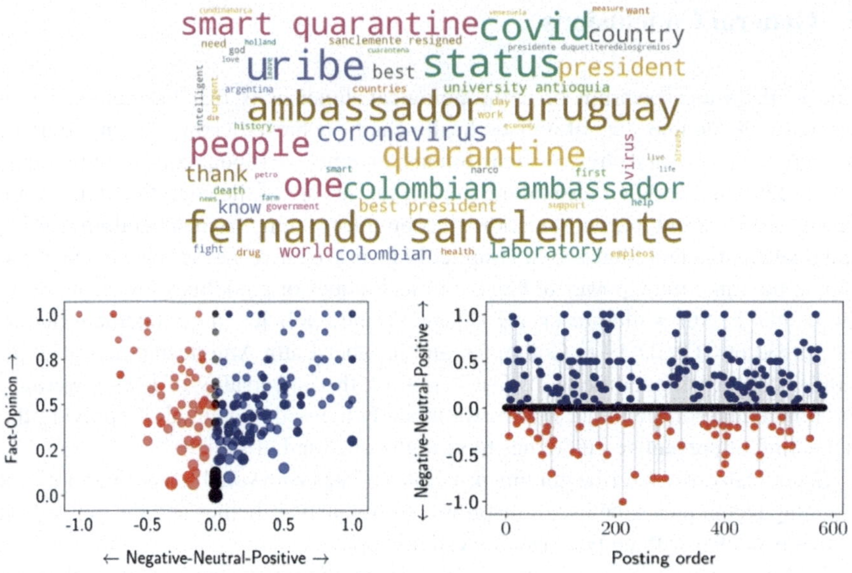

Fig. 7 Sentiment analysis of April 6 and 7

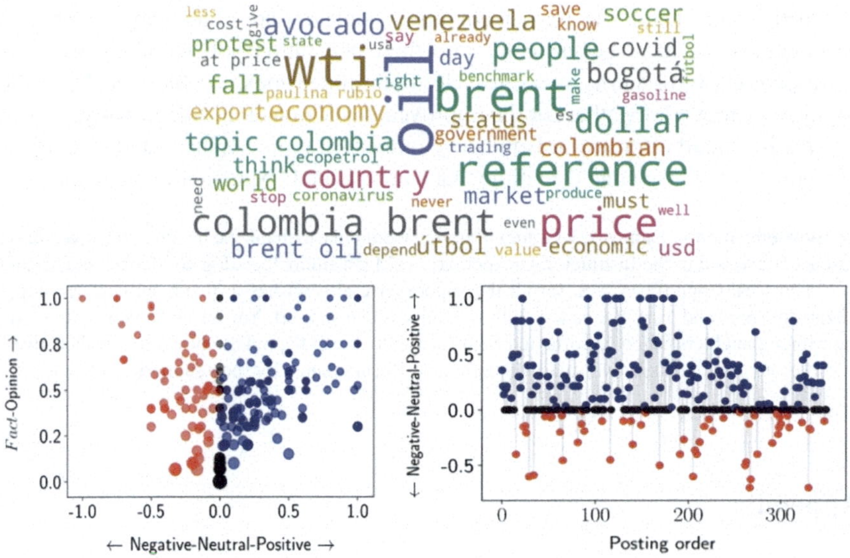

Fig. 8 Sentiment analysis of April 20 and 21

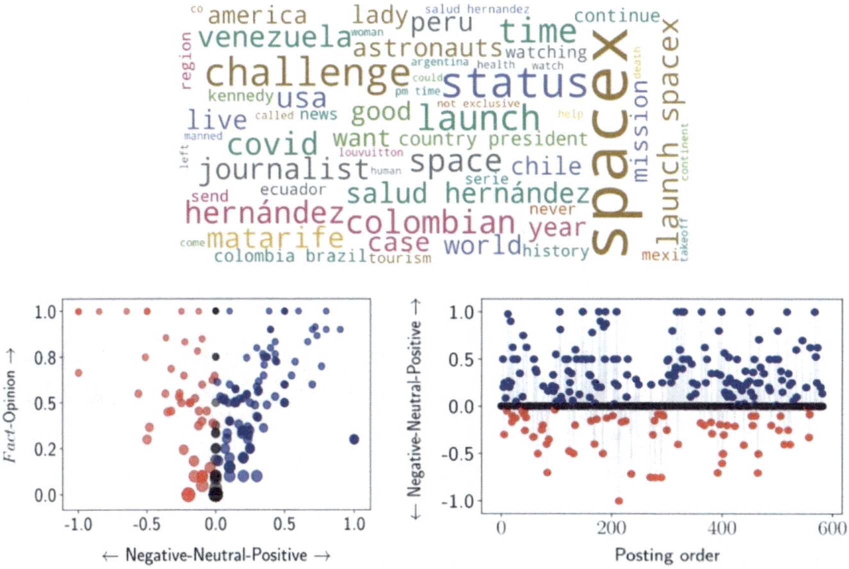

Fig. 9 Sentiment analysis of May 30 and 31

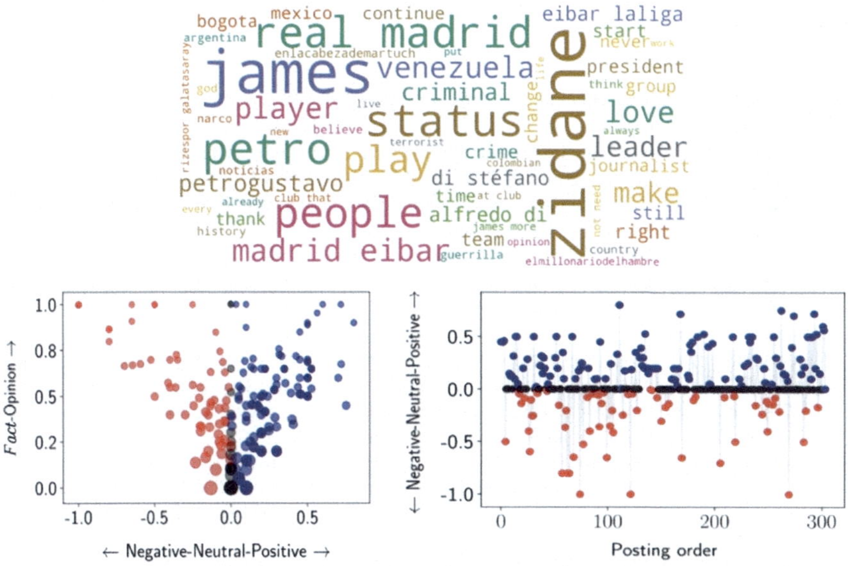

Fig. 10 Sentiment analysis of June 14 and 15

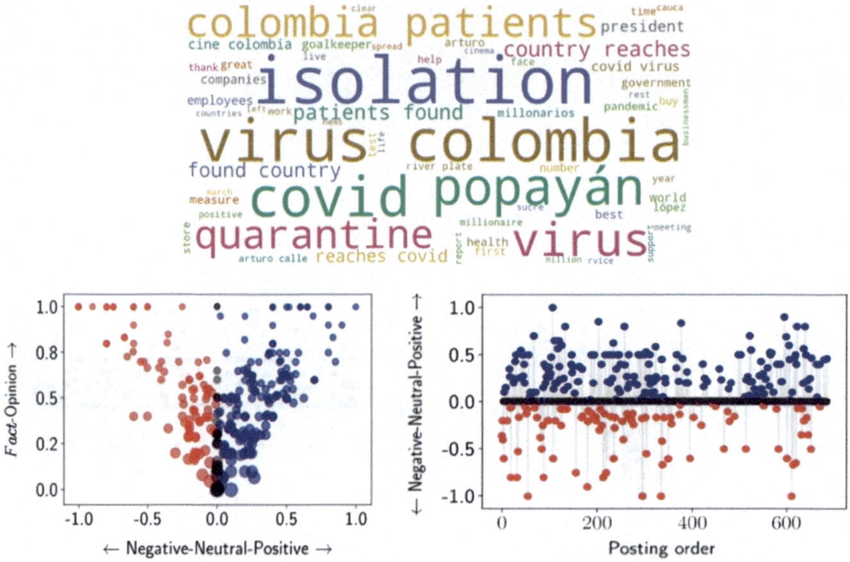

Fig. 11 Sentiment analysis of June 19 and 20

References

1. Alimadadi A, Aryal S, Manandhar I, Munroe PB, Joe B, Cheng X (2020) Artif Intell Mach Learn GHT covid-19
2. Bullock J, Pham KH, Lam CSN, Luengo-Oroz M et al (2020) Mapping the landscape of artificial intelligence applications against covid-19. arXiv:2003.11336
3. Ciolac EG (2012) High-intensity interval training and hypertension: maximizing the bene ts of exercise? Am J Cardiovasc Dis 2(2):102
4. Fauconnier G, Turner M (2003) The way we think. Basic Books
5. Gawain S (2016) Creative visualization-: use the power of your imagination to create what you want in your life. New World Library
6. Gómez Ramírez DAJ (2020) Artificial mathematical intelligence: cognitive, (meta)mathematical, physical and philosophical foundations. Springer International Publishing
7. Guo Y-R, Cao Q-D, Hong Z-S, Tan Y-Y, Chen S-D, Jin H-J, Tan K-S, Wang D-Y, Yan Y (2020) The origin, transmission and clinical therapies on coronavirus disease 2019 (covid-19) outbreak – an update on the status. Military Med Res 7(1):1–10
8. Hand DJ, Adams NM (2014) Data mining. Wiley StatsRef: Statistics Reference Online, pp 1–7
9. Hota S, Pathak S (2018) KNN classifier based approach for multi-class sentiment analysis of twitter data. Int J Eng Technol 7(3):1372–1375. https://doi.org/10.14419/ijet.v7i3.12656. ISSN 2227-524X
10. Hu B, Ge X, Wang L-F, Shi Z (2015) Bat origin of human coronaviruses. Virol J 12(1):221
11. Johnson J (1985) Exercise, aging and health. Occup Health Nurs 33(3):137–151
12. Laiton-Donato K, Villabona Arenas CJ, Usme Ciro JA, Franco Munoz C, Alvarez-Diaz DA, Villabona-Arenas LS, Echeverria-Londono S, Franco-Sierra ND, Cucunuba ZM, Florez-Sanchez AC, Ferro C, Ajami NJ, Walteros DM, Prieto-Alvarado FE, Duran-Camacho CA, Ospina-Martinez ML, Mercado-Reyes MM (2020) Genomic epidemiology of sars-cov-2 in colombia. https://doi.org/10.1101/2020.06.26.20135715
13. Lako G, Johnson M (1980) The metaphorical structure of the human conceptual system. Cogn Sci 4(2):195–208,

14. Le B, Nguyen H (2015) Twitter sentiment analysis using machine learning techniques. In Le Thi HA, Nguyen NT, Do TV (eds) advanced computational methods for knowledge engineering. Springer International Publishing, Cham, pp 279–289. ISBN 978-3-319-17996-4
15. Lee SY (2019) Document vectorization method using network information of words. PLOS ONE 14(7):1–13. https://doi.org/10.1371/jour-nal.pone.0219389
16. Pang B, Lee L (2008) Opinion mining and sentiment analysis. Found Trends Inf Retr 2(1–2):1–135. ISSN 1554-0669. https://doi.org/10.1561/1500000011
17. Richardson S, Hirsch JS, Narasimhan M, Crawford JM, McGinn T, Davidson KW, Barnaby DP, Becker LB, Chelico JD, Cohen SL et al (2020) Presenting characteristics, comorbidities, and outcomes among 5700 patients hospitalized with covid-19 in the New York City area. Jama
18. Rodriguez-Morales AJ, Gallego V, Escalera-Antezana JP, Mendez CA, Zambrano LI, Franco-Paredes C, Suarez JA, Rodriguez-Enciso HD, Balbin-Ramon GJ, Savio-Larriera E et al (2020) Covid-19 in latin america: the implications of the first confirmed case in brazil. Travel Med Infect Dis
19. Ross A, Thomas S (2010) The health benefits of yoga and exercise: a review of comparison studies. J Altern Complement Med 16(1):3–12
20. Sailunaz K, Alhajj R (2019) Emotion and sentiment analysis from twitter text. J Comput Sci 36:101003. ISSN 1877-7503. https://doi.org/10.1016/j.jocs.2019.05.009
21. Shereen MA, Khan S, Kazmi A, Bashir N, Siddique R (2020) Covid-19 infection: origin, transmission, and characteristics of human coron-aviruses. J Adv Res
22. Simbana-Rivera K, Gomez-Barreno L, Guerrero J, Simbana-Guaycha F, Fernandez R, Lopez-Cortes A, Lister A, Ortiz-Prado E (2020) Interim analysis of pandemic coronavirus disease 2019 (covid-19) and the sars-cov-2 virus in latin america and the caribbean: morbidity, mortality and molecular testing trends in the region. medRxiv
23. Sparkes AC, Smith B (2013) Qualitative research methods in sport, exercise and health: from process to product. Routledge
24. Vaishya R, Javaid M, Khan IH, Haleem A (2020) Artificial intelligence (ai) applications for covid-19 pandemic. Diabetes Metab Syndr: Clin Res & Rev
25. Yang J, Zheng Y, Gou X, Pu K, Chen Z, Guo Q, Ji R, Wang H, Wang Y, Zhou Y (2020) Prevalence of comorbidities and its effects in patients infected with sars-cov-2: a systematic review and meta-analysis. Int J Infect Dis 94:91–95
26. Yaqub U, Sharma N, Pabreja R, Chun SA, Atluri V, Vaidya J (2018) Analysis and visualization of subjectivity and polarity of twitter location data. In: Proceedings of the 19th Annual International Conference on Digital Government Research: Governance in the Data Age, dg.o '18, New York, NY, USA. Association for Computing Ma-chinery. ISBN 9781450365260. https://doi.org/10.1145/3209281.3209313
27. Ye Z-W, Yuan S, Yuen K-S, Fung S-Y, Chan C-P, Jin D-Y (2020) Zoonotic origins of human coronaviruses. Int J Biol Sci 16(10):1686

A Drone System for Detecting, Classifying and Monitoring Solid Wastes Using Computer Vision Techniques in the Context of a Smart Cities Logistics Systems

Adrian Ramirez-Lopez, Alberto Cortes-González, Gilberto Ochoa-Ruiz, Alberto Ochoa-Zezzatti, Lina Maria Aguilar-Lobo, Diego Moreno-Jacobo, and Christian Mata-Miquel

Abstract Currently, a titanic task in the solid waste collection is the location of this type of waste within a collection circuit as part of a Logistics 4.0 model, in our research proposal a drone is used to determine in a time horizon the solid waste through which it will pass in a determined time in a specific collection route, this will allow determining in a suitable form the necessary space of the different sections from the collection truck and will influence in determining patterns of solid waste together with which it will be able to specify the requirements of the vehicle fleet required for specific demands and contingencies determined by its high variability using an artificial vision system for detection, classification, and monitoring of solid waste in an intelligent drone under the Logistics 4.0 paradigm.

Keywords Solid waste management · Computer vision · Deep learning · Smart cities · Logistics 4.0

1 Introduction

In recent years, we have witnessed an astounding increase in population worldwide, and an enormous concentration of this population in big cities. In general terms, this has lead to a steady increment in production of goods that inevitably end as solid

A. Ramirez-Lopez · L. M. Aguilar-Lobo
Universidad Autónoma de Guadalajara, Zapopan, Mexico

A. Cortes-González · G. Ochoa-Ruiz (✉)
Tecnologico de Monterrey, Escuela de Ingeniería y Ciencias, Jalisco, Mexico
e-mail: gilberto.ochoa@tec.mx

D. Moreno-Jacobo · C. Mata-Miquel
Universitat Politècnica de Catalunya, Barcelona, Spain

A. Ochoa-Zezzatti
Universidad Autónoma de Ciudad Juárez, Ciudad Juárez, Mexico

© The Author(s), under exclusive license to Springer Nature Switzerland AG 2021
A. Ochoa-Zezzatti et al. (eds.), *Technological and Industrial Applications Associated with Intelligent Logistics*, Lecture Notes in Intelligent Transportation and Infrastructure,
https://doi.org/10.1007/978-3-030-68655-0_27

wastes that require a great deal of effort regarding their collection, management and storage. In many in-development countries, such measures are not always effectively implemented, which has adverse public health and the environmental implications, and in many cases poses hazards in public security as well. For instance, in many cities solid wastes laying in the streets can clog the city water drainage systems, leading to floods and other disastrous situations. Therefore, effective waste spotting mechanisms are very much required to avoid these situations.

In this paper, we present a system that seeks to mitigate some of the abovementioned problems. It represents the convergence of several technologies and disciplines, namely: Internet of Things, Artificial Intelligence (in the form of computer vision and machine learning) and approach based in logistics and Industry 4.0 principles. We have trained a deep learning-based object detection system based on TinyYolo for automatically spotting solid wastes in the wild using very large database of solid waste objects created by us and which is freely available. This model is amenable to mobile computing implementation and has been integrated within an embedded device mounting in a drone system. The main rationale of our system is to detect possibly hazardous solid waste piles, particularly in areas considered risky by the authorities; in order to effectively implementing this system, the locations the risk zones and of the detected solid waste piles are georeferenced. In this manner, a large database for reporting, follow and management can be created following principles of logistics and industry 4.0.

The rest of this paper is organized as follows: in Sect. 2 we further elaborate the motivation of this work, pointing out at the more pressing concerns in waste management and the problem of waste accumulation in risk zones. In Sect. 3 we carry out a thorough analysis of the state of the art, both for the automation spotting and classification of solid wastes in the wild using computer vision techniques, as well as the integration of artificial intelligence methods in the context of smart cities. In Sect. 4 we provide ample details about the proposed dataset, as well as for the training and validation of the implemented waste spotting and detection deep learning-based system. Subsequently, in Sect. 5 we provide ample details about the integration of the computer vision mobile system into the drone and logistics infrastructure. Finally, we carry out a discussion on the use and potential of drones and data analytics in this context, concluding the article.

2 Motivation

Significant development and increase in urbanization and the accompanying agglomeration of people in large cities around the world is leading to an ever-increasing waste generation [1]. According to several sources, in the past century only, waste production has risen tenfold and by 2025 it is expected to double again [2]. These often hazardous and solid waste are often not recycled properly and is usually either dumped in a landfill or incinerated as can be attested in very large landfills in many cities around the world [3]. In some cases, waste incinerators are being deployed to

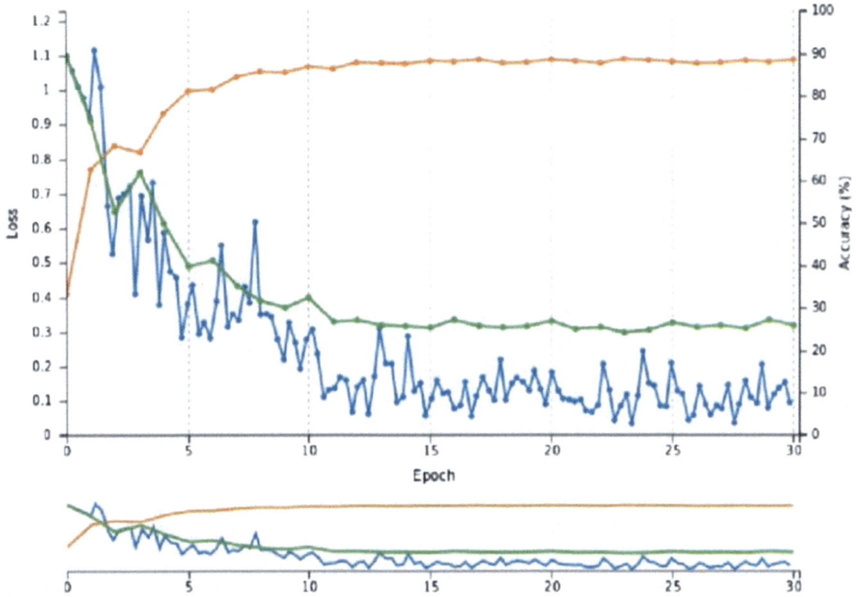

Fig. 1 Training and validation loss curves for the augmented dataset

cope with the increased amount of detritus, but there are a number of concerns over fine particle emission, which can lead to a number of associated health risks as well as contributing to global warming [4, 5]

Some previous works have addressed the problem through the smart cities and smart objects paradigm, by proposing systems able to monitor the status of the garbage can in real time. However, these bins do not have any waste classification mechanism, which can lead to a recycling bin becoming contaminated. General waste can be classified into the categories shown in Fig. 1 to allow for more efficient recycling. The most common contaminants in an urban environment are napkins, plastic food wrapping, plastic bags, coffee cups, coffee sleeves, rubber gloves and plastic medical waste. A number of schemes have been tried to inform and better educate the public on the categories and benefits of waste classification in different demographics [6]. An automated classification mechanism would solve this problem and encourage user engagement as they would not be worried about placing an item in a recycling bin that could possible contaminate it. Deep neural networks have shown increased accuracy in a number of recent benchmark competitions in machine learning and pattern recognition [7]. We can frame the waste classification problem as an image classification problem where we have a camera that takes an image of the waste and uses a DL model to identify to what class belongs to.

Also, researchers are thinking about how to use technology to achieve urban cleaning and urban street cleanliness assessment. Borozdukhin et al. [8] proposed a method to solve the optimization of garbage disposal in big cities. The method searches for the time-optimized dynamic route for garbage collection trucks by establishing a mathematical model of dynamic optimal paths, which can make the garbage collection trucks spend shorter time from the garbage collection area to the landfill area. However, the system only considers the route selection of garbage collection trucks and it does not consider the urban cleanliness assessment. Clean Street LA [9] is an initiative by the London city Mayor and the system uses ESRI and GIS tool to map and plot the street cleanliness status block by block. Multiple layers and grids are created to reflect different parts of the city. Cleanliness information on the streets with a cleanliness score is visualized on a map. This information is used to decide on the area that requires attention. However, the limitation of the system is that the monitoring is limited to garbage bins and cannot be extended to monitor the streets. In [10], a mobile app was developed to evaluate street cleanliness and waste collection. This app is based on a plan of indicators that can be used to evaluate the street cleanliness and waste collection service of Santander municipality. Specific methodologies for calculating and evaluating 59 indicators have been developed to obtain information regarding the status of the different elements of the service. Pearson correlation coefficient results suggest that an inverse relationship between the street cleanliness index values and the frequency street cleanliness services/population density ratio exists. In short, although researchers are concerned about street cleanliness, they have not yet applied mobile edge computing, cloud computing, and deep learning to assess street cleanliness in time.

3 State of the Art

Conversely, recent strides in deep learning techniques have enabled us to implement very sophisticated and resource efficient object detection systems in edge devices [11]. These progresses were possible due to the introduction of single stage detectors such as YOLO [12] or RetinaNet [13] (for a detailed discussion in this topic, see [14] for an excellent survey). Two-stage detectors such as Fast-RCNN could achieve high accuracies but with the cost of high inferences times, making them unsuitable for deployment on resource constrained devices [15]. On the other hand, one-stage detectors suffered from low accuracies but offered very low inference times; this problem has steadily been addressed by the DL community and recent work (both academic and commercial) have made use of these object detectors for implementing real-time acquisition systems on mobile devices [16].

There has been many theoretical projects and several experimental projects individually done by garbage companies as well. From the researches that have been done, the dataset is usually the problematic part in most of the cases. The need of a dataset that properly encompasses the problem set is of extreme importance. The recent developments in convolutional neural networks (CNN) has created an awesome environment for learning the features from images resulting in image classification, segmentation and detection [17]. Therefore, such way seems to be the best way to classify wastes on a whole. Awe et al. [18] propose an experimental project using a Faster R-CNN model to classify waste into three categories: paper, recycling, and landll that achieved a mean average precision of 68%. Thung and Yang [19] deployed support vector machine (SVM) and a convolutional neural network (CNN) to classify waste into six categories. It achieves an accuracy rate of 63% for SVM and 73% for CNN. Rad et al. [20] developed a GoogLeNet-based vision application to localize and classify urban wastes, but the classes the system is capable of detecting is restricted to tree leaves and cigarette buts; the study claims to have an accuracy rate ranging from 63% to 77% for different waste types. In more recent work, authors in [21] designed a project to detect whether an image contains garbage or not where he employs the pretrained AlexNet model and achieves a mean accuracy of 87.69%. This project, SpotGarbage, aims at segmenting garbage in an image without providing capabilities for more fine-grained waste classification, which can be useful in some situations but does not lead to adequate garbage sorting applications. This issue has been partially addressed by a more advanced computer vision approach to assess the recyclability of the detected garbage in a video stream. The dataset from Stanford clearly marks itself out due to its distinct features and the preprocessing; the dataset has been used by many of researchers in making more predictions and analysis as well. The size of the dataset being small needs to get some additional data in order to increase its accuracy as well as reliability [22].

In this project, we have made use of our deep learning model based on the YOLO (You Only Look Once) algorithm. We trained the model using an extension of the dataset introduced in [22], performing extensive data augmentation for battling overfitting, as it will be discussed in the next section.

4 Materials and Method

4.1 Dataset Collection

We have collected a large dataset from various sources (i.e. internet crawling, crowdsourced photos, among others) consisting of images of 5 classes, as depicted in Table 1.

The approach presented here is a multi-class classification problem, in which we take the 5 classes of the proposed dataset and train an optimized object detector model (Yolo, but it could be other as well) to implement a road damage acquisition system. However, training deep learning-based object detectors is not a trivial task:

Table 1 Number of trash instances in each class for: the original dataset and the augmented datasets

Dataset/Class	Glass	Paper	Plastic	Metal	Cardboard	Trash
TrashNet	501	594	403	482	410	137
Augmented	3,006	3,564	2,418	2,892	2,460	822

Table 2 Results of the validation of the model in the test set in terms of Mean Average Precision

Metric/Class	Glass	Paper	Plastic	Metal	Cardboard	Trash
mAP	0.92	0.95	0.88	0.95	0.93	0.95

one of the major issues of deep learning, when working with relatively small datasets such as the one used in this study, is the problem of overfitting.

Data augmentation is a manner for "synthetically" creating more training data from the existing examples in the datasets. They are generated by submitting images in the latter to a variety of image processing operations that yield believable looking images. The process of D.A. can significantly reduce the validation loss, as it is a form of model regularization. Using data augmentation clearly combats the overfitting problem, whose effect can be observed in the left side of the figure (this is the validation loss decreases and then increases again, whilst the training loss decreases for the training set).

Therefore for the purpose of this work, we performed six types of augmentation in the images of the dataset: flipping, rotation, shearing, translation, elastic transformation and histogram equalization. This process has the effect of multiplying the dataset by 6 fold (as it can be observed in the second row of Table 2) yielding a new dataset of 15,162 images.

4.2 *Model Training*

We carried out the training of the deep learning models described above using the following methodology. The labeled dataset is randomly shuffled and separated into training set and validation set, 70% percent of the images of each class was used for training, taking care that the model did not overfit; for this purpose, we have made use of dropout and data augmentation over the original dataset, as discussed above. Later, the other 30% of the images were used for validating the models.

The model description as well as the training phase were implemented on Keras and Tensorflow, running on Google Colab with a Tesla K80 GPU; the models were trained with batch size of 24 and number steps per epoch of 300, using Adam as optimizer, adapting the learning rate. The results of the training process are depicted on Fig. 1, where we can observe the monotonic decrease of the training error (blue line) as well as the validation error (green line) achieving a mean average procession of 90% for all the classes.

4.3 Model Validation

In our experiments, both the training and testing phases were carried out using Google Colab, which enabled us to have access to high performance computing (a Tesla K80 GPU and other resources) and to host our DB on the Google Cloud, all while avoiding any computational burden. In the evaluation phase, we also made use of this tool, importing the trained models and the 30% of the validation set; furthermore, we carried out experiments on real-time video to test the capabilities of the proposed models in a variety of scenarios using the Cartucho/mAP evaluation tool [23]. The results of this validation phase are summarized in Table 2.

Some examples of correctly identified images are shown in Fig. 2. These images comprise examples from the dataset (i.e. the metal and cardboard examples

5 Integration of the Computer Vision Model into the Drone System

The deep learning model based on YOLO described in the previous section is amenable for mobile computing applications and in this work, we investigate its use for implementing a drone based system for detecting, classifying and monitoring solid wastes in urban settings. The information gathered by the computer vision systems runs information that is fed to an analytics system for implementing a logistics model.

In the following Fig. 3, we describe our model associated with our project, first a group of drones are launched within a quadrant of the Smart City, with the information collected by the drones is considered a route for garbage collection and then generates a graph with it, then determines the fleet and labor required for this, considering that the drone establishes the approximate amounts of material to recycle.

First, the type of waste is identified and it is classified in: paper, aluminum, corrugated cardboard, glass and plastic. Once its characteristics are specified, the

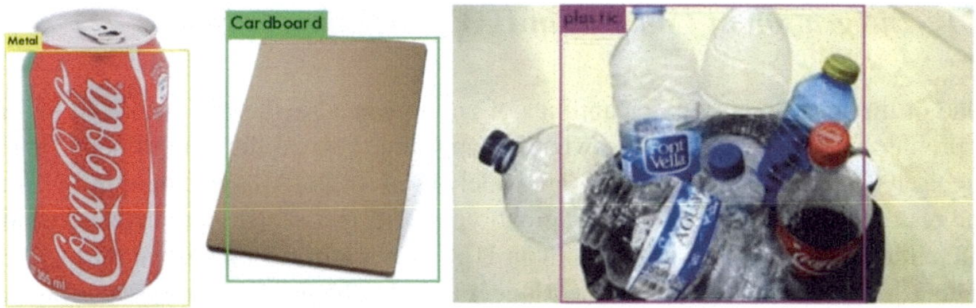

Fig. 2 Examples of instances in the dataset; the image in the right shows an example from the test set

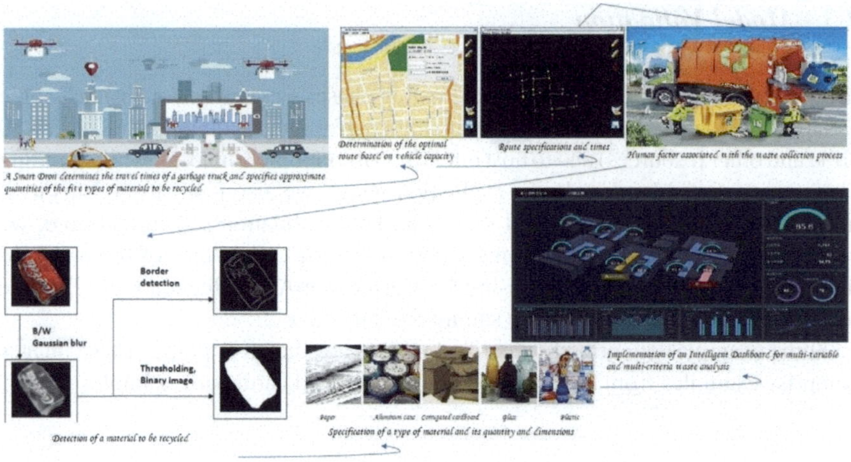

Fig. 3 Intelligent model associated with our research

social behavior regarding waste materials to be recycled can be seen in real time in an intelligent dashboard.

In Fig. 3, we can see in detail each component of our project, from the estimates made a cluster of drones were the routes where the vehicle fleet will pass for the collection of solid waste, then the information will be detailed between the physical locations of streets and avenues and then converted into a graph to determine if the route is optimized, correct and ideal, Together with the determination of the size of the team of workers that will carry out the solid waste collection, for this purpose they rely on the timely selection of each image that allows them to separate and determine the potential of each of the two five materials in order to be able to separate paper, aluminum, corrugated cardboard, glass and plastic. Finally, by using an Intelligent Dashboard it will be possible to determine recycling percentages, something of great importance for intelligent logistics 4.0 in a Smart City.

5.1 Integration of Cameras in UAV Navigation Systems

One of the most important capabilities of an unmanned aerial platform is to be in a privileged position to observe what is happening beyond our line of sight. It is precisely for this reason that UAVs are widely used for ISTAR-type missions where the platforms are equipped with different types of cameras and other detection and surveillance sensors. Recording from an UAV is not a trivial operation: during the flight, the plane moves, vibrates, is subject to light changes, the lenses are fragile etc. For this reason, the different specialized brands that we can find in the market offer increasingly robust products and allow us to capture stabilized images of high quality. The gimbal camera sector has evolved a lot during the last years, achieving

Fig. 4 Use of an intelligent UAV to determine tracks associated with a specific waste collection route

complex and high resolution mechanisms at a reduced size, which allows its easy integration in UAS systems. One of the challenges faced by gimbal manufacturers, and at the same time the manufacturers of platforms that incorporate these devices, is the stabilization of the camera, in order to obtain quality images for subsequent processing and extraction of data of interest. For this purpose, not only is it essential to incorporate good vibration damping systems, but also the compensation of movements and accelerations experienced during flight is key, as is possible see in Fig. 4.

To provide a solution to this situation, UAV Navigation proposes different levels in the integration of a gimbal camera with its autopilot so that our customers can get the maximum added value to their cameras when used in conjunction with our system. Thus, three different options for camera integration have been considered:

Level 1 integration of operatibility

In which the autopilot sends to the camera information of the flight dynamics (speed, GNSS position, accelerations…) allowing its correct stabilization and operation. As mentioned above, some of the cameras currently available on the market incorporate IMUs for calculating the attitude of the platform. However, the data obtained from the autopilot and its AHRS, systems designed for this purpose, have more advanced estimation algorithms and allow a more accurate stabilization. Moreover, the use of the data from the autopilot itself allows a reduction in the overall cost as it is not necessary to use cameras with integrated inertial systems and with a more economical cost and reduced weight.

Level 2 integration of tactical tasks

With this level of integration, the user can visualize the coordinates towards which the camera lens is pointing, expanding his situational awareness. In addition, the user can use these coordinates as a reference when navigating the platform. This level of integration allows the platform to therefore follow the coordinates provided by the camera (e.g. 'camera guidance' mode). This capability can be particularly useful in the tracking of moving targets. This level of integration can be achieved as long as the camera has the capability of geo-pointing. It is important to remember that the coordinates provided by the camera can either be acquired automatically or entered manually so users should be aware of the characteristics of their camera.

Level 3 Integration of strategies and stratagems associated with their nominal operation

In this last option, the complete integration of both systems (autopilot-camera) takes place. The ground station will provide the user with the option to control the place where he wants to direct the camera by simply clicking on the map and without the need for auxiliary and secondary applications. UAV Navigation, since its creation in 2004, has accumulated a great experience in the UAV sector and therefore has been very familiar with the evolution and technological progress of gimbals. The company has always maintained a relationship with the main camera manufacturers in the market with whom it has collaborated and tries to find joint solutions to the problems or challenges faced by manufacturers by providing alternatives that allow easier and higher quality use to its customers, allowing them to concentrate on the mission and not on the platform.

6 Discussion: Use of Drones into Smart City Logistics

Cities around the world are adding technology to improve the environmental, financial and social aspects of urban life. An entity that uses technology in such a way to improve the lives of its citizens and communication between residents and officials is known as a Smart City. Today many of the world's cities are smart, in a certain percentage, already considering themselves Smart Cities, which is a first step in digitally transforming many of their day-to-day processes. Although it is expected that with the arrival of the new generation networks worldwide this will be maximized, the truth is that many are already ready to use technologies such as the Internet of Things (IoT) to improve their services. In general, cities around the world are adding technology to improve the environmental, financial and social aspects of urban life. An entity that uses technology in such a way to improve the lives of its citizens and communication between residents and officials is known as a Smart City.

Next generation networks will serve as the foundation for small cell networks that will drive the next generation of wireless network infrastructure in a city. The connectivity and computing power they enable will make Smart Cities more of a reality, as city officials can adopt new technologies for Smart City solutions. With prices falling on sensors used for the Internet of Things, many cities say they are ready to change the way they operate in the next decade. For many industry specialists, the proliferation of IOT and the imminent development of 5G connectivity will open the floodgates for the first truly smart city.

The biggest impediment to moving the smart city from theory to practice was the high speed and bandwidth to handle the amount of data generated by sensors and process it in real time. This will be exemplified in everything from public transportation to law enforcement. The cost savings that will be generated by next generation networks will be the trigger for change. According to Accenture, smart city solutions applied to vehicle traffic management and power grids alone could yield $160 billion in benefits and savings through reductions in energy use, traffic congestion and fuel costs. This means that travel times will decrease, public safety will improve and significant efficiencies will be achieved on the smart grid. For example, in a smart city with next-generation networks, there are several layers that can be added to something as simple as a smart street light. Think about what can be added to it with a video camera or a gunshot detection sensor that will provide information to public safety officials so they can respond more quickly. Video analytics can be quickly evaluated and Artificial Intelligence can be added to make the city safer, as in Fig. 5.

Analyze that with only a medium investment you can get involved in these smart city drones that help with public safety. First responders can benefit from low-latency video cameras. A smart ambulance can move freely through a city, as it is connected to a network that communicates with traffic signals, and it can also do so with

Fig. 5 Intelligent UAV to determine the number of vehicles required for a waste collection route in a Smart City

the doctors in the trauma center of the hospital it is going to in order to quickly attend to a patient. Connected communities require a massive network ecosystem for information to flow, be processed and analyzed in just a few seconds. We are ready for this, and with the arrival of new generations of equipment and networks, the construction of smart cities will now move one hundred percent from theory to practice. This is already beginning to happen, but it will become more widespread. Many of us have heard about autonomous cars, and with this possibility will be real. Vehicles will talk to each other without latency. Cities will be able to use artificial intelligence to analyze the enormous amounts of data being collected. The resulting data can be used to automate processes that are currently done manually, and that is the most exciting thing about improving everyone's life. New generation networks allow technology to scale at low cost, so there is no reason why people and governments should not start joining this trend. Whether they are cities, mobile operators or innovators who have an idea for a service, they should start showing it and exploiting it in cities like Mexico. The factors that intervene in the operation of the system and from which they could affect or improve the performance of the same one are analyzed, for it, the following table is taken into account, in which the amount of space and what a photo in JPEG will occupy is deduced. In this table, it is possible to estimate an average occupation value under normal conditions, and therefore a card capacity.

A general use table could be the following (note: a compression ratio of 1:8 has been taken for such calculations). Specifications of an ideal group of components to a drone is analyzed in Tables 3 and 4.

Nowadays it is of vital importance to develop computer applications with different functionalities and high complexity in a fast and efficient way in all aspects. Performing tests to the software during the development process is a guarantee for the operation of the systems; besides corroborating the degree of reliability before being delivered to the final users; which reduces the defects by using appropriate techniques that enable effective software development processes, minimizing time and costs.

Table 3 How many photos can fit on a memory card?

Number of estimated photos					
Camera resolution					
SD camera target	12 MP	14 MP	16 MP	20 MP	24 MP
1	227	195	171	137	113
2	455	390	341	273	227
4	910	780	683	546	455
8	1.820	1.560	1.365	1.092	910
16	3.641	3.121	2.731	2.185	1.820
32	7.282	6.483	5.461	4.369	3.641
64	14.564	12.483	10.923	8.738	7.282

Table 4 System factors

Factors affecting system performance			
Factor description	Level 1	Level 2	Level 3
A. Battery life (min)	30	20	18
B. Flight time (min)	90	70	68
C. Processing (MB/s)	10	6	4
D. Storage	7.282	6.242	5461
E. Camera resolution (mp)	1080	720	720
F. Flight distance (km)	2	1	1.5
G. Speed without wind (km/h)	14	10	8
H. Height (m)	120	100	80

On the other hand, the orthogonal table will be applied as an alternative for the continuous improvement of the testing strategies and the decrease of the time required in its execution. In evaluating the literature on robust testing based on orthogonal arrays, emphasis was placed on the techniques employed by Dr. G. Taguchi for the improvement of product and process quality; applying the same in the engineering processes for software development. Planning and executing the test cases with the previous proposal guarantees the detection of the highest number of errors in the double and triple interactions of the system's functionalities. When calculating the distance of the equal points of the matrix for example m [0] [0] will be placed with a double greater than all the distances shown in Table 2, this modification was made in the distance function this in order to discard these points as minimum since if they remain as zero they would be taken as small values. The following Table 5 shows how it was formed.

In this way it is ensured that the minimum real values are taken, having the minimum of each point two situations necessary for a good result are processed, both situations happen when having the minimum value, one is when finding said value, that value is substituted in the matrix by the greater value this would mark the point as visited and it would be discarded that point be contemplated again, the next situation is to save the point of the visited matrix in an ArrayList which would be forming the optimal route. Our model consists in broad strokes, having initialized the distances between nodes, grouping the nodes or vertices into three types, the nodes marked, adjacent and distant. The initial node is a marked node, the adjacent nodes are nodes connected to a marked node and the distant nodes are nodes that have not yet been taken into account. It must be taken into account that each adjacent node saves which node was accessed. The procedure is as follows, the initial node is a node marked as visited and with zero cost, the adjacent nodes are marked to the current node and its cost is calculated from the current node, from this point the algorithm will cycle until all the nodes are marked nodes, from the marked nodes the lowest cost is taken and becomes the current node, the nodes connected to the current node that were adjacent to another node become adjacent nodes of the current node

Table 5 Distance matrix with maximum diagonal

	0	1	2		3	4	5	6	7	8
0	0	0.004754	0.008854		0.016618	0.008528	0.013256	0.005438	0.009781	0.008884
1	0.004754	0	0.004237		0.015505	0.007914	0.01495	0.00897	0.013555	0.013046
2	0.008854	0.004237	0		0.014227	0.008267	0.016203	0.01211	0.016556	0.017278
3	0.016618	0.015505	0.014227		0	0.008102	0.009119	0.014198	0.015547	0.024567
4	0.008528	0.007914	0.008267		0.008102	0	0.007995	0.007189	0.010359	0.016796
5	0.013256	0.01495	0.016203		0.009119	0.007995	0	0.008442	0.00738	0.018732
6	0.005438	0.00897	0.01211		0.014198	0.007189	0.008442	0	0.004585	0.010616
7	0.009781	0.013555	0.016556	0.010359	0.010359	0.00738	0.004585	0	0.012128	
8	0.008884	0.013046	0.017278	0.024567	0.016796	0.018732	0.010616	0.012128	0	

Fig. 6 Graph model associated with our UAV to determine the route with a specific quantity of each recyclable material

if its cost from the current node is smaller than they already have to reduce cost as is shown in next Fig. 6:

The factor model used to identify the best UAV associated with routing tasks and solid waste determination.

Assume we have a set of observed variables X1, X2,…, Xp and assume that in this underlying set* of k factors (the number of factors must be strictly less than the number of observed variables). In accordance with what we have stated in previous sections, the factors are latent variables* that explain the association between the manifested variables (in this case the X's); then, we can think of the factor model in a similar way to the line-*l regression model, in which this relationship between factors and variables is expressed, as follows

The factors f1, f2,…, fk, play the role of explanatory variables, and each one of the X's the role of response variables; the λ's are the coefficients associated to each factor, and they are called factorial loads; finally, the errors of the model are the U's. In this sense, the model is determined by the variables and not by the individuals. Factorial loads indicate the correlation between each variable and the corresponding factor; thus, a variable with a higher factorial load will be more representative of the factor. In this way, factorial loads serve to interpret the function that each variable performs in defining each of the factors.

In the factor model, f1, f2,…, fk are called common factors and u1, u2,…, up specific factors. The basic assumptions on which the model is built are as follows:

- The common factors fj $j = 1,2,\ldots, k$ are uncorrelated and have mean* zero and variance* one.
- The specific factors ui are not correlated and have mean zero and variance Ψi $i = 1, 2,\ldots, p$.
- The common factors are not correlated* with the specific factors.

Under these assumptions it is possible to decompose the variance of each of the observable variables of the model or indicator variables (Xi) into two uncorrelated components. On the one hand, the common variance, known as the communality* of the variable and which represents the variance of variable Xi that is explained by the common factors and, on the other hand, the specific variance* known as specificity and which is the variance not explained by these common factors.

The common factors and their associated characteristics (communalities, specificities, number, etc.) represent the object of interest in factor analysis.

- In order to be able to properly manage what factorial analysis is, it is important to know a series of fundamental concepts, which we present below:
- Variance: this is a measure of the dispersion of the set of scores that tells us how much the values of a test or measure are dispersed. In this way, the variance represents the arithmetic mean of the deviations from the mean squared.
- Factor: Factor is a hypothetical entity* or latent variable that is assumed to be the basis of tests, scales, and reagents. Factors also represent the underlying dimensions or constructs that summarize or justify the original set of variables under observation.
- Self-value: also known as characteristic value, latent root or characteristic root, it is a measure of the amount of variance contained in the correlation matrix such that the sum of the self-values equals the number of variables. It is calculated as the column sum of the square loads for a factor.
- Factorial load: Factorial loads are measures of the correlation between original variables and factors, and the key to understanding the nature of a specific factor (Hair and Cools 2004, p. 769).
- Communality: These are the sums of squares of the factorial loads of a test or variable; communality refers to its variance from the common factor.
- Rotation: It is a process of manipulation or adjustment of the factor axes, in this way what we want to achieve is a simpler and basically more significant factorial solution. In this way, factor rotation aims to transform the initial matrix into one that is easier to interpret.

Types of rotation: The two most common types of rotation, as pointed out by Kerlinger and Lee (2002), are the so-called

- orthogonal
- oblique.

Orthogonal rotations maintain factor independence, i.e. the angles between the axes are kept at 90°. Given a factorial matrix A, we want to find an orthogonal matrix

T so that the new factorial matrix $B = AT$ can define factors that have a simpler structure. An analytical criterion considers the function:

where Y is a parameter such that $0 < y \leq 1$.

The methods available within the orthogonal rotation are described and presented below:

- Varimax: This method is in charge of minimizing the number of variables that have high saturations in each factor

If $y = 1$ minimize G equivalent to maximize the sum of the variances of the squares of the saturation coefficients of each column of A. Similarly if we consider a2ij/h2i, the rotation is called normalized Varimax

- Quartimax: This method takes care of minimizing the number of factors needed to explain each variable.

If y=0 minimize G equivalent to maximize the variance of the squares of the p.m saturation coefficients. If each saturation a2ij is divided by the communality, i.e. a2ij/h2i is considered, the rotation is called normalized quartimax.

- Equamax: This method is a combination of the previous methods, so it minimizes the number of variables that saturate high in a factor as the number of factors needed to explain a variable

Oblique rotations are rotations in which the factorial axes allow acute obtuse angles to be formed. Factor rotation calculated so that the extracted factors are correlated (Hair and Cools 2004, p. 785).

The methods available within the oblique rotation are now described.

The new factorial matrix* $P = AL$ defines oblique factors that have a simpler structure. An analytical criterion on the factor structure matrix Q considers the function:

where and is a parameter such that 0 _< and < _ 1: There are three especially interesting criteria, which have an interpretation similar to the orthogonal case and which can also be formulated, more adequately, by dividing by the communalities.

This model postulates that the observable variables depend on some correlating factors F' 1…, F'm and P unique factors. Thus, for each variable Xi

1. Oblique factorial model matrix P = (pij)

   The saturation of the variable Xi by a factor of F0 j is very high:

2. Matrix of correlations between oblique factors

   ϕ = (p'ij) where p 'ij = cor(F' i; F' j):

3. Oblique factorial structure (reference structure) Q = (qij) where qij = cor(Xi; F0 j):

If we indicate F0 = (F'1 … F'm)' and write the model in matrix form X = PF0 + DU;

We easily test the relationship between the three P matrices, ϕy Q Q = Pϕ;
And Thurstone's version of the theorem for correlated factors
$R = P\phi P'' + D2$:
If the factors are orthogonal, the factorial model coincides with the factorial structure and we have
$P = Q$; $\phi = Im$:

- Direct Oblimin: This method is used when delta equals zero.
- Delta: The value of delta allows to control the degree of obliquity that can be reached by the factors of the solution.
- Promax: Oblique rotation that allows the factors to be correlated

It is important to verify if each one of the variables included in the analysis are explained acceptably by the model, this can be achieved using the estimation* of the communalities, since the communality represents the proportion of the indicator variance that is explained by the common factors of the model.

Once the factorial analysis is done, the final step is to assign the factorial scores to each individual in the sample. This score construction generates a new variable for each factor in the model

We proceeded to use this analysis because we seek to avoid all predictive methods, such as: discriminant analysis, analysis based on main components, among others. In other words, we do not seek to predict which variables will be affected in the future, but rather the main objective is to identify the most important variable or variables that are causing the greatest impact in order to obtain errors when making a change in the design level.

7 Conclusion

Being a project that improves a route, not only can it be used for schools, it can also be used for parcel companies, delivery companies or transport companies. Using a mobile device associated with a more detailed framework could even determine specific days with more transported objects associated with each student, as can be seen in Fig. 7.

An aspect of utmost importance will be to adapt in time, the specifications of each individual that allows it to be competitive, helping it with mobility in a Smart City, as can be seen in Fig. 8.

The route is formed in such a way that the next stop number is the combination with the previous stop number, in this way it is possible to know which is the point of start and departure, in order to perform the sum of the distance and corroborate which is the optimal route. An important aspect is to solve with less time the resolution of a new neighborhood with points of waste collecting with different locations and determine the correct locations of each stop of this route, as you can see in Fig. 9.

Fig. 7 Intelligent design of a mobile application based on ubiquitous computing that would allow each person to send information associated with the organization of waste transport and online specifications of each particular situation, of the dimensions and care of their respective trash classified

Fig. 8 The OECD considers that by 2027, seven of the 12 components of a Smart City will be associated with improving school performance, including student classification of trash

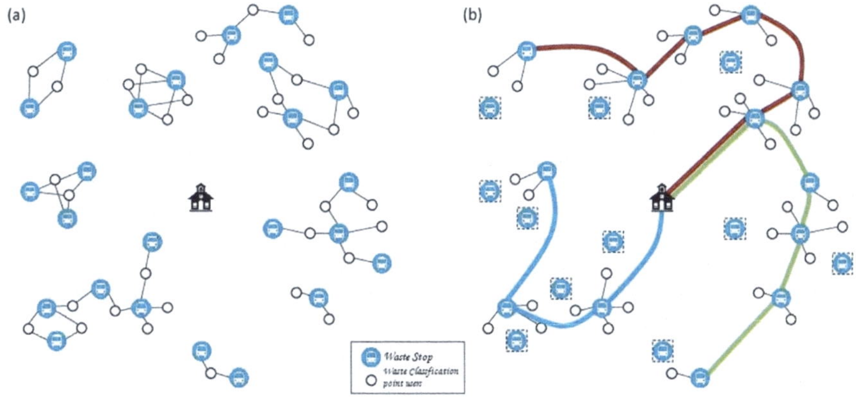

Fig. 9 Description of each group of students and their associated stops (**a**) and then the execution of waiting times (**b**)

References

1. Vethaak AD, Leslie HA (2016) Plastic debris is a human health issue
2. Hoornweg D, Bhada-Tata P et al (2012) What a waste: a global review of solid waste management, vol 15
3. Folianto F, Low YS, Yeow WL (2015) Smartbin: smart waste management system. In: 2015 IEEE tenth international conference on intelligent sensors, sensor networks and information processing (ISSNIP), Singapore, pp 1–2, https://doi.org/10.1109/issnip.2015.7106974
4. Naroznova I, Møller J, Scheutz C (2016) Global warming potential of material fractions occurring in source-separated organic household waste treated by anaerobic digestion or incineration under different framework conditions. Waste Manag 58:397–407
5. Zaman AU (2016) A comprehensive study of the environmental and economic benefits of resource recovery from global waste management systems. J Clean Prod 124:41–50
6. Mourad M (2016) Recycling, recovering and preventing food waste: competing solutions for food systems sustainability in the united states and france. J Clean Prod 126:461–477
7. White G, Clarke S (2020) Urban intelligence with deep edges. IEEE Access 8:7518–7530
8. Borozdukhin A, Dolinina O, Pechenkin V (2016) Approach to the garbage collection in the 'smart clean city' project. In: Proceedings of IEEE International Colloquium on Information Science and Technology (CiSt), October 2016, pp 918–922
9. LA (2017) Clean Streets LA Challenge. http://cleanstreetsla.com/clean-streets-challenge/. Accessed 5 Oct 2017
10. López I, Gutiérrez V, Collantes F, Gil D, Revilla R, Gil JL (2017) Developing an indicators plan and software for evaluating street cleanliness and waste collection services. J Urban Manag 6(2):66–79
11. Ochoa-Ruiz G, Angulo-Murillo AA, Ochoa-Zezzatti A, Aguilar-Lobo LM, Vega-Fernández JA, Natraj S (2020) An asphalt damage dataset and detection system based on RetinaNet for road conditions assessment. Appl Sci 10:3974
12. Adarsh P, Rathi P, Kumar M (2020) YOLO v3-tiny: object detection and recognition using one stage improved model. In: 2020 6th international conference on advanced computing and communication systems (ICACCS), Coimbatore, India, pp 687–694. https://doi.org/10.1109/icaccs48705.2020.9074315
13. Lin T, Goyal P, Girshick RB, He K, Dollár P (2017) Focal loss for dense object detection. http://arxiv.org/abs/1708.02002
14. Liu L, Ouyang W, Wang X, Fieguth P, Liu X, Pietikäinen M (2016) Deep learning for generic object detection: a survey. arXiv:1809.02165v2
15. Huang J, Rathod V, Sun C, Zhu M, Korattikara A, Fathi A, Fischer I, Wojna Z, Song Y, Guadarrama S, Murphy K (2018) Speed/accuracy trade-offs for modern convolutional object detectors. arXiv:1611.10012v3
16. Howard AG, Zhu M, Chen B, Kalenichenko D, Wang W, Weyand T, Andreetto M, Adam H (2017) Mobilenets: efficient convolutional neural networks for mobile vision applications. arXiv:1704.04861
17. Krizhevsky A, Sutskever I, Hinton GE (2012) ImageNet classification with deep convolutional neural networks. In: Proceedings of neural information processing system conference, Lake Tahoe, CA, USA, December 2012
18. Awe O, Mengistu R, Sreedhar V (2017) Smart trashnet: waste localization and classification. CS229 Project Report
19. Thung G, Yang M (2016) Classification of trash for recyclability status
20. Rad MS, Kaenel AV, Droux A et al (2017) A computer vision system to localize and classify wastes on the streets. In Computer vision system. Springer, Cham, p 195204
21. Mittal G, Yagnik KB, Garg M, Krishnan NC (2016) Spotgarbage: smartphone app to detect garbage using deep learning. In: Proceedings of the 2016 ACM international joint conference on pervasive and ubiquitous computing, ser. UbiComp 16. ACM, New York, NY, USA, p 940945

22. Gyawali D, Regmi A, Shakya A, Gautam A, Shrestha S (2004) Comparative analysis of multiple deep CNN models for waste classification. https://arxiv.org/abs/2004.02168
23. Cartucho J (2020) Mean average precision. https://github.com/Cartucho/mAP

Geo-Referenced Correlation for a Fire in a Smart City Urban Forest Using Hybrid Drone Data and Satellite Images

Alberto Ochoa-Zezzatti, Gilberto Ochoa-Ruiz, and Lina Maria Aguilar-Lobo

Abstract The current technology used to detect forest fires is: weather stations and satellite images, the latter is a great option, but, from space, fire detection has limitations to be large enough to be seen from the orbital altitude. That is why a drone can be used to monitor a forest for signs of fire before the satellite observes it, but a single drone to cover a large area of hectares is not optimal. A swarm of drones with automatic organization capability, equipped with atmospheric sensors that detect fire danger conditions or even a fire at an early stage, should be used to optimize coverage of the area. Implementing a heuristic algorithm for the automatic organization of the drone swarm applicable for the detection and warning of forest fires is a task still unfinished in the literature. Forest fires are a great environmental problem because they are detected mainly until they are very advanced and some square kilometers have already been burned. When they are detected in the advanced stage, the fire will be difficult to contain.

Keywords Data mining · Clúster of drones · Fires in a urban forest · Mobile devices

1 Introduction

According to Sethi [1], preventing deforestation and promoting afforestation have often been cited as strategies to curb global warming.

A. Ochoa-Zezzatti
Universidad Autónoma de Ciudad Juárez, Ciudad Juárez, Mexico

G. Ochoa-Ruiz (✉)
Tecnologico de Monterrey, Escuela de Ingenieria y Ciencias, Zapopan, Mexico
e-mail: gilberto.ochoa@tec.mx

L. M. Aguilar-Lobo
Universidad Autónoma de Guadalajara, Zapopan, Mexico

© The Author(s), under exclusive license to Springer Nature Switzerland AG 2021
A. Ochoa-Zezzatti et al. (eds.), *Technological and Industrial Applications Associated with Intelligent Logistics*, Lecture Notes in Intelligent Transportation and Infrastructure,
https://doi.org/10.1007/978-3-030-68655-0_28

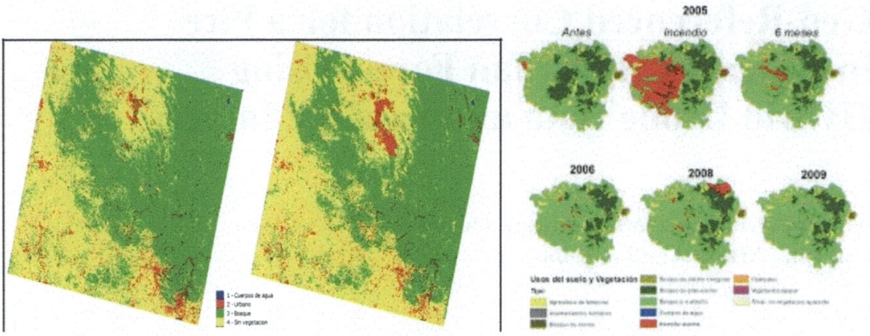

Fig. 1 Comparative map of the land use classification map for the years 2015 and 2019, and the past

In order to carry out the simulation with our proposal, we analyzed an endemic species of spotted salamander that inhabits in "La Primavera" forest. The high incidence of forest fires reduces the opportunity for ecosystems to recover in time and, therefore, the importance of this research is the use of drones that can help reduce this problem, as can be seen in Fig. 1.

2 Implementation of the Intelligent Application

We propose the use of a heuristic mechanism for the automatic organization in swarms of drones in a society of the type: "Smart City". According to the National Center for Forest Fire Control (Centro Nacional de Control de Incendios Forestales), from January 1 to October 19, 2017, there were 808 fires in Chihuahua that burned 37, 557.47 Hectares (Ha) [2]. Many parts of forest "La Primavera" that historically have the largest affected area is near to Zapopan [3]. For a drone, it is necessary to have a mechanism to carry out these activities efficiently and, mainly, to survey the area to optimize time. And the technological resources in Fig. 2 show the need to coordinate the drone swarm.

The detection of forest fires is in an initial state to implement a heuristic algorithm focused on the automatic organization of unmanned swarms, which poses some challenges:

(A) Drone technology. Current drone technology has a limited capacity to take long-range flights due to battery life, which is consumed mainly by the flight of drones. Adding a sensor data logger and transmitting data to a ground station represents a burden for battery life that must be taken into account. The sensor and data transmission electronics do not represent a load on the air vehicle.

(B) Radio signal using a mobile device. The data from the sensor data logging will be transmitted to the ground station. Due to the geography of the forest, not all areas are covered by radio signals. This can be solved by using some

Fig. 2 Implementation of a proposed solution to adapt a drone to prevent forest fires for a longer period of time in the forests of Chihuahua

drone swarm agents that serve as signal relays. This may present a challenge to the self-organizing algorithm.

(C) Altitude. Current drone technology is limited in the altitude it could reach depending on the orography where it will be implemented. This challenge is outside the scope of the self-organizing algorithm, but will be taken into account because the spherical coordinates will be used. Since we will be working with a very large area of forest, this area must be delimited by areas, which is the basis on which the algorithm will define the priorities that the drone swarm agents must visit.

The area to be monitored is considered in the algorithm to be used. In the following section, we describe which are the already known algorithms that will be referred to.

3 Test Development of a Mobile Device and Our Proposal Heterogeneous Drone Cluster

We considered monitoring a large geographical area using a swarm of drones to prevent forest fires. The area to be monitored is divided into well identified sub-areas. A swarm of drones is composed of a set of heterogeneous drones that are at a starting point [4–6]. Therefore, the problem we address is to create a program that contains the assignment of drones to geographic sub-areas for monitoring and detecting forest fires, so that the algorithm completion time is minimized, as in Fig. 3. A model planning system for organizing the visit of drones to geographic

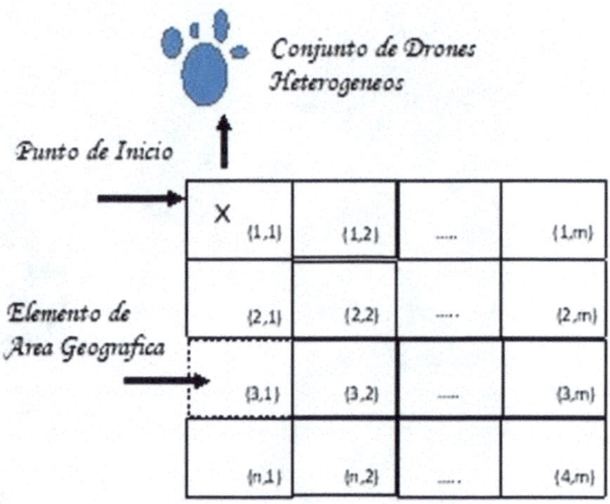

Fig. 3 Representation of the drone cluster coverage problem for a geographic area

sub-areas consists of the following elements: the geographic area, the drone swarm, and an objective function for programming the drone swarm, which is relevant even in satellite detection as in Fig. 4.

The Heterogeneous Drone Cluster (HDS) can be represented by a HDS DAG: (D, E). D represents the set of heterogeneous drones that make up the swarm. E is the set of directed arcs that connect different pairs of drones, so e (di, dj) denotes a precedence that indicates that dron dj cannot begin its mission until di completes its mission. For convenience, Pred (di) denotes the subset of drones directly preceding di and Succ (di) denotes the subset of drones directly following di. The input drones are those with: |

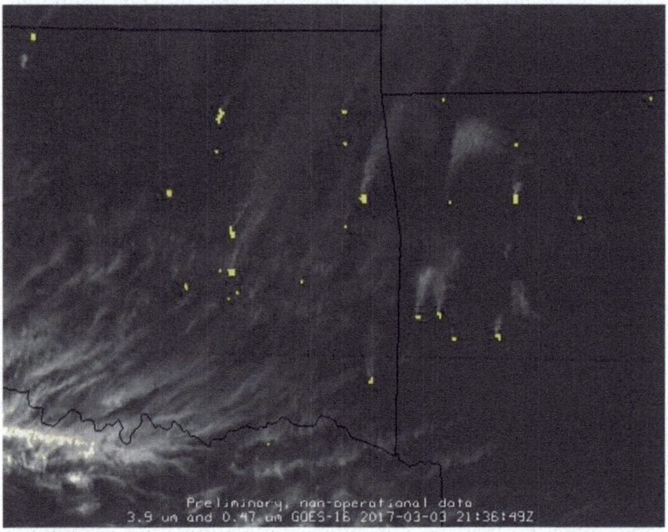

Fig. 4 Satellite detection of forest fires

Pred (di) | = 0 and the output drone is those with: | Succ (di) | = 0. For simplicity, in these cases we consider the use of dummy tasks so that the DAG contains only one input and one output drone. Recalling that drones are heterogeneous, we represent the estimated time of flight from base at a0 with EFT= $D \times A \rightarrow$ Int, where EFT (di, aj) denotes the time for a drone di to reach a geographic subarea aj. To simplify, we consider that the flight time to return to base in a0 is the same as the time to reach a particular area from a0. A drone can be assigned to different missions, but can only perform one mission at a time. Therefore, at time t we consider: avail: $D \rightarrow [0\ldots1]$, which captures the availability of each drone at time t. It should be noted that the mission time of a particular drone is given when it is running at full availability. W (di) denotes the time for a drone to perform a certain job once it reaches a secondary geographic area. The configuration (di) denotes the setup time for a drone to start a new mission. We assume that the flight and setup time information is provided in standard time units, compatible with our drone performance measures. Scheduling drones in geographic areas requires consideration of four events: (a) The moment the drone begins its mission. (b) The time for a drone to reach a particular geographic area. (c) The time for a drone to perform certain work once it reaches its geographic area, and (d) The time for a drone to return to its assigned base. Therefore, we must first predict the time a particular drone moves away from a0 to perform its mission in a particular sub-area and the time the drone returns to base. We must first define two mutually referential quantities. EDT (di, am) is the Estimated Drone Departure Time, the values of: (di, a, am) are calculated by Eq. 1:

$$\mathrm{EDT}(\mathrm{di}, \mathrm{am}) = \mathrm{Configuration}(\mathrm{di}) + \mathrm{maxdj} \in \mathrm{Pred}(\mathrm{di})\{\mathrm{ERT}(\mathrm{dj}, \mathrm{a0})\} \tag{1}$$

This is calculated by: ERT (di, a0), which denotes the estimated Return Time from dron di to base located at a0 and is calculated by Eq. 2:

$$\mathrm{ERT}(\mathrm{dj}, \mathrm{a0}) = \mathrm{EDT}(\mathrm{di}, \mathrm{am}) + (2 * \mathrm{EFT}(\mathrm{di}, \mathrm{am})) + \mathrm{W}(\mathrm{dj}) \tag{2}$$

The configuration (di) is the preparation time for a drone to start a new mission. It is added to the result of the maximum block in Eq. (3), which returns the estimated maximum return time at which each drone in Pred (di) returns to base. Once all drones have been programmed, the estimated program completion time is determined by the estimated return time of the departure drone. The estimated completion time is also known as the program schedule.

$$\mathrm{ERT} = (\mathrm{doutput},\ \mathrm{am} * \mathrm{n}). \tag{3}$$

Figure 5 shows the conceptual diagram of the research carried out.

The objective function of the drone programming is to create a program that contains the assignment of drones to geographical sub-areas, so that their scope is minimized, similar to what was proposed in [5–7].

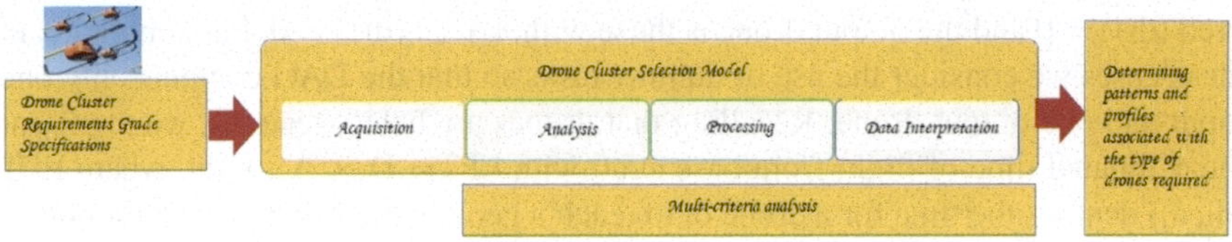

Fig. 5 Proposed solution of our research

4 DERT Algorithm

The DERT algorithm is based on the well-known list scheduling approach. Our interest in this approach is to explore low-computer complexity strategies and apply them to prevent and fight forest fires with the use of drones. Therefore, the DERT algorithm basically consists of two phases: The drone prioritization phase in which a priority ranking is established for each drone [8]. The geographic sub-area assignment phase where each drone is assigned to that geographic sub-area that optimizes a predefined cost function The performance of a drone cluster is linked to the configuration of the application domain, the better adapted, robust and validated, the larger the area it can cover to prevent forest fires. The DERT algorithm is shown in Fig. 6.

4.1 The Drone Prioritization Phase

We use DRu (di), an ascending range defined as the length of the critical path from dron di to the output drone. DRu (di) is calculated recursively in Eq. 4:

1. Set the drone flying time.
2. Set the drone setup time.
3. Set the drone work time.
4. Calculate DR_u for each dron by traversing the graph from the exit node to the entry node and keep the values in *L*.
5. Sort the drones in *L* in descending order of DR_u values.
6. Create a list *LSA* with the sub-areas composing A.
7. **while** there are unvisited areas in ***LSA*** do
8. Select the first sub-area a_m from *LSA*
9. **for** each available dron d_i ($avail(d_i)=1$) in L **do**
10. Compute *EDT* (d_i, a_m) value.
11. Compute *ERT* (d_i, a_0) value.
12. Assign dron d_i to the sub-area a_m that minimizes *ERT* of d_i
13. Set *avail* (d_i) = 0 from the time between *EDT* and *ERT*.
14. **end while**

Fig. 6 The DERT algorithm

$$\mathrm{DRu(di) = avg(FTi) + max\ vj \in\ Succ(vi)(DRu(vj))} \tag{4}$$

where: avg (FTi) is the average visit time for a drone (di) in all subareas, as in Eq. 5.

$$\mathrm{avg(FT(di))} = \sum\nolimits_{\mathrm{k=0}}^{\mathrm{nm}} \frac{\mathrm{(di, ak)}}{\mathrm{n, m}} \tag{5}$$

4.2 The Assignment Phase of a Sub-area

The DERT algorithm considers that a drone can be assigned to several missions, but can only be performed once at a time. A mission consists of leaving the base a0 to an area assigned am, perform a job once it reaches am and return to a0. In our case, the work that a drone does in a particular area is monitoring. The assignment phase for each drone to a geographic sub-area that offers the minimum estimated return time, takes the time complexity: O (d × e) for d drones and precedence.

5 Simulation

This research still needs to be complemented by further simulation and testing before implementing it in a real environment. In addition to the engineering challenges of a physical drone swarm, the developed algorithm will be simulated for tuning and validation as in Fig. 7.

For this, the forest and forest fires will be modeled to replicate as close as possible to the ecosystem where these drones will eventually be flying, see Fig. 8.

The first step is to get the coordinates where the user is located, and then sent them to the server, which calculates the number of traffic incidents within the closest radio specified in the configuration of the mobile device, the results are processed to determine a numerical index which will then be represented with a color for better end-user visualization. All this information is obtained from a transit database and analyzed with Data Mining. After the values are sent to the mobile device and interpreted to construct an URL which is then sent to the Google Maps API. Then a map will be ultimately display with the area indicators. The Fig. 9 represents the operation of our intelligent mobile device application, which is divided into two parts, the mobile application and Web services hosted on the server. The database, where the incidences data is stored, has 3 tables.

The structure of the database has a table called accidents, where each record contains the geographic coordinates expressed in decimal degrees, which determine the specific position of the accidents [1] and is visualized in our application as is analyzed in [7, 9–12]. To locate nearby points within a radius requires four parameters: latitude, longitude, and distance equatorial radius, the latter determines the

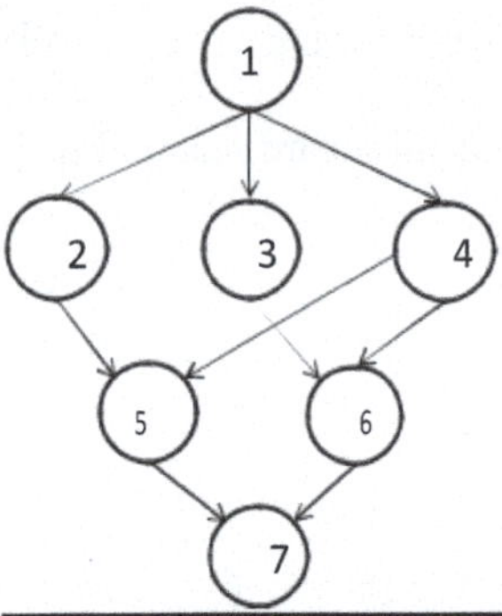

a1	a2	a3	a4
a5	a6	a7	a8
a9	a10	a11	a12

Dron	a1	a2	a3	a4	a5	a6	a7	a8	a9	a10	a11	12
1	7	3	9	4	9	4	3	3	1	2	9	9
2	4	7	18	3	2	7	7	2	7	8	7	5
3	8	2	19	4	7	1	2	8	4	4	4	1
4	9	9	17	3	2	7	9	1	2	2	3	5
5	4	8	10	5	9	9	8	8	9	1	4	9
6	5	5	9	8	2	3	5	4	5	8	8	4
7	2	3	11	2	3	3	3	7	3	2	7	3
8	7	5	14	4	7	5	5	5	5	3	3	4
9	3	4	20	3	4	4	4	4	2	4	1	4
10	1	5	2	7	2	5	5	5	10	7	3	2

Fig. 7 Proposed architecture for our drone cluster simulation model

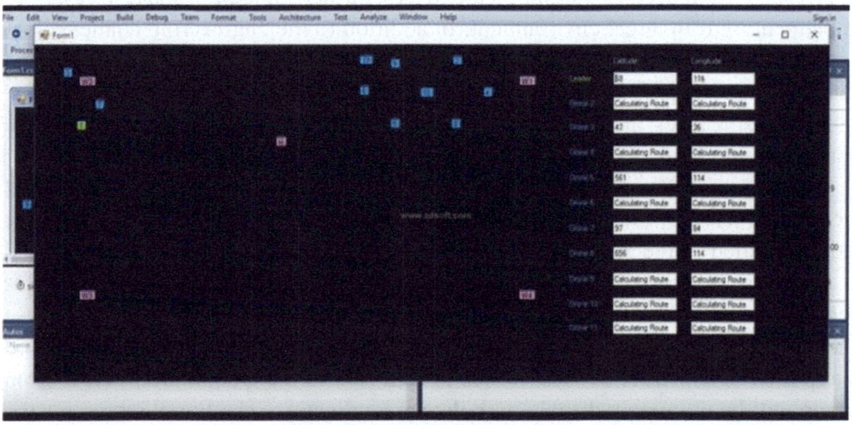

Fig. 8 Drone cluster simulation, using our smart tool for monitoring

maximum distance the search radius, these parameters are part of a method based on havesine formula, which is used to calculate great circle distances between two points on a sphere. Haversine function that is different from haversine formula is given by the function of semiversine where:

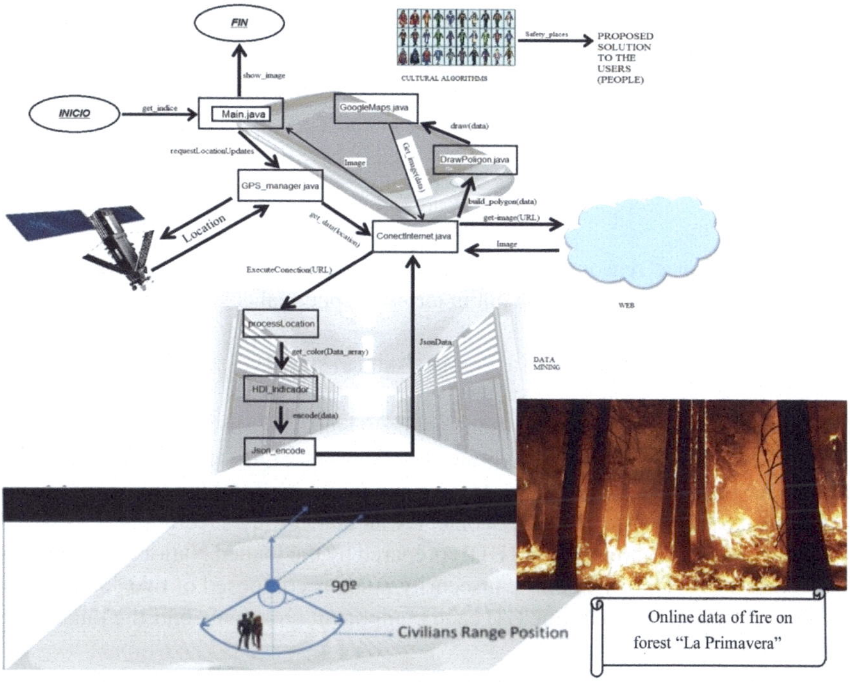

Fig. 9 Functional diagram of our Intelligent application including hybrid methodology

$$\text{semiversin}(\theta) = \text{haversin}(\theta) = \frac{\text{versin}(\theta)}{2} = \sin^2\frac{\theta}{2}$$

Haversine formula for any two points on a sphere:
d is the distance between two points (along a great circle of the sphere),
R is the radius of the sphere,
$\varphi 1$ is the latitude of point 1,
$\varphi 2$ is the latitude of point 2, and
$\Delta\lambda$ is the length difference

$$\text{haversine}\frac{d}{R} = \text{haversin}(\varphi 1 - \varphi 2) + \cos(\varphi 1)\cos(\varphi 2)\text{haversine}(\Delta\lambda)$$

Solving haversine formula can calculate the distance, either by applying the function inverse haversine or by using the arcsine where:

$$\text{h} = \text{haversine}\left(\frac{\text{h}}{\text{R}}\right)$$

$$\text{d} = \text{R haversine}^{-1}(\text{h}) = 2\text{R} \arcsin(\sqrt{\text{h}})$$

With this formula we can construct the following instructions, taking into account the role of semiversine we have:

$$h = \sin^2\left(\frac{\varphi 1 - \varphi 2}{2}\right) + \cos(\varphi 1)\cdot\cos(\varphi 2)\cdot\sin^2\left(\frac{\Delta\lambda}{2}\right)$$

$$d = 2R\arcsin(\sqrt{h})$$

For performance reasons when haversine implements the function using SQL statements used a similar formula but in terms of spherical cosine law called cosine.

$$\cos(c) = \cos(a)\cos(b) + \sin(a)\sin(b)\cos(c)$$

This is only an approximation when applied to land, since the Earth is not a perfect sphere radius, as we approach the poles it varies. The SQL statement based on the formula haversine find all locations that are within the range of the variable $ radius and groups them by type of traffic accident. Calculate of the uncertainty is modeled on the human development index (HDI) prepared by the United Nations Program for Development (PNDU). The traffic insecurity index is composed of two components Number of Events (Q) and time (H) each component represents half the total value of the index.

Each expressed with a value from 0 to 1 for which we use the following general formula:

$$\text{Component index} = \frac{(\text{Value} - \text{minimum})}{\text{Maximum} - \text{minimum}}$$

The following point to take into account is the modification of the position of the civilians when they converge with the player once obtained with the previous formula of distance. If this position is not modified, the civilians tried to converge in our position being caused a graphical collision. In order to be able to resist this, function of control is necessary to add one, so that it allows us to locate to the civilians in a radius very near the position of the player. The following figure illustrates as they must comply with the civilians given the position of the user.

In order to obtain the previous thing, the following equations were used (6) and (7):

$$\left\{\left[\left(y - y_p\right) - \left(x - x_p\right)\tan\left(\frac{\theta}{2}\right)\right]\right\} < 0 \tag{6}$$

$$\left\{\left[\left(y - y_p\right) + \left(x - x_p\right)\tan\left(\frac{\theta}{2}\right)\right]\right\} < 0 \tag{7}$$

where:

x and y are each civilian's position on the plane. x_p and y_p are the player's current position on the plane.

θ is the angle where an arc are described to place the civilians. This cold is taken as a constant with a value of 90º.

6 Conclusions and Future Research

A relevant aspect of this research is the early detection of recurrent fires in areas of high forest potential and therefore the impact on the population that depends on this type of wood material. A relevant map of the drought and therefore of the possible forest fires can be seen in Fig. 10.

Illegal tree cutting detection is one of the future works to be developed in the state of Chihuahua, one of the following works related to the heuristic algorithm presented for the automatic organization of drone swarms is to monitor forests to detect illegal tree cutting. Not only inspecting forests with UAVs, but equipping these swarm agents with image recognition based on an automatic learning model to be developed. The authors of this research are now beginning to get involved in this

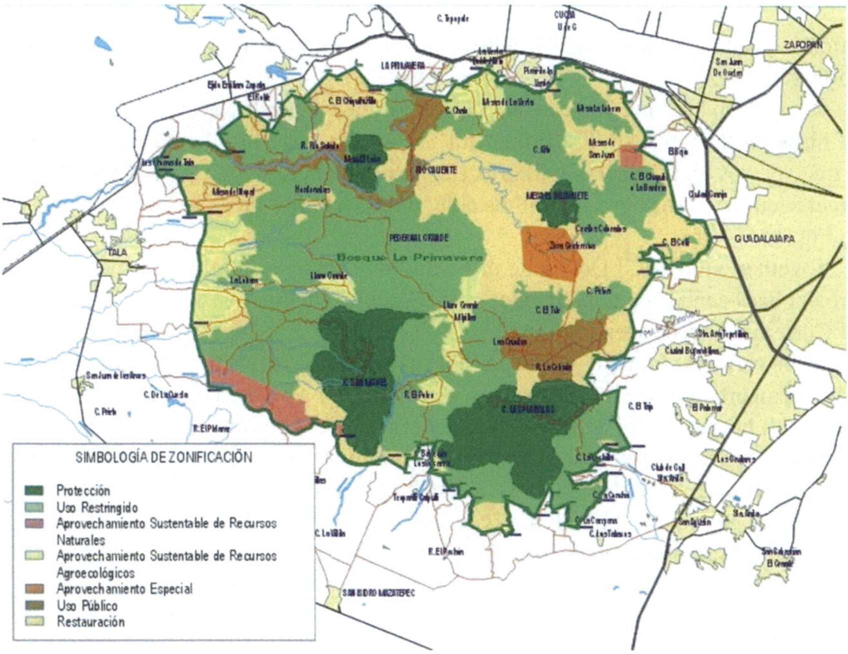

Fig. 10 Map associated with the drought in the spring forest and its correlation with forest fires

Fig. 11 A portion of illegally logged timber identified by an automatic learning model

effort, as can be seen in Fig. 11, which shows the stratification of illegal logging in forests.

Using deep learning you can quickly determine when a fire starts and with photos of birds in the forest of "La Primavera" what effect the fire damage will have on a tree species and how this will affect its population ecology by determining with great accuracy if that species will be able to recover after those fires, as you can see in Fig. 12.

Together with UACJ Doctorate in Technology students, attached to the UABC, a project has been set up to help redefine the Vaquita's habitat on the coasts of the Sea of Cortez, as can be seen in Fig. 13.

In the very near future it will be possible to determine in real time the side effects of a fire in an urban forest in a Smart City, especially for a particular species of tree bird for it will be used mobile devices, social networks and complementary information from drones and satellite information, however this type of technology must increase its use and make it scalable before in 2037 half of the 11,787 species of birds are doomed to extinction. In addition is possible used this technology associated with Intelligent Smart Farming as is visible on Fig. 14.

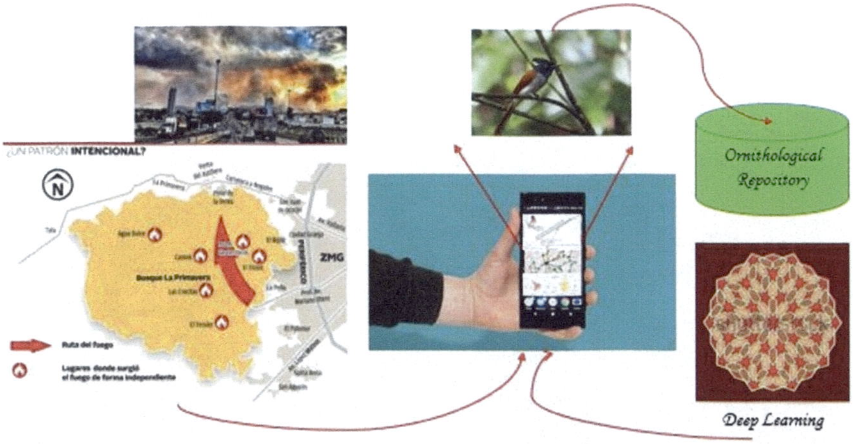

Fig. 12 Identifiably representation of a special issue of arboreal bird and calculate the effects of a fire in "La Primavera" urban forest implemented on an intelligent dashboard associated with preservation of bird's species

Fig. 13 Future project linked to the redefinition of the Vaquita's habitat through the use of a cluster of specialized drones

Fig. 14 A Smart farming model using intelligent cluster of drones

References

1. Sethi V (1986) Natural language interfaces to databases: MSI impact, and survey of their use and importance. University of Pittsburgh
2. Malhi Y, Roberts JT, Betts RA, Killeen TJ, Li W, Nobre CA (2008) Climate change, deforestation, and the fate of the Amazon. Science 319(5860):169–172
3. ClimaTemps (2017) Chihuahua climate & temperature, November 2017, de ClimaTemps Sitio web. http://www.chihuahua.climatemps.com/
4. Köppen climate classification (2017) Köppen climate classification, November 2017, de Köppen climate classification Sitio web. http://hanschen.org/koppen/
5. Gómez C, Cruz-Reyes L, Rivera G, González J, Quiroz-Castellanos M (2012) Impact of initial tuning for algorithm that solve query routing. In: Management intelligent systems, pp 315–323
6. Gómez C, Cruz-Reyes L, Schaeffer E, Meza E, Rivera G (2010) Local survival rule for steer an adaptive ant-colony algorithm in complex systems. In: Soft computing for recognition based on biometrics. Springer, Berlin, pp 245–265
7. Gómez C, Cruz-Reyes L, Schaeffer E, Meza E, Rivera G (2011) Adaptive ant-colony algorithm for semantic query routing. J Autom Mob Robot Intell Syst 5:85–94
8. Bosch (2019). https://www.bosch-mobility-solutions.com/media/global/highlights/connected mobility/smart%20agriculture/bosch_connected_mobility_smart_agriculture_text_image_n evonex.jpg. Accessed 4 Dec 2019
9. Martınez-Carranza J, Marquez F, Garcia EO, Munoz-Meléndez A, Mayol-Cuevas W (2015) On combining wearable sensors and visual SLAM for remote controlling of low cost micro aerial vehicles. In: IEEE 3rd workshop on research, education and development of unmanned aerial
10. Sudhakar S, Vijayakumar V, Sathiya Kumar C, Priya V, Logesh Ravi, Subramaniyaswamy V (2020) Unmanned Aerial Vehicle (UAV) based Forest Fire Detection and monitoring for reducing false alarms in forest-fires. Comput Commun 149: 1–16
11. Mahmoud MAI, Ren H (2019) Forest fire detection and identification using image processing and SVM. JIPS 15(1):159–168
12. Indu V, Thampi SM (2019) A nature - inspired approach based on forest fire model for modeling rumor propagation in social networks. J Netw Comput Appl 125:28–41

Evaluation of Drones for Inspection and Control in Industry 4.0

Diego Moreno-Jacobo, Gustavo Toledo-Nin, Alberto Ochoa-Zezzatti, Vianey Torres, and Fernando Estrada-Otero

Abstract This research presents a brief idea of drones, their technology and can be used for different purposes. New technologies such as machine learning, Artificial Intelligence (AI), the Industrial Internet of Things (IIoT) and drones are transforming the workforce. Major aim of this project is for surveillance, inspection. The most important feature of the drone is the transmission of images and video in real time (FPV). They highlighted the safety areas, as risky tasks such as thermometry and specific tool maintenance can now be performed without exposing workers. Finally, more than 40 cases of use have been validated, mainly for the security of assets, for the areas of exploration, projects, engineering, construction, planning and logistics, maintenance, quality and environment, production and above all, personnel safety.

Keywords Surveillance drone · FPV drone · Energy efficient drone · Less noise · Industry 4.0

1 Introduction

The 2016 was a drone's year, a true revolution occurred in terms of the hardware and software used in both the manufacturing process and in controlling drones. This has made these devices easier to maneuver, able to fly more, be safer and able to obtain video footage and high-resolution photos at a professional level. In recent research it is clearly observed that drones are about to create a revolution in human life and industry. The most important aspect is that it can be used for effective surveillance at places where human being can't reach. To reduce the cost of the drones, Low cost avionics System Prototypes are made. It makes the structure light as well as strong.

The paper deals with a drone which has its location being directed by GPS giving its exact location. The most important aspect of this paper deals with the detection of unidentified obstacles which tend to create disturbance in the normal environment. Capturing and reporting all minute details as the camera associated with it moves as

D. Moreno-Jacobo · G. Toledo-Nin · A. Ochoa-Zezzatti (✉) · V. Torres · F. Estrada-Otero
Juarez City University, Ciudad Juárez, Mexico
e-mail: alberto.ochoa@uacj.mx

© The Author(s), under exclusive license to Springer Nature Switzerland AG 2021

A. Ochoa-Zezzatti et al. (eds.), *Technological and Industrial Applications Associated with Intelligent Logistics*, Lecture Notes in Intelligent Transportation and Infrastructure,
https://doi.org/10.1007/978-3-030-68655-0_29

per our direction. Using of GPS for automatic mission planning, the stabilization of the drone for proper surveillance and the FPV part are the advantageous where we can control it through the remote.

In our capture of models and alternatives for our company or industry we are going to raise a comparative of certain models of Drones with certain characteristics and using TOPSIS we will be able to arrive at our ideal solution. The TOPSIS method has been selected because it is one of the most widespread and applied methods of operational research in the industry. It was developed by Hwang and Yoon (Technique for Order Preference by Similarity to Ideal Solution, 1995) based on the concept that it is desirable for a given alternative to be located at the shortest distance from an ideal positive solution and at the longest distance from an ideal negative solution.

The Ministry of Environment and Water in Dubai has started using drones for monitoring the work of crushers and quarries. Nokia plans to use flying drones as a public safety system for cities. PrecisionHawk [1] has been offering remote sensing and data processing services using drones for various applications such as infrastructure monitoring and search/rescue. Recent advances in sensor and embedded device technologies are also pushing the pervasive data acquisition and processing capabilities in different city infrastructures such as roads, traffic signals, sidewalks and bridges to monitor.

2 Models of the Interface

Each of the screen will have relationship with the different module of the intelligent dashboard. The next window will show the first module called "Agregar" in this module you can add an alternative by the name of the product and 5 of their characteristics included the price, this is with the purpose to add the most characteristics for compare the alternatives (Fig. 1).

The variables of each Self. Entry is declared as StringVar(). When you introduce the information, they go to the DataBase.

In the second module called "Consultar" you can view all the alternatives saved in the BD separated by categories. If you need to check any data these modules are the indicated (Fig. 2).

The third module is called "Modificar/Actualizar" in this module you can modify or update the attributes of you alternatives only if you need to make some corrections before the program create the TOPSIS method. The correct form to use this module you need to write the code generated when you add any alternative on the BD, when you write the code all the fields gonna be full with the information (Fig. 3).

Fourth module is very simple due you only need to write the code of the alternative to eliminate by 1. Be careful of digit the correct code (Fig. 4).

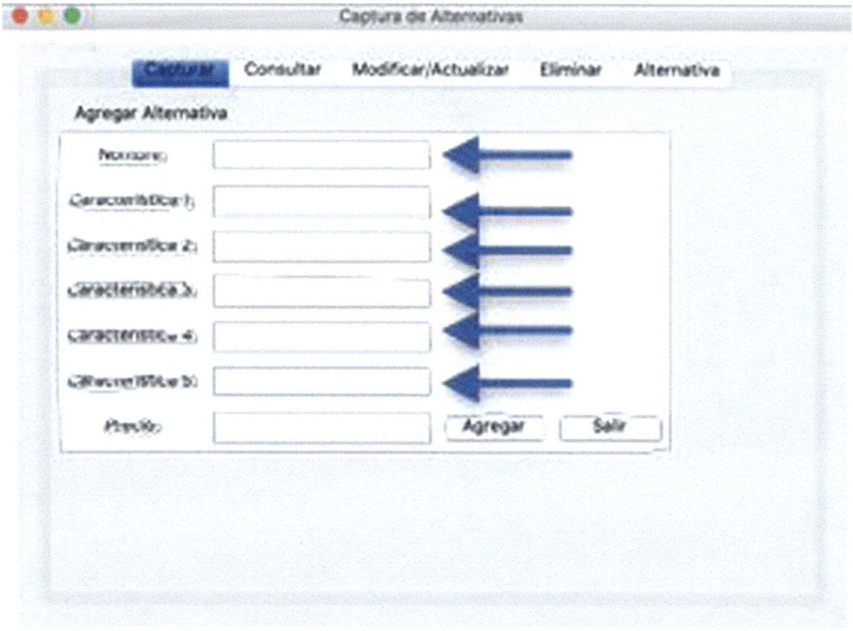

Fig. 1 Initial module of our intelligent dashboard

3 Drone Applications

Drones can be used for many applications but we will focus mainly on those in Industry 4.0. We will analyze Thermography, Photogrammetry, Scanning (point cloud), Inspection of power lines, Inspection of infrastructure, Inspection of wind turbines, Inspection of chimneys and gas leaks, Inspection of pipelines, tanks and reservoirs, Inspection of oil platforms, Energy efficiency [2].

The objective of the use of this new technology is to facilitate the work of companies in the industrial sector as well as to minimize expenses and risks, reducing the use of personnel, the time of execution of the work and as a consequence the cost.

Thermography

Thermography is a technique that allows calculating and determining the surface temperature of objects at a distance, without the need for physical contact with the object being studied. Thermography allows the capture of infrared radiation of the electromagnetic spectrum, using thermographic or thermovision cameras. We can carry out both qualitative (by comparing temperatures) and quantitative (by giving numerical values) inspections, incorporating qualified thermographs in the template (Fig. 5).

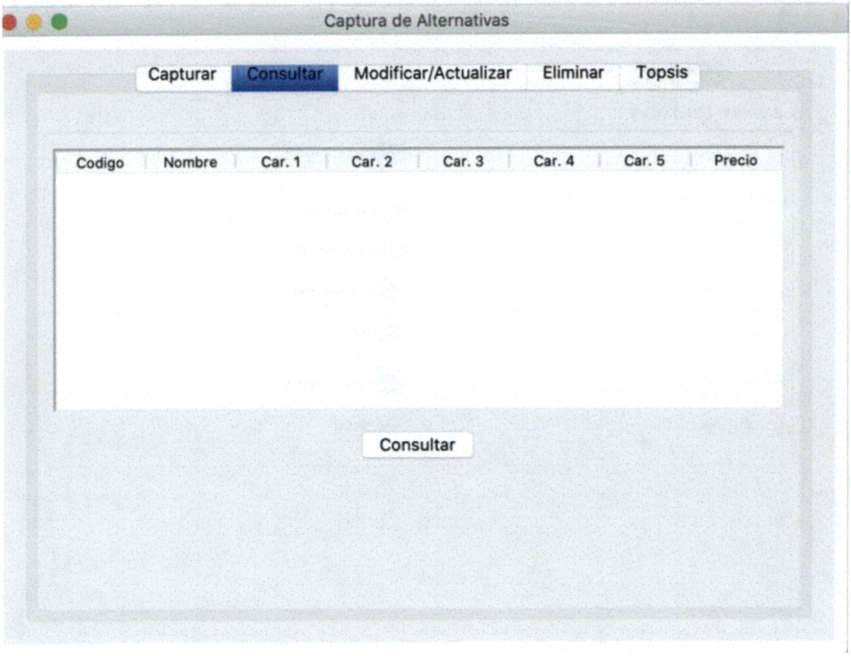

Fig. 2 Secondly module of our intelligent dashboard

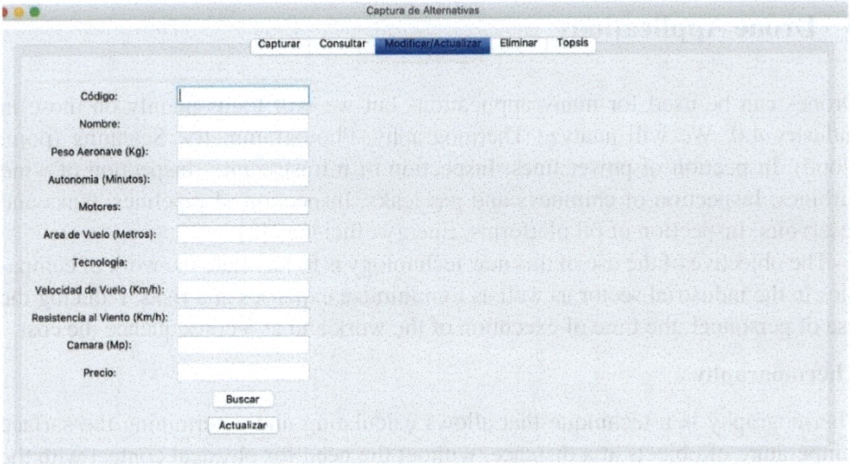

Fig. 3 Third module of our intelligent dashboard

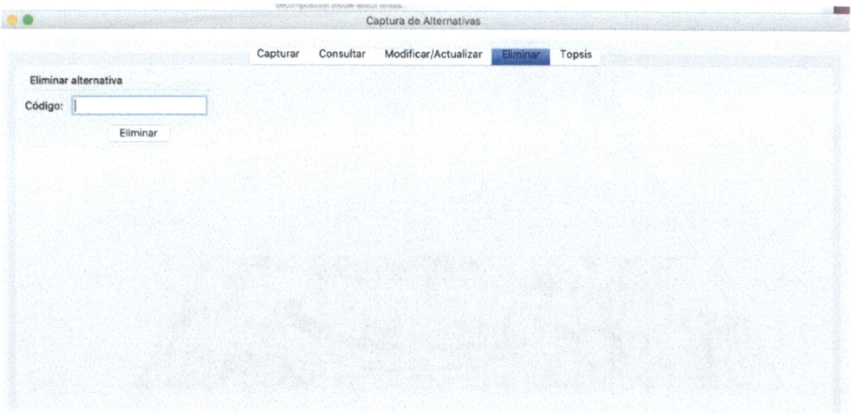

Fig. 4 Fourth module of our intelligent dashboard

Fig. 5 Thermography with drones, where others can't reach

Photogrammetry

Photogrammetry is a technique to determine the geometric properties of objects and spatial situations from photographic images by obtaining 3D models and/or maps from the generated point clouds. With it and with our drones equipped with GPS, the georeferencing of the objects and points scanned is obtained. In the same way, by exchanging different cameras we can generate thermal or visible spectrum maps (Fig. 6).

Fig. 6 Photogrammetry with drones

Scanning (point cloud)

By combining traditional scanning techniques, such as three-dimensional laser scanners, and the point clouds generated by our drones, we obtain a data file with a totally complete point cloud, avoiding blind spots in the facilities (Fig. 7).

Inspection of power lines

Exhaustive visual and thermographic inspection of the electric, low and high voltage lines. The drones prevent workers from climbing the tower and traveling long distances for the execution of our projects, delivering higher quality data by not running the risk of taking information. Inspections are speeded up and power cuts are avoided, executing this, right at the time of the execution of the work (Fig. 8).

Fig. 7 Lifting of buildings by applying point clouds

Fig. 8 Expansion of power line inspection with drone technology infrastructure inspection

Safety inspection of buildings, structures and cranes, by means of a periodic review. It allows us to detect and measure corrosion, deformation and cracks. Traditional methods are often risky, slow, expensive and involve closing lanes, using truck cranes, walkways, entre otras actividades (Fig. 9).

Inspection of wind turbines

Inspection of shovels, main hub and blades. Inspection of mast, hub and blades in both visual and thermal spectrum (Fig. 10).

Inspection of chimneys and exhaust gases.

Visual inspection for corrosion, deformation, surface treatment and possible cracks. Detect hot spots in an exhaust stack and cold spots from gas leaks. It is not necessary to stop production to carry out the inspection (Fig. 11).

Inspection of oil platforms

Visual inspection for corrosion, deformation, surface treatment and possible cracks. Inspection of structures at great heights and areas close to contact with the sea (Fig. 12).

But before we go into detail, let's see how a drone works

Fig. 9 Inspection and monitoring of infrastructure

Fig. 10 Wind power and renewable energies: new inspection system for wind turbine blades

Fig. 11 Minimum height of industrial chimneys

Fig. 12 Oil platforms

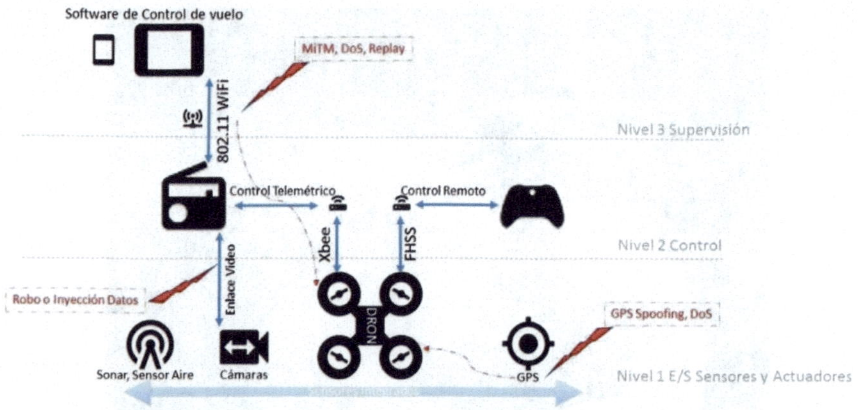

How it can be shown in the image we see that the drone can be operated in different ways, remotely with a control or with the application for mobile devices.

The only disadvantage that the drones have, is the WIFI connectivity, they need to have a very good antenna to not have interference when receiving images or live video.

Here is a graph showing the drone market by sector (Fig. 13).

According to Gartner market research, the market size for commercial drones in 2016 was 2.8 billion with only 110 billion units sold. In 2017, it is estimated that commercial drone sales will grow by about 60% to 170,000. While personal drones dominate unit sales in 94% of the market, they comprise only 40% of the market's

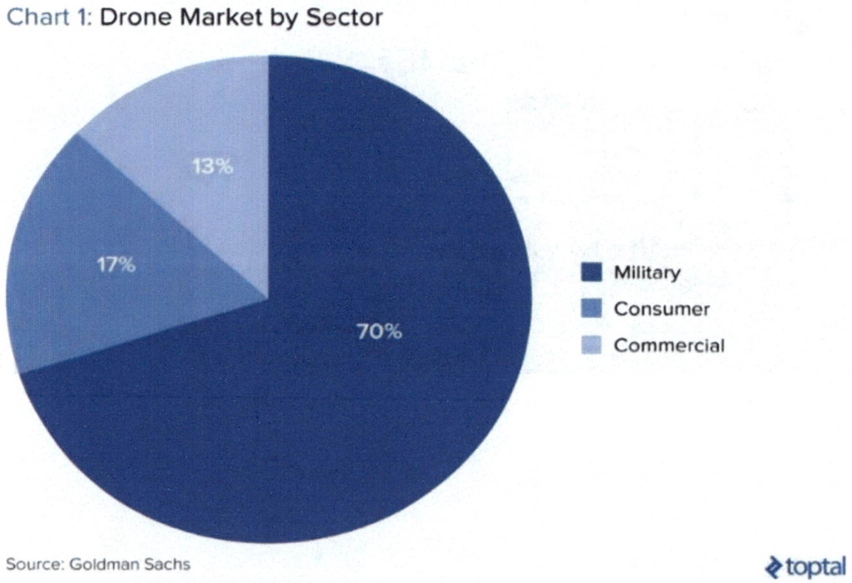

Fig. 13 Graph by Goldman Sachs

revenue share. Commercial drones account for only 6% of the market but their price tags of around $100,000 are projected to account for 60% of the industry's revenue.

Qualcomm Ventures, the most active venture capitalist, making six investments to date in companies focused on mapping, pipeline inspection, delivery, 3D mapping, stand-alone solutions and business solutions Qualcomm also acquired Kmel Robotics to expand its cellular technologies within the drone operations, launching its own commercial robotics solutions. Google Ventures, which has also been actively investing in long-distance delivery, commercial end-to-end solutions and 3D mapping. Google Ventures has developed 63 patents related to drones through various drone capabilities. Intel Capital, which has invested in a complete end-to-end business solution, an analytics solution and a hardware company. They have also acquired two companies, including Ascending Technologies, which is developing "sense and avoid" algorithms, and MaVinci, which is developing flight planning software. These acquisitions and investments have helped Intel develop its own solutions for commercial drone applications.

4 The TOPSIS Method

In our application to use the TOPSIS method [3] we are going to break it down part by part.

Let's first select our alternatives and then separate them by attributes, giving a weight to each attribute.

Alternatives: These are the options which are to be evaluated for selection of the best.

Attributes: These will impact the selection of alternatives. Weight: These estimates relative importance of attributes.

Attribute	Weight on kgs	Autonomy	Motors	Distance	Technology	Velocity	Resistance	Camera	Price
Weight	5	8	4	8	7	4	8	7	6

Rating Scale 1 Not Important—10 Very Important

Attribute	Weight	Rating scale
Weight on kgs	5	1 None—10 Very Important
Autonomy	8	1 None—10 Very Important
Motors	4	1 None—10 Very Important

Attribute	Weight	Rating scale
Distance	8	1 None—10 Very Important
Technology	7	1 None—10 Very Important
Velocity	4	1 None—10 Very Important
Resistance	8	1 None—10 Very Important
Camera	7	1 None—10 Very Important
Price	6	1 None—10 Very Important

5 MOORA Versus TOPSIS

The Multi-Objective Optimization by Radius Ratio Analysis (MOORA) method [4] starts with the response matrix of the other alternatives and different criteria. It then constructs the decision matrix, which contains n rows representing the A1 alternatives, an in the evaluation, and J + L the columns representing the criteria under evaluation (J quantitative criteria and L qualitative criteria). For this step we will obtain the final decision matrix (FDM) with the following formula.

$$MDF = [VO, VST]$$

$$= \begin{matrix} A^1 \\ A^2 \\ \vdots \\ A^n \end{matrix} \begin{bmatrix} x_1^1 \cdots x_j^1 \; x_{j+1}^1 \cdots x_{j+L}^1 \\ x_1^2 \cdots x_j^2 \; x_{j+1}^2 \cdots x_{j+1}^2 \\ \vdots \; \ldots\ldots \; \ldots \; \ldots \; \ldots \\ x_1^2 \cdots x_j^n \; x_{j+n}^n \cdots x_{j+n}^n \end{bmatrix}$$

where Ai represents the alternatives, for i = 1… n, and xji represents the entries for alternative i with respect to criterion j.

Once we select our attributes and the weight of each of them, we will proceed to the decision matrix for each of the alternatives. When we have the ideal matrix, we will use the TOPSIS methodology. Technique for Order Preference by Similarity to Ideal Solution. TOPSIS selects the alternative that is the closest to the ideal solution and farthest from negative ideal solution.

In this Method two artificial alternatives are hypothesized:

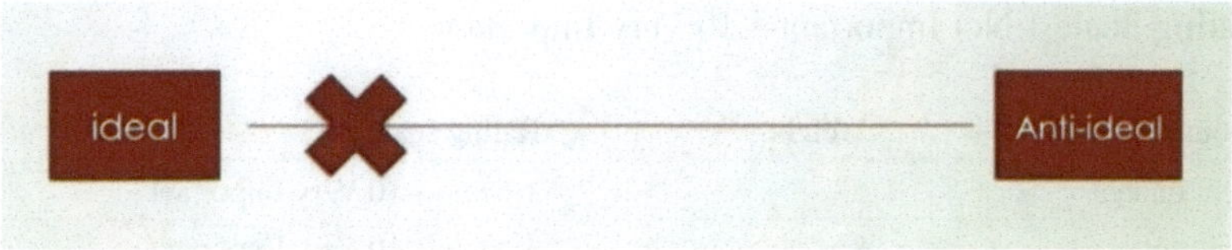

Calculate the standardized decision matrix.

It is feasible that the qualification criteria are expressed in different units or scales of measurement; therefore, normalization is carried out, where each response xij is compared to a denominator, which is representative of all the alternatives involving the criterion.

$$|X_j| = \sqrt{\sum_{1}^{n} x_i^2}$$

Therefore, the normalization of each FDM input is

Ideal Alternative: One which has the best attributes values.

Negative Ideal Alternative: One which has the worst attribute. carried out according to the equation.

Calculate the weighted standardized decision matrix.

Taking into account the different importance of the criteria, the WNxij normalized weighted scores are calculated with the equation.

Calculation of the contribution of each alternative.

The contribution of each alternative is obtained using the equation.

The previous steps were for the Moora method, as a conclusion we can say that the most significant method is that of TOPSIS because taking into account the number of variables it is difficult to obtain a concrete result with MOORA. This is because of the weighting of each attribute.

6 Carbon Footprint

The dictionary says that the footprint is the sign left by the foot of the man or the animal on the ground through which it passes. More precisely, the carbon footprint is defined as the totality of greenhouse gases emitted directly or indirectly by an individual, organization, event or product.

Since we are talking about greenhouse gases, let's identify the main ones:

- Water vapour (H_2O): Odourless and colourless gas obtained by evaporation or boiling of liquid water or by sublimation of ice. It is the one that contributes most to the greenhouse effect due to the absorption of infrared rays.
- Carbon dioxide (CO_2): Also called carbon dioxide.

- Methane (CH_4): It is the simplest alkane hydrocarbon. It is colourless and odourless and hardly soluble in water in its liquid phase. In nature it is produced as a final product of the anaerobic decomposition of plants. It is a GHG about 23 times more powerful than CO_2.
- Nitrogen oxides (NOx): These are several binary gaseous chemical compounds formed by the combination of oxygen and nitrogen. The most common process of formation of these inorganic compounds is combustion at high temperatures, a process in which air is usually the oxidizer.
- Ozone (O_3): A substance whose molecule is composed of three oxygen atoms.
- Chlorofluorocarbons (CFCs): Derivatives of saturated hydrocarbons obtained by replacing hydrogen atoms with mainly fluorine and chlorine atoms. They have been widely used as refrigerant gases, extinguishing agents and propellants for aerosols.

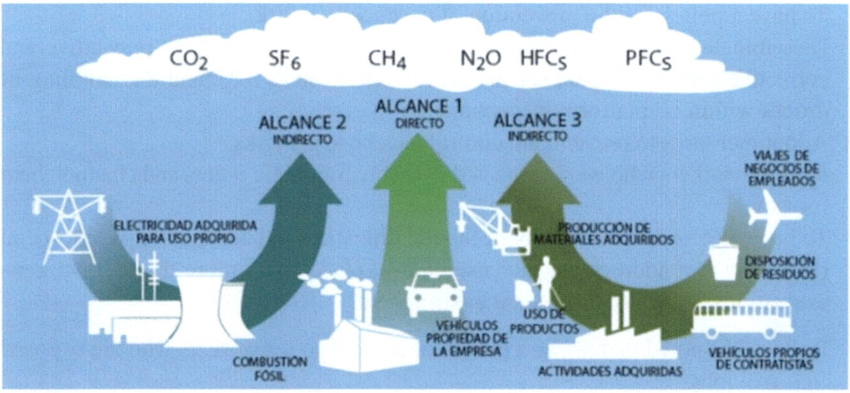

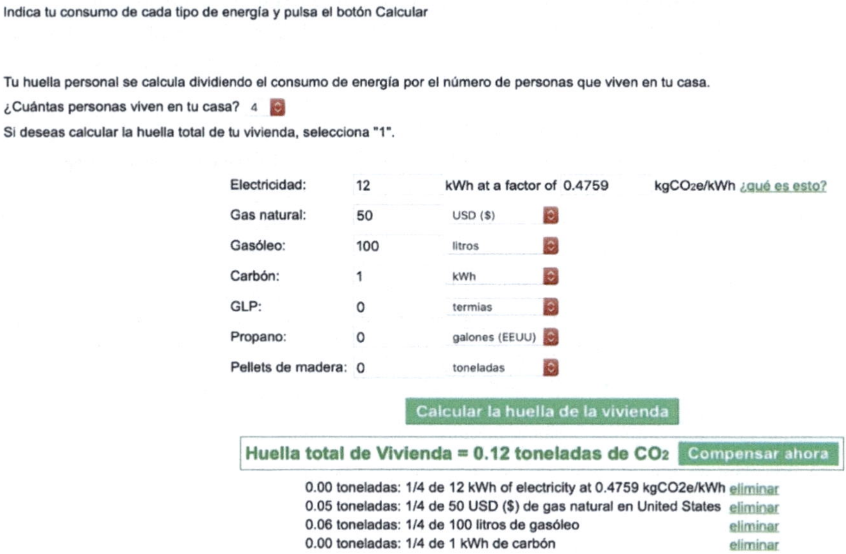

These are my registered data according to the website "https://calculator.carbon footprint.com."

7 NOM-035 STPS 2018

NOM-035 STPS 2018 aims to establish the elements to identify, analyze and prevent psychosocial risk factors, as well as to promote a favorable organizational environment in the work centers. NOM-035 is applied to the oil platform in the inspection of its installations, to carry out visual or data surveys in order to obtain a detailed vision of the areas where anomalies are presumed, whether for corrective or predictive maintenance purposes [5]. In order to comply with the psychosocial risk factors Identification, analysis and prevention will take into account the following points:

- To have a policy for the prevention of psychosocial risk.
- Disseminate and provide information to workers on the policy, preventive and corrective actions adopted for the prevention of these risks and the handling of drones within oil platform inspections.
- Adopt measures to prevent and control psychosocial risks.
- Identify workers who were subjected to severe traumatic events and channel them for care.
- To have a record of the results of the identification and analysis of psychosocial risk factors, in addition, in the case of workplaces with more than 50 workers, evaluations of the organizational environment.

It should be noted that this is a benefit or advantage for workers who are exposed to high-risk jobs.

Depending on the type of sensor used, it is possible to detect cracks and corrosion, so that previous evaluations can verify its evolution. The lowering of the risk index is another point in favors since with our proposal the exposure of the human element is avoided, at least for this inspection stage. This norm is understood by two very important aspects that we will mention next:

Favorable organizational environment: In order to identify a favorable organizational environment, it is necessary to take into account that the workers must have the sense of belonging to the company, as well as the adequate communication of the work regulated by the LFT. Psychosocial risk factors: These are all those that can cause anxiety disorders, not organic of the sleep or wake cycle, of serious stress and of adaptation, derived from the nature of the functions of the job, the type of working day and the exposure to severe traumatic events or acts of violence in the workplace to the worker, due to the work carried out. This regulation is only applicable in country's territory and in all work centers. There are three categories, depending on the number of workers working there.

In this project it will be catalogued in the work center where less than 15 workers work which consists of:

- To establish, implement, maintain and disseminate in the workplace a policy for the prevention of psychosocial risks that contemplates: the prevention of psychosocial risk factors; the prevention of violence in the workplace and the promotion of a favorably organizational environment.
- To adopt measures to prevent psychosocial risk factors, to promote a favorably organizational environment, of this way as well as to address practices opposed to a favorably organizational environment and acts of violence in the workplace.
- Identify workers who were subjected to severe traumatic events during or on the occasion of work and channel them to care.

Disseminate and provide information to workers.

A course is implemented to implement a prevention against risks that may have the workers course that is supported by the STPS this course leads to a few hours of training for those workers selected for such tasks of handling the drones this in order to accredit the skills of them. The advantages of implementing drones in the inspection of oil platforms is to lower the level of risk to the lives of workers who are dedicated to the inspection of anomalies found on them. The prevailing regulations in Mexico for the operation of drones are known, as well as the basic principles for the flight mechanics of multi-rotors under a safe environment and expressly arranged for working hours.

References

1. Behzadian M, Otaghsara SK, Yazdani M, Ignatius J (2012) A state-of the-art survey of TOPSIS applications. Consultada por internet, Mayo 2020. Direccion de internet. http://www.elsevier.com/locate/es
2. Phillips AJ (2019) Impacto de los drones en la era de la Industria 4.0. Consultada por internet, Mayo 2020. Dirección de internet. http://www.logisticamx.enfasis.com/articulos/84628-impacto-los-drones-la-era-la-industria-40
3. Byun HS, Lee KH (2005) A decision support system for the selection of a rapid proto-typing process using the modified TOPSIS method. Int J Adv Manuf Technol 26:1338–1347
4. Brauers WK, Zavadskas EK, Peldschus F, Turskis Z (2008) Multi-objetive decision-making for road design. Transport 23(3):183–193
5. Secretaría del Trabajo y Previsión Social (2019) Norma Oficial Mexicana NOM-035-STPS-2018. Factores de riesgo psicosocial en el trabajo-Identificación, análisis y prevención. Consultada por internet, Mayo 2020. Dirección de internet. https://www.gob.mx/stps/articulos/norma-oficial-mexicana-nom-035-stps-2018-factores-de-riesgo-psicosocial-en-el-trabajo-identificacion-analisis-y-prevencion?fbclid=IwAR1UK6rzSOt5GZeJgsRCVRE9VH4KBw5igrouWa1uOrFAtAHHDkij0kDxt0

Uncertain Analysis Based on Milk-Runs Systems Using Bayesian Networks

Roberto Murillo-Ramirez and Giovanni Lizarraga-Lizarraga

Abstract For several years, the role of original equipment manufacturers (OEM) has been changing. Now, their primary operations focus on design, assembly systems, and marketing. Considering that component suppliers provide more than 95% parts of a light vehicle, the complexity of the logistics is high. Seeking to comply with the Just in Time (JIT) philosophy and thus avoid the shortage of components on assembly lines, the companies use the inbound logistics to perform these operations from suppliers to OEMs. The Milk-run concept refers to carrying out the consolidated supply of components to OEMs through the planning of routes considering xed and cyclical time windows. Currently, Milk-run (MR) systems suffer from interruptions in component deliveries; this can produce monetary losses per minute of delay in the assembly line of OEM. Problems in MR have been approached from the perspective of optimization models with variations within them, and there are a few studies about variability and uncertainty in MR operations. The results of this research are favorable because not only uncertainty is quantified; the causes of optimistic and pessimistic scenarios are also quantified.

Keywords Milk-runs · Bayesian networks · Uncertainty

1 Introduction

The automotive industry has been a forerunner in important areas such as global economic structures and manufacturing systems. This industry has moved its manufacturing plants to different parts of the world, which benefits the economy of the countries in which they move [1, 2].

R. Murillo-Ramirez (✉) · G. Lizarraga-Lizarraga
Universidad Autonoma de Nuevo Leon, San Nicolas de los Garza, Mexico
e-mail: roberto.murillorm@uanl.edu.mx

G. Lizarraga-Lizarraga
e-mail: giovanni.lizarragalz@uanl.edu.mx

© The Author(s), under exclusive license to Springer Nature Switzerland AG 2021

A. Ochoa-Zezzatti et al. (eds.), *Technological and Industrial Applications Associated with Intelligent Logistics*, Lecture Notes in Intelligent Transportation and Infrastructure,
https://doi.org/10.1007/978-3-030-68655-0_30

Another significant contribution has been the implementation of lean manufacturing systems, and its objective is to eliminate processes that do not add value to a company. One of the most representative tools of these systems is the Just In Time (JIT) manufacturing philosophy, which supports the correct coordination of the actors in a supply chain system [3, 4].

The role of Original Equipment Manufacturers (OEM) has changed over time; in the past, car manufacturing was carried out entirely within OEM's. Today, their main functions focus on the design, marketing, and assembly systems of vehicles. The current models of light vehicles assembled by the OEM's are composed of an average of 3,500 to 7,000 parts and components. The suppliers' status takes total relevance since they are responsible for supplying 95% of the vehicle components [5]. Inbound logistics in OEM's refers to the articulation of the supply of components, following established parameters, correct quantities, and specific components in the right order. The modes of transportation that used to carry out the logistics may be the following:

(a) point-to-point transportation, (b) expedited transportation, (c) Milk-run round-trip and (d) Milk-run open [6]:

(a) This kind of transportation is planned and based on the monthly requirement of components.
(b) This mode is activated by a variation in the demand for components.
(c) Component collection system that begins at the plant taking empty equipment to suppliers.
(d) Component collection system that begins at a supplier on the route.

In the literature review, these systems are called Milk-Runs (MR) and present specific characteristics such as cyclical and xed scheduling of routes and collection times previously defined and coordinated by the OEMs [7].

MR freight transport operations are carried out by outsourcing services companies, which seek to establish efficient and pro table transportation routes, thus minimizing losses of time and money [6].

2 Problem Statement

It is known that variations and interruptions may occur within the operations of an MR. Disruptions in freight transport are associated with errors in daily work and their execution. However, minimal they may seem, these interruptions can cause changes in everyday practices and, therefore, lead to disturbances in a network [8]. Thus, the disruptions presented in transport are showed. We explain how these interruptions can affect the mobility of goods, in such a way that the entire flow of a transport network suffer alterations. In [9], the author mentions the effects caused by the uncertainty in such operations.

One of OEM's main goals is to speed up the entry of components to the assembly lines, thus reducing the number of delays and low inventory levels. Delays result in monetary losses per unit of time [3]. Interruptions in MR generate uncertainty and to face the uncertainty presented in component delivery delays, expedited freight transport is used. This is an alternative currently used by OEMs to avoid delays in production lines. Consequently, this represents a higher inherent cost and more complexity in logistics operations.

This uncertainty has not yet been incorporated and analyzed within the MR scenarios, decisions to either to expedite or not to expedite components are made empirically or by intuition without considering a decision statement-making [10].

In this research, we identify, describe, and document the problems related to MR's uncertainty and how these factors affect decision-making regarding expedited component transportation. We analyze the operating scenarios to observe the uncertainty involved in the MR.

3 Literature Review

This section describes the problems that some authors have associated with MR. MR's have pick-up and delivery time windows. There are also a limited number of cargo units with specific capacities that make them a candidate to be solved by a vehicle routing problem (VRP) [11]. It is essential to mention that the MR's have specific characteristics and the VRP models only contribute in a part of the design and planning of the MRs, but do not adequately contemplate the logistics operations involved.

Usually, the main target of optimizing VRP models is to obtain the minimum total travel distance using heuristic algorithms or linear programming. For component pick-up systems MR, a tool known as the vehicle routing problem with MRVRP dairy routes is used. In this model, two approaches are considered: minimizing the distance or the total costs [12]. According to [13], VRP models can present one of the following strategies: minimize aspects such as total cost of operation, full transport time, total distance traveled, and waiting time. You can also look for pro t maximization, customer service, vehicle utilization, etc.

In [6], the authors solved the Periodic Vehicle Routing Problem with Service Choice (PVRPSC) using mixed-integer programming. Among its results, they provided a basis for the decision regarding the selection of the transport mode and a 15% reduction in annual operating costs. In another paper, [3] explains the VRP model and its implications for the management of parts collecting systems. They mentioned that this model has variants that may consider factors such as time windows or dynamic order processing, and for both cases, there are modifications to the model. The author also mentions that as the size of the problem grows, the calculation time will be longer.

These authors modify the classic VRP problem and seek to adapt certain needs of their case studies, as can be seen in Table 1, in which the VRP can include factors of time windows, dynamism and include CRS costs. These models present specific objectives to be achieved. Still, models do not contemplate any behavior that implies uncertainty, which means that the values in established parameters are not usually the most appropriate.

There is a stochastic approach of the VRP's. This variant is known in the literature as SVRP and includes non-deterministic factors since, in real life, the variables considered present some degree of uncertainty. These variables include, usually, service and travel times, customers, and the most crucial factor is the demand. These variants are widely studied in the literature [14], and can help make partial decisions because the real performances of these variables are not frequently measured.

The generation of cost structures is one of the approaches found in the literature. These investigations contemplate implicit costs in the design, planning, and execution of MRs, and are mostly used to examine hidden costs and future scenarios. Another perspective of the investigations associated with the term Milk-Runs is the MR in the plant (MRIP), which refers to the design of component pick-up systems. The main target is to design and plan the implementation of MR tours in manufacturing plants, with the help of automatic or semi-automatic equipment, to reduce operating times and human mistakes.

In our literature review, we did not found published works related to levels of uncertainty in MR. In the next section, we introduce some of the tools used to analyze scenarios where uncertainty is incorporated as a preamble to understand the objective of this research. Uncertainty refers to the least possible or null understanding of a scene that requires inquiries or answers that the standard models do not work contemplate. There are tools used in the quantification of uncertainty [15]. One of them is the Monte Carlo method, which consists of generating a series of numbers using a random sampling of probability distributions and can give approximate solutions by experimenting with real phenomena [16]. Within these tools, there are also

Table 1 Overview of problems related with Milk-run systems

Authors	Problems					
	VRP	VRPTW	DVRP	Cost	MRIP	Uncertainity
Du 2007	x	x	x	–	–	–
Aragao 2016	x	–	x	–	–	–
Meyer 2018	x	–	–	x	–	–
Kluska 2018	–	–	–	–	x	–
De 2016	–	–	–	x	–	–
Guner 2017	x	–	x	–	–	–
Staab 2016	–	–	–	–	x	–
Bocewicz 2019	x	–	–	–	x	–
Fedorko 2018	–	–	–	–	x	–
Lin 2015	x	–	–	–	–	–
Ma 2013	x	–	–	–	–	–
Novaes 2015	x	–	x	x	–	–
Ranjbaran 2020	–	–	–	x	–	–

fuzzy sets in which a linguistic label is used to determine a membership function and helps to represent the ambiguity or vagueness of real phenomena [4]. One more quantification method for uncertainty is the polynomial chaos method.

In this method, an uncertainty propagation is performed on the model predictions and reduces the complexity in computational resolution compared to the Monte Carlo method [17]. The last uncertainty method we reviewed was the Bayesian model, based on the conditional probabilities between variables according to Bayes' theorem. This tool has proven to be efficient in the correct quantification of uncertainty in highly complex scenarios.

4 Methodology

The selected tool to quantify the uncertainty, among the existing methods, was the Bayesian networks because they have the following advantages:

- Ideal for situations where there is a few data or imperfect knowledge, which cannot be solved by rigorous probabilistic models [18].
- This tool is based mainly on intuition, which is formalized in certainty factors [18].
- General reasoning is used under uncertainty with in-depth knowledge [18].

Due to these aspects, Bayesian networks promise to be a good option to quantify the uncertainty in MR and also provide an important guideline in terms of literature contribution.

Bayes' method is the inverse effect of the total probability. Given two mutually independent events, the result in the calculation of the total probability is the probability of the intersection of one event to the other and is expressed in the following way:

$$P(A \cap B) = P(A) * P(B) \tag{1}$$

Where $P(A \cap B)$ is the probability of the intersection of event A and event B.

In this way, it is easier to understand how Bayes' theorem works; it is the inverse form of the previous equation since prior evidence of one event is required. This is the probability of occurrence of an event since another has happened and is expressed as follows:

$$P(A_i|B) = \frac{P(A_i|B)}{P(B)} \tag{2}$$

Where: $P(A_i|B)$ is the probability that event A occurs knowing that event B occurred.

As we mentioned before, a scenario is structured to better outline the MR operations, where we exemplify the behavior of the independent variable A, concerning a dependent variable B.

We try to explain how realiced the estructure of MR scenario with the Fig. 1, where:

C = Complete scenario of operation MR.

A_i = Independent variables of MR, the subscript i show the count to each event.

B = Dependent variable that represents the probability of expediting components.

The developed methodology is based on MR scenarios that are schematized for their analysis using Bayes' theorem. The next step is establishing the variables type Ai to proceed with the creation of the directed acyclic graph and the calculation of the conditional probabilities. Bayesian networks have two parts, one qualitative and the other quantitative. The qualitative component is made based on the experience of the people involved in MR operations [19]. The methodology developed arises from the application of models that use Bayesian networks. The authors [20–22] adjusted the variables of their models to be able to predict the events in their corresponding scenarios. The following is the assignment of variables to the scenario.

These variables require a probabilistic value based on the possible states they can take. For example, the states that these variables can have can be expressed as follows:

- True—false
- Complete—incomplete
- In time—delated.

These statements are analyzed by the expert people consulted, in order to obtain the values of the probability of occurrence of each sentence according to their experience. It is essential to mention that the sum of the percentages in the possible states of a variable must be equal to one hundred percent in all cases (see Fig. 2).

Once the conditional probabilities for the scenario are determined, we will continue with the creation of the graphical model that will help in the probabilistic reasoning and is represented by the directed acyclic graphs formed by nodes and arcs. Each variable represents the nodes with its respective probabilities, and the arches are the direct dependency relationships between variables. The purpose of Bayesian networks is to know the posterior probability of the variables of the network. These probabilities arise when specific evidence is given that some variable has taken a certain state, while no evidence is given to the variables, the network will take values of the a priori probability distribution, in which only the probability of an event occurring without prior evidence is considered. The following steps are required to create the graph:

- Determination of the relationship between variables, identify the parent nodes and the child nodes.
- Validate the interdependence or conditional dependence between variables.

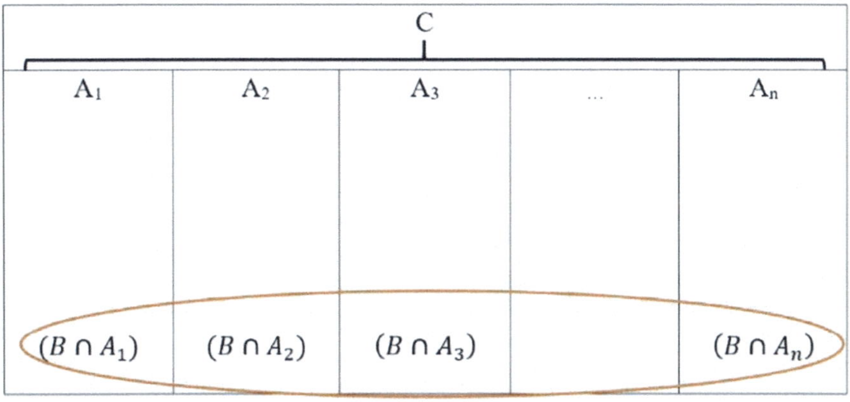

Fig. 1 Venn diagram with respect to the Bayes theorem scenario [20]

Fig. 2 Values of variables

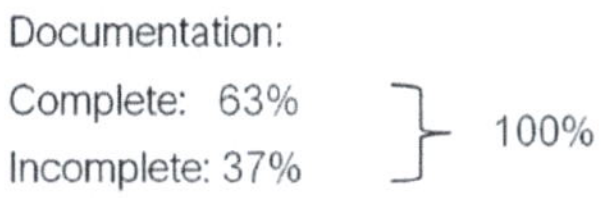

The first part is determined by calculating the dependency probabilities, which are developed by the following equation:

$$P(B|A_1 \ldots A_n) = (B|A_i) \tag{3}$$

The resulting probability for each variable $(B|A_i)$ produced by this equation serves as a statement for the creation of the graph. If they determine the dependency or independence relationship for the nodes, these relationships are expressed within the graph, as shown in Fig. 3.

These graphical representations are used in the creation of a graph based on the structure of Bayesian networks. Within the graphs, the relationships between nodes can have several behavior patterns: convergent behavior, that happens when a parent node has two children (see Fig. 4); divergent behavior, when a variable has two parents; and sequential behavior, in which there is a relationship between a parent node and a child node that in turn is the parent of another node.

Once the graph is created, the next step is the calculation of the conditional probabilities. This step is di cult and extremely laborious because each probability is calculated with the following equation:

Fig. 3 Structure of graph model

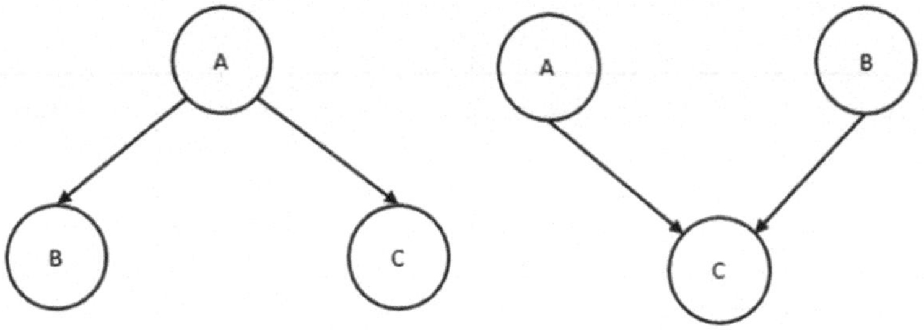

Fig. 4 Types of structure in graph models

After the creation of the graph, the calculation of the conditional probabilities is difficult and the manual process is extremely laborious, because include each probability spread following equation:

$$P(A_i|B) = \frac{P(A_iB)}{P(B)} = \frac{P(B|A_i)}{\sum_{i=1}^{n} P(A_i)P(B|A_i)} \quad (4)$$

Where $P(A_i|B)$ is the conditional probability of an event A_i given that there was an event B of an expedition of components. The calculations are made for each relationship between parent and child node, in such a way that they can be shown in result matrices that will help to inference the causes and probabilities of the operational scenarios. Considering the last part, this research pretends to obtain a basis given the calculation of the probabilities and causes that influence the events where component expediting occurs. The inferences about the event studied will serve as a statement for decision-makers regarding two types of planning:

- Seek the possible integration of the expediting of components in the MR planning knowing the probability that these events will happen given certain findings and
- Identify the processes that negatively influence the MR's execution and seek their improvement; these MR operations involved can be seen in Table 2.

5 Case of study

We worked in collaboration with a company dedicated to the management of type C components. This classification includes materials such as adhesives, plugs, fuses, indicators, led accessories, security seals, wires, hardware, and terminals for wires, among others. Within the manufacturing industries, all components' importance is kept in mind, no matter their size and volume, they are essential for a manufacturing process. At this point, the importance of managing these components is rectified.

The company carries out management to guarantee its clients solutions based on the challenges that manufacturing companies present. These challenges are the reduction of total ownership costs without risking factors such as product quality and inventory levels in parts consumption.

It is valuable for the company to contemplate that hidden costs are commonly omitted in the management of class C components. Only 15% of the total value of a component is known, while the other 85% is unknown or ignored. These hidden costs impact areas such as logistics, inventory management, engineering, quality, and finance.

It will be advantageous for the company to provide a specific supply chain for class C components, which will improve the consolidation of suppliers in a single operating figure, reduce the degree of uncertainty, decrease delivery times, and increase productivity. The following are the company's value-added services: risk minimization, inventory optimization, operations improvement, and quality improvement.

It is important to mention that the company carries out projects according to the requirements of each specific client. These requirements can be diverse and varied because the company has clients from different industries, such as automotive and household appliances, among other sectors.

6 Results and Analysis

For this investigation, the company provided information related to twenty-one MR routes. The creation of the scenario was based on the operations in Table 2. These operations were taken as the scenario variables. First, we assign probabilities of occurrence to these variables. These probabilities were determined by consulting the company's expert group through an online form that was shared. We sent the form to the experts so they can familiarize them-selves with the tool, then an online meeting was held with all the participants in which the probabilities were assigned by consensus.

In Fig. 5, we can see the structure of the directed acyclic graph and how the variables are related to each other. We can see that the network has a total of twelve nodes, whose output variable is the node "Expediting", which has as parent nodes "Load time", "Road accident", and "Change model" variables.

The propagation of probabilities in the network is shown in Fig. 6 and we can see that values showed in each node represent the "a priori" probabilities of the

Table 2 Variables of MR scenario

Variables		
Supplier	3PL transport	OEM
Shipping documentation	Vehicle arrives	Forecast production
Load time	Vehicular traffic	Stowage plan
Complete pallets	Maintenance	Change of model
Quality	Road accident	

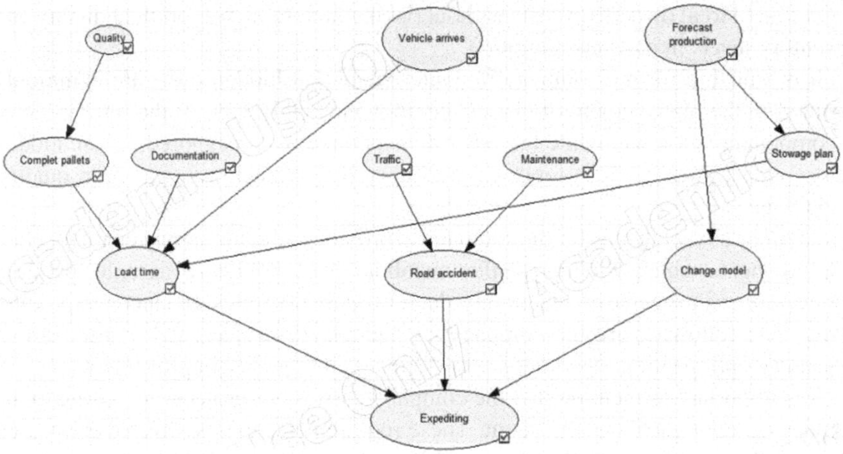

Fig. 5 Graphic model of MR systems

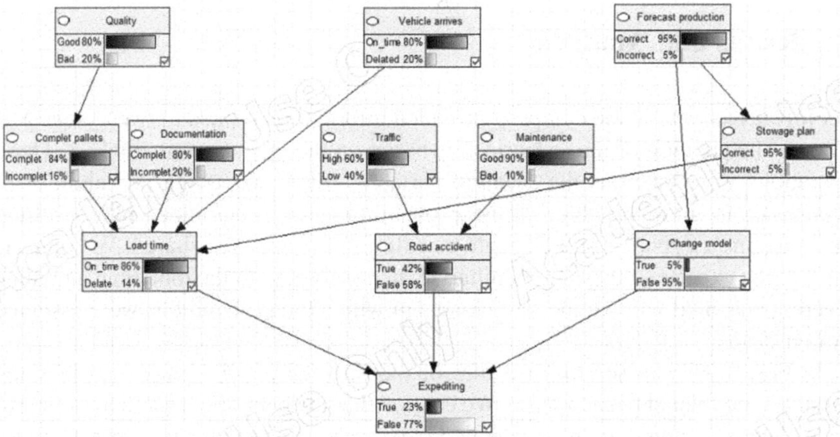

Fig. 6 Graphic model of probabilities a priori of MR systems

network. When we introduce the probabilities values gived from experts the target node "Expediting" take the value 77%, means that it is not required to expedite components versus 23% otherwise.

When obtaining the probability of occurrence, the acyclic directed graphic was created using the Genie version 2.4 software. The calculation of the "a priori" probabilities distributions and their respective conditional probability tables were obtained. In this sense, by providing evidence to the network's nodes, the conditional probabilities spread is carried out throughout the network, giving an estimation of the "a posteriori" probabilities.

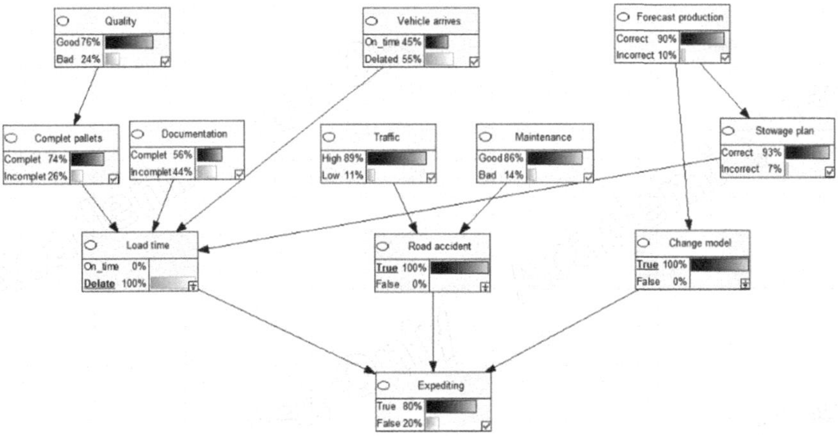

Fig. 7 Graphic model of negative scenario of MR systems

In the first part of the "a priori" probabilities, it is probability of occurrence is shown based on the experts' values. The second part, where realiced inferences, is made when certain evidence is given to the middle nodes, the probabilities spread are not only shown on the target node, but also affect the nodes probabilities on top and bottom positions creating an upstream and downstream effect. To exemplify this part, two scenarios were elaborated, one with favorable conditions for the MR's execution, and the other with negative conditions.

The first scenario is shown in Fig. 7, in which negative evidence was given to the "Load time", "Road accident", and "Change model" middle nodes and in this way, the network's values were modificated. The "a posteriori" probability changes its value, and the target node "Expediting" changes its probabilities of occurrence, now the probability that will be occurr as 80% true against 20% false.

As shown in Fig. 7, it can be observed that now the probability that the "Vehicle arrives" is delayed is 55% and that the "Documentation" of MR's could be 44% wrong. In comparison, there is an 89% probability that the traffic is high. With these observations, one of the advantages of Bayesian networks is fulfilled. It not only shows the probability of occurrence of the target variables but also quantifies the possible causes for these states in the variables, this prove the upstream and downstream effect.

The behavior of this network is similar to the previous one, by giving positive evidence to the network the upstream and downstream effect will repeat, now the probability that components will be expedited is 90% false (view Fig. 8), which is favorable for the company. Besides, it is also observed how should the values be so that the probability that the "Load time" is on time should present a "Vehicle arrives" with 86% on time, and that the "Documentation" of the SRC should be correct by 84%. In comparison, there is a 61% probability that the traffic was low. This scenario

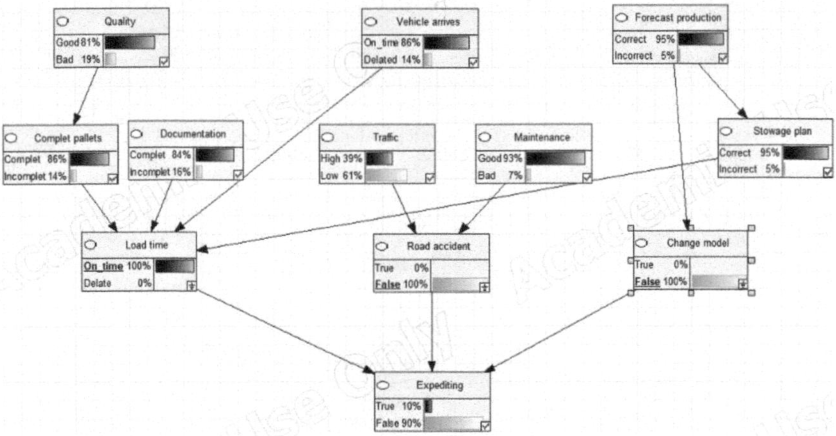

Fig. 8 Graphic model with positive evidence of MR systems

could be as a strategic guide for decision makers because shown the occur probability use minime of expediting transport and how to reduce the use of this.

7 Conclusions

This paper addresses a logistics issue related to the business structures commonly used by the automotive industry, the Milk-run systems. We reviewed the literature, and by studying the applications of these systems in the cases of study, we found that MRs are not exclusive to the automotive industry. Industries such as household appliances have adopted these systems to reduce inventories and achieve more agile and lean manufacturing. Also, we observed that in the current approaches in the literature related to MR, there are many guidelines to develop these systems. We can conclude that MRs is a topic that has not been studied so extensively, since these systems can be associated with other types of transport structures.

In the uncertainty part in the MR, the closest that could be found to the literature is the approach to the vehicle routing problem using stochastic variables. As mentioned before, the VRPs only manage to solve a part of the planning and design of MR, while there is still uncertainty in the other parts. It is convenient to mention that by relating the term" Uncertainty" with the term “milk-runs” the literature search is reduced to a minimum, which indicates that more studies can be carried out on this topic.

In the part of the uncertainty quantification tool, the Bayesian networks showed their effectiveness, not only in their maximum expressiveness of the probability distributions in the nodes using the directed acyclic graph, but it was also possible to weight causality of nodes given some evidence. It was possible to show to the company the probability of an event of expedited transport of components to happen and what are the possible causes that propitiate it.

7.1 Future Work

As mentioned before, as we did not have historical data on MR's execution, it was decided to consider the company's staff's expert opinion to calculate the probabilities of the variables. For this, the company was informed that the network could be fed in the future with historical operating data. The Bayesian network has an advantage over other probabilistic tools since it can work with little information or incomplete information.

Acknowledgements This research was made possible with the support of the software company, Bayesfusion, who provided a free academic license for GeNIe Modeler software in version 2.4.

To Universidad Autonoma de Nuevo Leon through the PICYT 2020 program for support to the project "Simulation model of the supply chain under disruptions."

References

1. Basurto Alvarez R (2013) Estructura y recomposicion de la industria auto-motriz mundial. Oportunidades y perspectivas para Mexico. Econom a UNAM, 75–92
2. Sabbagha O, Nizam Ad Rahman MN, Rosmanira Ismail W, Wan Hussain WM (2016) Impact of quality management systems and after-sales key performance indicators on automotive industry: a literature review. Procedia - Soc Behav Sci 68–75
3. Du T, Wang FK, Lu PY (2007) A real-time vehicle-dispatching system for consolidating milk runs. Transp Res Part E: Logistics Transp Rev 565–577
4. Gónzalez Díaz MC (2013) "SISTEMAS DIFUSOS" UNA HERRAMIENTA PARA LA ADMINISTRACION DE INVENTARIOS. Obtenido de UNAM: https://www.ptolomeo.unam.mx:8080/6536/Tesis.pdf
5. Bocewicz G, Nielsen P, Zbigniew B (2019) Milk-run routing and scheduling subject to different pick-up/delivery profiles and congestion-avoidance constraints. IFAC-PapersOnLine
6. Meyer A, Amberg B (2018) Transport concept selection considering supplier milk runs{an integrated model and a case study from the automotive industry. Transp Res Part E: Logistics Transp Rev 147–169
7. Baundin M (2004) Lean Logistics: the nuts and bolts of delivering materials and goods. Productivity Press
8. Jayaram J, Das A, Nicolae M (2010) Looking beyond the obvious: unraveling the Toyota production system. Int J Prod Econ 280–291.
9. Díaz Madroñero M, Mula J, Peidro D (2017) A mathematical programming model for integrating production and procurement transport decisions. Appl Math Model 527–543
10. Yin Z, Wang C, Yin Q (2018) Coordinating overseas and local sourcing through a capacitated expediting transportation policy. Transp Res Part E: Logistics Transp Rev 258–271
11. Borgstedt P, Neyer B, Schewe G (2017) Paving the road to electric vehicles a patent analysis of the automotive supply industry. J Cleaner Prod
12. Ma J, Sun G (2013) Mutation ant colony algorithm of milk-run vehicle routing problem with fastest completion time based on dynamic optimization. Discrete Dyn Nat Soc

13. Ulin Hernandez EJ (2019) Optimizacion de la red de distribucion en el servicio de paqueteria empleando una tecnologia emergente. Obtenido de UANL: https://eprints.uanl.mx/17868/1/1080288731.pdf
14. Oyola J, Arntzen H, Woodru DL (2018) The stochastic vehicle routing problem, a literature review, part I: models. EURO J Transp Logistics 7(3):193–221
15. Cardenas Monsalve JJ, Ramírez Barrera AF, Delgado Trejos E (2018) Evaluación y aplicación de la incertidumbre de medicion en la determinacion de las emisiones de fuentes jas: una revisión. TecnoLógicas, 231–244
16. Lopez Gomez A (2012) UNAM. Obtenido de Analisis del combustible de un Reactor Nuclear modular de helio con turbina de gas: https://www.ptolomeo.unam.mx:8080/xmlui/handle/132.248.52.100/720
17. Beltran FR (2002) Métodos para obtener conocimiento utilizando redes bayesianas y procesos de aprendizaje con algoritmos evolutivos. Tesis Doctoral, Universidad de Sevilla
18. Larrañaga P, Moral S (2011) Probabilistic graphical models in artificial intelligence. Applied Soft Computing
19. Abolbashari MH, Chang E, Hussain OK, Saberi M (2018) Smart buyer: a Bayesian network modelling approach for measuring and improving procurement performance in organisations. Knowl-Based Syst 142:127–148
20. Moreno Garza OH (2006) Redes Bayesianas una Apliacion de Control de Tra co
21. Pascual MB, Martínez AM, Alamillos AM (2014) Redes bayesianas aplicadas a problemas de credit scoring. Una aplicacion practica, Cuadernos de Economía, 37(104):73–86
22. Kent J, Dowling R, Maalsen S (2017) Catalysts for transport transitions: Bridging the gap between disruptions and change. J Transp Geogr 200–207
23. Kaneko J, Nojiri W (2008) The logistics of Just-in-Time between parts suppliers and car assemblers in Japan. J Transp Geogr 155–173

Implementation of an Intelligent Visual Recognition System for the Proper Classification of Solid Waste Using a Mobile Application in a Smart City

Diego Moreno-Jacobo, Gilberto Ochoa-Ruiz, Alberto Ochoa-Zezzatti, Lina Maria Aguilar-Lobo, Adrian Ramirez-Lopez, and Daniel Angeles-Herrera

Abstract A preponderant aspect in the world's megacities that wish to become pioneers of a Smart City model is the adequate collection of solid waste and, to a large extent, adequate classification and separation, since there are specific recycling models derived from correct classification. In the present research, we propose the use of a mobile application that allows determining a correct classification between cardboard, glass, organic waste, metal (especially aluminum) and lime waste, this will allow the human team to make a correct decision on the type of waste to be classified, for its correct separation using an intelligent system of artificial vision for the competitive environment of a Smart City.

Keywords Solid waste management · Computer vision on mobile device · Deep learning · Smart cities · Logistics 4.0

1 Introduction

In recent years, we have witnessed an astounding increase in population worldwide, and an enormous concentration of this population in big cities. In general terms, this has led to a steady increment in production of goods that inevitably end as solid wastes that require a great deal of effort regarding their collection, management and storage. In many in-development countries, such measures are not always effectively

D. Moreno-Jacobo · A. Ochoa-Zezzatti
Universidad Autónoma de Ciudad Juárez, Chihuahua, Mexico

G. Ochoa-Ruiz (✉)
Tecnologico de Monterrey, Escuela de Ingeniería y Ciencias, Zapopan, Mexico, Monterrey, Mexico
e-mail: gilberto.ochoa@tec.mx

L. M. Aguilar-Lobo · A. Ramirez-Lopez
Universidad Autónoma de Guadalajara, Zapopan, Mexico

D. Angeles-Herrera
TecNM/ITS de Tantoyuca, Subdireccion de Posgrado e Investigación, Tantoyuca, Mexico

© The Author(s), under exclusive license to Springer Nature Switzerland AG 2021
A. Ochoa-Zezzatti et al. (eds.), *Technological and Industrial Applications Associated with Intelligent Logistics*, Lecture Notes in Intelligent Transportation and Infrastructure,
https://doi.org/10.1007/978-3-030-68655-0_31

implemented, which has adverse public health and the environmental implications, and in many cases poses hazards in public security as well [1]. For instance, in many cities solid wastes laying in the streets can clog the city water drainage systems, leading to floods and other disastrous situations. Therefore, effective waste spotting mechanisms are very much required to avoid these situations, but also for improving the recycling efforts by many municipalities.

2 Motivation

Significant development and increase in urbanisation and the accompanying agglomeration of people in large cities around the world is leading to an ever increasing waste generation. The basis of the recycling movement is the 3Rs rule, which refers to the initials Reduce, Reuse and Recycle. It was devised by Koizumi Junichiro, the then prime minister of Japan, during the G8 Summit in 2004. For this reason, it seeks to develop an application that is capable of determining the type of material to be recycled and its correlated effectiveness [2].

3 State of the Art

Do you have doubts about where to throw away some containers because of their composition or their appearance? To help to solve this problem, Ecoembes has created the Intelligent Recycling Assistant, A.I.R-e, a chatbot that, according to the company, “was born to give immediate answer to all the doubts and consultations of the citizen in the matter of recycling". Created by TheCircularLab, the center of innovation in the matter of circular economy of Ecoembes, and developed by Accenture, A.I.R-e makes use of the Artificial Intelligence to give interactive information, at the moment, on different questions related to the recycling of packages, like in which container should be deposited each package and the material of which it is composed, but, also with the recycling of another class of residues, like furniture, kitchenware and textile. A.I.R-e not only works through voice and text recognition, but it is also capable of recognizing, through photographs, the type of waste about which the user has doubts. Thus, the citizen will be able to send him images of those he wants to recycle and does not know where to do it and will receive “an immediate answer".

4 Materials and Method

With the advance of new technologies, we have found ourselves in the need to involve artificial intelligence in our daily lives, in turn trying to make this implicit in most of our computers and mobile devices. The problem arises when waste has to be sorted

Table 1 Number of trash instances in each class for: the original dataset and the augmented datasets

Dataset/Class	Glass	Paper	Plastic	Metal	Cardboard	Trash
Trash Net	501	594	403	482	410	137
Augmented	3,006	3,564	2,418	2,892	2,460	822

and it is possible to see if anything can be recovered. Because of this, the need to implement a solid waste classification system. It is based on the development of a graphic interface. Due to the complexity of a machine's ability to interpret images, we must consider that a machine has to learn skills for proper classification. We call this a test and training step.

4.1 Dataset Collection

We have collected a large dataset from various sources (i.e. internet crawling, crowd-sourced photos, among others) consisting of images of 5 classes, as depicted in Table 1.

The approach presented here is a multi-class classification problem, in which we take the 5 classes of the proposed dataset and train an optimized object detector model (Yolo, but it could be other as well) to implement a road damage acquisition system. However, training deep learning-based object detectors is not a trivial task: one of the major issues of deep learning, when working with relatively small datasets such as the one used in this study, is the problem of overfitting.

4.2 Dataset Preprocessing

We carried out the training of the deep learning models described above using the following methodology. The labeled dataset is randomly shuffled and separated into training set and validation set, 70% percent of the images of each class was used for training, taking care that the model did not overfit; for this purpose, we have made use of dropout and data augmentation over the original dataset, as discussed above. Later, the other 30% of the images were used for validating the models [3]. The model description as well as the training phase were implemented on Keras and Tensorflow, running on Google Colab with a Tesla K80 GPU; the models were trained with batch size of 24 and number steps per epoch of 300, using Adam as optimizer, adapting the learning rate. The results of the training process are depicted on Fig. 1, where we can observe the monotonic decrease of the training error (blue line) as well as the validation error (green line) achieving a mean average procesion of 90% for all the classes.

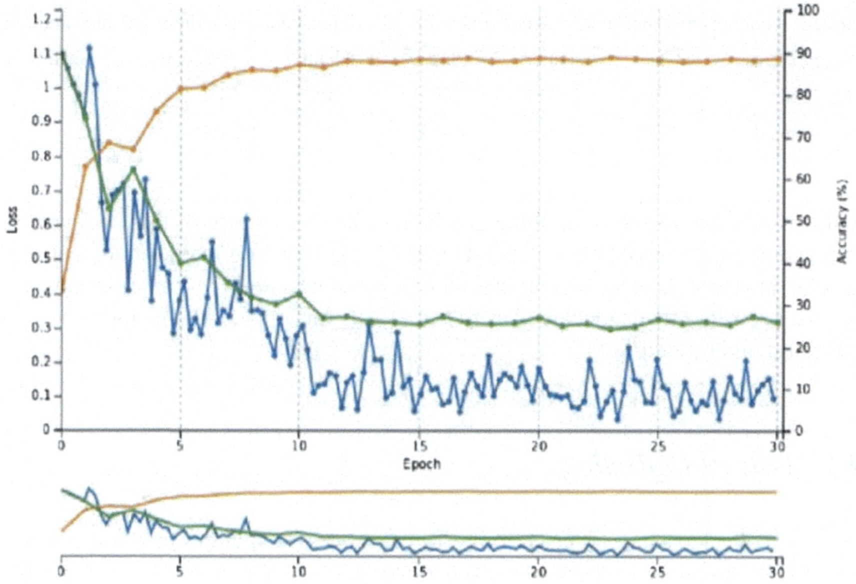

Fig. 1 Training and validation loss curves for the augmented dataset

4.3 *Model Training*

Here is a very simple example of how our research works, in next Fig. 2, we explain the first approach to functionality:

Thanks to the new tools provided by google we can make the training of our models are even easier, in the following images we will show how to use the tool, as is possible see in Fig. 3.

In this part of our intelligent mobile application we are allowed to upload images or through a webcam capture the images of the model we want to process, in next Fig. 4 is possible determine a specific image of a material.

Fig. 2 Functionality of our model

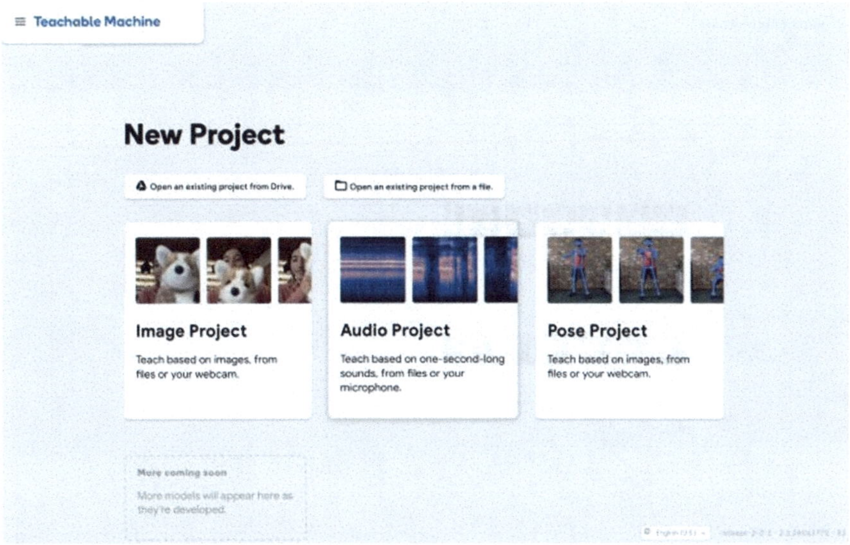

Fig. 3 Select the option Image Project

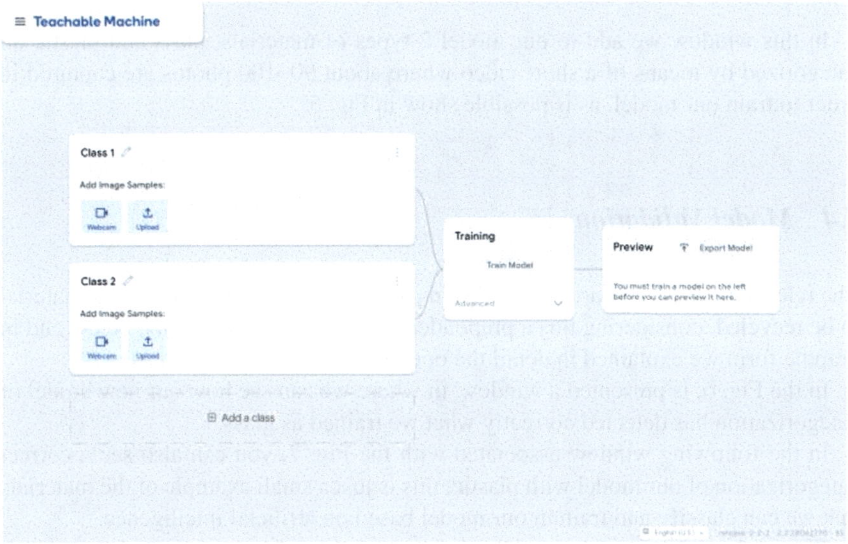

Fig. 4 Selection of a specific image of a material to classify

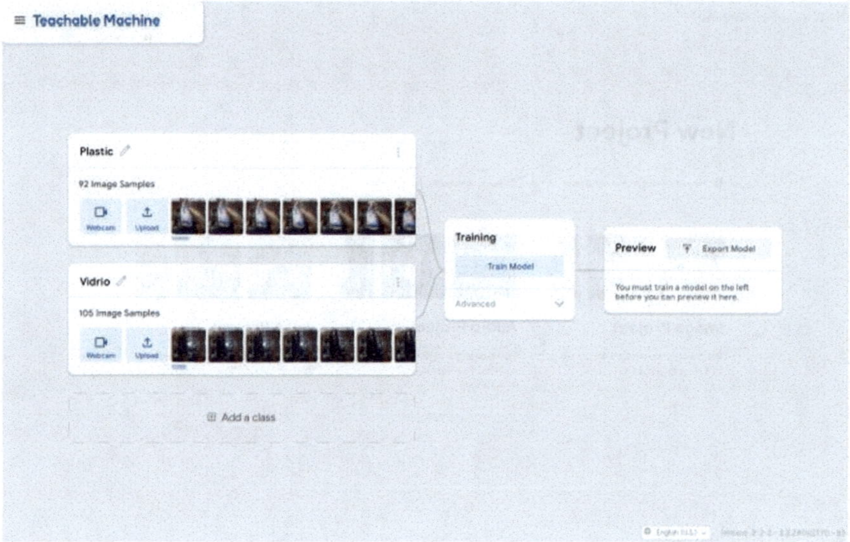

Fig. 5 Classification of a specific material using an image dataset

In this window we add to our model 2 types of materials, glass and plastic all categorized by means of a short video where about 90–100 photos are captured in order to train our model, as is possible show in Fig. 5.

4.4 Model Validation

The relevance of our research consisted in properly determining the type of material to be recycled, considering this a preponderant aspect in a Smart City, below and in graphic form we explained in detail the operation of our application as in [4].

In the Fig. 6, is presented a window, in where we can see how our new model of categorization has detected correctly what we trained as glass.

In the following window associated with the Fig. 7, you can also see a correct categorization of our model with plastic, this is just a small example of the materials that we can classify and train in our model based on artificial intelligence.

This is the turn of another application provided by google to train the model. It consists of adding a series of images in order to correctly classify the material, the only drawback is that you have to add all the images manually, model by model. In Fig. 8, we show the corresponding model associated with a specific material.

Once we capture in this case two models are processed to select the model or as we want it to compile. We select the model to train. In Fig. 9, is shown a complementary model associated with the dataset of each material.

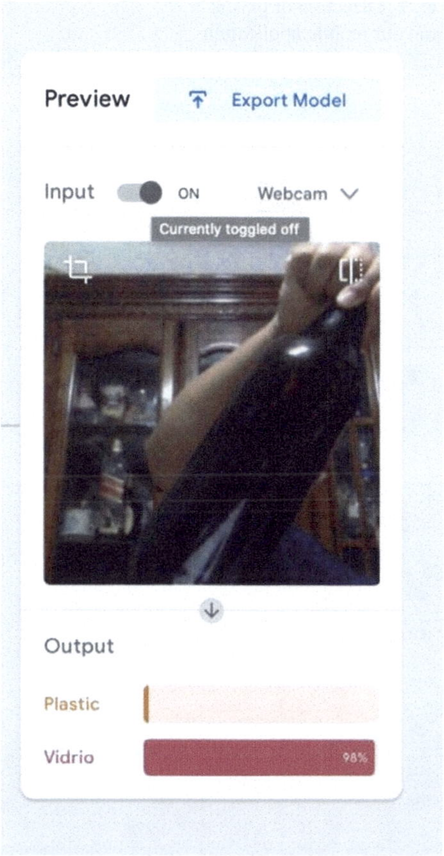

Fig. 6 Detection of glass from our mobile application

Once we train the model he will give us an estimate, the more images we add of our categories we will have a better ranking by our neural network. In the next part if we want to test our model otherwise we proceed to download the model to put it in our app. Once this is done we select the lower right button to download our model. Now it's MIT App Inventor turn to develop the application. In this image we can see the main screen of our application. In Fig. 10 is shown an application with the features of select and classify corresponding material to recycle.

In the image below we selected our screen and also on it we will upload our model previously trained.

Where, in the box marked in red you must upload the selected model already with its respective training. This is show in Fig. 11.

This is the secondary screen, where they show 3 buttons, when we press "sort video", we will be asked for permission to access camera and in real time we will be able to point out the objects and see their categorization, in this case we are going to use the same objects with those of us who trained our model. In Fig. 12 is possible identify specific modules of our mobile application.

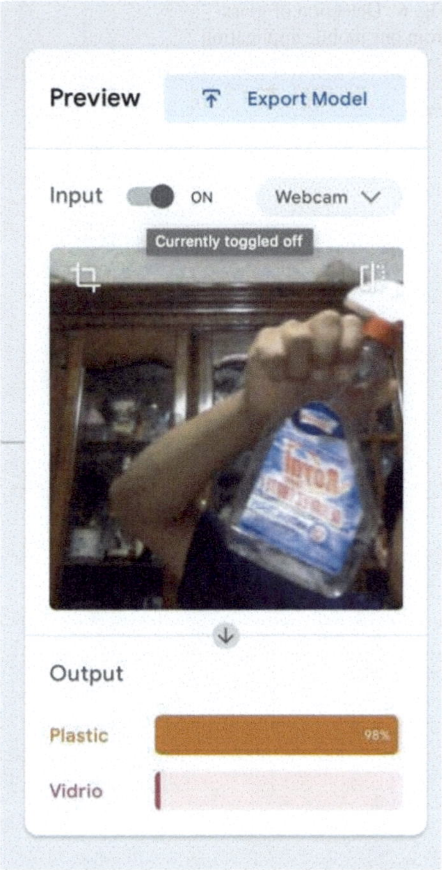

Fig. 7 Detection of plastic from our mobile application

Here we see how we have a greater number on the plastic this because of the neural network our maximum would be 1 and the minimum 0, we see that we when training our model it we categorize as plastic and at the time of focus with the camera we see which one has more weight on this categorization. In Fig. 13 is possible identify correctly a specific recyclable material as plastic.

Now if we focus our second object of the model we can see how quickly the numbers change and now the Glass over Plastic, likewise being 1 maximum and 0 as a minimum. In Fig. 14 is possible identify correctly solid waste associated with glass.

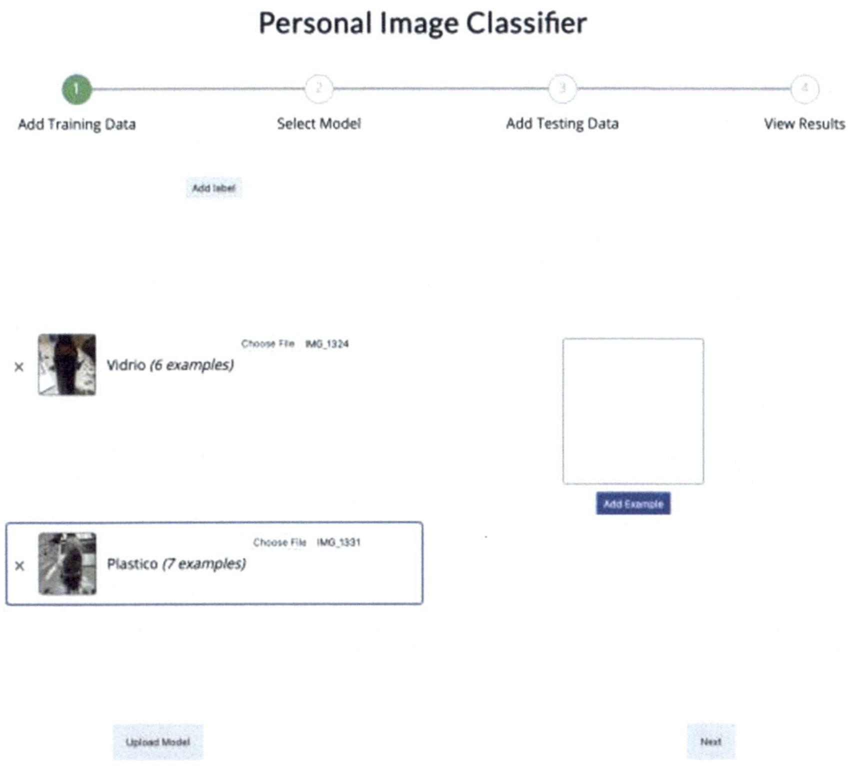

Fig. 8 Selection of a specific material to determine its correct classification

5 Integration into of an Intelligent Mobile Device Application

Properly specifying the type of material to be recycled will be timely due to the immense future opportunities that smart cities will have to determine what type of reuse could be given to it, in the present many clothes are developed using recycled materials or use to improve the life of citizens in a Smart City [5], the relevance of our ineligible model is presented in Fig. 15.

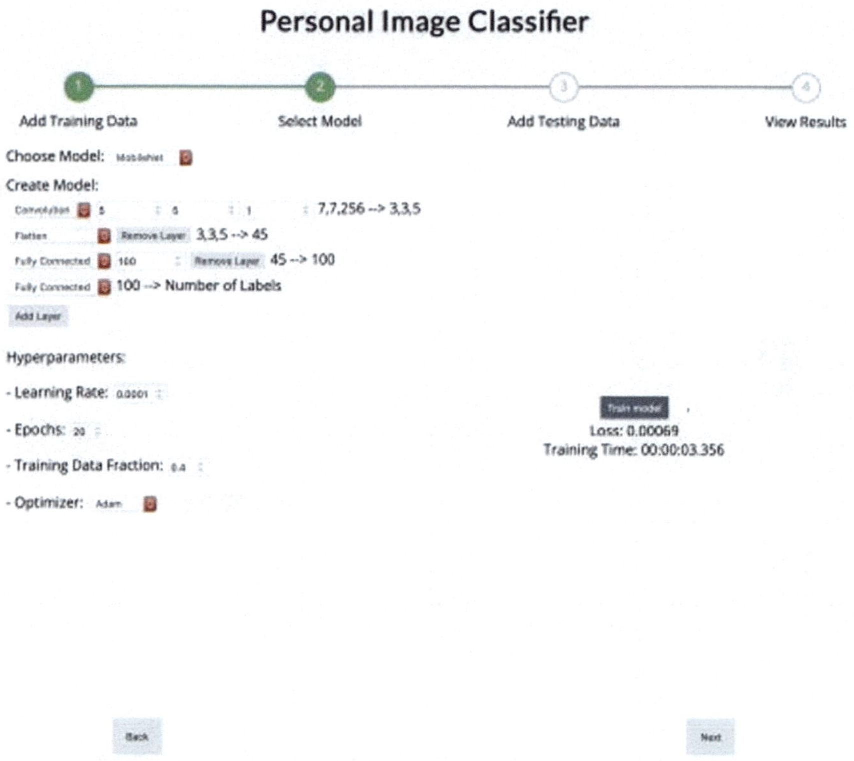

Fig. 9 Addition model of each material with its respective dataset

6 Discussion: Use of Intelligent Mobile Device Application to Improve the Competitiveness into Life Citizens on a Smart City

In order to understand the importance of the use of mobile applications that the Smart Cities will get in a couple of decades, we first need to know about the technological revolution that the Internet of Things (IoT), an increasingly important concept, brings with it. When we talk about the Internet of Things, we are referring to the sensors or objects of our daily lives that, through networks connected between them, send millions of data in real time to interpretation and analysis centers for their subsequent study and decision making. It is not uncommon to hear more and more talk about the term IoT, since more and more electronic or non-electronic objects are interconnected. To understand the development of IoT, three important and decisive aspects of its implementation must be considered:

(a) Miniaturization by which the components are becoming smaller and smaller, which is making it easier to connect almost anything, from anywhere and at any time.

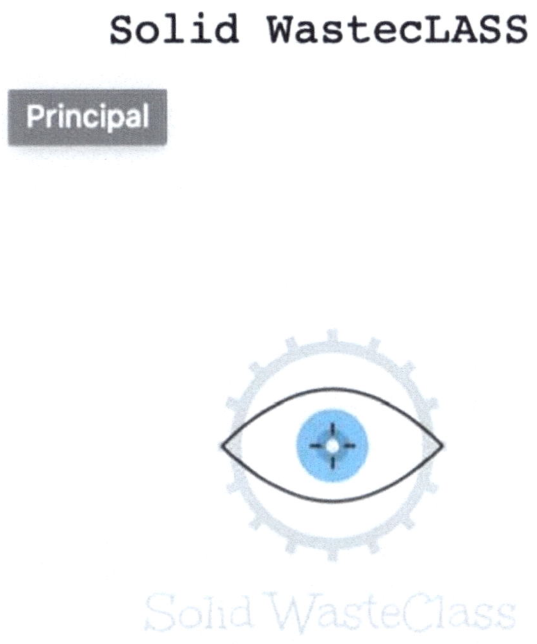

Fig. 10 An application with similar features to classify recyclable materials

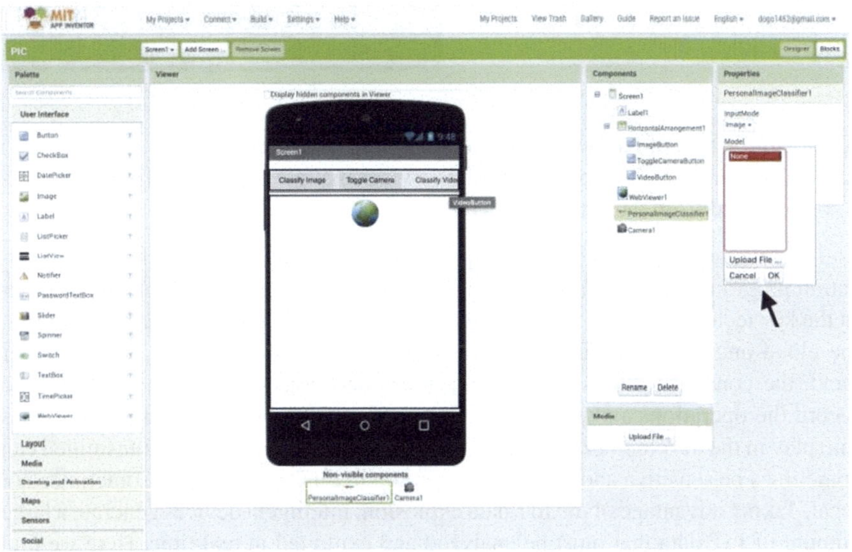

Fig. 11 Training set of our model to classify a recyclable material

Fig. 12 Modules mobile application used to our research

(b) Elimination of barriers in the mobile phone infrastructure.
(c) Dissemination of applications and services that put into use and value the data obtained from the Internet of Things.

The Internet of Things depends on a number of technologies, such as the application programming interfaces (APIs) that connect devices to the Internet. Another of the key technologies for IoT are Big Data management tools, predictive analytics, the cloud and radio frequency identification (RFID), among others. On the other hand, the communications between machines and objects need the blockchain to record the operations and protect their data safely. Thus, cybersecurity also comes into play in the development of this technology since the information transmitted can sometimes be sensitive and human intervention in this type of tool is sometimes not legal. Taking advantage of the IoT data explosion, intelligent devices generate a large amount of IoT data that must be analyzed and exploited in real time. Here we find tools such as predictive analytics and Big Data. IoT becomes the Internet of industrial things (IIoT), industry 4.0. The IIoT uses M2M technology, is used to support all kinds of things, from remote monitoring and telemetry, to predictive maintenance. The term M2M (machine to machine) refers to machine to machine communication

Fig. 13 Correctly identification of plastic material with our intelligent mobile application

without the need for people through, either a communication by proximity, being the 2 objects within the same network or remotely connected by an application or software for the purpose of monitoring, controlling or programming a particular device in a closed environment [6]. At the level of industrial development, the Internet of things in the Smart City has driven the development of new production models that collect data on how a city, and a society in particular, moves, to establish efficient methods with which to make proper use of the city's resources, the aim is to develop sustainable models of consumption and production to boost the economy. In intelligent cities the IOT is implemented through sensors, giving importance to public services such as transport, lighting, irrigation systems, industrial waste collection as in our research, among others. so it will achieve an efficient management that reduces public spending. Smart cities must opt for self-sustaining models in terms of energy and resource use, through the IoT. Thanks to this, the quality of life of both citizens and the environment in which they live will be improved.

Fig. 14 Correctly identification of glass material with our intelligent mobile application

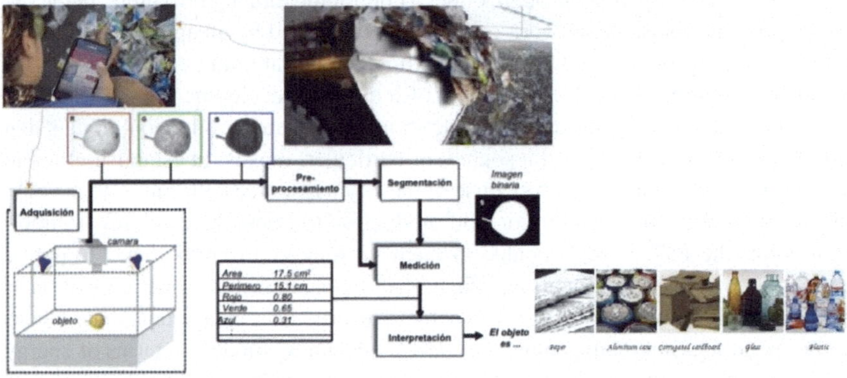

Fig. 15 Intelligent model of an image analysis process: image acquisition, pre-processing, segmentation, measurement including feature extraction, interpretation and finally classification of the five types of recyclable material

7 Conclusion and Future Research

Being a project that improves a route in which is very important determine the quantity of waste residues, not only can it be used for a Smart City, it can also be used for parcel companies, delivery companies or transport companies [7]. Using a mobile device associated with a more detailed framework could even determine specific days with more transported objects associated with each student, as can be seen in Fig. 16.

The route is formed in such a way that the next stop number is the combination with the previous stop number, in this way it is possible to know which is the point of start and departure, in order to perform the sum of the distance and corroborate which is the optimal route. An important aspect is to solve with less time the resolution of a new neighborhood with points of waste collecting with different locations and determine the correct locations of each stop of this route, as you can see in Fig. 18.

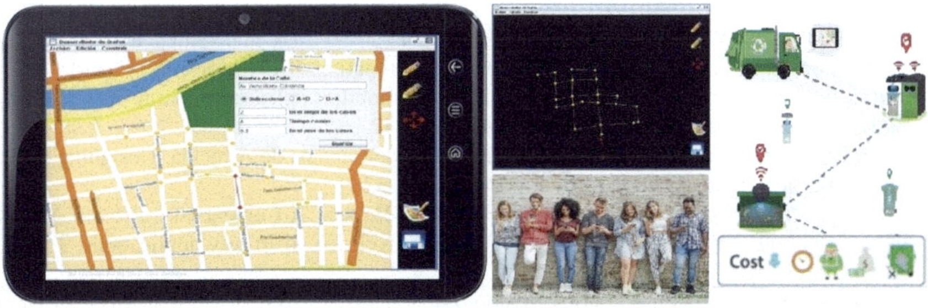

Fig. 16 Intelligent design of a mobile application based on Ubicuro Computing that would allow each person to send information associated with the organization of waste transport and online specifications of each particular situation, of the dimensions and care of their respective trash classified by recyclable materials in a Smart City. An aspect of utmost importance will be to adapt in time, the specifications of each individual that allows it to be competitive, helping it with mobility in a Smart City, as can be seen in Fig. 17

Fig. 17 The OECD considers that by 2027, seven of the 12 components of a Smart City will be associated with improving school performance, including classification of trash to their citizens, considering reuse, recycle, reinvent, reach 287 different products which were made from recycled materials

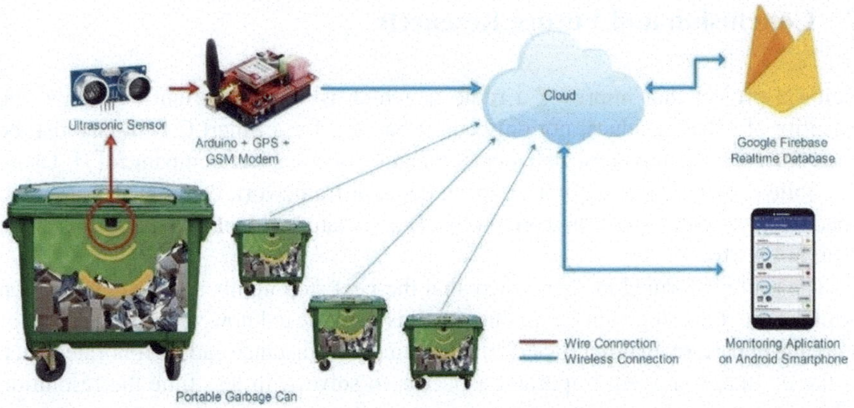

Fig. 18 Description of each group of people and their associated kind of solid waste materials and then the execution of waiting times to specify a quantity of solid waste residues

References

1. Deng Y, Chen Z, Yao X, Hassan S (2019) Jia Wu (2019) Task Scheduling for Smart City Applications Based on Multi-Server Mobile Edge Computing. IEEE Access 7:14410–14421
2. Blut C, Blut T, Blankenbach J (2019) CityGML goes mobile: application of large 3D CityGML models on smartphones. Int J Digital Earth 12(1):25–42
3. Maria RSFR, de Souza JD, McGregor, (2019) Eduardo Santana de Almeida: Designing smart city mobile applications. Emp Softw Eng 24(6):3255–3289
4. Habbal A (2019) Swetha Indudhar Goudar, Suhaidi Hassan: A Context-aware Radio Access Technology selection mechanism in 5G mobile network for smart city applications. J Netw Comput Appl 135:97–107
5. Holguin L, Ochoa-Zezzatti A, Larios-Rosillo VM, Cossio E, Maciel R (2019) Gilberto Rivera Zarate: Small steps towards a smart city: Mobile application that provides options for the use of public transport in Juarez City. ISC2 2019: 100–105
6. Idwan S, Mahmood I (2020) Junaid Ahmed Zubairi, Izzeddin Matar: Optimal Management of Solid Waste in Smart Cities using Internet of Things. Wireless Pers Commun 110(1):485–501
7. Topaloglu M, Yarkin F, Kaya T (2018) Solid waste collection system selection for smart cities based on a type-2 fuzzy multi-criteria decision technique. Soft Comput 22(15):4879–4890

Logistics on the Designing of an Electronic Colorblindness Application for Early Colorblindness Detection in Children by Using a Modified Ishihara Test

Martín Montes Rivera, Alejandro Padilla Díaz, Julio César Ponce Gallegos, Juana Canul-Reich, Alberto Ochoa-Zezzatti, and Miguel A. Meza de Luna

Abstract Colorblindness is a condition that affects the cones in the eyes; it can be congenital or acquired and is considered an average disability that affects about 10% of the world's population. Children with colorblindness have difficulties when they enter to an educational environment because the educational materials are mostly developed for people with normal vision. Unfortunately, they must deal with the common materials, classes, and not inclusive environments, because colorblindness is usually diagnosed until adulthood. The main reason that makes difficult the earlier detection of colorblindness is related with the tests for its detection, because they are too complex for being applied in children. Moreover, preschool children cannot read numbers that are commonly used in the Ishihara test, and other types of tests are too challenging and require high attention, making them inadequate for children. This work focuses on modifying the Ishihara test by changing the numbers to more familiar figures and reducing the number of plates required for the diagnosing of preschool children, while maintaining the logistics of the original test, the proposed test also ads audio support, so that the kids understand what they should see and do, even if

M. M. Rivera (✉)
Universidad Politécnica de Aguascalientes, Aguascalientes, México
e-mail: martin.montes@upa.edu.mx

A. P. Díaz · J. C. P. Gallegos · M. A. M. de Luna
Universidad Autónoma de Aguascalientes, Aguascalientes, México
e-mail: apadilla2004@hotmail.com

J. C. P. Gallegos
e-mail: julk_cpg@hotmail.com

M. A. M. de Luna
e-mail: meza_capacitacion@hotmail.com

J. Canul-Reich
Universidad Juárez Autónoma de Tabasco, Villahermosa, México
e-mail: juana.canul@ujat.mx

A. Ochoa-Zezzatti
Universidad Autónoma de Ciudad Juárez, Ciudad Juárez, México
e-mail: alberto.ochoa@uacj.mx

© The Author(s), under exclusive license to Springer Nature Switzerland AG 2021
A. Ochoa-Zezzatti et al. (eds.), *Technological and Industrial Applications Associated with Intelligent Logistics*, Lecture Notes in Intelligent Transportation and Infrastructure,
https://doi.org/10.1007/978-3-030-68655-0_32

they cannot read the instructions in the test. The proposed digital application helps to identify children who suffer from colorblindness, is designed for being applied with simplicity by every teacher, and its use helps him to identify the students that require special attention and materials in their education process, because they suffer from a diminished perception of color.

1 Introduction

The sense of sight in human beings as in other organisms depends on their eyes; these use two types of cells for the perception of images, rods and cones [1].

The rods allow to identify the luminosity, i.e. the amount of light received from the environment and the cones allow to identify the color or frequency in the spectrum of light received [2].

In most people, there are three types of cones, each one to perceive a basic color, these can be red, green, or blue, and the other colors that are generated are the result of various combinations that are received amounts of light in tune with the frequencies of these basic colors [3–5].

The world around us is designed to work with the colors that are perceived with three cones, since most people, can perceive the environment with three basic colors, i.e. they are trichromats, however, there are data of people with a fourth type of cone, which allows them to perceive more colors than the average person visualizes, however, these people often have problems describing the environment and tones they perceive, since the world is not made contemplating their sensory perceptions [6].

On the other hand, there are also cases of people with a lower color perception, this condition is called colorblindness and is considered a medium disability, since the colors with trichromat perception are used in various activities, such as to identify objects in a conversation, to identify dangerous situations, know when to advance in a traffic light, to decide what clothes to buy and to enjoy art forms such as painting or photography [7, 8].

Colorblindness can be catalogued in four variants according to the cones available to perceive the environment, these variants can be anomalous trichromacy, dichromacy and monochromacy or achromatopsia [2].

The most common variant of colorblindness is anomalous trichromacy, in which there are all the cones for color perception, but there is a deficiency in some of them. Anomalous trichromacy can be separated depending on its severity into mild anomalous trichromacy, medium or strong and depending on the color in which the deficiency is presented, trichromacy can be divided into deuteranomalia or deficiency in the perception of green, protanomalia in the perception of red and tritanomalia in the perception of the blue [9].

Another variant of colorblindness that occurs less frequently, but more severely is the dichromacy with an absent cone, i.e., the person cannot perceive one of the basic colors, which causes problems with all colors that have this tone in their constitution,

Table 1 Prevalence of colorblindness in the world [2]

Type	Variant	Prevalence H/M	
Monochromacy	Acromatopsia	0.00003%	
Dichromacy	Deuteranopia	1.27%	0.01%
	Protanopia	1.01%	0.02%
	Tritanopia	0.0001%	
Anomalous Tri	Deuteranomalia	4.63%	0.36%
chromacy	Protanomalia	1.08%	0.03%
	Tritanomalia	0.0002%	

for example, a person who has problems with the green cone, will have problems with all forms of green, but also with yellow and brown because they are constituted with green as a base. Dichromacy can also be classified depending on the absent cone, it is deuteranopia when the green cone is absent, protanopia when the red cone is absent, and tritanopia when the blue cone is absent [2].

Monochromacy or achromatopsia, is the rarest condition of colorblindness, in this, all the cones are absent, therefore, only the environment is perceived in gray scales or luminosity and although it is very rare, represents a major difficulty for people who suffer from it, since they cannot live a normal life without the assistance of healthy people. These people cannot drink a liquid from a bottle without first looking for a label confirming its contents, they cannot identify if a food is in good condition before eating it, they cannot choose their clothes, identify a parking place, among other difficulties [2].

About 10% of people suffer from some deficiency or blindness to color, that is, about 700 million people suffer from colorblindness, considering that the world population exceeds 7,000 million inhabitants. Table 1 shows the percentages of incidence in the world for men and women with each of the variants of colorblindness [2].

As often as in adults worldwide, 1 in 10 children is born colorblind, facing a world that is not designed for him, generating various difficulties even in their learning and school performance [10]. Mexico has a similar situation in colorblindness as detailed in the study of prevalence of colorblindness in public school children in Mexico [11].

Children with visual difficulties associated with colorblindness may have school problems when performing activities that involve color identification, such as using educational material with colored content, for example, relating content in their textbooks and participating in games, among other difficulties related to visual tasks [10].

Despite all the difficulties that children with colorblindness are exposed to, this condition is one of the vision anomalies that takes longer to be detected by parents and teachers, because by the intelligence or support they receive from other people they manage to get ahead with their education [10].

Teaching materials are often designed for people with normal vision, so it is important to detect vision before school age so that children receive appropriate assistance in their educational processes [11].

The detection of colorblindness can be done through the application of various tests that are designed according to the variant that is presented and the colors that are confused in this condition, for this, it is important to reproduce the perception of a colorblind people so that the tests be correctly designed. There are several algorithms that adjust the model of a digital image represented with red, green and blue parameters (RGB), to simulate colorblindness and then allow people with a normal vision to see as people with this medium disability do [12].

Figure 1 shows the color spectrum seen by an average trichromat, Fig. 2 by a person with protanopia, Fig. 3 by deuteranopia, and Fig. 4 by tritanopia, all of them obtained applying colorblindness simulation models.

The most common test used by ophthalmologists, based on the principle of detecting confusing colors, is the Farnsworth-Munsell test, in which the patient is presented with discs with colors of the entire color spectrum and is asked to make their arrangement in the correct order, this test is highly accurate in identifying any variant

Fig. 1 Full color spectrum

Fig. 2 Full color spectrum perceived by a protanope

Fig. 3 Full color spectrum perceived by a deuteranope

Fig. 4 Full color spectrum perceived by a tritanope

of colorblindness and the severity with which it is presented, however, its application is highly complicated and to carry it out correctly, specific ambient conditions are required, such as a specific brightness of 25 candles at 6740 K, which describe the lighting conditions at midday [13]. Figure 5 shows a simplified Farnsworth-Munsell test manufactured by Lea Color Vision Testing.

The most commonly used tests for the detection of colorblindness are Ishihara plates, designed for the detection of protanopia, protanomalia, deuteranopia, and deuteranomalia, however, any other variant of colorblindness cannot be detected by Ishihara plates as they are designed to detect the colors that confuse people with these variants of colorblindness [15]. Figure 6 shows plate 42 that is seen by people with protanopia or strong protanomalia as a 2 and by deuteranopia and strong deuteranomalia as a 4.

The tests to perform the Ishihara test are widely known and simple to evaluate, because the difficulties to perceive the color are observable when the individual who is submitted to the test, cannot see the number inside or it is difficult to visualize it, in addition, depending on the plate with which you have problems, it is identified which variant of colorblindness is presented. Table 2 is used as a comparison list to identify the type of colorblindness variant presented with the Ishihara test.

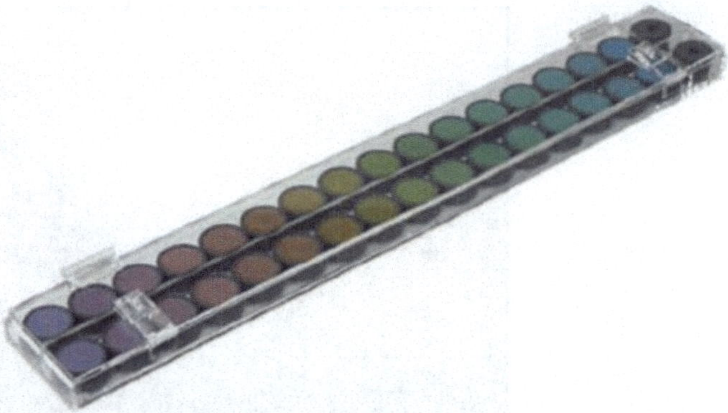

Fig. 5 Farnsworth-Munsell Test manufactured by Lea Color Vision Testing [14]

Fig. 6 Ishihara Test Plate 42 [15]

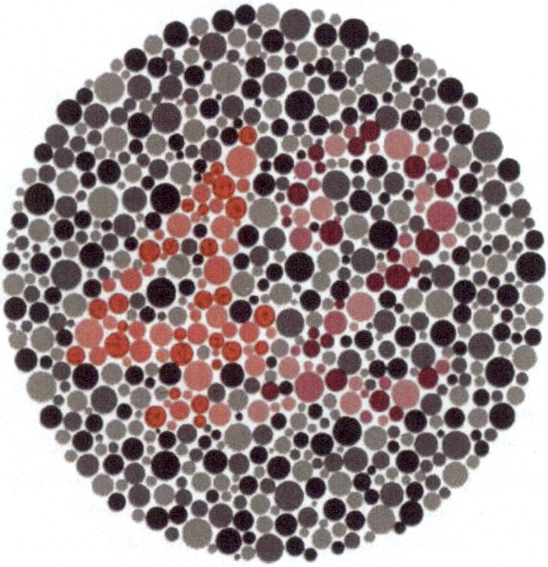

The initial problem with this type of test is the purchase of the same, however, currently there are several websites of organizations, which allow a rapid assessment when there are suspicions of colorblindness.

An ageappropriate colorblindness test in [16] includes the possibility of detecting colorblindness in toddlers when dealing with figures; however, given the options presented for each plate, the test can still be complicated for a preschooler. One of the images presented in this test is shown in Fig. 7. In this slide, the child is asked whether he sees a letter B behind the mesh, a star behind the mesh, or the mesh itself.

Developing a test that can be done at the preschool level with a group of children would make it possible to raise awareness of the difficulties presented by some of

Table 2 Checklist for evaluation with Ishihara test of 17 plates, where the X mark that cannot be read [15]

Plate Number	Plate perceived trychromats	Plate perceived by protanopes or deuteranopes		Plate perceived by monochramats	
1	12	12		12	
2	8	3		X	
3	29	70		X	
4	5	2		X	
5	3	5		X	
6	15	17		X	
7	74	21		X	
8	6	X		X	
9	45	X		X	
10	5	X		X	
11	7	X		X	
12	16	X		X	
13	73	X		X	
14	X	5		X	
15	X	45		X	
		Protanope		Deuteranope	
		Strong	Mild	Strong	Mild
16	26	6	(2)6	2	2(6)
17	42	2	(4)2	4	2(4)

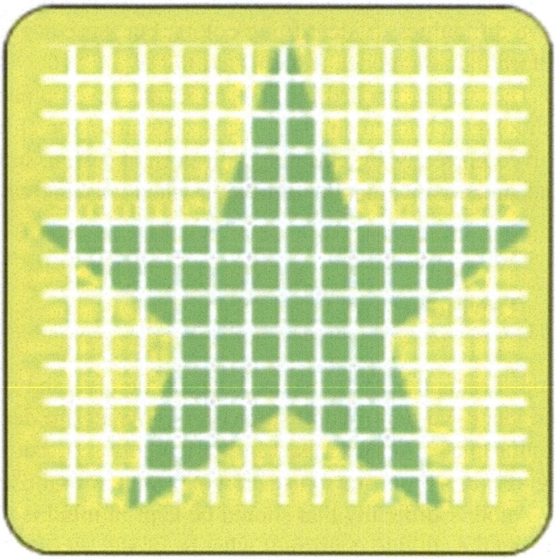

Fig. 7 Star test plate in [16]

the peers in that group and allow the preschool teacher to identify students who have problems with color perception, to adjust the activities to children who might face this type of condition in a preschool group.

The aim of this work is to review the issue of detecting colorblindness in preschool children, the difficulties and guidelines, while proposing images that can be identified through sight by groups of healthy children and disagree with those perceived by people with some form of colorblindness, in this chapter focusing primarily on the identification of protanopia, protanomalia, deuteranopia and deuteranomalia in the classromg, so that the teacher can recommend to parents going with an specialist to confirm a positive evaluation in their son and then, present a more extensive diagnosis. The tests are also mounted on an application so that they can be taken from home or in their classroom with the help of parent or tutor.

2 Backgrounds

The concern for colorblindness in children is not an issue that has begun to study recently and there are several papers that are presented around this average disability.

Several of these papers focus on identifying the incidence of colorblindness in children in certain parts of the world, such as at [11], where colorblindness is detected in schoolage children in Mexico. The same incidence is being studied in eastern Nepal [17]. In [18] a similar study is conducted on children between the ages of 6 and 15 in Jalandhar, in all cases using Ishihara plates.

Other works identify problems that children with colorblindness have and their detection.

One of these studies shows that children with colorblindness problems often have difficulties in identifying colors, but these are less than those expected in the colorblindness model, as indicated in [19].

The work proposed in [20] shows the development of a computer interface for schoolage children, it uses images suitable for children in a game that can be solved correctly by children with normal vision, however, they need instructions and supervision, making it difficult when working with a group of children.

3 Development

Considering that the most used diagnostic tests and simple to evaluate are linked to the identification of confusing colors, images are designed with these colors using similar plates to those used in the Ishihara test, but instead of using numbers and figures, it is proposed to use drawings similar to those recognized at an early age.

Another difficulty that should be kept in mind is to keep the indications simple, asking the child to confirm if what he or she sees is correct or not. For this purpose, the Microsoft Net Assembly speech interface is used to make the computer tell the child what it would have to see in the developed images.

However, in the proposed test logistics of the original Ishihara test must be maintained, so that its quality and methodology can be used in the but being adapted to children by modifying the plates and the quantity of them, so that they can be focused during their diagnose. In that sense, the proposed test follows the original instructions in the Ishihara test, and therefore, the first thing shown in the application is a description addressed to the applicator or the teacher, which indicates the variants of colorblindness that can be detected, the purpose of the test and the Ishihara instructions, but adapted to digital application for diagnosing preschool children.

Once the applicator clicks the button that starts the test, the images are presented to the children for three seconds, as it is proposed in the original Ishihara test, for avoiding the use of contrast and brightness to try to identify the images in the test, but in the proposed children test the digital application plays in the speaker at the same time the instructions, so that the child understand what he should see and the application identify whether or not the child can see the figure.

As in the original test the presented plates are already associated with some variant of colorblindness depending on the perception, in that sense, children feedback to the applicator telling him if they see the presented plate correctly or not, and depending on their answer, the application will catalogue the answer with some variant of colorblindness.

After ending the test, the total number of associations are counted and compared with the total number of plates, which in the case of the children application is reduced to six plates for reducing the required attention, helping the teacher to apply the test with simplicity to the children, this reduction of plates does not affect the aim of detecting children with possible problems associated with a diminished perception of color, because after that the children are identified they must be further evaluated by an specialist.

It is hoped that, with this test, the applicator or teacher can identify which children have problems with the color perception, and then he should inform the parents, so that they can make the child diagnosed by a specialist. After a positive diagnose, the teacher should consider the special condition of the affected children when preparing the material for his classes.

The flow diagram presented in Fig. 8 describes the logistics inside the application with the proposed test designed for identifying colorblindness in preschool children.

4 Results

The plates developed are based on the colors used in the Ishihara plates, with designs that are known by preschool children, including a face, a three, a candy, a boat a sun and a house, which are shown from Figs. 9, 10, 11, 12, 13, 14.

Before applying the test to children, the plates must be evaluated for verifying that they will be not correctly seen by children with this affection. A good alternative for

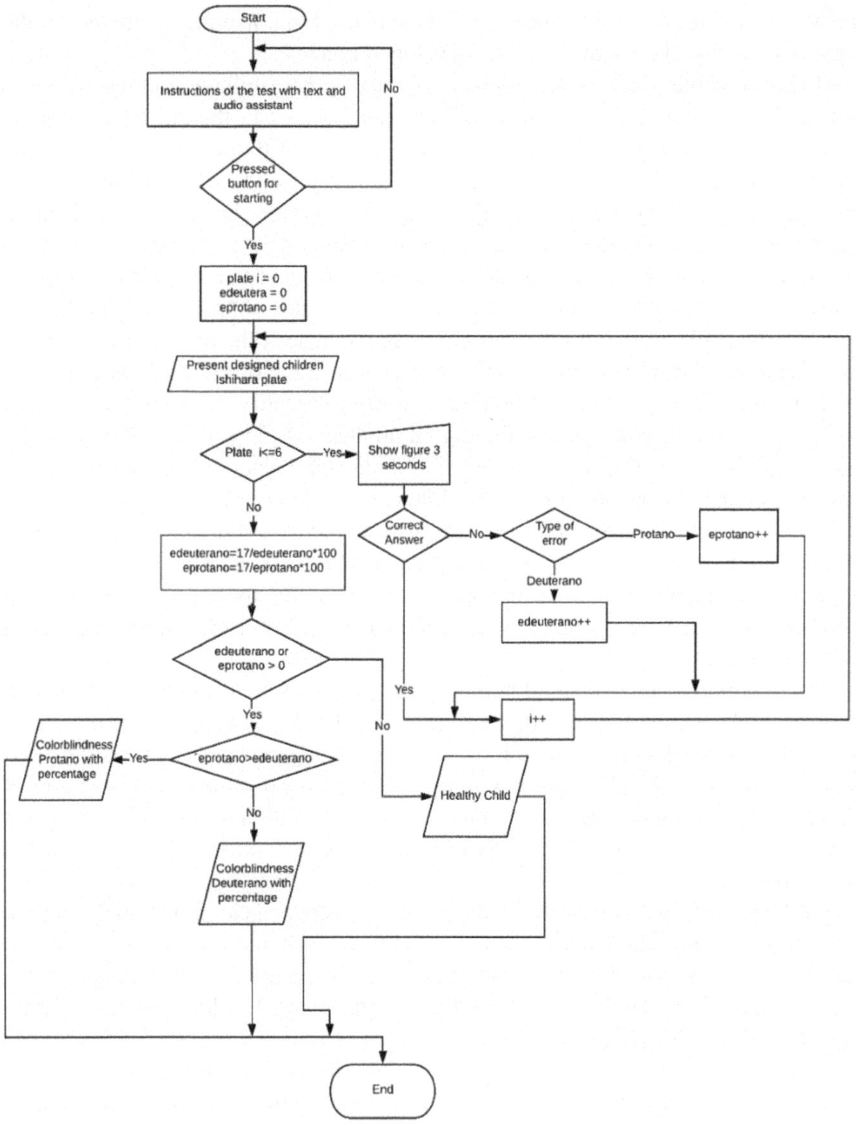

Fig. 8 Flow diagram with the application logistics for detecting colorblindness in preschool children

perceiving the plates as people with colorblindness does is by using a digital camera and processing the images with a colorblindness simulation model as described in [21].

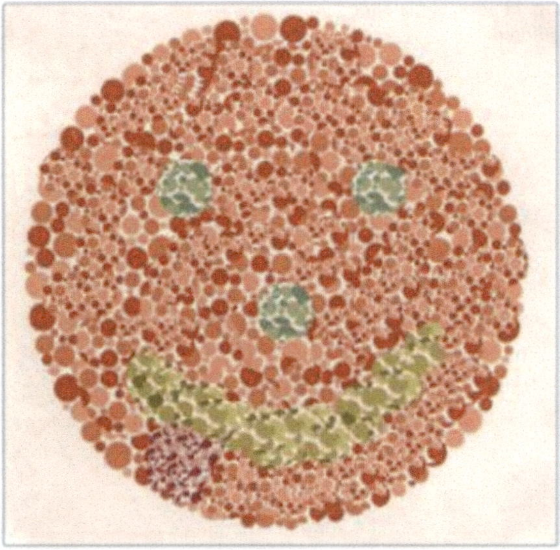

Fig. 9 Face plate for children

Fig. 10 Tree plate for children

The Table 3 shows the images obtained using a dichromacy colorblindness simulation model (protanopia, deuteranopia, and tritanopia), so that each plate can be perceived like in each of the dichromacy critical colorblindness variants.

The application interface during all the moments of the test are shown from Figs. 15, 16, 17, 18, 19, 20, 21, 22.

The Fig. 15 shows the instructions are shown and read when opening the application to start the test, the translated instructions say:

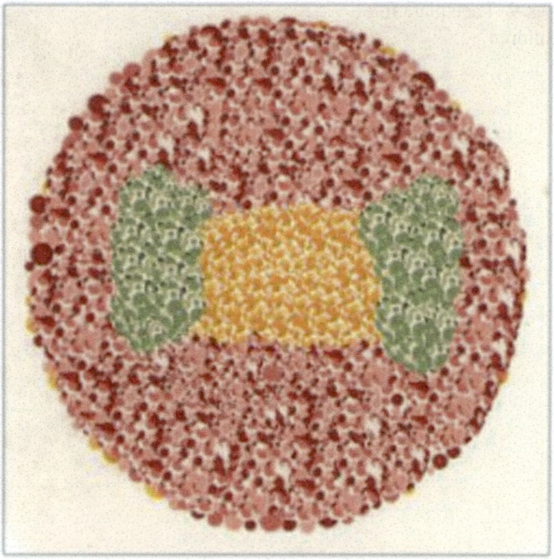

Fig. 11 Sweet plate for children

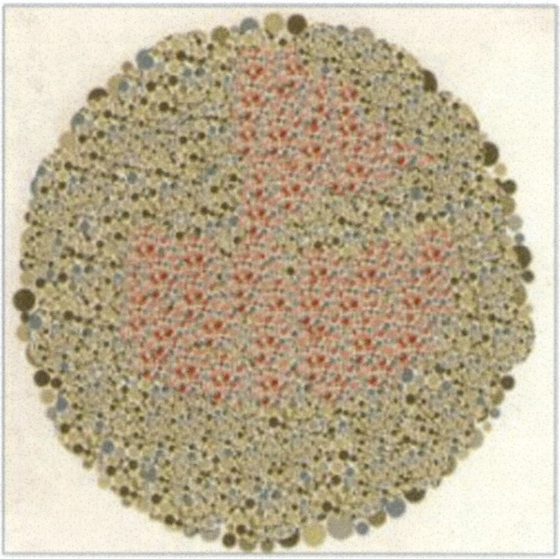

Fig. 12 Boat plate for children

Ishihara automated test.

Instructions: The Ishihara test is a useful tool for diagnosing people with colour blindness in its variants: protanopia, deuteranopia, protanomalia and deuteranomalia.

This test is specially designed for children, only has 6 plates that are child friendly, the test can to a child or group, the teacher should only mark if the child or the whole

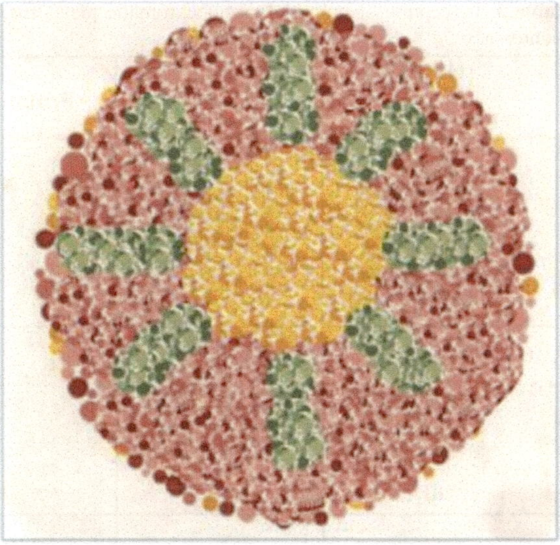

Fig. 13 Sun plate for children

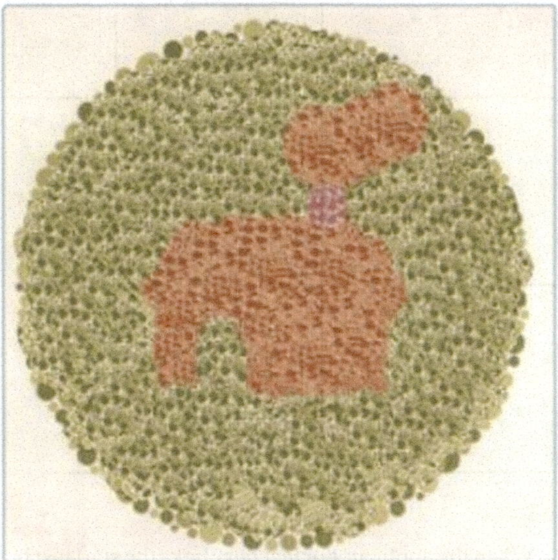

Fig. 14 House plate for children

group of children managed to see the figures in a way right or wrong, a computer will indicate what the child should see and children should indicate whether they see it or not.

To start the test, click on Start Test, display the images that are presented and select from the drop-down list the option that corresponds to what you see.

Table 3 Perception of proposed pseudochromatic plates for children with different variants of dichromacy

Normal perception	Deuteranopia	Protanopia	Tritanopia
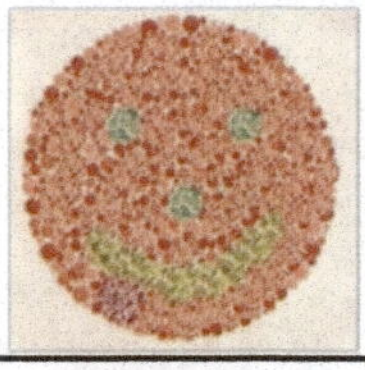	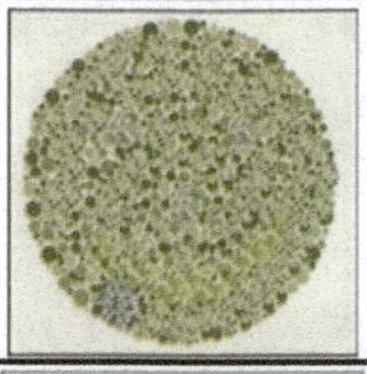	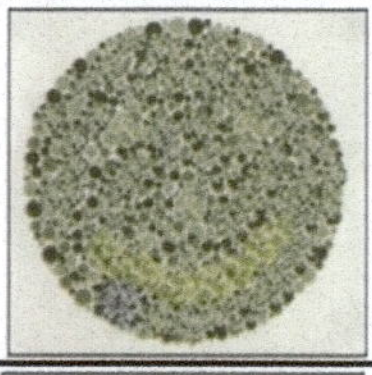	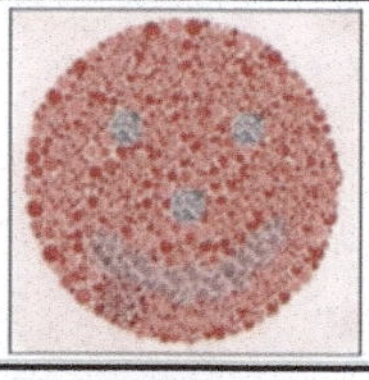
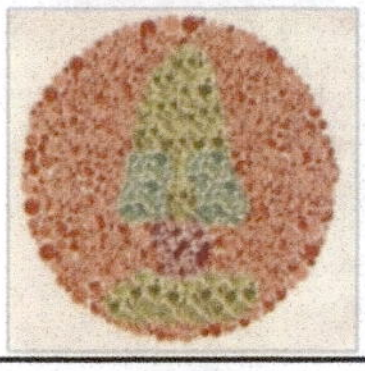	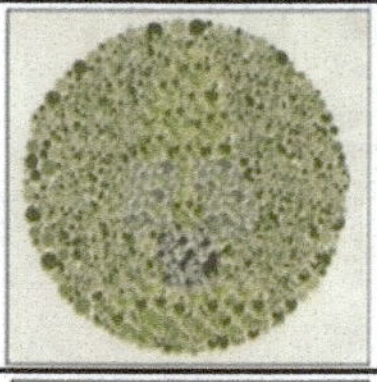		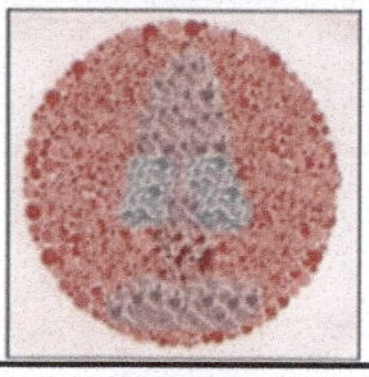
	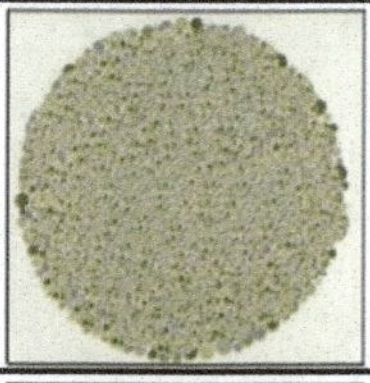		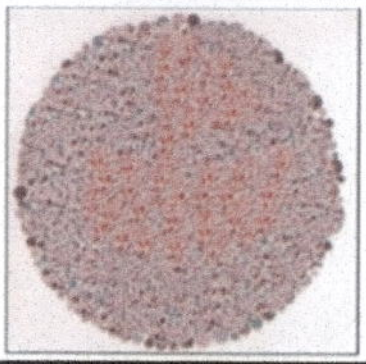
	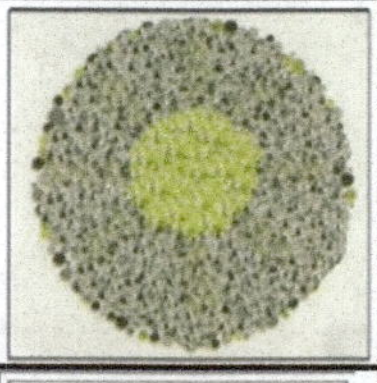	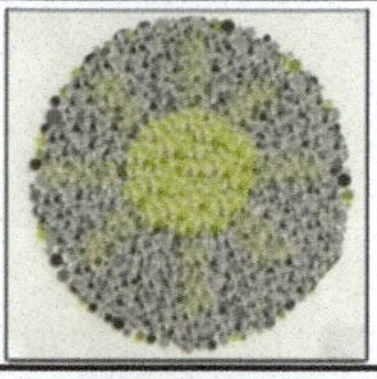	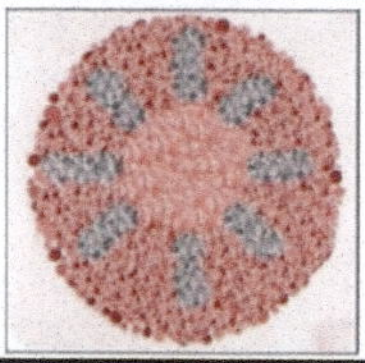
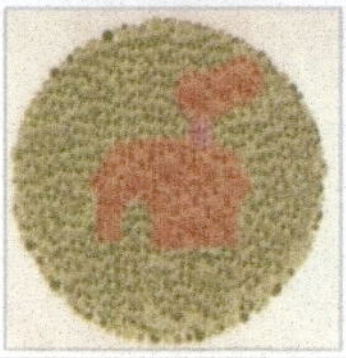		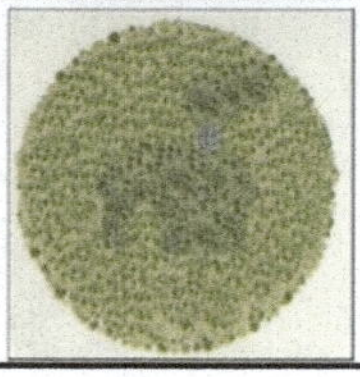	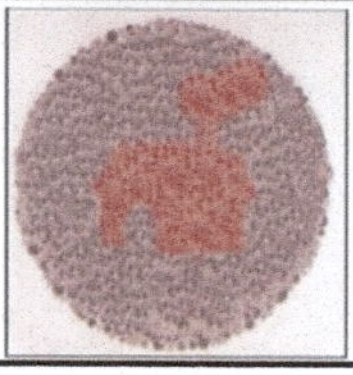

When the applicator or teacher clicks the start test button, images begin to be shown Figs. 16, 17, 18, 19, 20, 21, 22 as the sound produced with Microsoft Net Assembly tells the children which image they should see and then the applicator selects from the drop-down list whether the plate was viewed correctly or incorrectly.

When the test is completed after reporting the percentage of failure, a chart is received indicating that a doctor should be visited visually and by audio, in case of failure in the identification of the plates. The user has the possibility to do the test again.

The diagnosing of the test translated says:

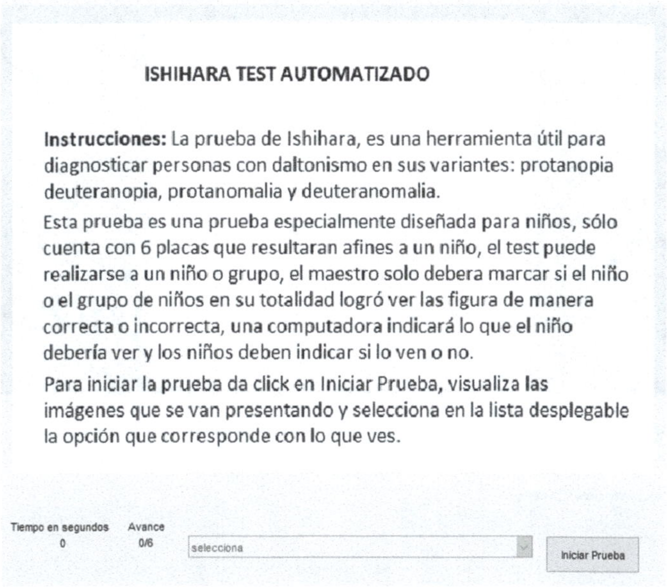

Fig. 15 Instructions shown in the application when opening the test

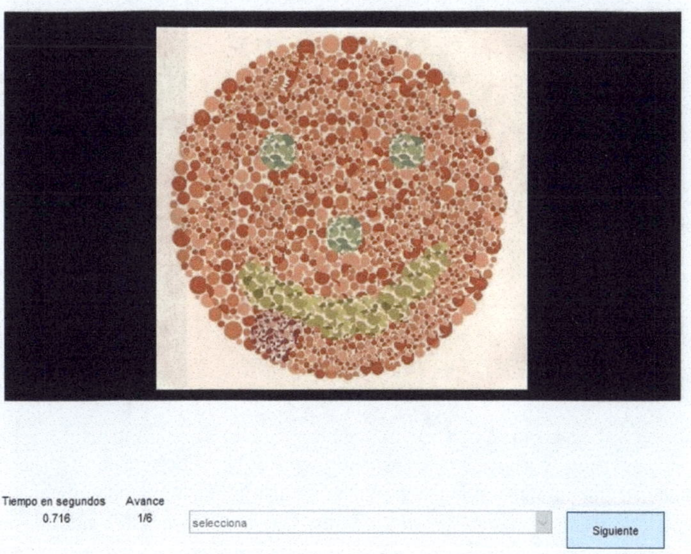

Fig. 16 The face shown in the application when the Microsoft Voice Assistant says "You should see a Face"

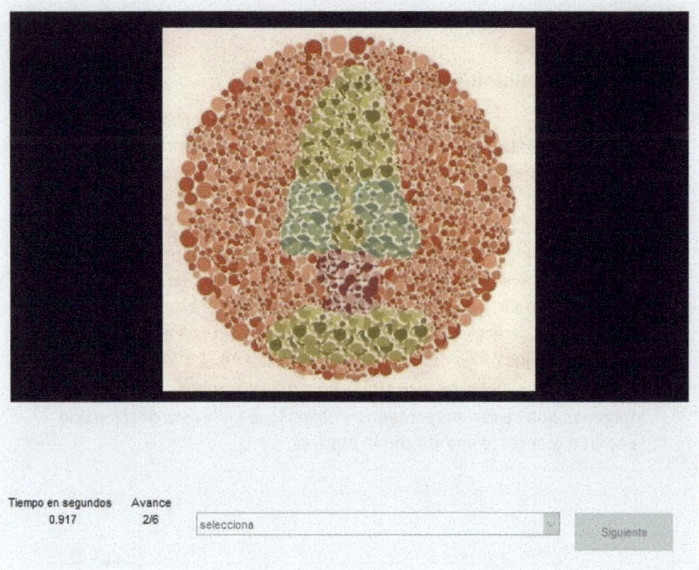

Fig. 17 Tree shown in the application when the Microsoft Voice Assistant says "You should see a Tree"

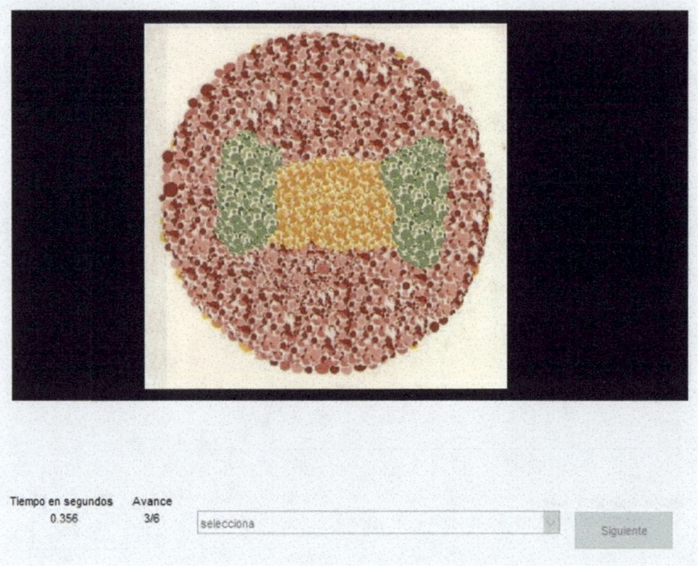

Fig. 18 Candy shown in the application when the Microsoft Voice Assistant says "You should see a Candy"

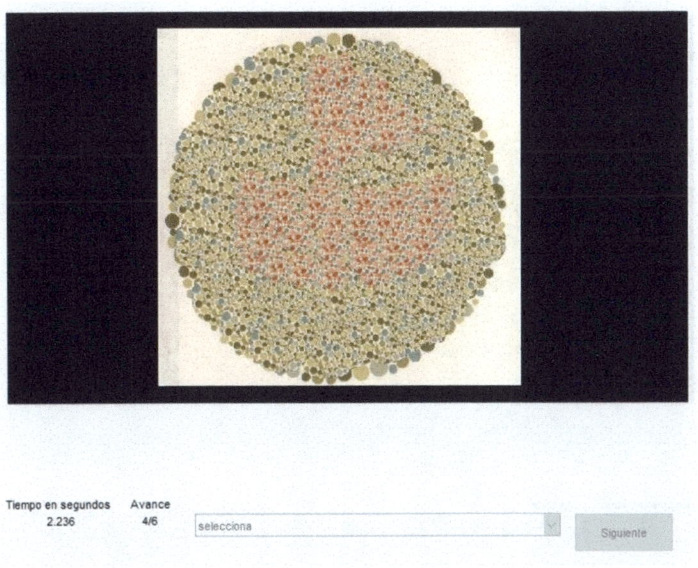

Fig. 19 Ship shown in the application when the Microsoft Voice Assistant says "You should see a Ship"

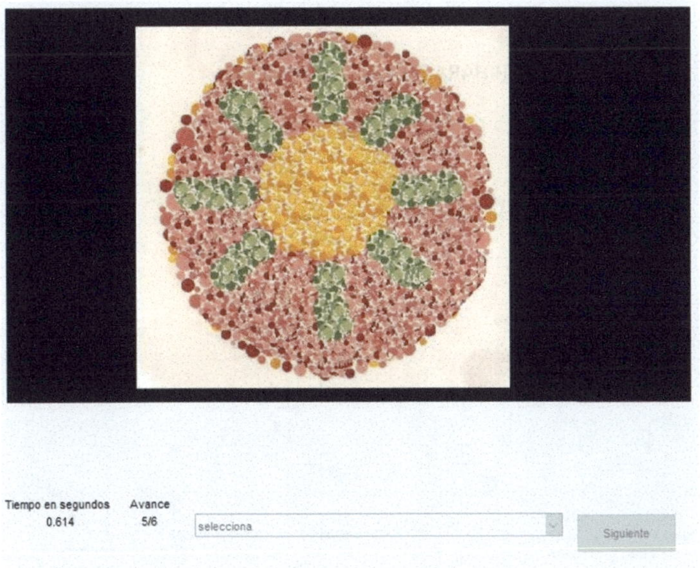

Fig. 20 Sun shown in the application when the Microsoft Voice Assistant says "You should see the Sun"

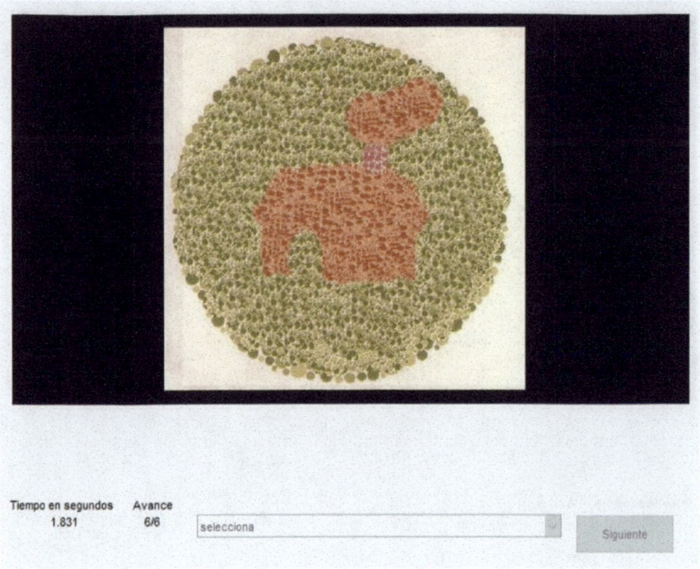

Fig. 21 House shown in the application when the Microsoft Voice Assistant says "You should see a House"

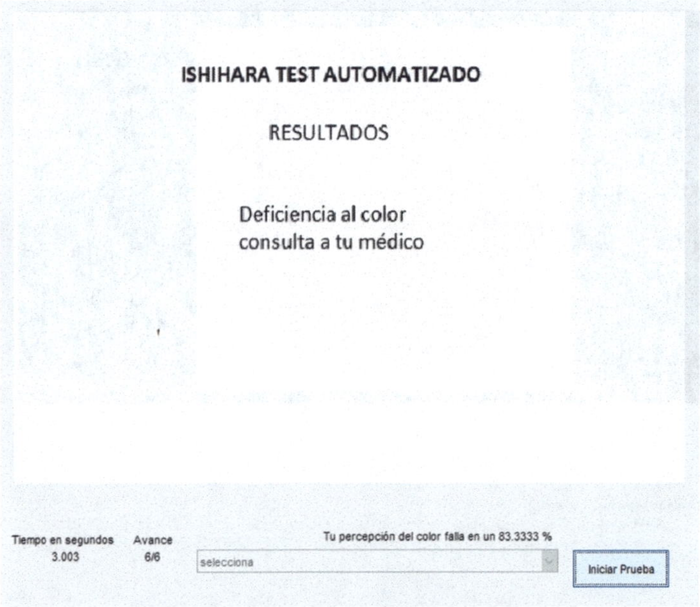

Fig. 22 Conclusion of the test indicating that there is deficiency in color perception and a doctor should be visited

Ishihara automated test results

Color deficiency.

see your doctor.

5 Conclusions

In this work we manage to design plates with figures that preschool children can identify, as well as using them in an application that presents an audio aid that tells children what they should see on each plate and with the help of an applicator that could well be the teacher can apply the test. The proposed test is simplified for reducing the required attention in the children and changing the numbers in the original Ishihara test by familiar figures, but the logistics in the original test are retained so that a diminished perception of color can be diagnosed.

The design of each of the plates is intended to make it difficult for people with problems in the different variants of colorblindness that are detected by the Ishihara test, i.e., protanopia, protanomalia, deuteranopia, and deuteranomalia.

The graphs show different figures depending on the variants of colorblindness that are presented as shown in the results section.

6 Future Work

Apply the pilot test with preschool children to verify that they are familiar with the figures, as well as seek to detect real cases of colorblindness using the proposed test. The results reported by the test must be evaluated in the presence of specialist in colorblindness and child psychologist, for improving the test in quality of the detection of colorblindness and simplicity.

The plates proposed for detecting colorblindness has familiar figures for a child, but the user interface in the application still uses the windows general styles, in order to improve the attention of the children, a new user interface must be designed with a more attractive appearance for preschool kids.

In the same way the Microsoft Voice is not the most attractive to children, with that in mind is recommended to generate special records with instructions using more characterized voices that can be better accepted by children.

References

1. Richmond Products: Color Vision Deficiency: A Concise Tutorial for Optometry and Ophthalmology. Richmond Products (2012)
2. Colorblindor: Color Blind Essentials (2020) https://www.color-blindness.com/color-blind-essentials/. Accessed 11 Aug 2020
3. Deeb SS (2004) Molecular genetics of color-vision deficiencies. Vis Neurosci 21:191–196
4. Meng M, Tanaka G (2019) Proposal of minimization problem based lightness modification method considering visual characteristics of protanopia and deuteranopia. In: 2019 Asia-pacific signal and information processing association annual summit and conference, APSIPA ASC 2019. pp. 1417–1422. Institute of Electrical and Electronics Engineers Inc. (2019). https://doi.org/10.1109/APSIPAASC47483.2019.9023294.
5. Rivera MM, Díaz AP, Reich JC, Gallegos JCP, Zezzatti AO (2019) Augmented Reality labels for security signs based on color segmentation with PSO for assisting colorblind people. Int J Comb Optim Probl Inform 10:7–20
6. Robson D (2016) Las mujeres con una visión superhuman. https://www.bbc.com/mundo/noticias/2014/09/140911_vert_fut_mujeres_vision_superhumana_finde_dv
7. Kato C (2013) Comprehending color images for color barrier-free via factor analysis technique. In: 2013 14th ACIS international conference on software engineering, artificial intelligence, networking and parallel/distributed computing. pp. 478–483. IEEE (2013). https://doi.org/10.1109/SNPD.2013.39
8. López de Nava A, Salini UB B (2020) Physiology, Vision. StatPearls Publishing (2020) from https://www.ncbi.nlm.nih.gov/pubmed/30860728. Accessed 11 Aug 2020
9. Huang C-R, Chiu K-C, Chen C-S (2011) Temporal Color Consistency-Based Video Reproduction for Dichromats. IEEE Trans Multimed 13:950–960. https://doi.org/10.1109/TMM.2011.2135844
10. Pardo Fernández, P.J., Gil Llinás, J., Palomino, M.I., Pérez Rodríguez, A.L., Suero López, M.I., Montanero Fernández, M., Díaz González, M.F.: Daltonismo y rendimiento escolar en la Educación Infantil. Rev. Educ. ISSN 0034–8082, No 330, 2003, págs. 449–462. 449- 462 (2003).
11. Jimenéz Pérez AA, Hinojosa García L, Peralta Cerda EG, García García P, Flores Peña Y, M, CVV, Cerda Flores RM (2013) Prevalencia de daltonismo en niños de escuelas pú-blicasde México: detección por el personal de enfermería. CIENCIAUANL 16, 140–144
12. Tomoyuki O, Kazuyuki K, Kajiro W, Yosuke K (2010) Proceedings of SICE annual conference 2010. Presented at the (2010)
13. Cranwell MB, Pearce B, Loveridge C, Hurlbert AC (2015) Performance on the Farnsworth-Munsell 100-Hue Test Is Significantly Related to Nonverbal IQ. Investig Opthalmol Vis Sci 56:3171. https://doi.org/10.1167/iovs.14-16094
14. Vision LC (2018) Color vision test color vision test clinical evaluation of color vision. www.good-lite.com1.800.362.38601.888.362.2576Faxwww.good-lite.com. Accessed 23 Dec 2018
15. Ishihara S (1973) Test for Colour-Blindness, 24 Plates Edition, Kanehara Shuppan Co. Ltd., Tokyo
16. Jouannic J (2019) color blindness test (free and complete). https://www.opticien-len-tilles.com/daltonien_beta/new_test_daltonien.php. Accessed 17 Dec 2019
17. Niroula DR, Saha CG (2010) The incidence of color blindness among some school children of Pokhara. Western Nepal Nepal Med Coll J 12:48–50
18. Moudgil T, Arora R, Kaur K (2016) Prevalance of Colour Blindness in Children. Int J Med Dent Sci 5:1252. https://doi.org/10.19056/ijmdsjssmes/2016/v5i2/100616
19. Lillo J, Davies I, Ponte E, Vitini I (2001) Colour naming by colour blind children. Anu Psicol 32:5–24

20. Nguyen L, Lu W, Do EY, Chia A, Wang Y (2014) Using digital game as clinical screening test to detect color deficiency in young children. In: Proceedings of the 2014 conference on Interaction design and children - IDC '14. pp. 337–340. ACM Press, New York, New York, USA (2014). https://doi.org/10.1145/2593968.2610486
21. Montes Rivera M, Padilla Díaz A, Ponce Gallegos JC, Canul-Reich J, Ochoa Zezzatti A, Meza de Luna MA (2019) Performance of Human proposed equations, genetic program-ming equations, and artificial neural networks in a real-time color labeling assistant for the colorblind. In: Martínez-Villaseñor L, Batyrshin I, and Marín-Hernández A (eds) Advances in soft computing. pp 557–578. Springer International Publishing, Cham. https://doi.org/10.1007/978-3-030-33749-0

The manufacturer's authorised representative in the EU is Springer Nature Customer Service Centre GmbH, Europaplatz 3, 69115 Heidelberg, Germany. If you have any concerns regarding our products, please contact ProductSafety@springernature.com

Printed and bound by CPI Group (UK) Ltd, Croydon, CR0 4YY
05/10/2026
02228339-0008